STRONG INTERACTIONS IN LOW DIMENSIONS

Physics and Chemistry of Materials with Low-Dimensional Structures

VOLUME 25

The titles published in this series are listed at the end of this volume.

STRONG INTERACTIONS IN LOW DIMENSIONS

Edited by

D. Baeriswyl

Department of Physics,
University of Fribourg,
Fribourg, Switzerland

and

L. Degiorgi

Solid State Physics Laboratory,
ETH Zürich, Switzerland

KLUWER ACADEMIC PUBLISHERS
DORDRECHT / BOSTON / LONDON

A C.I.P. Catalogue record for this book is available from the Library of Congress.

ISBN 978-90-481-6512-4 (PB)
ISBN 978-1-4020-3463-3 (e-book)

Published by Kluwer Academic Publishers,
P.O. Box 17, 3300 AA Dordrecht, The Netherlands.

Sold and distributed in North, Central and South America
by Kluwer Academic Publishers,
101 Philip Drive, Norwell, MA 02061, U.S.A.

In all other countries, sold and distributed
by Kluwer Academic Publishers,
P.O. Box 322, 3300 AH Dordrecht, The Netherlands.

Printed on acid-free paper

Printed in the Netherlands.

TABLE OF CONTENTS

Chapter 1

STRONG INTERACTIONS IN LOW DIMENSIONS: INTRODUCTORY REMARKS

D. Baeriswyl
Département de Physique, Université de Fribourg, Pérolles, CH-1700 Fribourg, Switzerland.
dionys.baeriswyl@unifr.ch

L. Degiorgi
Laboratorium für Festkörperphysik, ETH Zürich, CH-8093 Zürich, Switzerland.
degiorgi@solid.phys.ethz.ch

The physical properties of low–dimensional systems have fascinated researchers for a great part of the last century, and have recently become one of the primary centers of interest in condensed matter research. At the beginning, this field appeared much more like a playground for creative theorists than a serious domain of solid–state physics. In fact, the exact treatment of the one–dimensional Lenz–Ising model and Bethe's ingenious diagonalization of the antiferromagnetic Heisenberg chain were considered at most as first steps towards a theory of electronic and magnetic properties of real, three–dimensional crystals. Similarly, Tomonaga presented his study of sound waves in a one–dimensional system of interacting fermions as mathematically interesting but physically not very useful. A collection of important early developments has been assembled and commented upon by Lieb and Mattis [1], who emphasized that while exact solutions of one–dimensional models provide useful tests for approximate methods, "in almost every case the one–dimensional physics is devoid of much structure, and describes a colorless universe much less interesting than our own". This point of view was based on the observation that "in one dimension bosons do not condense, electrons do not superconduct, ferromagnets do not magnetize, and liquids do not freeze". Fortunately, it has since been demonstrated that the one–

D. Baeriswyl and L. Degiorgi (eds.), Strong Interactions in Low Dimensions, 1–19.

dimensional world has its own richness, and that there are real materials which to a good approximation may be considered as consisting of uncoupled chains, at least for temperatures or frequencies which are not too low. Some of the specific features predicted for interacting one–dimensional electron systems, such as charge– and spin–density waves, have indeed been observed in many quasi–one–dimensional materials [2], and under particular circumstances, such as Fermi–surface nesting, can also be found in higher dimensions.

In this book we attempt to convey the colorful facets of condensed matter systems with reduced dimensionality. We are of course aware of the fact that many important aspects must be left aside in such a collection of specific subjects; some of the most regretful omissions will be mentioned later. The following introductory remarks are intended as an aid to identifying some of the essential concepts which will reappear at several places in the subsequent chapters. At the same time this introduction may help to connect the different topics treated in the book, some of which might at first sight appear rather disparate.

1. Ordering in low dimensions

In our three–dimensional world we are accustomed to the spontaneous appearance of order at sufficiently low temperatures. Liquids condense to form periodic solids, magnetic moments are aligned in ferromagnetic or antiferromagnetic configurations, and Fermi liquids turn into superfluids. At zero temperature the state of lowest energy determines the stable configuration of a system, but at finite temperatures it is the minimum of the free energy which determines whether the order parameter – the magnetization in a ferromagnet, the condensate fraction in a superfluid or the intensity of Bragg peaks in a periodic solid – remains finite or is completely suppressed due to thermal fluctuations. The destabilizing effects of temperature are particularly strong in one dimension. While in two or higher dimensions the Ising model exhibits long–range order below a finite critical temperature, this is no longer true in one dimension, where thermal fluctuations destroy the spin correlations beyond a finite correlation length. These fluctuations are even more effective in the case of a continuous order parameter. Thus, according to the Mermin–Wagner theorem, the classical two–dimensional XY model has no true long–range order at finite temperatures [3]. Nevertheless this model – which can be used for representing the phase fluctuations of a complex order parameter or the spin configurations in an easy–plane Heisenberg ferromagnet – shows a transition from a disordered high–temperature phase with exponentially decaying correlations to a phase with quasi–long–range order below the Kosterlitz–Thouless temperature T_{KT}. The appropriate quantity describing this transition is the phase stiffness or superfluid density, which is finite below T_{KT} and shows a universal discontinuity at T_{KT} [4].

At zero temperature classical spin models on a bipartite lattice have long–range order, but this is not necessarily true for the corresponding quantum models. Thus for both the spin–$\frac{1}{2}$ quantum XY and Heisenberg models the spin correlations decay algebraically, and even the Ising chain becomes disordered under a sufficiently strong transverse field. The latter case is very illuminating [5, 6] as the external field allows one to drive the system through a quantum critical point, a second–order phase transition at $T = 0$ where quantum fluctuations are relevant [7]. For dimensions higher than 1 the quantum XY model has long–range order for all values of the spin, at least on a hypercubic lattice [8]. This result has also been proven for the quantum Heisenberg antiferromagnet for 3 or higher dimensions (and any spin) [9]. For dimension D=2 a rigorous proof still seems not to be available for $S = \frac{1}{2}$, but both numerical simulations and analytical calculations indicate that long–range order does exist, albeit with a reduced moment [10].

Low–dimensional systems not only experience strong quantum and thermal fluctuations, but also admit ordering tendencies which are difficult to realize in three–dimensional materials. Prominent examples are spin– and charge–density waves in quasi–one–dimensional organic compounds and spontaneous circulating currents (leading to "orbital antiferromagnetism") in two dimensions. The competition among several possible order parameters leads to rich phase diagrams and an enhanced sensitivity to disorder or applied external fields. Some of these order parameters are very difficult to observe directly. As an example, while phases with spontaneous spin or charge currents around the plaquettes of a square lattice occur naturally in models of interacting fermions in two dimensions [11], their unambiguous detection appears to be very difficult. Thus it is at present not clear whether the so–called "pseudogap phase" in the layered cuprates is related to such a hidden order parameter [12].

2. Dimensional crossover

A real material is not truly one–dimensional (1D), but at most quasi–one–dimensional, *i.e.* a collection of weakly coupled chains. (Notable exceptions include quantum wires and nanotubes.) Thus one of the important questions will be the extent to which the coupling between chains is relevant. An illustrative example is the Ising model on a square lattice with different ferromagnetic exchange constants $J_{\|}$ and $J_{\perp}$ in the two directions. According to Onsager's exact solution [13], the critical temperature T_c for the transition from the paramagnetic to the ferromagnetic phase is given by

$$\sinh(2\beta_c J_{\|})\ \sinh(2\beta_c J_{\perp}) = 1\ , \tag{1.1}$$

where $\beta_c = 1/k_B T_c$. For an array of weakly coupled chains ($J_{\|} \gg J_{\perp}$), Eq. (1.1) becomes

$$2J_{\perp}\xi_{\|}(T_c) = k_B T_c\ , \tag{1.2}$$

where $\xi_{\|}(T)$ is the correlation length of the 1D Ising model. From this relation one may be tempted to conclude that the critical temperature is of the order of the interchain coupling $J_{\perp}$. However, this is not correct because the correlation length is strongly temperature–dependent, $\xi_{\|}(T) = \frac{1}{2}\exp(2\beta J_{\|})$. Inserting this relation into Eq. (1.2) and taking the logarithm yields

$$k_B T_c = \frac{2J_{\|}}{\ln(k_B T_c/J_{\perp})}\ . \tag{1.3}$$

This logarithmic dependence is very weak, and therefore the critical temperature remains of the order of $J_{\|}$, unless the ratio $J_{\perp}/J_{\|}$ is vanishingly small.

The behavior of Eq. (1.3) is rather special as it depends on the exponential temperature dependence of the correlation length. For a 1D Heisenberg antiferromagnet one finds instead $\xi_{\|}(T) \approx J_{\|}/(2k_B T)$ (up to some logarithmic corrections [14]). In this case the use of Eq. (1.2) would again predict a finite critical temperature for a 2D array of coupled chains, but this is not correct because the 2D Heisenberg model orders only at zero temperature. Eq. (1.2) is therefore not universally valid, and in particular cannot be applied for models with gapless excitations.

A dimensional crossover is expected to occur at a temperature $T_{\rm cr} > T_c$. Below $T_{\rm cr}$ correlations between chains (or planes) extend beyond the distance between the structural units, and thus a continuum theory of the Ginzburg–Landau type,

$$F = F_0 + \int d^3r\ [\ a|\Psi|^2 + b|\Psi|^4 + \sum_{\alpha} c_\alpha|\partial_\alpha\Psi|^2\ ]\ , \tag{1.4}$$

should be applicable, where $\Psi(\mathbf{r})$ is a real or complex order–parameter field, the coefficient a changes sign at the mean–field temperature, $b > 0$ and the relative sizes of c_α, $\alpha = 1, 2, 3$, describe the nature of the spatial anisotropy. For a quasi–1D situation one of the coefficients c_α is significantly larger than the two others, while for a quasi–2D case one coefficient is significantly smaller. A simple rescaling, $x_\alpha = x'_\alpha \sqrt{c_\alpha}$, transforms Eq. (1.4) into an isotropic functional in which all the critical properties are equal to those of the isotropic Ginzburg–Landau theory. Such a rescaling can sometimes be justified even for quasi–2D superconductors where the coupling to the electromagnetic field must be included [15]. One may conclude that characteristic low–dimensional effects emerge above $T_{\rm cr}$. For strongly anisotropic situations, as in certain layered materials, $T_{\rm cr}$ may be so close to T_c that for all practical purposes the continuum description of Eq. (1.4) is never valid. Instead, the discrete sequence of stacks of planes, or of arrays of chains, must be described explicitly, as for example in the Lawrence–Doniach model of layered superconductors [15, 16].

A key issue, which will play a rôle in several chapters of this book [17, 18], is the dimensional crossover observed as a function of frequency in dynamic correlation functions. Consider for example a quasi–one–dimensional conductor in which an electron moves preferentially along a chain (parameter $t_\|$) and hops occasionally to a neighboring chain (parameter $t_\perp \ll t_\|$). The response of this system to an external stimulus of frequency ω is expected to be three–dimensional at low frequencies but essentially one–dimensional for sufficiently large frequencies. If the quasiparticle description of Fermi–liquid theory is applicable, the characteristic frequency separating the two regimes is expected to be on the order of $t_\perp/\hbar$. However, if this description is not valid, the discussion becomes much more subtle [19, 20, 21]. For dynamic order–parameter correlation functions the crossover is expected to occur for $\hbar\omega_{\rm cr} \approx k_B T_{\rm cr}$.

3. Magnetism in insulators and metals

Magnetism has been a subject of amazement since antiquity [22], and remains one of the most active fields of solid–state physics. Magnetic moments appear, according to Hund's rules, in atoms or ions with a partially filled shell. In an ionic crystal where charge delocalization is small, the picture may be modified by crystal–field effects or by the Jahn–Teller distortion [23]. The Heisenberg model,

$$H = -J \sum_{\langle i,j \rangle} \mathbf{S}_i \cdot \mathbf{S}_j \ , \tag{1.5}$$

in which the spins $\mathbf{S}_i$ occupy the sites of a lattice and the exchange interaction acts only among nearest neighbors, is often sufficient to describe accurately the magnetic properties of materials with local moments, such as the magnetic susceptibility, the magnetic contribution to the specific heat, or magnetic neutron scattering. Depending on the context, the coupling J may represent the (ferromagnetic) direct exchange interaction between electrons occupying orthogonal orbitals [24] or the (antiferromagnetic) kinetic exchange across non–magnetic ions in transition–metal compounds [25].

The localization of electronic states is not necessarily the result of a small overlap beween atomic wave functions, but can also be produced by disorder or Coulomb correlations. Materials undergoing a metal–insulator transition are thus particularly interesting for their rich magnetic properties [26, 27, 28]. The combined effects of strong disorder and Coulomb interactions may even produce local moments in the metallic phase [29, 30]. Because both disorder and correlation effects are strong in low dimensions, it is not surprising that the interplay between localization and magnetism is pronounced in organic chain compounds [31] and in the layered cuprates [12].

Despite the simple form of the Heisenberg model (1.5) it is very difficult to obtain closed analytical solutions. A famous exception is the spin–$\frac{1}{2}$ Heisenberg chain, where Bethe found an ingenious method for calculating the energy eigenstates [32]. The spin correlation function for the antiferromagnetic case has been calculated using both field–theoretic [33, 34, 35] and numerical techniques [36], and found to behave asymptotically as

$$\langle \mathbf{S}_i \cdot \mathbf{S}_{i+n} \rangle \sim \frac{(-1)^n}{n} \sqrt{\ln(cn)} \tag{1.6}$$

for $n \to \infty$, where c is a constant. This implies that the 1D antiferromagnetic Heisenberg chain has quasi–long–range order. The elementary excitations are spinons, which can be viewed as traveling domain walls between regions of opposite (staggered) order parameter [37], and have a gapless energy spectrum. In contrast, as suggested by Haldane, the spin–1 chain behaves like a spin liquid, with exponentially decaying spin correlations and a gap in the excitation spectrum [38, 39]. This conjecture has since been confirmed both numerically [40, 41] and experimentally [6]. A closely related phenomenon is observed in spin–$\frac{1}{2}$ ladders, where a spin gap appears for an even number of legs, while odd–leg ladders are gapless [42]. Several materials are now available which can be considered as consisting of weakly coupled spin ladders [43].

New phenomena can occur in the presence of frustration, as for example in the antiferromagnetic Heisenberg chain when an additional next-nearest–neighbor antiferromagnetic exchange constant J' is included. For the particular ratio $J' = J/2$ the exact ground state is known to be a valence–bond solid, *i.e.* a simple product of nearest–neighbor singlet pairs [44, 45]. This state has a broken translational symmetry with long–range order, but the range of spin correlations is limited to nearest–neighbor sites. In two dimensions, frustration can be produced either by the geometry of the lattice [46] or by competing exchange couplings [47], but the rich variety of possible phases between Néel order in unfrustrated systems and spin liquids in strongly frustrated systems has not yet been fully explored. For many materials, magnetic frustration appears to be the origin of a spin gap [6].

Magnetic order is not restricted to insulators with localized magnetic moments, but occurs also in some alloys of non–magnetic metals. A microscopic approach in terms of electronic band structure and Coulomb matrix elements is required to treat such cases. The problem of ferromagnetism in itinerant electron systems was addressed in the early days of solid–state physics. According to Stoner, ferromagnetism occurs if $U\rho(\varepsilon_F) > 1$, where U represents the electron–electron interaction strength and $\rho(\varepsilon_F)$ is the electronic density of states at the Fermi energy [48].

A more explicit treatment of metallic ferromagnetism was proposed by Gutzwiller, Hubbard and Kanamori [49, 50, 51] in terms of the Hubbard Hamiltonian,

$$H = -t \sum_{\langle i,j\rangle,\sigma} \left(c_{i\sigma}^{\dagger} c_{j\sigma} + c_{j\sigma}^{\dagger} c_{i\sigma} \right) + U \sum_{i} n_{i\uparrow} n_{i\downarrow} \,, \tag{1.7}$$

where $c_{i\sigma}^{\dagger}$ ($c_{i\sigma}$) creates (annihilates) an electron with spin σ at site i and $n_{i\sigma} = c_{i\sigma}^{\dagger} c_{i\sigma}$ expresses the electron density.

For a half–filled band the fully polarized state is an eigenstate of the Hamiltonian (1.7), with energy $E = 0$. However, it is easy to see that this cannot be the true ground state, at least for bipartite lattices, where a variational wave function with alternating values of $\langle n_{i\uparrow}\rangle - \langle n_{i\downarrow}\rangle$ on the two sublattices has lower energy. The amplitude of such a spin–density wave is small for small U but tends to 1 as $U \to \infty$ (Néel state). In one dimension the fully polarized eigenstates are easy to determine also for other fillings, because they correspond to those of non–interacting spinless fermions, but the true ground state is found to be a spin singlet [52]. This is no longer true if next–nearest–neighbor hopping is included, in which case ferromagnetism has been found numerically for the Hubbard model in both one [53] and two dimensions [54] for large values of U. However, it remains quite generally true that the conditions for the appearance of ferromagnetism are considerably more stringent than the simple Stoner criterion.

4. Charge order

The conventional Hubbard model (1.7), originally conceived for describing ferromagnetism, has played (and still plays) a major role in the study of the Mott transition, *i.e.* the metal–insulator transition produced by strong electronic correlations [26, 27, 55]. At zero temperature this transition occurs at half–filling as a function of U, or at a (sufficiently large) given value of U as a function of density. Increasing U results in a suppression of the double occupancy $d_i = \langle n_{i\uparrow} n_{i\downarrow} \rangle$, and at half–filling d_i decreases from 1/4 for $U = 0$ to 0 for $U \to \infty$ in any dimension. Correspondingly, the magnetic moment,

$$\langle \mathbf{S}_i^2 \rangle = \frac{3}{4} \hbar^2 (1 - 2d_i) \; , \tag{1.8}$$

evolves from $\frac{3}{8}\hbar^2$ for $U = 0$ to $\frac{3}{4}\hbar^2$, the value for a localized spin $\frac{1}{2}$, for $U \to \infty$. [We recall that for large values of U, $U \gg t$, the low–energy eigenstates of the Hubbard Hamiltonian at half–filling are equal to those of the Heisenberg model (1.5) with $J = 4t^2/U$.] The genuine Mott transition (not masked by antiferromagnetic ordering, which leads to a "Slater insulator" [56]) occurs in the region where U is similar in magnitude to the bandwidth. The double occupancy or the magnetic moment, which change smoothly in this region, are therefore not suitable for characterizing the transition. A better quantity is the fluctuation of polarization [57], which has values on the order of the system size in the metallic phase, but of order unity in the insulating phase.

Because of its conceptual simplicity, the Hubbard model has become very popular for describing materials, such as La_2CuO_4, which would be metals from the point of view of their (LDA) band structure but turn out to be insulators with an energy gap for charge excitations. However, as already recognized by Hubbard [50], such a description assumes implicitly an efficient metallic screening, which is of course not available on the insulating side, where a full account of the long–range nature of Coulomb forces is required. At half–filling, charge fluctuations are strongly suppressed by the on–site interaction, and so the long–range part of the interaction plays a minor role. For other fillings, however, charge fluctuations may be strongly enhanced in order to reduce the long–range part of the Coulomb interaction, which can lead to inhomogeneous charge patterns such as charge–density waves, Wigner crystals, or charge stripes [58].

The concept of charge– (or spin–)density waves is usually associated with a small–amplitude spatial modulation of the charge (or spin) density. These density waves occur preferentially in low–dimensional metals, for which a single wave vector $\mathbf{Q}$ connects large portions of the Fermi surface ("nesting") [2, 17, 18]. They may be considered as condensed electron–hole pairs, similar to the way in which superconductivity can be interpreted as a condensation of Cooper pairs. This Fermi–surface instability must be contrasted with the appearance of periodic charge patterns in the classical limit, where the kinetic energy is negligible and the electrons arrange themselves to minimize the total potential energy, forming a Wigner crystal in the continuum [59] and a "generalized Wigner lattice" in a periodic solid [60]. A simple model which illustrates both the charge–density–wave instability of the metallic phase and the crossover towards a generalized Wigner lattice as the interaction strength is increased is a 1D system of spinless fermions with an average number density of 1/2 and an interaction between nearest–neighbor sites. The Hamiltonian is

$$H = -t\sum_i \left(c_i^\dagger c_{i+1} + c_{i+1}^\dagger c_i\right) + V\sum_i n_i n_{i+1} \ , \tag{1.9}$$

where $n_i = c_i^\dagger c_i$ measures the density at site i. This model can be mapped to the XXZ Heisenberg Hamiltonian with exchange couplings $J_x = J_y = 2t$, $J_z = V$ [61], which has an easy–plane (XY–type) region for $V < 2t$ and an easy–axis (Ising–type) region for $V > 2t$. Correspondingly, the fermion system is metallic (or, more precisely, a Luttinger liquid [62]) for $V < 2t$ and insulating with a charge–ordered ground state for $V > 2t$. The amplitude of the charge modulation begins from 0 at $V = 2t$ and evolves continuously to 1 for $V \to \infty$, where the fermions occupy every other site in order to minimize the nearest–neighbor repulsion.

Charge ordering is a widespread phenomenon [63]. It has been observed in organic chain compounds such as TTF–TCNQ and the Bechgaard salts [18], in transition–metal dichalcogenides, in layered nickelates and cuprates [58], and in manganites, and it has also been used to explain transport anisotropies in quantum Hall systems. Very often, structural distortions accompany an inhomogeneous charge distribution, and in some cases, such as the Peierls instability, the electron–phonon coupling is even the driving force for charge–density–wave formation.

Electrons carry both charge and spin, and it is therefore natural to consider whether charge and magnetic order are correlated. A simple picture is available for some organic chain compounds, in particular for $(TMTTF)_2X$ ($X=PF_6$, SbF_6, AsF_6) [64] and $(DCNQI)_2Ag$ [65], where charge ordering occurs at a relatively high temperature (c. 100 to 200K), whereas magnetic ordering occurs at much lower temperatures (of the order of 1 to 10K). A lattice model with dominant long–range Coulomb interactions is able to explain this separation of energy scales: the Coulomb interaction leads to the formation of a generalized Wigner lattice with a typical energy scale of 1eV [60], while the effective exchange constant between the nearest–neighbor spins is two orders of magnitude smaller for these materials [66].

A rich variety of different phases can emerge in systems where the energy scales for charge and magnetic orderings are comparable. Layered cuprates appear very likely to belong to this category: they show an intricate competition between antiferromagnetism, charge order, superconductivity and maybe other, more exotic, broken symmetries. A wealth of phenomena, which are still only partly understood, may be linked directly to this interplay of charge and spin degrees of freedom, in both normal and superconducting phases [12, 67]. One of the most intriguing new collective states is the stripe phase, where doped charges are located on spontaneously generated domain walls between antiferromagnetic regions [58]. In such a phase the motion of charges is essentially 1D, notwithstanding the 2D band structure of the CuO_2 plane. The consequences of this "dynamical dimension reduction" [68] have not been fully explored, and even the basic mechanism governing stripe formation is not yet firmly established.

5. The fate of the Fermi liquid in low dimensions

One of the cornerstones of solid–state physics is the Landau theory of the Fermi liquid [69], which explains why in ordinary metals the thermal and transport properties of strongly interacting electrons can be described in terms of weakly interacting (fermionic) quasiparticles. The fundamental reason for the success of this theory is the rarity of scattering events in the vicinity of the Fermi surface. The state of a normal Fermi liquid breaks down if the residual interactions between quasiparticles lead to a collective bound state, such as a superconducting or charge–ordered state. In a strict sense, the ground state of an interacting electron system is always expected to differ from that of a normal Fermi liquid, because superconductivity will always occur if it is not dominated by another instability [70]. However, this may occur at temperatures so low that for conventional experimental conditions the Fermi–liquid theory does provide a valid description.

Charge– and spin–density waves or superconductivity remove electronic states in the vicinity of the Fermi energy, either completely in the case of s–wave symmety or only partially for an order parameter $\Delta(\mathbf{k})$ with nodes on the Fermi surface. This mean–field picture, beautifully confirmed by tunneling and thermal experiments on superconductors [71, 72], is based on symmetry–breaking below the (mean–field) critical temperature. As discussed above, thermal fluctuations are so strong in one and two dimensions that a continuous symmetry is not broken at finite temperatures [3]. A simple example of the way in which strong order–parameter fluctuations determine the electronic structure around the Fermi energy was illustrated by Lee, Rice, and Anderson for the Peierls instability [73]. The replacement of new Bragg peaks (generated by a static order parameter) by a broadened structure factor representing the order–parameter fluctuations removes the gap between electronic bands, but if the correlation length is sufficiently large a pronounced pseudogap remains. This result has been confirmed recently by more detailed calculations [74, 75, 76], and may be relevant for the interpretation of photoemission experiments on one–dimensional conductors above the critical temperature of the (3D) Peierls transition [18, 77]. Whether the so–called pseudogap phase in layered high–temperature superconductors may be described similarly in terms of (superconducting) order–parameter fluctuations (preformed pairs) remains an open issue [12].

A more subtle breakdown of a normal Fermi liquid occurs in 1D systems of electrons with (short–range) interactions. In this case the very existence of fermionic quasiparticles is questionable, because the state resulting from the addition or removal of an electron may decay quickly into a charge and a spin excitation which propagate with different velocities (spin–charge separation) [21, 38]. This result, derived in the framework of the 1D electron gas [78], has its correspondence in lattice models such as the 1D Hubbard model where the elementary excitations are spinons, carrying spin but no charge, and holons, which carry charge but no spin [79]. A 1D metal described by a model of interacting electrons, not coupled to phonons, is thus predicted to behave quite differently from a Fermi liquid, namely as a so–called "Luttinger liquid" [21, 80]. However, the extent to which the experimental signature of a Luttinger–liquid state has been identified clearly in real materials remains controversial, although high–resolution photoemission [18] and optical absorption [17] measurements on quasi–one–dimensional conductors, as well as transport and tunneling experiments on quantum wires [81], have been used extensively for clarifying the issue.

It is worthwhile to add that non–Fermi liquid features are not a privilege of one–dimensional systems, but also appear routinely in heavy–fermion materials. Among other mechanisms, low–energy fluctuations due to the proximity of a magnetic phase transition near zero temperature have been invoked for interpreting the experimental findings in these systems [82]. The extent to which a similar mechanism may be responsible for the observed non–Fermi liquid behavior of optimally doped, layered cuprates [83], remains an open issue.

6. *Ab initio* calculations and effective models

Most electronic and magnetic properties of solids could be described accurately if the basic Hamiltonian of electrons and nuclei coupled by Coulomb interactions were tractable. This is unfortunately not the case without resort to a number of (often uncontrolled) approximations. A simple and widely used scheme is the Local Density Approximation (LDA) of Density Functional Theory, which has been applied successfully to many types of solids, at least for describing their ground–state properties [84] and lattice dynamics [85]. An extended version of LDA which allows for spin–polarized ground states, the Local Spin–Density Approximation, has been quite successful in describing the magnetism of transition metals and their alloys [28], but there are prominent exceptions, including the layered cuprates, where strong correlations must be treated more thoroughly than is possible within a local exchange–correlation potential. Some progress has been made recently for strongly correlated systems by combining LDA calculations with methods used previously for model Hamiltonians, such as the Dynamical Mean–Field Theory [86].

A different approach exploits the fact that one is often not concerned with all of the details embodied in the interacting electron–nucleus system, but rather in its low–energy behavior. Therefore one attempts a down–folding of the full Hamiltonian to a more simple, effective model which contains the essential variables associated with the low–energy degrees of freedom [87]. A good example of such a downfolding process is the BCS theory of conventional superconductors [88]. Instead of treating explicitly the interacting electron–phonon system, which itself should be considered as an effective model of Landau quasiparticles interacting with lattice vibrations, one uses the fact that the relevant electronic excitations have an energy lower than the phononic energy scale $\hbar\omega_D$. Second–order perturbation theory then allows one to calculate the effective attraction arising due to phonon exchange. The resulting BCS Hamiltonian is purely electronic, and is valid for low–energy excitations (with an energetic cut–off on the order of $\hbar\omega_D$). A further downfolding can be performed by proceeding from the microscopic BCS Hamiltonian to the Ginzburg–Landau functional, which describes the low–energy fluctuations of the order parameter close to the mean–field critical temperature [89].

A systematic elimination of high energy scales and a corresponding transformation of the original model into an effective low–energy Hamiltonian is known as renormalization (or also as the Renormalization Group). Originally developed in quantum field theory, this method was applied in the 1970s to (classical) phase transitions, where it not only provided specific tools for calculating critical exponents, but also demonstrated why seemingly different systems show the same behavior close to a critical point. More recently the Renormalization Group method has also been applied to many–electron systems. In one dimension the problem may be formulated in terms of a small set of coupling constants for the scattering processes between the two Fermi points, two for forward scattering (g_2, g_4), one for backward (g_1) and one for Umklapp scattering (g_3). The renormalization of the Hamiltonian can then be visualized as a flow in the space of the coupling parameters $g_1, ..., g_4$[78, 90]. In two dimensions the problem is more complicated because the Fermi surface is a line, and therefore the scattering vertex must be treated as a function of momenta rather than of a small set of coupling constants. While in the isotropic case the problem can still be treated analytically (to lowest order in perturbation theory) [91, 92], a more realistic Fermi surface, such as that of a partially filled tight–binding band, has been investigated mostly numerically [93, 94]. Often some couplings grow strongly upon renormalization. This is usually interpreted as the signal of an instability (superconductivity, charge–density waves, flux phases, and others), but at the same time the perturbative Renormalization Group approach becomes invalid. Field–theoretic methods applicable close to a quantum–critical point [7], or duality transformations to collective variables [5], may in some cases help to overcome this difficulty, but quite generally the physics of interacting electrons on a 2D lattice, which is genuinely a strong–coupling problem, leaves still a lot of scope for future studies.

7. About this book

Many important developments in solid–state physics during the last three decades have been associated with materials or structures of reduced dimensionality, including quasi–one–dimensional conductors, layered cuprates, the 2D electron gas at semiconductor interfaces, surfaces, and surface adsorbates. Both the controlled preparation of new materials and advances in experimental and theoretical techniques have been essential for this rapid progress. The aim of this book is not to review the most important results that have been accumulated recently, but rather to explain thoroughly certain selected topics. The emphasis of the more experimental chapters is on the application of dynamical probes, such as neutron scattering [6], optical absorption [17, 67], and photoemission [12, 18], as well as on transport studies, both electrical [81] and thermal [43]. Some of the more theoretical chapters are directly relevant for experiments, such as optical spectroscopy [87], transport in one–dimensional models [80], and the phenomenology of charge inhomogeneities in layered materials [58], while others discuss more general topics and methods, for example the concept of a Luttinger liquid and bosonization [21] or duality transformations, both promising tools for treating strongly interacting many–body systems [5].

Many important topics are treated only marginally or not at all. Thus the ubiquitous electron–phonon interaction is mentioned only briefly [18], although it is of primary importance for the Peierls instability in one–dimensional conductors. Similarly, the conjugated polymers are not discussed, although the concepts of neutral and charged solitons [95, 96] are intimately related to those of spinons and holons. Magnetic fields are known to induce spectacular effects in low–dimensional systems, such as the Quantum Hall Effect in the 2D electron gas [97], the appearance of different forms of vortex matter – solid, liquid, glass – in layered superconductors [15] or the field–induced spin–density waves in the Bechgaard salts [98]. Regretfully a detailed discussion of these beautiful phenomena has been omitted. On the experimental side two important dynamical probes have not been discussed, light scattering and magnetic resonance. Light scattering, in particular magnetic Raman scattering, has recently been used successfully for investigating the excitations in low–dimensional spin systems [99]. Nuclear magnetic resonance spectroscopy gives important insight into local magnetic fields and spin dynamics, while nuclear quadrupole resonance spectroscopy can provide valuable information about the charge distribution, which is generally very difficult to measure by other means [100]. On the theoretical side two important recent developments are not included, namely the extension of the Renormalization– Group method to many–fermion systems in two dimensions (discussed briefly above), and the progress in numerical techniques, most notably the Density–Matrix Renormalization Group [101] and new Quantum Monte Carlo algorithms [102].

For those topics which have been omitted, we trust that the interested reader may find due information and inspiration from the references cited. For those which have been included, we hope that the following 12 chapters serve to provide a coherent overview of the fundamental issues in contemporary studies of strongly interacting systems in low dimensions.

Acknowledgment

We are grateful to Bruce Normand for helpful discussions and for a critical reading of the manuscript.

References

[1] E. H. Lieb and D. C. Mattis, *Mathematical Physics in One Dimension*, Academic Press, New York (1966).

[2] For a pedagogical introduction see G. Grüner, *Density Waves in Solids*, Addison–Wesley, Reading (1994).

[3] N. D. Mcrmin and H. Wagncr, Phys. Rev. Lett. **22**, 1133 (1966).

[4] J. M. Kosterlitz and D. J. Thouless, J. Phys. C **6**, 1181 (1973).

[5] See the contribution of M. P. A. Fisher in this book.

[6] See the contribution of C. Broholm and G. Aeppli in this book.

[7] S. Sachdev, *Quantum Phase Transitions*, Cambridge University Press, Cambridge (1999).

[8] T. Kennedy, E. H. Lieb and B. S. Shastry, Phys. Rev. Lett. **61**, 2582 (1988).

[9] T. Kennedy, E. H. Lieb and B. S. Shastry, J. Stat. Phys. **53**, 1019 (1988).

[10] For a review see E. Manousakis, Rev. Mod. Phys. **63**, 1 (1991).

[11] B. Binz, D. Baeriswyl and B. Douçot, Eur. Phys. J. B **25**, 69 (2002).

[12] See the contribution of J. C. Campuzano in this book.

[13] L. Onsager, Phys. Rev. **65**, 117 (1944).

[14] K. Nomura and M. Yamada, Phys. Rev. B **43**, 8217 (1991).

[15] G. Blatter, M. V. Feigel'man, V. B. Geshkenbein, A. I. Larkin and V. M. Vinokur, Rev. Mod. Phys. **66**, 1125 (1994).

[16] W. E. Lawrence and S. Doniach, *Proc. 12th Int. Conf. on Low Temperature Physics*, edited by E. Kanda, Keigaku, Tokyo (1971), p. 361.

[17] See the contribution of L. Degiorgi in this book.

[18] See the contribution of M. Grioni in this book.

[19] P. W. Anderson, Phys. Rev. Lett. **67**, 3844 (1991).

[20] J. Voit, Rep. Prog. Phys. **58**, 977 (1995).

[21] See the contribution of K. Schönhammer in this book.

[22] For a historical introduction see D. C. Mattis, *The Theory of Magnetism I*, Springer Series in Solid–State Sciences **17** (1981).

[23] K. Yosida, *Theory of Magnetism*, Springer Series in Solid–State Sciences **122** (1996).

[24] W. Heisenberg, Z. Phys. **49**, 619 (1928).

[25] P. W. Anderson, Phys. Rev. **115**, 2 (1959).

[26] N. F. Mott, *Metal–Insulator Transitions*, Taylor and Francis Ltd., London (1974).

[27] F. Gebhard, *The Mott Metal–Insulator Transition*, Springer Tracts in Modern Physics **137** (1997).

[28] P. Fazekas, *Electron Correlation and Magnetism*, World Scientific, Singapore (1999).

[29] M. Milovanović, Phys. Rev. Lett. **63**, 82 (1989).

[30] V. Dobrosavljević and G. Kotliar, Phys. Rev. B **50**, 1430 (1994).

[31] For a review see D. Jérome, in *Strongly Interacting Fermions and High T_c Superconductivity*, edited by B. Douçot and J. Zinn–Justin, Elsevier, Amsterdam (1995), p. 249.

[32] H. Bethe, Z. Physik **71**, 205 (1931).

[33] I. Affleck, D. Gepner, H. J. Schulz and T. Ziman, J. Phys. A: Math. Gen. **22**, 511 (1989).

[34] R. R. Singh, M. E. Fisher and R. Shankar, Phys. Rev. B **39**, 2562 (1989).

[35] T. Giamarchi and H. J. Schulz, Phys. Rev. B **39**, 4620 (1989).

[36] K. A. Hallberg, P. Horsch and G. Martinez, Phys. Rev. B **52**, R719 (1995).

[37] L. Faddeev and L. Takhtajan, Phys. Lett. A **85**, 375 (1981).

[38] F. D. M. Haldane, Phys. Rev. Lett. **50**, 1153 (1983).

[39] I. Affleck, J. Phys. Cond. Matt. **1**, 3047 (1989).

[40] S. R. White and A. A. Huse, Phys. Rev. B **48**, 3844 (1993).

[41] O. Golinelli, Th. Jolicoeur and R. Lacaze, Phys. Rev. B **50**, 3037 (1994).

[42] For a review see E. Dagotto and T. M. Rice, Science **271**, 618 (1996).

[43] See the contribution by A. V. Sologubenko and H. R. Ott in this book.

[44] C. K. Majumdar and D. K. Ghosh, J. Math. Phys. **10**, 1388 (1969).

[45] A. Auerbach, *Interacting Electrons and Quantum Magnetism*, Springer–Verlag, Berlin (1994).

[46] See G. Misguich, D. Serban and V. Pasquier, Phys. Rev. B **67**, 214413 (2003), and references therein.

[47] S. Sachdev and K. Park, Ann. Phys. (N. Y.) **298**, 58 (2002).

[48] E. C. Stoner, Proc. Roy. Soc. London A **165**, 372 (1938).

[49] M. C. Gutzwiller, Phys. Rev. Lett. **10**, 159 (1963).

[50] J. Hubbard, Proc. Roy. Soc. London A **276**, 238 (1963).

[51] J. Kanamori, Prog. Theor. Phys. **30**, 275 (1963).

[52] E. Lieb and D. Mattis, Phys. Rev. **125**, 164 (1962).

[53] S. Daul and R. M. Noack, Phys. Rev. B **58**, 2635 (1998).

[54] R. Hlubina, S. Sorella and F. Guinea, Phys. Rev. Lett. **78**, 1343 (1997).

[55] P. W. Anderson, in *Frontiers and Borderlines in Many–Particle Physics*, edited by R. A. Broglia and J. R. Schrieffer, North–Holland, Amsterdam (1988), p. 1.

[56] J. C. Slater, Phys. Rev. **82**, 538 (1951).

[57] C. Aebischer, D. Baeriswyl and R. M. Noack, Phys. Rev. Lett. **86**, 468 (2001).

[58] See the contribution of A. Castro Neto and C. Morais Smith in this book.

[59] E. Wigner, Trans. Faraday Soc. **34**, 678 (1938).

[60] J. Hubbard, Phys. Rev. B **17**, 494 (1978).

[61] E. Lieb, T. Schultz and D. Mattis, Ann. Phys. **16**, 407 (1961).

[62] F. D. M. Haldane, Phys. Rev. Lett. **45**, 1358 (1980).

[63] For a recent collection of papers see S. Brazovskii, N. Kirova and P. Monceau, eds., *Proc. Int. Workshop on Electronic Crystals*, J. Phys. IV **12**, Pr9 (2002).

[64] D. S. Chow, F. Zamborszky, B. Alavi, D. J. Tantillo, A. Baur, C. A. Merlic and S. E. Brown, Phys. Rev. Lett. **85**, 1698 (2003).

[65] K. Hiraki and K. Kanoda, Phys. Rev. Lett. **80**, 4737 (1998).

[66] B. Valenzuela, S. Fratini and D. Baeriswyl, Phys. Rev. B **68**, 045112 (2003).

[67] See the contribution of D. van der Marel in this book.

[68] V. J. Emery, S. A. Kivelson and M. Tranquada, Proc. Natl. Acad. Sci. **96**, 8814 (1999).

[69] D. Pines and P. Nozières, *The Theory of Quantum Liquids*, Addison–Wesley, Redwood City (1966).

[70] W. Kohn and J.M. Luttinger, Phys. Rev. Lett. **15**, 524 (1965).

[71] M. Tinkham, *Introduction to Superconductivity*, McGraw–Hill, New York (1975).

[72] J. R. Waldram, *Superconductivity of Metals and Cuprates*, IOP Publishing, Bristol (1996).

[73] P. A. Lee, T. M. Rice and P. W. Anderson, Phys. Rev. Lett. **31**, 462 (1973).

[74] A. J. Millis and H. Monien, Phys. Rev. B **61**, 12496 (2000).

[75] L. Bartosch and P. Kopietz, Phys. Rev. B **62**, R16223 (2000).

[76] H. Monien, Phys. Rev. Lett. **87**, 126402 (2001).

[77] L. Bartosch, Phys. Rev. Lett. **90**, 076404 (2003).

[78] For an early review see J. Sólyom, Adv. Phys. **28**, 201 (1979).

[79] F. H. L. Essler and V. E. Korepin, Phys. Rev. Lett. **72**, 908 (1994).

[80] See the contribution of X. Zotos and P. Prelovšek in this book.

[81] See the contribution of A. Yacoby in this book.

[82] G. R. Stewart, Rev. Mod. Phys. **73**, 797 (2001).

[83] S. Sachdev, Rev. Mod. Phys. **75**, 913 (2003).

[84] R. O. Jones and O. Gunnarsson, Rev. Mod. Phys. **61**, 689 (1989).

[85] R. Car and M. Parrinello, Phys. Rev. Lett. **55**, 2471 (1985).

[86] For a recent review of the LDA+DMFT method see K. Held *et al.*, in *Quantum Simulations of Complex Many–Body Systems: from Theory to Algorithms*, edited by J. Grotendorst, D. Marks and A. Muramatsu, NIC Series **10**, 175 (2002).

[87] See the contribution of A. J. Millis in this book.

[88] J. R. Schrieffer, *Theory of Superconductivity*, Addison–Wesley, Redwood City (1964).

[89] P. G. de Gennes, *Superconductivity of Metals and Alloys*, Addison–Wesley, Redwood City (1966).

[90] For a recent discussion see C. Bourbonnais, B. Guay and R. Wortis, in *Theoretical Methods for Strongly Correlated Electrons*, edited by A. M. Tremblay, D. Sénéchal, A. Ruckenstein and C. Bourbonnais, Springer–Verlag, Berlin (2003).

[91] R. Shankar, Rev. Mod. Phys. **66**, 129 (1994).

[92] T. Chen, J. Fröhlich and M. Seifert, in *Fluctuating Geometries in Statistical Mechanics and Field Theory*, edited by F. David, P. Ginsbar and J. Zinn–Justin, Elsevier, Amsterdam (1996).

[93] D. Zanchi and H. J. Schulz, Z. Phys. B **103**, 339 (1997); Phys. Rev. B **61**, 13609 (2000).

[94] C. J. Halboth and W. Metzner, Phys. Rev. B **61**, 7364 (2000).

[95] A. J. Heeger, S. Kivelson, J. R. Schrieffer and W. P. Su, Rev. Mod. Phys. **60**, 781 (1988).

[96] D. Baeriswyl, D. K. Campbell and S. Mazumdar, in *Conjugated Conducting Polymers*, edited by H. Kiess, Springer Series in Solid–State Sciences **102**, 7 (1992).

[97] For a pedagogical introduction see S. M. Girvin, in *Topological Aspects of Low–Dimensional Systems*, edited by A. Comtet, T. Jolicoeur, S. Ouvry, and F. David, Springer–Verlag, Berlin (2000).

[98] W. Kang, S. T. Hannahs and P. M. Chaikin, Phys. Rev. Lett. **70**, 3091 (1993).

[99] P. Lemmens, G. Güntherodt and C. Gros, Phys. Rep. **376**, 1 (2003).

[100] For a review on the application of these techniques on both normal and superconducting phases of layered cuprates see A. Rigamonti, F. Borsa and F. Carretta, Rep. Prog. Phys. **61**, 1367 (1998).

[101] I. Peschel *et al.*, eds., *Density Matrix Renormalization: a New Numerical Method in Physics*, Springer–Verlag, Berlin (1999).

[102] H. G. Evertz, Adv. Phys. **52**, 1 (2003).

Chapter 2

DYNAMIC CORRELATIONS IN QUANTUM MAGNETS

Collin Broholm

Department of Physics and Astronomy, Johns Hopkins University, 3400 North Charles Street, Baltimore, MD 21218, and NIST Center for Neutron Research, Gaithersburg, MD 20899, USA

broholm@jhu.edu

Gabriel Aeppli

NEC Research Institute 4 Independence Way, Princeton, NJ 08540, USA, and London Centre for Nanotechnology, Gower Street, London WC1E 6BT, UK

gabe@research.nj.nec.com

Abstract Dynamic correlations in spin systems near quantum phase transitions where static long range order vanishes are explored via magnetic susceptibility measurements, carried out using inductance coils and inelastic neutron scattering. Experiments on the Ising ferromagnet $Li(Ho, Y)F_4$ in a transverse field introduce the concept of a quantum critical phase transition with and without disorder. Quantum disordered systems examined include the alternating spin chain copper nitrate, the Haldane spin-1 chain Y_2BaNiO_5, and frustrated $Cu_2(C_5H_{12}N_2)_2Cl_4$ (CuHpCl). The uniform spin-1/2 chains $KCuF_3$ and copper benzoate, and the frustrated spinel $ZnCr_2O_4$ are studied as examples of nearly quantum critical systems. Finally, two square lattice spin-1/2 antiferromagnets, La_2CuO_4 and $Cu(DCOO)_2 \cdot 4D_2O$ (CFTD), exemplify renormalized classical systems with strong fluctuations.

Keywords: Quantum Magnetism, Neutron Scattering, Transverse Field Ising Model, Quantum Spin Chains, Frustrated Magnets, Impurities in Quantum Magnets, Two-dimensional Hubbard model.

D. Baeriswyl and L. Degiorgi (eds.), Strong Interactions in Low Dimensions, 21–61.

Introduction

Conventional magnetic systems develop some form of static order at a critical temperature of order the spin-spin interaction strength. However, reduced dimensionality[1], competing interactions[2], and interactions with mobile electrons and holes, can stabilize an essentially quantum mechanical many- body state in the limit $T \to 0$. These systems are of great interest because they represent unique cooperative states of matter that can be explored in detail with a range of experimental techniques[3]. Understanding them might also provide the key to understanding several of the main problems of modern condensed matter physics, most notably those of high-temperature superconductivity and non-Fermi liquids. In this article we shall review experiments that probe dynamic correlations at low temperatures in quantum magnets. The aim is to give an overview of the variety of phenomena that have been found recently in this field rather than to give an exhaustive guide to the literature.

1. Neutron Scattering

This article focuses on the dynamical magnetic susceptibility. The most common probes of this quantity are mutual inductance measurements of the bulk ($Q = 0$) susceptibility and magnetic neutron scattering. The neutron scattering cross section[4] arising from electromagnetic interactions between the neutron magnetic dipole moment and the spin and orbital magnetic dipole moment of electrons in solids is as follows:

$$\frac{d^2\sigma}{d\Omega dE'} = \frac{k_f}{k_i}(\gamma r_0)^2 |\frac{g}{2}F(\mathbf{Q})|^2 \sum_{\alpha\beta}(\delta_{\alpha\beta} - \hat{Q}_\alpha \hat{Q}_\beta)\mathcal{S}^{\alpha\beta}(\mathbf{Q},\omega). \tag{2.1}$$

Here $\mathbf{Q} = \mathbf{k_i} - \mathbf{k_f}$ and $\hbar\omega = E_i - E_f$ are the wave vector and energy transfer to the sample in the scattering event, $r_0 = 0.28179 \cdot 10^{-12}$ cm is the classical electron radius, $\gamma = 1.913$ and $g \approx 2$ are the spectroscopic g-factors of the neutron and the magnetic atom respectively, and $F(\mathbf{Q})$ is the magnetic form factor[5]. The interesting part of Eq. (2.1) is the Fourier transformed two point dynamic spin correlation function:

$$\mathcal{S}^{\alpha\beta}(\mathbf{Q},\omega) = \frac{1}{2\pi\hbar}\int dt e^{i\omega t}\frac{1}{N}\sum_{ij}\langle S_i^\alpha(t) S_j^\beta(0)\rangle e^{-i\mathbf{Q}\cdot(\mathbf{R}_i - \mathbf{R}_j)} \tag{2.2}$$

$\mathcal{S}^{\alpha\beta}(\mathbf{Q},\omega)$ can be related to the generalized spin susceptibility through the fluctuation dissipation theorem[4].

$$\mathcal{S}(\mathbf{Q},\omega) = \frac{1}{1 - e^{-\beta\hbar\omega}}\frac{\chi''(\mathbf{Q},\omega)}{\pi(g\mu_B)^2} \tag{2.3}$$

where $\beta = 1/k_B T$ and χ'' denotes the imaginary part of the susceptibility.

The neutron scattering experiments reported in this article were performed using instruments at the NIST Center for neutron research, the now defunct DR3 research reactor of Risø National Laboratory, Denmark, and at the ISIS facility in the United Kingdom. The NIST and Risø facilities are continuous neutron sources based on fission and the instruments used monochromators based on Bragg reflection from graphite [6]. The ISIS facility is a pulsed spallation neutron source where time of flight is used to determine the energy of the incident and scattered beam[7].

2. Experiments on Insulating Quantum Magnets

Low energy theories for insulating magnets are typically based on a spin hamiltonian of the form.

$$\mathcal{H} = -\sum_{\langle ij \rangle} J_{ij}^{\alpha\beta} S_i^\alpha S_j^\beta - g\mu_B H \sum_i S_i^z + \sum_i G(\mathbf{S}_i) \tag{2.4}$$

$J_{ij}^{\alpha\beta}$ are potentially anisotropic exchange constants for spin pairs $\langle ij \rangle$. We have included the Zeeman term describing the effects of an applied magnetic field, H, which can induce quantum phase transitions between different ground states at $T = 0$. For spin quantum numbers greater than 1/2 there is also a single ion anisotropy term represented by $G(\mathbf{S}_i)$.

The simple ferro-magnetic state with all spins parallel is an eigenstate of $\mathcal{H}$ and it is the ground state for isotropic and uniaxial systems with $J_{ij} > 0$. For a three dimensional bi-partite lattice with inter-sublattice interactions $J_{ij} < 0$, the Néel state with antiparallel nearest neighbor spins is an excellent approximation to the ground state and the basis for the Néel-Anderson spin wave theory of antiferromagnetism. However, there are quantum corrections to the Néel state, which diverge and signal its irrelevance when the manifold of putative soft spin waves becomes sufficiently large[8]. A well known example is the one dimensional antiferromagnet where Néel order is replaced by different quantum many body states for half odd integer and integer spins[1]. However, we shall see that quantum disordered states are also possible in higher dimensions when competing interactions disfavor static order.

2.1 Ising Model in Transverse Field

The simplest quantum spin model is the Ising magnet in a transverse field. The corresponding Hamiltonian is

$$\mathcal{H} = -\sum_{ij} J_{ij}^{zz}\sigma_i^z\sigma_j^z + \Gamma_e \sum_i \sigma_i^x, \tag{2.5}$$

where the σ's are Pauli spin matrices, J_{ij}^{zz} are longitudinal exchange constants, and Γ_e is an effective transverse field, perpendicular to the Ising axis[9]. In the classical limit where $\Gamma_e = 0$, the commutator of $\mathcal{H}$ and σ vanishes, such that any spin configuration is dynamically stable as long as there are no couplings to other degrees of freedom, such as phonons. As soon as Γ_e becomes non-zero, the commutator becomes non-zero, with the result that Heisenberg's equation of motion,

$$i\hbar\frac{d\sigma_i}{dt} = [\sigma_i, \mathcal{H}], \tag{2.6}$$

becomes non-trivial. Figure 2.1 provides a very dramatic illustration of the quantum speed-up, which occurs[10] on application of a transverse field. The material in this case is not low-dimensional, but is instead the prototypical Ising insulator $Li(Ho,Y)F_4$, which is a diluted ferromagnet [11] where the magnetic moments are carried by Ho^{3+} ions. The key point is the two orders of magnitude increase in the characteristic(peak) relaxation frequency on applying a field of 0.8 T perpendicular to the Ising (c) axis of this body centered tetragonal material.

Once it was realized that $Li(Ho,Y)F_4$ provides an excellent realization of the transverse field Ising model, a series of experiments was performed to probe various quantum mechanical aspects of both pure and disordered magnets. One particular attraction of this material is that classical quenched disorder and quantum fluctuations are independently tuneable, the former via random substitution of Y for Ho, and the latter via an external magnetic field, H_t. The pure material, $LiHoF_4$, has the phase diagram[12] shown in Figure 2.2. There is a phase boundary separating the ferromagnet with a spontaneous moment along c from the paramagnet. For $H_t = 0$, the boundary terminates at a classical Curie point at $T_c = 1.53$ K, while for $T = 0$, it ends at a quantum critical point at $H_c = 5$ T. The solid line through the data corresponds to a mean field description, which takes account of both the real crystal field level scheme for the electrons attached to the Ho^{3+} ions as well as the Ho nuclear spins, coupled via the hyperfine interaction to the electrons. It is the latter interaction that gives rise to the low-temperature upturn of the phase boundary. Mean field theory not only accounts for the phase boundary, it also gives an excellent account of the critical properties, which were determined with unprecedented accuracy for the quantum critical point. Most notably, as shown in Figure 2.3, the susceptibility diverges along both thermal (T) and quantum (H_t) trajectories with exponent indistinguishable from the mean field value of unity.

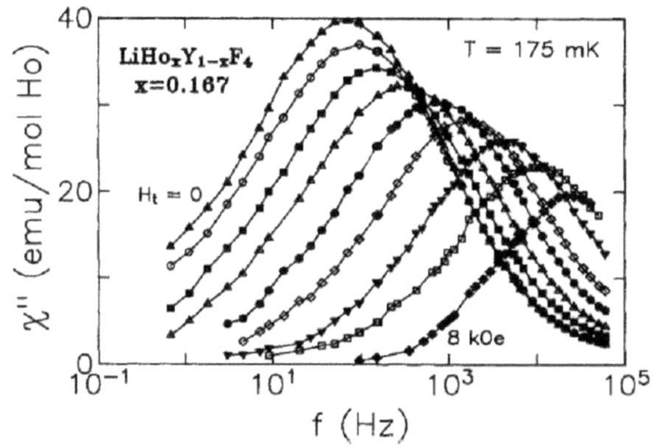

Figure 2.1. AC susceptibility data for the Ising magnet $LiHo_{0.167}Y_{0.833}F_4$ as a function of a transverse field (oriented perpendicular to the easy axis). From ref. [10].

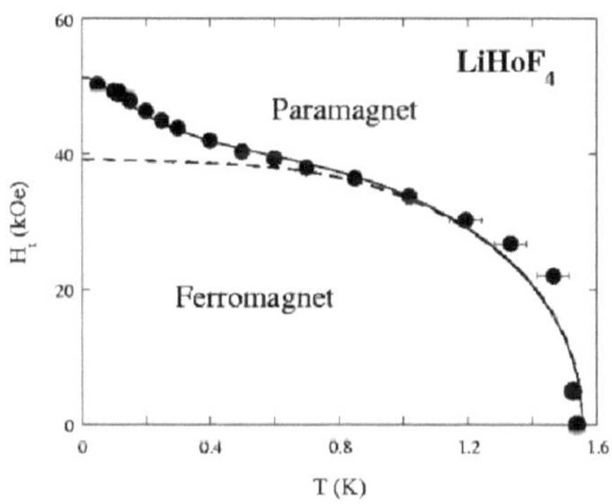

Figure 2.2. Phase diagram for pure $LiHoF_4$ in a magnetic field H_t applied perpendicular to the Ising easy axis. Dashed line is from mean field theory accounting only for electronic moments; solid line includes nuclear moments as well. From ref. [12].

The experimental findings for pure $LiHoF_4$ are in accord with longstanding theory[13], which equates $T = 0$ quantum phase transitions in d dimensions with thermal phase transitions in $d + 1$ dimensions. Actually, due to the long range nature of dipolar interactions which dominate the physics of $LiHoF_4$, the effective dimensionality[14] for the thermal phase transition is four, already implying mean field exponents (with logarithmic corrections) for the thermal transition.

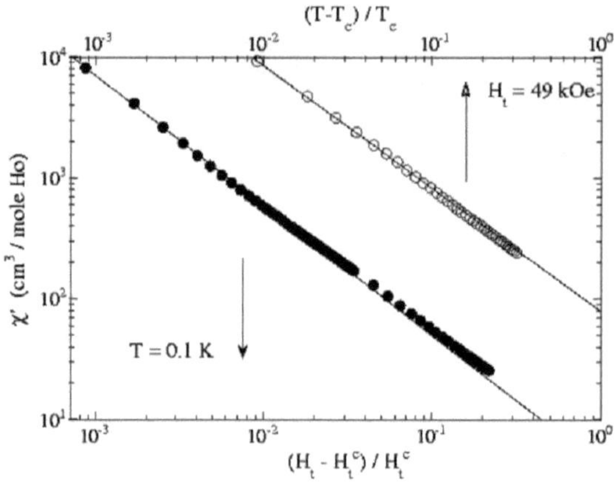

Figure 2.3. Magnetic susceptibility along thermal(T) and quantum(H_t) trajectories near the quantum critical point for $LiHoF_4$. From ref. [12].

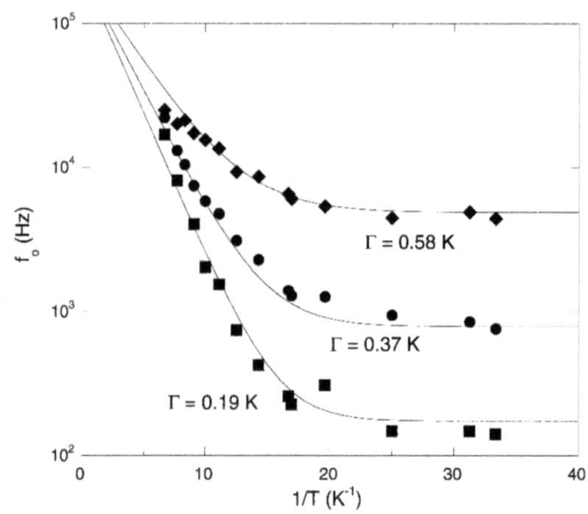

Figure 2.4. Characteristic relaxation rate f_0 for magnetization in the disordered ferromagnet $LiHo_{0.44}Y_{0.56}F_4$ as a function of $1/T$ where T is temperature and the effective transverse field Γ, computed from the laboratory transverse field H_t. For high T(at left), the relaxation rate follows a thermal activation law, whereas for low T, it is dominated by quantum tunneling and becomes T-independent, but susceptible to speed-up as quantum fluctuations are enhanced with rising Γ. From ref. [20].

Upon substitution of a non-magnetic Y for Ho, the behavior of the system is still largely in agreement with mean field theory, as long as the Y fraction, $1-x$, is not too large and no quantum disorder is inserted via a transverse field. Specifically, there is ferromagnetic order which sets in sharply at $T_c(x) = xT_c(x=1)$, which for smaller x (e.g. $x = 0.167$) gives way to spin glass order appearing at a not much lower temperature, due to the random sign of the dipolar interaction linking nearest Ho neighbors placed at random. At lower dilution in zero transverse field, and for finite transverse field even in the ferromagnetic regime, there are many effects entirely at odds with the mean field approach:

- The quantum critical point occurs at a transverse field which is unexpectedly low, i.e. Γ_c does not scale with x in the way in which $T_c(x)$ does.
- Near its quantum critical point, the disordered ferromagnet(with $x = 0.44$) acquires a broad spectrum of relaxation times not seen near its classical critical point[15].
- The paramagnetic-glass boundary[10] for the material which in zero transverse field behaves as a classical spin glass, seems to be characterized by an exponent for the non-linear susceptibility[16] that vanishes as the quantum glass transition is approached.
- For a surprisingly high Ho fraction, $x = 0.04$, the spin glass phase gives way to an antiglass state[17, 18] for which the distribution of barriers to relaxation narrows rather than broadens on cooling. This is completely unexpected classically, where for arbitrarily small concentrations of dipoles, spin glass behavior always obtains [19].

A full discussion of the above points would require a separate review article. We mention these results here simply to emphasize how the interplay of disorder and quantum mechanics produces entirely unexpected phenomena even in a high-dimensional insulator. Before leaving the subject, however, we point out that $Li(Ho, Y)F_4$ in a transverse field has also permitted two experiments which in many ways provide key justifications for work on quantum magnetism. The first, illustrated in Figure 2.4, revealed tunable quantum tunneling[20] of ferromagnetic domain walls, while the second[15], illustrated in Figure 2.5, showed that quantum processes can be used to anneal complex systems, with outcomes different from the classical, thermal annealing proposed by Kirkpatrick, Gelatt, and Vecchi[21]. Quantum routes to solving general optimization problems are now being applied to problems not only in

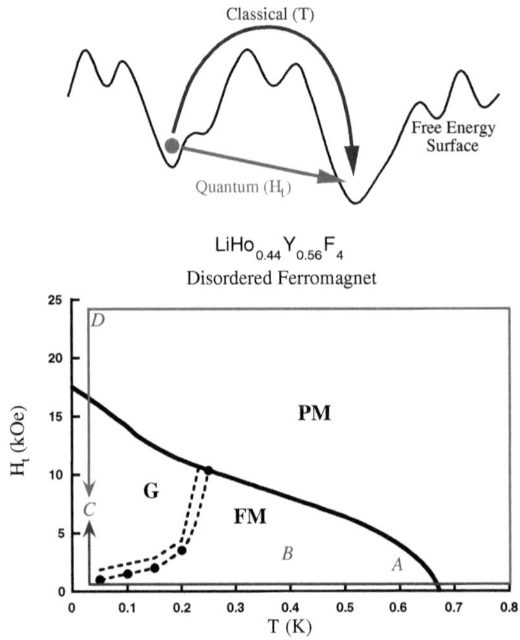

Figure 2.5. The concept(upper frame)underlying quantum annealing is that a complex state space can be explored using quantum tunneling instead of or in addition to the thermal hopping paradigm promoted by Kirkpatrick, Gelatt, and Vecchi [21]. Brooke and collaborators[22] have tested this concept for the disordered ferromagnet $LiH_{0.44}\ Y_{0.56}\ F_4$, for which ferromagnetic domain wall tunneling is tuneable, as described in Figure 2.4 and its caption. They explored the consequences of approaching the same low-T point in $H_t - T$ plane via routes(lower frame) where the primary settling from a disordered paramagnetic state was due to thermal(blue) and quantum(red) relaxation, respectively. The experiments showed that the states reached along the two different trajectories are different, with the quantum trajectory favoring a state with intrinsically more rapid fluctuations.

physics but also in biology, and already appear to have certain advantages[22].

2.2 Magnets with Gapless or nearly gapless spectra

From the highly anisotropic Ising model based on rare earth ions, we turn to quasi-isotropic spin systems based on transition metal ions[23]. Isotropic magnets with static correlations necessarily have a gapless spectrum due to long wave length excitations that twist the order parameter through isotropic spin space[25, 26]. Quantum and thermal fluctuations of these (putative) Goldstone modes can destabilize static correlations in low dimensional[8] and frustrated systems[2]. In this section we discuss renormalized classical and near quantum critical systems[9] where the excitation spectrum remains gapless on the scale of the exchange constant, J, at low temperatures.

2.2.1 Spin-1/2 Antiferromagnetic Chain. The uniform antiferromagnetic Heisenberg spin-1/2 chain is gapless and quantum critical at $T = 0$. The exact Bethe ansatz ground state has dynamic quasi-long-range order associated with resonating valence bonds on all length scales[27, 28]. The excited states are topological excitations called spinons[29, 30], uncharged Fermions that carry spin-1/2. Neutrons can only create and annihilate spinons in pairs and consequently the experiment probes the matrix element weighted joint spinon density of states rather than the spinon dispersion relation[31, 32, 33].

Orbitally anisotropic, the Kramers doublet Cu II ion is naturally predisposed to forming low-dimensional Heisenberg spin-1/2 systems. Figure 2.6 shows high quality neutron scattering data from the antiferromagnetic spin-1/2 chain $KCuF_3$ obtained using time of flight instrumentation at the ISIS facility[34, 35, 36]. The important qualitative feature is that there is a continuum at every wave vector, not a resonant mode that would have produce a resolution limited ridge. This confirms that quasi-particles cannot be individually created or anihilated as magnons when there is static long range order. Previous data from $KCuF_3$ has been found[34, 35] to be consistent with an approximate two spinon scattering cross section due to Müller *et al*[31, 32]. Recently, the exact two-spinon scattering cross section was calculated[37, 33] and it was shown that the distinction from the approximate expression is quite subtle[33, 38].

The Jordan-Wigner transformation maps the spin-1/2 chain to a one-dimensional lattice of spin-less Fermions. For the easy plane spin-1/2 chain there are no interactions between Fermions so that system maps

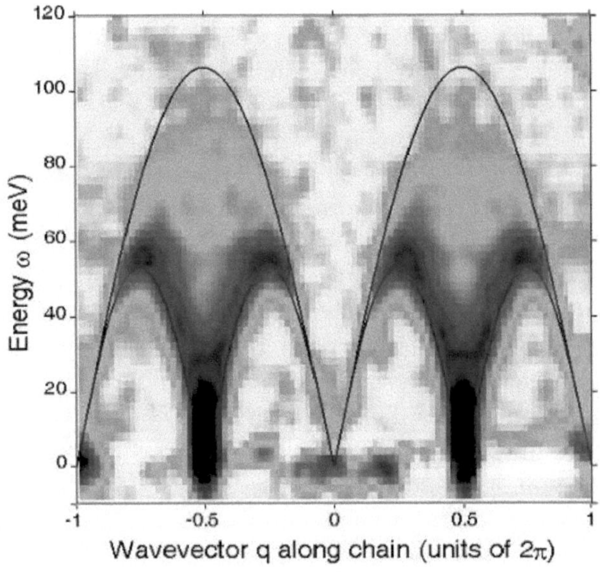

Figure 2.6. Inelastic magnetic neutron scattering from a single crystal of the antiferromagnetic spin-1/2 chain $KCuF_3$. The data was acquired on the time of flight spectrometer MAPS at the ISIS spallation neutron source with the chain axis oriented perpendicular to the incident beam. Reproduced from Ref. [35, 36].

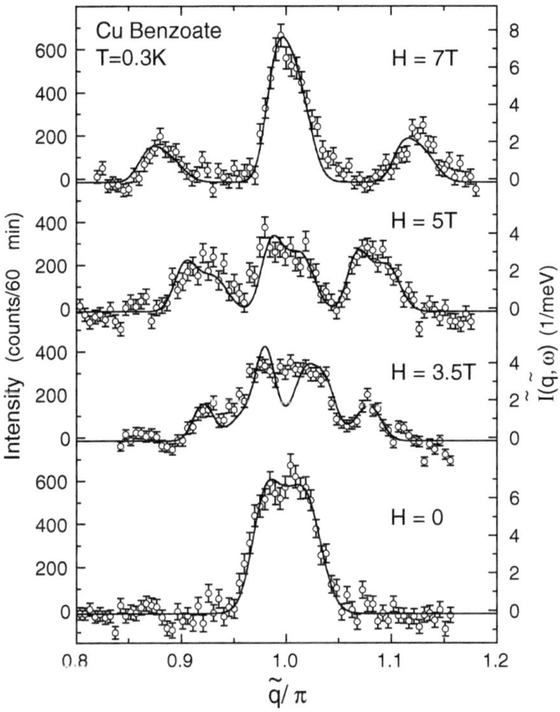

Figure 2.7. Constant energy scans for $\hbar\omega = 0.21$ meV for various magnetic fields at $T = 0.3$ K in the uniform antiferromagnetic spin-1/2 chain, copper benzoate. The data provide evidence for incommensurate soft modes in the magnetized state. Solid lines show a resolution convoluted model calculation as described in Ref.[42].

to a degenerate Fermi gas[39]. The longitudinal term that is added for Heisenberg spins corresponds to strong interactions between Fermions. Nonetheless, the quasi-particles remain Fermions in a so called Luttinger liquid state. Evidence of Fermionic quasi-particles in a spin-1/2 chain can be obtained from neutron scattering experiments that probe the system in a strong magnetic field. An applied field should shift the Luttinger liquid away from half filling so low energy excitations with wave vector transfers that connect the Fermi points become incommensurate [40, 41].

While the exchange constant is too large to observe such high field effects in $KCuF_3$, there are organo-metallic spin-1/2 systems with exchange constants of order 10 K, where a reduced field as large as, $g\mu_B H/J \approx 1$, can be achieved[38]. Figure 2.7 shows low energy constant energy cuts through $\mathcal{S}(\mathbf{Q},\omega)$ for the spin-1/2 chain copper benzoate at various applied fields. Upon application of a field, incommensurate peaks

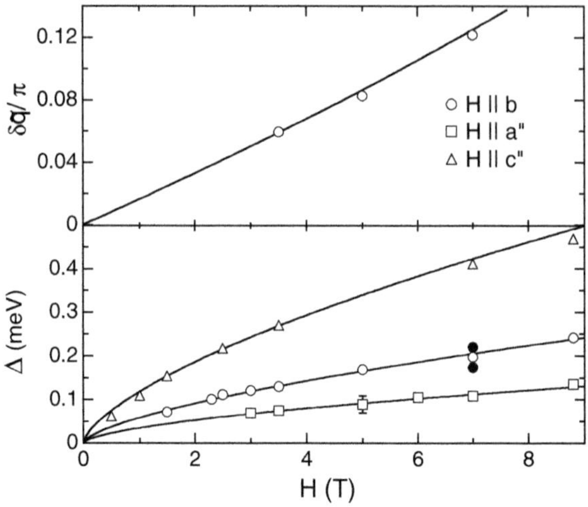

Figure 2.8. (a) Field dependence of the incommensurate wave vector for low energy spin excitations in the spin-1/2 chain copper benzoate. (b) Field dependence of the gap in the excitation spectrum of same material from neutron scattering and specific heat data. Reproduced from Ref. [42].

indeed split off from the $q = \pi$ high symmetry point. The rate of shift (see Fig. 2.8(a)) is consistent with the Fermi velocity that can be derived from the zero field excitation spectrum[24] or from low temperature specific heat data[43].

Surprisingly, the field also induces a gap in the excitation spectrum for copper benzoate[42]. The gap is due to a residual transverse staggered field that grows in proportion to the applied field as a results of a staggered Landé g-tensor and staggered Dzyaloshinskii-Moriya interactions. A staggered field is a relevant operator for the Luttinger liquid and drives a phase transition from the zero field quantum critical state, to a finite field state with uniform- and staggered-magnetization and a gapped excitation spectrum. To lowest order the gap is expected to rise as a power-law and it can be shown that in the limit $H \to 0$ the power is 2/3[44, 45, 46]. The data in Fig. 2.8(b) bears out this parameter free results.

2.2.2 Square Lattice S=1/2 Antiferromagnet. The square lattice antiferromagnet has been the subject of perennial study because this model is realized by a variety of lamellar materials containing transition metal ions such as Ni and Cu. Additional impetus is due to the fact that the high-temperature superconductors are derivatives of materials that are, for many purposes, well-described as two-dimensional

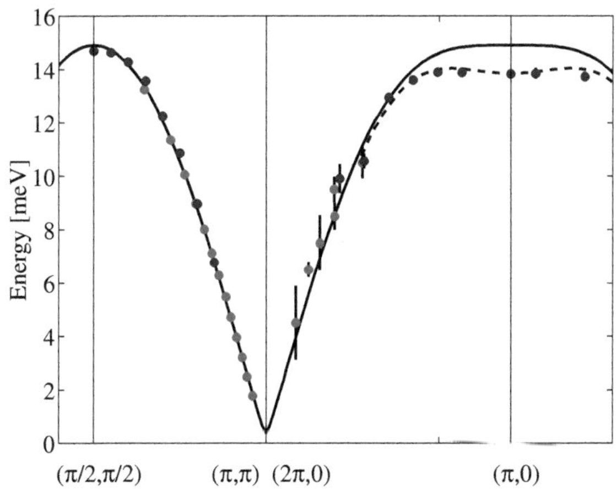

Figure 2.9. Measured dispersion for spin waves in the model two-dimensional Heisenberg antiferromagnet copper formate tetra-deuterate, from ref. [51].

$S = 1/2$ Heisenberg antiferromagnets. We discuss these materials - more properly thought of as Mott-Hubbard insulators - in the next section, and describe here recent progress on compounds that come closer to the Heisenberg ideal. One disappointment has been how well renormalized classical theory describes the experiments for these materials. In particular, there is a substantial ordered moment ($\langle S_z \rangle = 60$ % of the maximum value for $S = 1/2$), the dominant excitations at $T = 0$ are spin waves, and warming leads to an exponentially activated - rather than linear in $T-$form for the temperature ($T-$)dependent magnetic correlation length[47, 48, 49, 50].

One material that has been extensively characterized within the last years is the deuterated (to remove the large incoherent background associated with neutron scattering from hydrogen) analog (CFTD) of copper formate tetrahydrate, an organometallic for which the exchange constants are sufficiently low so that the magnetic properties can be easily mapped up to energies and temperatures of order the underlying nearest neighbor exchange constant, $J = 6.3$ meV. We focus[51] first on the excitations at zero temperature, namely the spin waves, whose dispersion is shown in Fig. 2.9. The solid line through the data corresponds to the classical prediction. What is especially significant is that since the controlling factor in the classical theory is the spatial Fourier transform of the pattern of exchange interactions, and this quantity is independent of position along the magnetic zone boundary, joining e.g. $(\pi, 0)$ and $(2\pi, \pi)$, the spin waves should also be dispersionless along the magnetic zone boundary as well. However, when quantum corrections are inserted, this degeneracy is lifted. This follows because the classical Néel state is not the actual ground state for the system, and the real ground state wavefunction contains corrections involving spins flipped relative to simple unit cell doubling. If a spin displaced along the diagonal from such a defect is flipped to build the wavefunction for a spin wave, the energy cost remains 4J, as for the undisturbed Néel state. On the other hand, if it is displaced along a square edge, the energy cost will be lower because reversal lowers the energy of the bond along the displacement by $2J$ (see Fig. 2.10). Spin waves should therefore be more energetic at $(\pi/2, \pi/2)$ than at $(\pi, 0)$, as is indeed observed for CFTD. The line of reasoning just given can be followed through quantitatively in a series expansion [52] away from the Ising limit, and yields the quantum renormalized spin wave dispersion indicated by the dashed line in Fig. 2.9. The agreement of theory and experiment is excellent, as it is also[53] for a square sublattice of the much more complex antiferromagnet, $Sr_2Cu_3O_4Cl_2$.

Other results for the CFTD include relatively complete images of the melting of the magnetic order on warming. Fig. 2.11 summarizes how the spin waves soften and broaden on warming. There is no need to invoke cross-over to a quantum critical regime at high temperature to account for these data[51].

2.2.3 Two-dimensional Hubbard Model at half filling.

The Hubbard model contains key features of the high-temperature superconductors and reduces to the Heisenberg model in the limit when the on-site Coulomb interaction U becomes infinite. For these reasons, the two-dimensional $S = 1/2$ Heisenberg model has become very popular, and most studies of the parents of the cuprate superconductors have been interpreted in the context of the Heisenberg model. Because these studies have been extensively reviewed elsewhere[54], we concentrate here on a neutron scattering experiment which brings out the difference between the Hubbard and Heisenberg models.

Fig. 2.12 shows the spin wave dispersion[55], together with the spin wave intensities, along the major symmetry directions for La_2CuO_4. We see conventional spin waves with conventional amplitudes, in agreement with theory (solid line). Note the diverging amplitude as the magnetic zone center at $(\pi\pi)$ is approached; it is this divergence which distinguishes antiferromagnets from ferromagnets, and which results in the infrared catastrophe for the former as $d \to 1$. Equally interesting, though, is that the spin waves disperse along the zone boundary between $(\pi 0)$ and $(3\pi/2\pi)$, but with sign opposite to that for the Heisenberg model and CFTD - i.e. the spin waves along the diagonal are less energetic

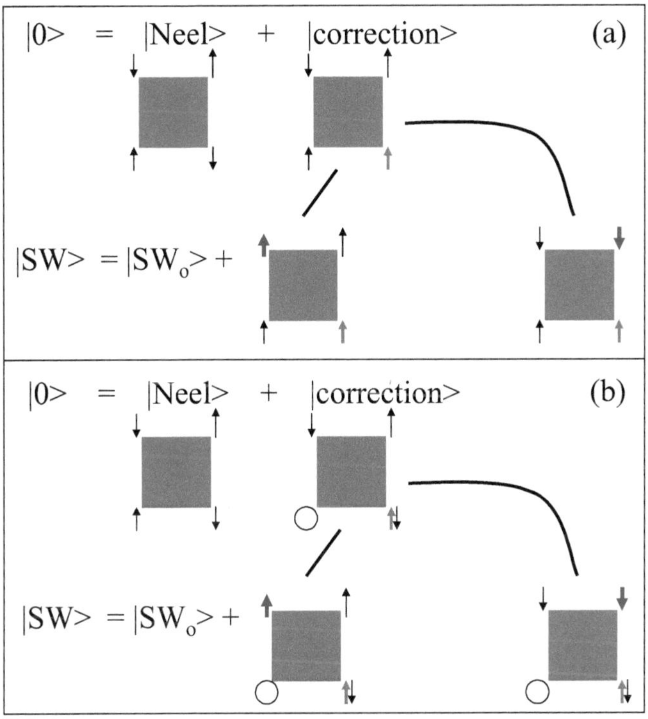

Figure 2.10. Rationale for why spin waves disperse along magnetic zone boundaries in quantum cases even though they are not allowed to do so classically for the simple Heisenberg antiferromagnet with nearest-neighbor coupling only. In (a) we consider the $S = 1/2$ Heisenberg antiferromagnet with completely localized electrons. The true ground state $|0\rangle$ is the classical Néel state $|0\rangle$ with quantum corrections, which can be considered as a sum of properly phased spin flips relative to $|0\rangle$. The wavefunctions for the magnons are then the classical magnons formed from the Néel state via a set of phased spin flips, plus a series of correction terms, the first of which are simply the same spin flips applied to the corrections to the ground state. The energy cost for the latter will depend on whether the spin wave is moving along (1,0) or (1,-1) through the defective spin in the correction to $|0\rangle$; in the former case it will be lower by J because one of four "bad" bonds for the single spin flip relative to $|0\rangle$ will be healed due to the correction to $|0\rangle$. The upshot is that spin waves will be softer at the (1/2,0) zone boundary than at (1/4,1/4). In (b), we apply similar reasoning to the Hubbard model. Here, as long as U is finite, the Néel state is corrected by terms which entail pairs of unoccupied and doubly occupied sites. The cost of flipping a spin separated along the diagonal from the doubly occupied site is lower than for such a flip adjacent to the doubly occupied site on account of the fact that the flipped electron can move to the vacant site in the former case, but it would be trapped at its site in the latter instance. The discrepancy in quantum confinement energy therefore leads to a hardening of the spin waves at (1/2,0)-type zone boundaries relative to those at (1/4,1/4), a result opposite to the effects of quantum corrections on the simple Heisenberg ($U \to \infty$) model.

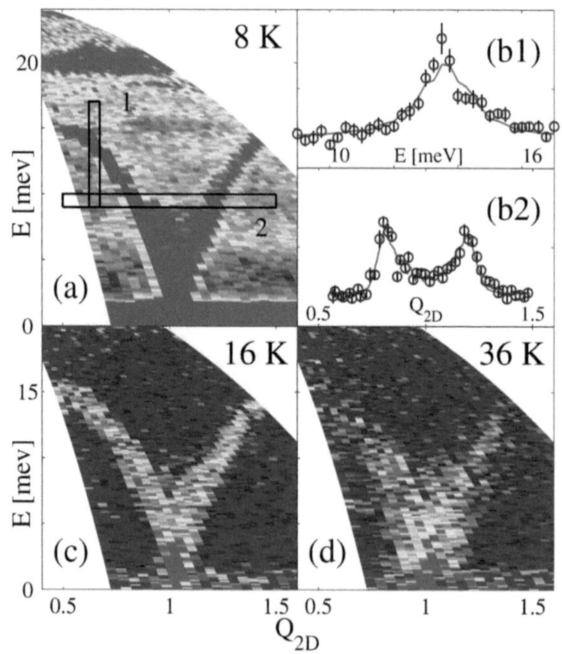

Figure 2.11. Images in momentum-energy space of the inelastic structure factor for copper formate tetra-deuterate, measured on warming using the HET instrument at the ISIS proton-driven pulsed neutron source. (b1) and (b2) show constant$-Q$ and constant$-E$ cuts respectively through the low temperature data in frame (a). From ref[51].

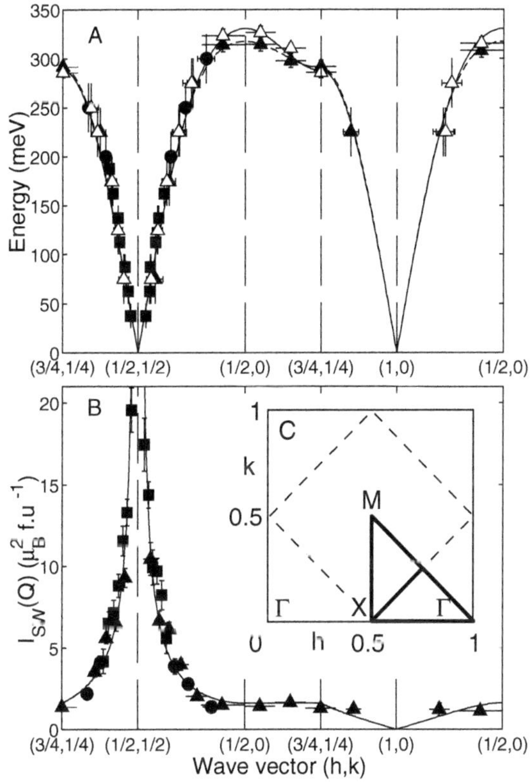

Figure 2.12. Spin wave energies and amplitudes measured for La_2CuO_4 using the position sensitive detector bank of HET at ISIS. From ref[55].

than along the edge. Several explanations may be provided, including large interactions between pairs of further neighbor copper atoms. However, quantum chemical considerations as well as comparison to other oxides of copper, including[53] $Sr_2Cu_3O_4Cl_2$ and the ladder compound [56], $Sr_{14}Cu_{24}O_{41}$, indicate that the most likely origin are cyclic four-spin exchange interactions[57] around Cu plaquettes. Such interactions J_c, in turn, are derivable from a Hubbard model, and come about because of fluctuations of the charge state of the magnetic ions. Because the on-site Coulomb repulsion, U, is finite, such fluctuations are allowed, and give rise to a substantial $J_c = 61(8)$ meV. Indeed, from the measured spin wave dispersion relation, it is possible to derive[58] the two parameters for a single-band Hubbard model to describe the CuO_2 planes in La_2CuO_4. The outcome is a bandwidth $t = 0.33(2)$ eV and $U = 2.2(4)$ eV, in agreement with values obtained from photoemission[59] and optical[60] spectroscopies. Thus, a charge neutral probe - spin wave spectroscopy using neutrons - provides two key parameters describing charge fluctuations[61].

Fig. 2.10 illustrates why the zone boundary dispersion for the Hubbard model is expected to be of opposite sign to that for the Heisenberg Hamiltonian. The correction terms to the Néel state in this case must also account for doubly occupied sites with unoccupied neighbors. To gain understanding, we follow the procedure used to infer the relative energies of zone boundary spin waves for the Heisenberg case along the zone diagonal and edge, and compare the costs of spin flips next to a doubly occupied/unoccupied pair. Attempts to flip the red spin displaced from the hole along the (1,0) direction will be forbidden due to the Pauli principle, while flipping spins along the (1,1) direction (as also along the (0,1) direction) will cost $3J$. Thus, the zone boundary modes along $(\pi, 0)$ should be harder than along $(\pi\pi)$.

2.2.4 Frustrated Magnets. While the Mermin-Wagner theorem[8] precludes static long range order at finite T in a two dimensional square lattice Heisenberg antiferromagnet, the correlation length grows exponentially for $T < J$ and long range order therefore develops for $T \sim O(J)$ in materials with finite inter-plane interactions. However, if the dominant interactions are frustrated in the sense that no static spin configuration can satisfy all interactions[63] then a phase transition to long range order may be absent or T_c may be much less than J even in three dimensional systems. The experimental signature of a highly frustrated magnet is therefore a large value of the frustration index[2], $f = |\Theta_{CW}|/T_N$, and correlations that span two or three dimensions.

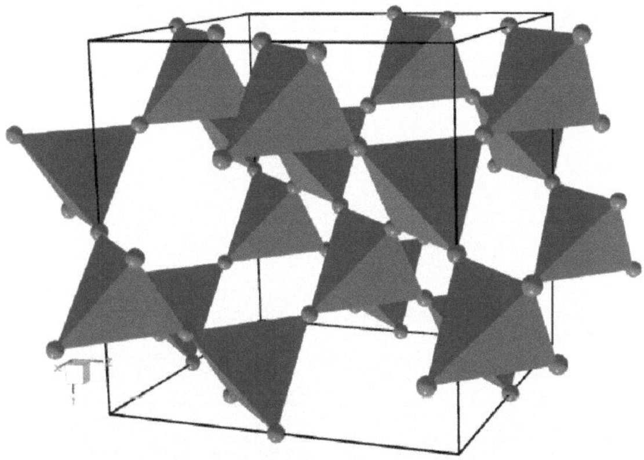

Figure 2.13. Octahedrally coordinated "B" sites in a normal spinel AB_2O_4. The space group is Fd$\bar{3}$m with chromium atoms in the parameter free 16d positions. The magnetic atoms lie on the vertices of a network of cornersharing tetrahedra. Mangetic atoms on this type of lattice are also found in the pyrochlore structure[62].

Here Θ_{CW} is the Curie-Weiss temperature derived by fitting high temperature susceptibility data to $\chi(T) = C/(T - \Theta_{CW})$.

An intriguing example is provided by spinel oxides where magnetic ions can occupy the vertices of a three dimensional network of corner-sharing tetrahedra as shown in Fig. 2.13. The spin hamiltonian for antiferromagnetically interacting spins on this lattice can be written as the sum over tetrahedra of the total spin on each tetrahedron. Each spin is part of two tetrahedra and this prevents simultaneous singlet formation on all tetrahedra. Still it appears that for spin-1/2 the ground state is a cooperative singlet[64]. For $S \to \infty$ there is a degenerate manifold of classical spin configurations all characterized by zero magnetization on all tetrahedra[65]. For intermediate spin, there is as yet no theoretical consensus on the nature of the low temperature spin state.

Zinc-chromite ($ZnCr_2O_4$) has quasi-isotropic spin-3/2 degrees of freedom on a lattice of corner-sharing tetrahedra with nearest neighbor antiferromagnetic interactions. The Curie-Weiss temperature is $\Theta_{CW} = -390$ K[2] and yet the material does not undergo a phase transition until $T = 12.5$ K[66]. The quasi-elastic excitation spectrum with a relaxation rate of order $k_B T/\hbar$ indicates that the material is close to a quantum critical point. The fluctuations in the quantum critical phase of $ZnCr_2O_4$ feature a distinct wave vector dependence that is consistent with short range antiferromagnetic correlations within hexagonal rings formed by

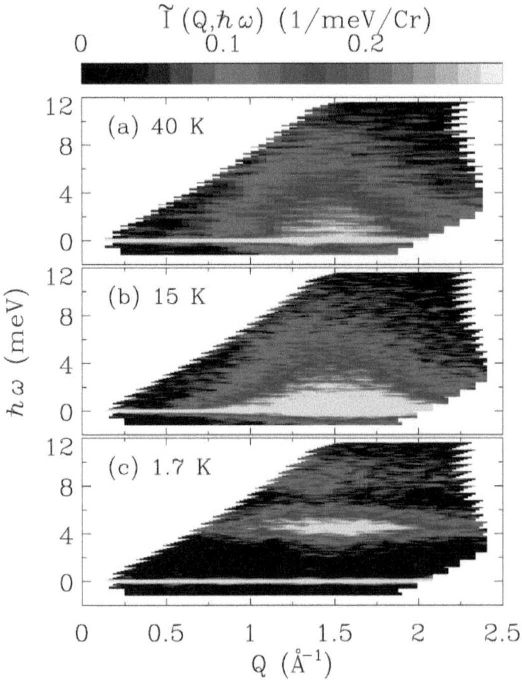

Figure 2.14. Inelastic magnetic scattering from a powder sample of $ZnCr_2O_4$ at three temperatures surrounding a magneto-elastic transition at $T_c = 12.5$ K. The transition entails a comprehensive rearrangement of the excitation spectrum from quasi-elastic scattering for $T > T_c$ to a local spin resonance at $\hbar\omega \approx J$ and spin waves with an anisotropy gap of approximately 0.5 meV for $T < T_c$. From Ref. [66]

adjoining tetrahedra[67]. Thus it appears that one can describe the quantum critical phase in terms of weakly interacting hexagonal spin directors rather than strongly interacting spins.

As illustrated in Fig. 2.14, a gap of order the exchange constant, J, abruptly opens in the quasi-elastic excitation spectrum at the low T phase transition. The first order nature of the transition and a tetragonal lattice distortion indicate that this is an inherently magneto-elastic phase transition, where frustration associated with the highly symmetric cubic phase is relieved at the expense of elastic energy to enable a lower energy spin configuration[68, 69]. There is an interesting analogy between this phase transition and the spin-Peierls transition for the uniform spin-1/2 chain[70]. In both cases a near quantum critical phase is replaced by a lower energy spin state by deforming the lattice and in both cases there is a prominent finite energy spin resonance. There are also important differences, most notably $ZnCr_2O_4$ achieves long range order while the low temperature spin state is quantum disordered in the dimerized spin-1/2 chain. If these differences can be accepted as variants, then the spin-Peierls phenomenon can be generalized to a much wider class of quantum critical spin systems on compliant lattices.

2.3 Magnets with Gapped Excitation Spectra

Contrasting with renormalized classical spin systems and separated from them by a quantum critical point, are isotropic spin systems with an isolated singlet ground state[9]. We discuss experiments probing the fundamental excitations in quantum disordered spin systems ranging from weakly coupled dimers to a frustrated three dimensional spin-liquid. Owing to the finite correlation length, quantum disordered systems have a localized response to impurities. The intricate structure of the spin polaron that forms around impurities can be probed by inelastic magnetic neutron scattering and an example of such an experiment on a hole doped Haldane spin liquid is presented.

2.3.1 Alternating spin-1/2 chain. As the uniform spin-1/2 chain is quantum critical, there is a strong response to any so-called relevant perturbation. We saw an example in section 2.2.1 where an effective staggered field induces staggered magnetization and a spin-gap. Bond alternation, that is doubling the unit cell by adjusting the ratio of even to uneven bonds, $\alpha = J'/J$, is a relevant perturbation that creates a gap in the excitation spectrum without static spin correlations. Bond alternation can occur spontaneously in a "standard" spin-Peierls transition where lattice energy is expended to stabilize a quantum disordered

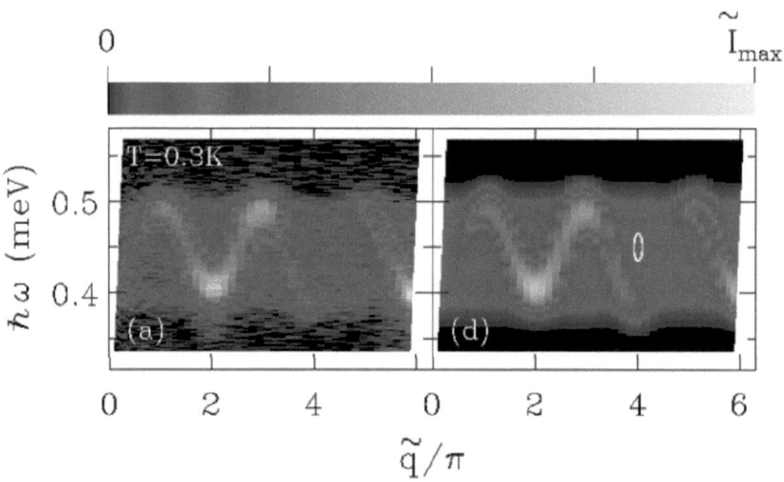

Figure 2.15. (a) Inelastic magnetic neutron scattering from copper nitrate at $T = 0.3$ K compared to (d) a single mode approximation that takes into account the crystal structure, the instrumentation resolution and the first moment sum-rule for $\mathcal{S}(\mathbf{Q}\omega)$. The ellipsoid in (d) indicates the FWHM of the resolution function. The experiment was performed on the IRIS backscattering time of flight spectrometer at the ISIS pulsed spallation source. Adapted from ref. [71]

state[70], as observed for example in $CuGeO_3$[72]. However, there are also spin-1/2 chains with purely structural reasons for bond alternation. In the limit of strong bond alternation ($\alpha << 1$) the origin of the spin gap is easy to understand via perturbation theory. Neglecting at first the weaker bonds, J', in comparison to the stronger bonds, J, the spin system can be described as a collection of strongly coupled spin pairs that do not interact with each other. Each pair of spins with an antiferromagnetic Heisenberg exchange interactions, J, has a singlet ground state with a triplet excited state at an energy J above the ground state. Introducing intra spin pair interactions, J', yields a cosine dispersion relation with bandwidth J' as localized triplet excitations become wave packets that can propagate along the chain.

Neutron scattering experiments were performed on the alternating spin-1/2 chain copper nitrate[71] ($Cu(Ni_3)_2 \cdot 2.5D_2O$) and low temperature data for the dynamic spin correlation function are shown in Figure 2.15. The small ratio of single magnon bandwidth to mean singlet-triplet transition energy indicates a strongly dimerized system ($\alpha = 0.24$) that should be amenable to perturbation theory from the isolated dimer limit. Also shown is the calculated intensity distribution derived from perturbation theory and the single mode approximation[83], which is in

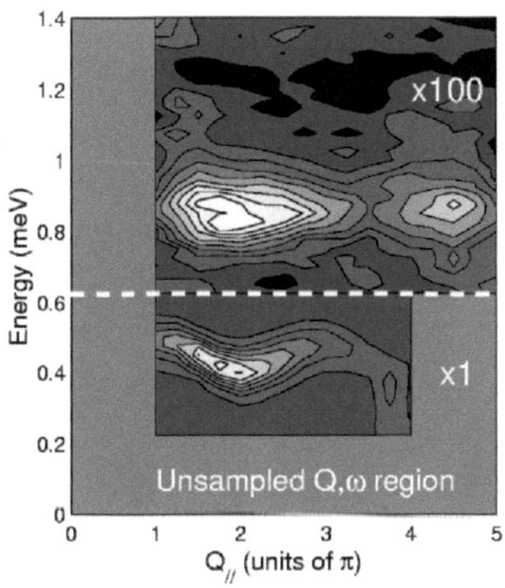

Figure 2.16. Inelastic magnetic neutron scattering from the alternating spin-1/2 chain copper nitrate at $T = 0.3$ K. The data shows both one magnon and two magnon scattering. Note the factor one hundred change in scale between the two contributions to $\mathcal{S}(\mathbf{Q}\omega)$. The data were acquired on the SPINS cold neutron triple axis spectrometer at NIST using a horizontally focusing analyzer and a composite deuterated $Cu(NO_3)_2 \cdot 2.5D_2O$ sample with a total mass of 14.1 g. From Ref. [73]

good agreement with the experimental data. Note in particular that the "incommensurate" wave vector dependence of the intensity is a simple consequence of structural dimerization[71].

Looking with greater sensitivity at higher energies, a two-magnon contribution to the excitation spectrum was identified[73] and it is shown in Fig. 2.16. The two magnon cross section is approximately 2 % of the one-magnon cross section in copper nitrate, which is also consistent with $O(\alpha^2)$ perturbation theory. Magnons do have short range interactions and it has been predicted that they can form singlet and triplet two-magnon bound states for certain ranges of center of mass momentum [74]. The experimental data is consistent with a predicted bound state at $Q_{//} = 3\pi$ though a bound state is not required to account for the relatively low-resolution data that is presently available[73].

2.3.2 Haldane spin-1 Chain. Experimentally, the distinctions between a Néel ordered magnet and the spin-1/2 chain are actually quite subtle. This is however not the case for the Haldane spin-1 chain. This system has an isolated singlet ground state[75, 76, 1] and the gap, Δ, to excited states produces exponentially activated behavior at low temperatures ($T << \Delta$). Despite its prominent effects on all thermo-

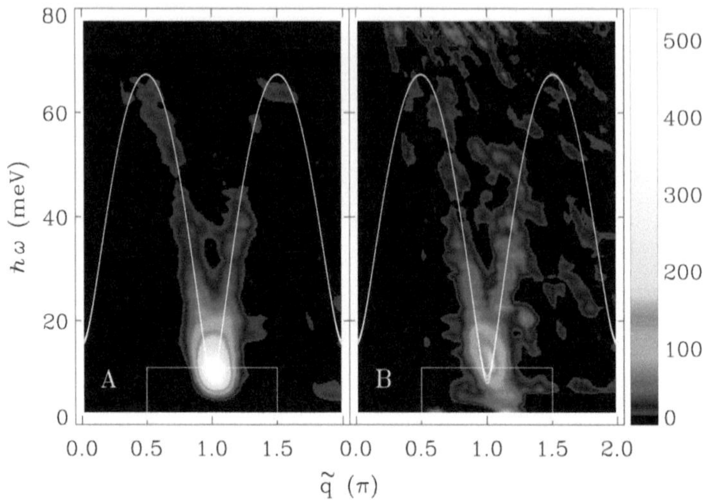

Figure 2.17. Dynamic spin correlation function for spin-1 chains in pure and doped Y_2BaNiO_5. frame A shows data for a pure sample while frame B is data for $Y_{2-x}Ca_xBaNiO_5$ with $x = 9.5$ %. There is a gapped coherent mode in the pure sample and much as in a semiconductor, doping introduces bound states in the gap. Boxes indicate regions examined in Fig. 2.20. The color bar shows values for $\frac{ki}{kf}\frac{d^2\sigma}{d\Omega dE_f}$ in units of mbarn meV^{-1} per Ni. From Ref. [80]

magnetic properties, the gap was first discovered theoretically by F. D. M. Haldane[75, 76]. The original work was based on a large S mapping of the spin hamiltonian to a continuum field theory, the non-linear sigma model. A similar mapping can be undertaken for the spin-1/2 chain but for half odd integer spins the Lagrangian has a topological term that gives rise to gapless "instanton" excitations (spinons). The nature of the Haldane ground state was elucidated through the discovery by Affleck, Kennedy, Lieb, and Tasaki, that the spin hamiltonian for the spin-1 chain is closely related to a total spin pair projection operator on the spin-2 sector[77]. The ground state for that model is a valence bond solid, where the two spin-1/2 degrees of freedom that make up each spin-1 form singlets with their counterparts on the two neighboring sites. The valence bond solid states can be generalized to higher dimensions and may form a useful basis for understanding magnetization plateaus in more complicated systems[78, 79].

A direct view of the dynamic spin correlation function for the antiferromagnetic spin-1 chain was obtained through neutron scattering on Y_2BaNiO_5[80] and is shown in Fig. 2.17. The Haldane gap in this material is approximately 9 meV with a splitting of 2 meV due to crystalline anisotropy[81]. A resonant mode extends over much of the Brillouin zone and it has been shown to carry most of the spectral weight[82]. Consequently the wave vector dependence of the intensity is well accounted for by the single mode approximation[83]. Conceptually this mode can be described as a triplet bound state that propagates coherently through the singlet ground state. It was recently shown that finite temperature properties of the spin-1 chain can be accounted for through a semi-classical approximation to interacting triplet wave packets[84].

Returning to the absolute zero temperature, for smaller wave vector transfer there is a continuum[85] that was recently detected experimentally[86] through neutron scattering experiments on $CsNiCl_3$. Neutron scattering experiments have also detected a continuum at higher energies for $q \approx \pi$ in this material[87]. This continuum is not expected in the one dimensional limit and there are experimental indications that inter-chain interactions play a role in producing it[88].

2.3.3 Frustrated Magnets with a spin gap. There are many organo-metallic spin systems with complex patterns of interacting spin-1/2 degrees of freedom and a singlet ground state[89, 90, 91, 92, 38]. These moment free magnets are of interest for exploring quantum critical phase transitions as the spin gap can be closed using a 10 T superconducting magnet. The common explanation for the zero field gap is singlet formation due to some form of dimerization as in copper nitrate (see section 2.3.1). However, closer examination with neutron scattering indicates that frustration often plays an important role in suppressing Néel order and stabilizing the quantum disordered phase in such materials.

One example is CuHpCl ($Cu_2(C_5H_{12}N_2)_2Cl_4$), which had an early life as a spin-1/2 ladder[93]. The ladder model however turned out to be inconsistent with inelastic neutron scattering experiments from a powder sample[96, 95] (see Fig. 2.18). A model free analysis of the inelastic neutron scattering data based on the first moment sum-rule[83] revealed that the ground state features frustrated bonds (see Fig. 2.19). Co-analysis of powder and single crystal inelastic scattering data revealed a highly complex three dimensional magnetic lattice in CuHpCl[95], which can be described as interleaving sheets of distorted Shastry-Sutherland [97] spin planes. The molecular spin-1/2 pair that is the central motif is frustrated in the ground state, that is the corresponding bond energy raises rather than lowers the ground state energy[95]. The molecular bond is the third leg in a total of six frustrated spin triangles. Apparently localizing frustration to this bond is worth the energy gain from singlet

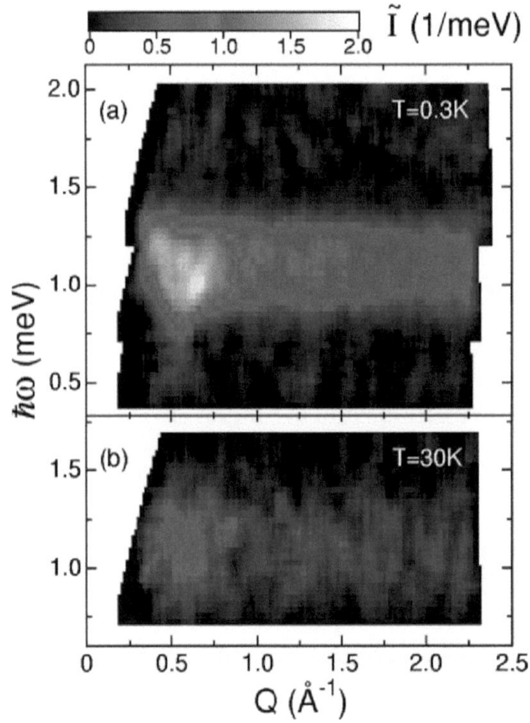

Figure 2.18. Neutron scattering from a powder sample of CuHpCl at (a) $T = 0.3$ K and (b) $T = 30$ K. The low temperature gap in the excitation spectrum is consistent with expectations based on specific heat and susceptibility data that can be accounted for by the spin ladder model[93, 94]. However, the wave vector dependence of the neutron scattering data is inconsistent with the spin ladder model, which predicts a global maximum in $\mathcal{S}(Q)$ at $Q = 1.3$ Å^{-1}. From Ref. [95]

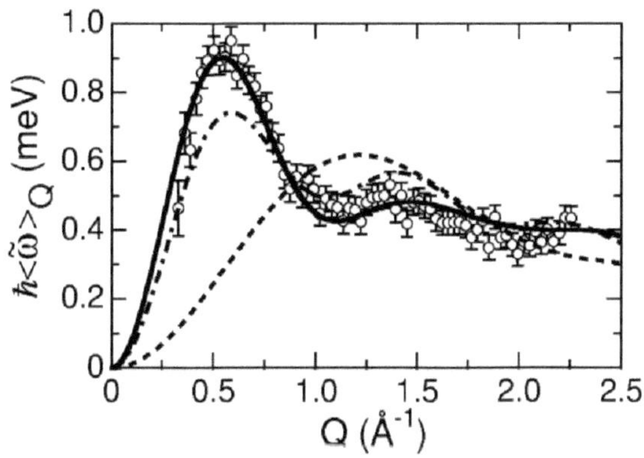

Figure 2.19. Wave vector dependence of the first moment $\hbar\langle\omega\rangle_{\mathbf{Q}} \equiv \hbar^2\int_{-\infty}^{\infty}\omega\mathcal{S}^{\alpha\alpha}(\mathbf{Q},\omega)d\omega$ of inelastic magnetic powder neutron scattering from CuHpCl. For a spin system described by a Heisenberg spin hamiltonian, this quantity is a lattice fourier transform of inter-spin bond energies[83, 82]. The dashed line is the prediction for a spin ladder. The dashed dotted line is the best fit with a single dominant bond. It fails to account for the large peak to high $Q-$plateau ratio. This feature of the data can be directly related to bond frustration. The solid line is the best self consistent fit to powder and single crystal data, which incorporates several frustrated bonds. From Ref. [95]

formation on the surrounding spin pairs. An interesting consequence of this is that each molecule in CuHpCl is an effective spin-1 degree of freedom that goes on to form a cooperative singlet with neighboring molecules.

PHCC (Piperazinium hexachlorodicuprate) is a quasi-two-dimensional example of a complex frustrated singlet ground state system[98]. While further theoretical and experimental work is needed to understand organic spin-1/2 singlet ground state systems, it is clear that they form a considerably more interesting class of materials than previously recognized where frustration can play an important role despite low symmetry.

2.3.4 Impurities in Gapped Quantum Magnets. Gapped quantum magnets are in many ways the magnetic analogue of semiconductors. In the pure state at low temperatures the magnetic susceptibility is exponentially activated as is the conductivity of a semiconductor. And just as impurities enhance the conductivity of a semiconductor, impurities generally enhance the susceptibility of a gapped quantum magnet. The analogy springs from the cooperative singlet nature of both

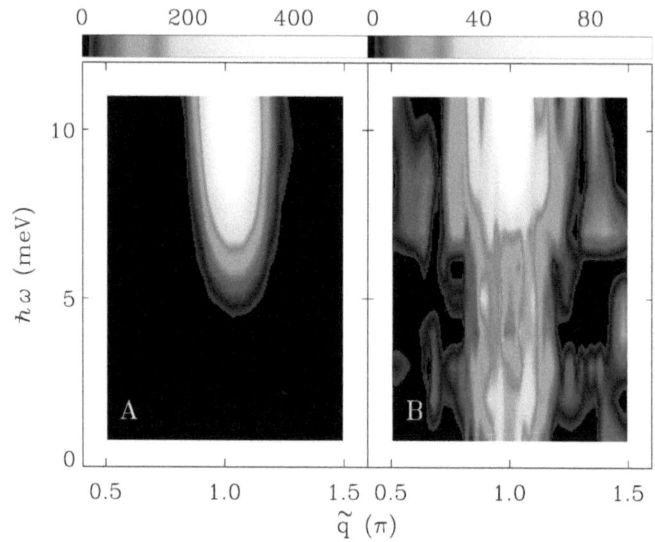

Figure 2.20. Low energy detail of magnetic excitations in (A) pure and (B) 9.5 % calcium doped Y_2BaNiO_5. (A) shows time-of-flight data at T=10K (MARI spectrometer, ISIS pulsed neutron source) while (B) shows data collected at T=1.5K using triple-axis spectrometers (SPINS and BT2 at the NIST steady state neutron source with final energies 5 meV and 14.7 meV respectively). The color bar shows values for $\frac{ki}{kf}\frac{d^2\sigma}{d\Omega dE_f}$ in units of mbarn meV^{-1} per Ni. From Ref. [80]

systems and the fact that doping offsets a delicate balance that cancels magnetism and charge transport respectively in the clean limit.

One of the interesting aspects of doped quantum magnets is that impurities create complex isolated spin degrees of freedom, not simple atomic spins. This cannot be appreciated with standard susceptibility measurements where the signature of impurity spin is a low temperature $1/T$ up-turn[99] typically denoted a "Curie-tail". While such measurements provide access to the overall density of impurity spin, magnetic neutron scattering is required to unravel the detailed structure of the corresponding spin polarons. An interesting example was provided by experiments on $Y_{2-x}Ca_xBaNiO_5$. Replacing Ca^{2+} for Y^{3+} creates holes on superexchange mediating oxygen atoms that become potentially mobile ferromagnetic impurity bonds in the Haldane spin-1 antiferromagnet[100, 80]. Figure 2.20 provides the low energy detail of $\mathcal{S}(Q\omega)$ for pure and Ca-doped Y_2BaNiO_5. As was already apparent from Fig. 2.17, calcium doping introduces magnetic excitations below the clean limit Haldane gap. In addition, the greater detail of Fig. 2.20, reveals a doubly peaked wave vector dependence of the scattering cross section.

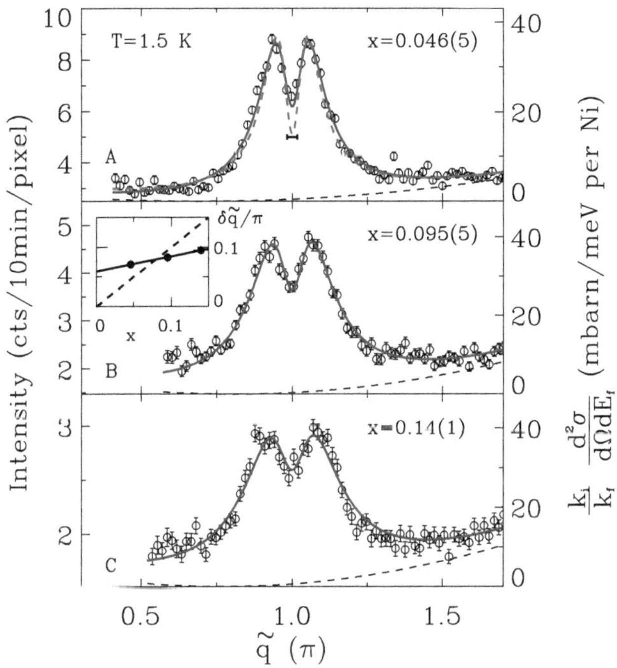

Figure 2.21. $Q-$scans, collected using SPINS at NIST, through the sub-gap inelastic scattering that develops on Ca doping Y_2BaNiO_5. Average energy transfer $\hbar\omega = 4.5$ meV and the energy resolution of the spectrometer was 2 meV full width at half maximum. The dashed green line in A shows a single impurity model convoluted with the instrumental resolution (solid bar in frame A). The red lines take into account that neighboring impurities truncate the spin polarization cloud around an impurity. Inset in B shows half the distance $\delta\tilde{q}(x)/\pi$ between the peaks of two Lorentzians superposed to fit the data. From Ref. [80]

Fig. 2.21 shows the wave vector dependence of the sub gap scattering for three different samples. The peak positions do not depend on the level of doping which indicates a single impurity effect. The proposed structure of a single spin polaron is shown in figure 2.22. Surrounding the hole is an antiferromagnetic droplet with a central phase shift of π. While the antiferromagnetic correlations call for a peak at $q = \pi$ with a half width at half maximum that equals the inverse correlation length, $\kappa = \xi^{-1}$, the central inversion symmetry disallows scattering for $q \equiv \pi$, resulting in a doubly peaked structure factor. In the experiment the intensity does not actually vanish between the two peaks. Reasons for this include instrumental resolution, neighboring impurities, and the hole spin. All these effects can be taken into account and the solid lines in Figure 2.21 show the result. The primary parameters are the Haldane length and an overall scale factor. The impurity concentration is known from neutron activation analysis. Within the proposed model, the data place a lower limit of 2 lattice spacings on the hole-spin localization length, a number that is consistent with estimates of the hole-charge localization length from transport data.

3. Beyond the metal-insulator transition

Beyond the metal-insulator transition, the most dramatic phenomenon that occurs in two-dimensional magnets is high-temperature superconductivity. There is a vast literature on the magnetic fluctuations in the high-T_c materials, and we concentrate here only on aspects which are illuminating for quantum magnetism in general. Probably the most important result is that doping towards the metallic state actually converts the renormalized classical behavior of the insulator into quantum behavior in a way that is in surprising agreement with the theory of quantum phase transitions for two-dimensional insulators. Indeed, this is the only case we are aware of where two of the quantum critical exponents expected for a two-dimensional antiferromagnet have actually been observed.

To give more detail, we refer to the schematic phase diagram[101], shown in Fig. 2.23, for the cuprates. Here, the variables are doping x, chemical pressure y, and temperature, T. There are several phases, including the parent antiferromagnet (AFM), the high temperature superconductor (SC), spin glass (SG), and a longer period antiferromagnet commonly referred to as 'striped'. There are then several possible quantum critical points, involving for example the cross-over between the SG and SC states, and also between the SC and 'striped' states. The first evidence for quantum critical behavior was obtained[102] for

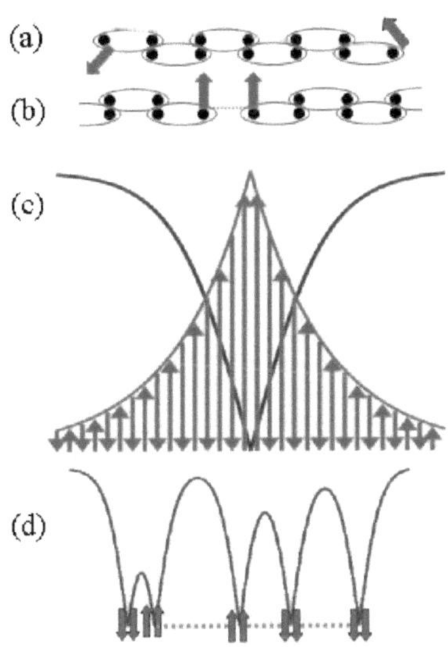

Figure 2.22. Schematic of the spin polaron surrounding a hole in a Haldane spin-1 chain. (a) shows the chain-end composite spin degree of freedom. (b) shows such composite chain-end spins coupled ferromagnetically through a hole impurity. (c) shows the proposed spin structure corresponding to (b). (d) shows polarons at finite concentration. The blue lines in (c) and (d) indicate the strength of a local observable (such as a near neighbor singlet operator) inherent to the bulk.

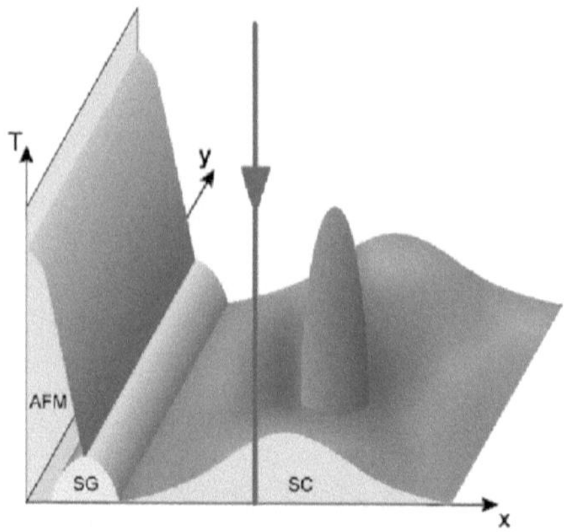

Figure 2.23. Schematic phase diagram for La_2CuO_4, where the variables are hole doping (e.g. via substitution of Sr for La) x, chemical pressure (e.g. via substitution of Nd for La) y, and temperature T. The possible phases are simple antiferromagnetic (AFM), spin glass (SG), and superconductor (SC). The dark bulge in the center of the diagram is the locus long-period striped ordering. From ref[101].

the former, and is shown in Fig. 2.24. The key conclusion is that for this $x = 0.05$ sample, which eventually undergoes spin freezing, the imaginary part of the magnetic susceptibility is well described (for $T > 75$ K) using the simple function,

$$\chi''(\omega) = \arctan(\omega/\Gamma_0), \tag{2.7}$$

where $\Gamma_0 = (k_B T/2\hbar)$. This function manifestly displays "E/T scaling" - i.e. on warming the only energy scale is temperature[9]. "E/T scaling" generically happens below the upper marginal dimensionality for a quantum phase transition, and has been seen for a few rare earth and actinide compounds in addition to both Ba- and Sr-doped[103] La_2CuO_4, as well as oxygen deficient $YBa_2Cu_3O_{6+\delta}$[104, 105].

On increasing the doping to enter the superconducting regime, the magnetic fluctuations become incommensurate. The amplitude and width in reciprocal space of the corresponding peaks in the magnetic structure function have been measured[101] as a function of temperature above the superconducting transition. Fig. 2.25 shows the outcome for a sample with doping $x = 0.14$, slightly below optimal. The important results are that the magnetic fluctuations here are nearly singular, undergoing a nearly two decade rise on cooling from room temperature to the superconducting $T_c = 35$ K. Furthermore, $k_B T$ and E added in quadrature (with no adjustable relative pre-factor) give the inverse

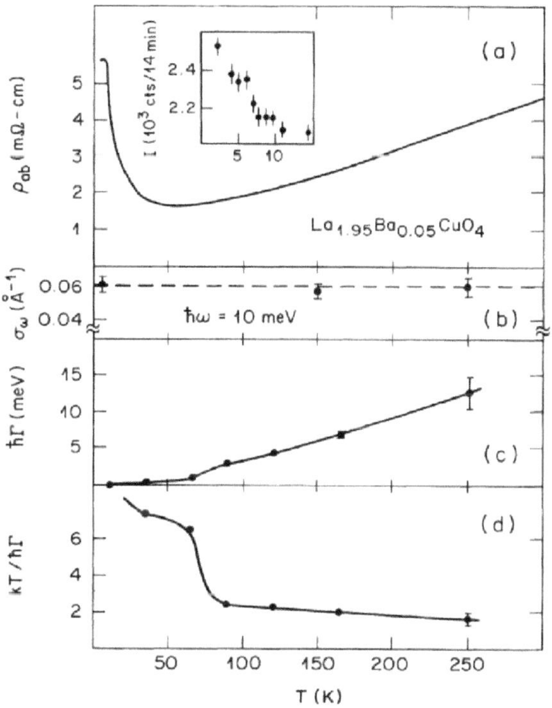

Figure 2.24. Temperature dependence of various quantities related to spin correlations in $La_{1.95}Ba_{0.05}CuO_4$. (a) Basal-plane resistivity (inset: elastic neutron scattering intensity proportional to the spin glass order parameter), (b) Gaussian width (standard deviation), σ_ω, of constant energy scans measured at $\hbar\omega = 10$ meV, (c) magnetic relaxation rate $\hbar\Gamma$, and (d) calculated ratio $k_B T/\hbar\Gamma$, which is proportional to the nuclear spin-lattice relaxation rate. From ref. [102].

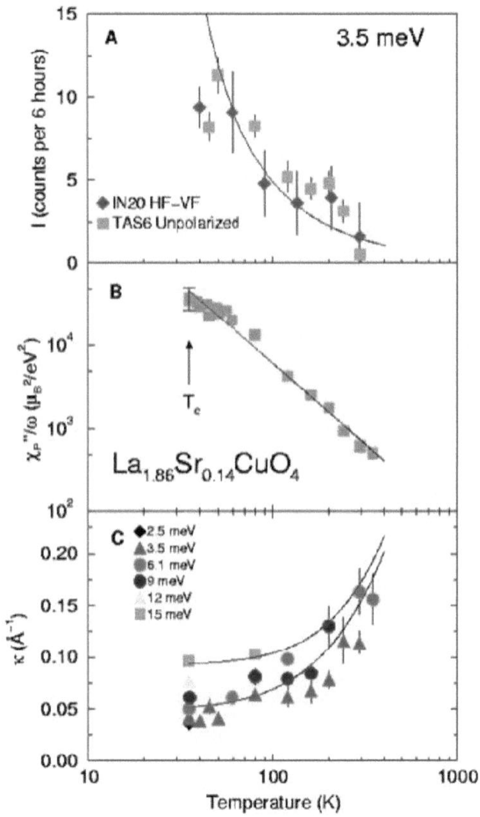

Figure 2.25. Temperature dependence of a variety of parameters characterizing the incommensurate peaks for paramagnetic $La_{1.86}Sr_{0.14}CuO_4$: (A) measured peak intensities for polarized and unpolarized neutrons, (B) ratio of imaginary susceptibility to frequency obtained at incommensurate positions from data analysis which takes instrumental resolution into account, and (C) and resolution-corrected peak widths at a variety of energy transfers. From ref. [101].

correlation length, or peak width, as shown in Fig. Fig. 2.26. Therefore, the quantum critical point that is regulating the correlation length in $La_{1.86}Sr_{0.14}CuO_4$ is consistent with a dynamical exponent $Z = 1$. We have performed a more detailed analysis, where Z was treated as an adjustable parameter, with the outcome that $Z = 1.0(2)$. Having identified one exponent, Z, characterizing the quantum critical point controlling the magnetic fluctuations in $La_{1.86}Sr_{0.14}CuO_4$, we attempt to identify another. In particular, $\chi''(\omega, T)/T$ should be proportional to $\kappa(\omega = 0, T) - \delta$ where $\delta = (2 - \eta + Z)/Z$. Thus, re-plotting χ''/ω against κ should yield d, and therefore the second critical exponent, η, characterizing the quantum spin fluctuations in $La_{1.86}Sr_{0.14}CuO_4$. The inset to Fig. 2.26 shows the result of this procedure: $\delta = 3.0(3)$, which together with our result for Z implies $\eta \approx 0$. The conclusion is that metallic and (eventually) superconducting $La_{1.86}Sr_{0.14}CuO_4$ has magnetic fluctuations with the exponents $\eta = 0$ and $Z = 1$ associated with quantum critical points in two-dimensional *insulating* magnets[106]. Thus its seems as if the mobile carriers increase the quantum fluctuations in the same fashion as, for example, frustration brought about via next nearest neighbor interactions; the mobile, Fermionic character of the carriers seems not to matter.

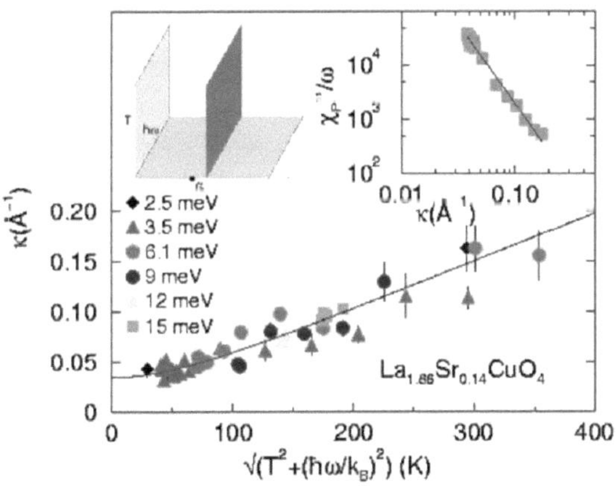

Figure 2.26. Resolution-corrected peak widths plotted against T and energy transfer added in quadrature. The solid line corresponds to a Z=1 quantum critical point. The upper left inset illustrates why this is a sensible way to plot the data: the current experiment resides on the dark plane where the quantum control parameter α is fixed by the x=0.14 Sr content of our sample, and Euclidean distances to the quantum critical point indicated by the solid circle are the regulator of the dynamic susceptibility. The upper right inset shows how the peak response depends on the extrapolated zero frequency inverse coherence length. From ref. [101].

4. Conclusions

We have given a broad, but by no means exhaustive, survey of experiments on quantum magnetism, especially in less than three dimensions, performed over the last decade. In spite of the large activity on many fronts, there are still many future opportunities in this field. Firstly, ever improving single crystal samples and neutron instruments, and within a few more years, the new Oak Ridge Spallation Neutron Source(SNS), will provide the definitive overview of magnetic fluctuations in the high temperature superconductors. This overview, especially performed as a function of variables such as magnetic field and pressure, will help to pin down (finally) whether the mechanism for high-temperature superconductivity is genuinely magnetic. While the resolution of the cuprate superconductivity problem will certainly be important, we anticipate even more exciting results to emerge from new materials built from other (i.e. not copper oxide planes) low-dimensional or frustrated building blocks, especially if mobile charge carriers can be inserted into them. This is a field with great room for ingenuity and surprise, in disciplines ranging from chemical synthesis to quantum many body physics.

Acknowledgments

We are grateful to our many colleagues who have contributed to the work reported in this article through collaboration on experiments and analysis, or through stimulating discussions. We are also grateful to S. E. Nagler for allowing reproduction of Figure 2.6, which has yet to appear in print. The US National Science Foundation supported Work at JHU through DMR-0074571. and work at NIST through DMR-9986442.

References

[1] I. Affleck, J. Phys. Cond. Matt. **1**, 3047 (1989).

[2] A.P. Ramirez, in Handbook of Magnetic Materials North-Holland, Amsterdam, **13**, 423 (2001).

[3] J. W. Allen, M. Aronson, G. S. Boebinger, C. Broholm, S. Lance Cooper, J. E. Crow, P. C. Hammel, G. Lander, Physica B **318**, 12-23 (2002).

[4] S.M. Lovesey, *Theory of Thermal Neutron Scattering from Condensed Matter*, (Clarendon Press, Oxford) 1984.

[5] P. J. Brown, in "International Tables for Crystallography", Volume C, edited by A. J. C. Wilson and E.Prince, Kluwer Academic Publishers Boston (1999).

[6] J. M. Tranquada, G. Shirane, S. M. Shapiro, "Neutron Scattering with a Triple-Axis Spectrometer", Cambridge University Press (2002).

[7] K. Skjöld and D. L. Price (Eds.), "Neutron Scattering". Volume **23** of "Methods of Experimental Physics", Orlando, Academic Press, (1987).

[8] N. D. Mermin and H. Wagner, Phys. Rev. Lett. **17**, 1133 (1966).

[9] S. Sachdev, "Quantum Phase Transitions", Cambridge University Press (2001).

[10] W. Wu et al. , Phys. Rev. Lett. **67**, 2076 (1991).

[11] J. A. Griffin, M. Huster, and R. J. Folweiler, Phys. Rev. B **22**, 4370 (1980).

[12] D. Bitko, T. F. Rosenbaum, and G. Aeppli, Phys. Rev. Lett. **77**, 940 (1996).

[13] M. Suzuki, Prog. Theor. Phys. **56**, 2076 (1976).

[14] A. Aharony and B. I. Halperin, Phys. Rev. Lett. **35**, 1308 (1975).

[15] J. Brooke, D. Bitko, T. F. Rosenbaum, G. Aeppli, Science **284**, 779-781 (1999).

[16] W. Wu et al. , Phys. Rev. Lett. **71**, 1919 (1993).

[17] D. H. Reich et al. , Phys. Rev. Lett. **59**, 1969 (1987).

[18] S. Ghosh, R. Parthasarathy, T. F. Rosenbaum, G. Aeppli, Science **296**, 2195-2198 (2002).

[19] A. Aharony and M. J. Stephen, J. Phys. C **14**, 1665 (1981).

[20] J. Brooke, T. F. Rosenbaum, G. Aeppli, Nature **413**, 610 - 613 (2001).

[21] S. Kirkpatrick, C. D. Gelatt Jr. , and M. P. Vecchi, Science **220**, 671 (1983).

[22] Giuseppe E. Santoro, Roman Martonak, Erio Tosatti, and Roberto Car Science **295:** 2427-2430 (2002).

[23] J. B. Goodenough, "Magnetism and the Chemical Bond", New York, Interscience Publishers, (1963).

[24] D. C. Dender, D. Davidovic, D. Reich, C. Broholm, K. Lefmann, and G. Aeppli, Phys. Rev. B **53**, 2583 (1996).

[25] B. I. Halperin and P. C. Hohenberg, Phys. Rev. **188**, 898 (1969).

[26] Halperin B. I. and Saslow W. M., Phys. Rev. B, **16**, 2154, (1977).

[27] H. Bethe, Z. Phys. **71**, 205 (1931).

[28] M. Karbach and G. Müller, Comp. in Phys. **11**, 36-43 (1997); M. Karbach, K. Hu, and G. Müller, ibid **12** 565-573 (1998); M. Karbach, K. Hu, and G. Müller cond-mat/0008018 (2000).

[29] J. Lowenstein, in *Recent Advances in Field Theory and Statistical Mechanics* Les Houches Summer School 1982 session XXXIX, edited by L. B. Zuber and R. Stora, North Holland, (1984).

[30] L. Fadeev, in *Recent Advances in Field Theory and Statistical Mechanics* Les Houches Summer School 1982 session XXXIX, edited by L. B. Zuber and R. Stora, North Holland, (1984).

[31] G. Müller, H. Beck, and J. C. Bonner, Phys. Rev. Lett. **43**, 75 (1979).

[32] G. Müller, H. Thomas, H. Beck, and J. C. Bonner, Phys. Rev. B **24**, 1429 (1981).

[33] M. Karbach, G. Müller, A. H. Bougourzi, A. Fledderjohann, and K. H. Mutter Phys. Rev. B **55**, 12510 (1997).

[34] D. A. Tennant, T. G. Perring, R. A. Cowley, S. E. Nagler Phys. Rev. Lett. **70**, 4003 (1993).

[35] D. A. Tennant, R. A. Cowley, S. E. Nagler, A. M. Tsvelik, Phys. Rev. B **52**, 13368 (1995).

[36] B. Lake et al. to be published (2003).

[37] A.H. Bougourzi, M. Couture, and M. Kacir, Phys. Rev. B **54**, R12669 (1996).

[38] C. Broholm, G. Aeppli, Y. Chen, D. C. Dender, M. Enderle, P. R. Hammar, Z. Honda, K. Katsumata, C. P. Landee, M. Oshikawa, L. P. Regnault, D. H. Reich, S. M. Shapiro, M. Sieling, M. B. Stone, M. M. Turnbull, I. Zaliznyak, and A. Zheludev, p 211-234 in High Magnetic Fields applications in condensed matter physics and spectroscopy C. Berthier, L. P. Lvy, and G. Martinez, Eds. Springer Verlag (2002).

[39] A. Luther and I Peschel, Phys. Rev. B **12**, 3908 (1975).

[40] E. Pytte, Phys. Rev. **10**, 4637 (1974).

[41] N. Ishimura and H. Shiba, Prog. Theor. Phys. Jpn. **57**, 6 (1977); **57**, 1862 (1977); **64**, 479 (1980).

[42] D. C. Dender, P. R. Hammar, Daniel H. Reich, C. Broholm, and G. Aeppli, Phys. Rev. Lett. **79**, 1750 (1997).

[43] P. R. Hammar, M Stone, D. H. Reich, C. Broholm, P. J. Gibson, M. M. Turnbull, C. P. Landee, and M. Oshikawa, Phys. Rev. B **59**, 1008 (1999).

[44] I. Affleck and M. Oshikawa, Phys. Rev. B **60**, 1038 (1999).

[45] F. H. M. Essler and A. M. Tsvelik, Phys. Rev. B **57**, 10592 (1998).

[46] M. Oshikawa and I Affleck, Phys. Rev. Lett. **79** 2883 (1997).

[47] S. Chakravarty, B. I. Halperin, D. R. Nelson Phys. Rev. B **39** (1989) and Phys. Rev. Lett. **60** (1988).

[48] B. Keimer et al., Phys. Rev. B **46**, 14 034 (1992).

[49] M. Greven et al., Z. Phys. B **96**, 465 (1995).

[50] H. M. Ronnow, D. F. McMorrow, and A. Harrison A Phys. Rev. Lett. **82** (1999).

[51] H. M. Ronnow et al. , Phys. Rev. Lett. **87**, 0372021 (2001)

[52] R. R. P. Singh et al. , Phys. Rev. B **52**, R15695 (1995).

[53] Y. J. Kim et al. , Phys. Rev. Lett. **83**, 852 (1999).

[54] See e. g. M. A. Kastner, R. J. Birgeneau, G. Shirane, and Y. Endoh, Reviews of Modern Physics **70** , Page: 897-928 (1998)

[55] R. Coldea, S. M. Hayden, G. Aeppli, T. G. Perring, C. D. Frost, T. E. Mason, S. -W. Cheong, Z. Fisk, Phys. Rev. Lett., **86**, 5377-5380 (2001).

[56] R. S. Eccleston et al. , Phys. Rev. Lett. **81**, 1702 (1998).

[57] S. Brehmer et al. , Phys. Rev. B **60**, 329 (1999).

[58] A. H. MacDonald, S. M. Girvin, and D. Yoshioka, Phys. Rev. B **41** 2565 (1990) and **37**, 9753 (1988).

[59] C. Kim et al. , Phys. Rev. Lett. **80**, 4245 (1998).

[60] H-B. Schuttler and A. J. Fedro, Phys. Rev. B **45**, 7588 (1992).

[61] N. M. R. Peres and M. Araujo, Phys. Rev. B **65**, 2402 (2002).

[62] J. E. Greedan, J. Mat. Chem. **11**, 37 (2001).

[63] J. Villain, Z. Phys. B Cond. Matter **33** 31 (1979).

[64] B. Canals and C. Lacroix, Phys. Rev. Lett. **80**, 2933 (1998).

[65] R. Moessner and J. T. Chalker, Phys. Rev. Lett. **80**, 2929 (1998); Phys. Rev. B **57**, R5587 (1998); **58**, 12049 (1998).

[66] S.-H. Lee, C. Broholm, S-W. Cheong, T.H. Kim, and W. Ratcliff II, cond-mat/9908433, Phys. Rev. Lett. **84**, 3718 (2000).

[67] S.-H. Lee, C. Broholm, W. Ratcliff II, G. Gasparovic, Q. Huang, T. H. Kim, and S.-W. Cheong, Nature **418**, 856 (2002).

[68] O. Tchernyshyov, R. Moessner, and S. L. Sondhi Phys. Rev. Lett. **88**, 067203 (2002).

[69] O. Tchernyshyov, R. Moessner, and S. L. Sondhi, Phys. Rev. B bf 66, 064403 (2002).

[70] E. Pytte, Phys. Rev. B **10**, 4637 (1974); M. C. Cross and D. S. Fischer, Phys. Rev. B **19**, 402 (1979); J. W. Bray, L. V. Interrante, I. S. Jacobs, and J. C. Bonner, in *Extended Linear Chain Compounds*, edited by J. S. Miller (Plenum, New York, 1982), Vol. 3, pp. 353-415.

[71] G. Xu, C. Broholm, D. H. Reich, and M. A. Adams, Phys. Rev. Lett. **84**, 4465 (2000).

[72] M. Hase, I. Terasaki, and K. Uchinokura, Phys. Rev. Lett. **70**, 3651 (1993).

[73] D. A. Tennant, C. Broholm, D. H. Reich, S. E. Nagler, G. E. Granroth, T. Barnes, G. Xu, B. C. Sales, and Y. Chen, to appear in Phys. Rev B (2002).

[74] G. S. Uhrig and H. J. Schulz, Phys. Rev. B **54**, 9624 (1996).

[75] F. D. M. Haldane, Phys. Lett. **93A**, 464 (1983).

[76] F. D. M. Haldane, Phys. Rev. Lett., **50**, 1153 (1983).

[77] I. Affleck, T. Kennedy, E. H. Lieb, and H. Takaki, Phys. Rev. Lett. **59**, 799 (1987).

[78] D. P. Arovas, A. Auerbach, and F. D. M. Haldane Phys. Rev. Lett. **60**, 531 (1988).

[79] M. Oshikawa, Phys. Rev. Lett. **84** 1535 (2000).

[80] Guangyong Xu, G. Aeppli, P. Bischer, C. Broholm, J. F. DiTusa, C. D. Frost, T. Ito, K. Oka, H. Takagi, and M. Treacy, Science 289, 419-422 (2000).

[81] Guangyong Xu, J. F. DiTusa, T. Ito, H. Takagi, K. Oka, C. Broholm and G. Aeppli, Phys. Rev. B **54**, R6827 (1996).

[82] S. Ma, C. Broholm, D.H. Reich, B.J. Sternlieb and R.W. Erwin, Phys. Rev. Lett., **69**, 3571 (1992).

[83] P. C. Hohenberg and W. F. Brinkman, Phys. Rev. B **10**, 128 (1974).

[84] K. Damle and S. Sachdev, Phys. Rev. B **57** 8307 (1998).

[85] I. Affleck and R. A. Weston, Phys. Rev. B **45**, 4667 (1992).

[86] I. A. Zaliznyak, S. H. Lee, and S. V. Petrov, Phys. Rev. Lett. **87**, 017202 (2001).

[87] M. Kenzelmann, R. A. Cowley, W. J. L. Buyers, R. Coldea, J. S. Gardner, M. Enderle, D. F. McMorrow, and S. M. Bennington, Phys. Rev. Let. **87**, 017201 (2001).

[88] M. Kenzelmann, R. A. Cowley, W. J. L. Buyers, Z. Tun, R. Coldea, and M. Enderle Phys. Rev. B **66**, 024407 (2002).

[89] L. J. deJongh and A. R. Miedema Adv. Phys. **23** 1 (1974).

[90] B. Leuenberger et al., Phys. Rev. B **30**, 6300 (1984).

[91] Y. Sasago, K. Uchinokura, A. Zheludev, and G. Shirane Phys. Rev. B **55** 8357 (1997).

[92] N. Cavadini, C. Ruegg, A. Furrer, H. U. Gudel, K. Kramer, H. Mutka, P. Vorderwisch Phys. Rev. B **65**, 132415 (2002).

[93] G. Chaboussant, M. H. Julien, Y. Fagot-Revurat, M. Hanson, L. P. Levy, C. Berthier, M. Horvatic, O. Piovesana Eur. Phys. J. B **6**, 167 (1998).

[94] P. R. Hammar and D. H. Reich J. App. Phys. **79** 5392-5394 (1996).

[95] M. B. Stone, J. Rittner, Y. Chen, H. Yardimci, D. H. Reich, C. Broholm, D. V. Ferraris, and T. Lectka, Phys. Rev. B 65, 064423 (2002).

[96] P. R. Hammar, D. H. Reich, C. Broholm, and F. Trouw Phys. Rev. B **57**, 7846 (1998).

[97] B. S. Shastry and B. Sutherland, Physica Amsterdam **108B**, 1069 (1981).

[98] M. B. Stone, I. A. Zaliznyak, Daniel H. Reich, and C. Broholm, Phys. Rev. B **64**, 144405 (2001).

[99] K. Kojima, A. Keren, L. P. Le, G. M. Luke, B. Nachumi, W. D. Wu, Y. J. Uemura, K. Kiyono, S. Miyasaka, H. Takagi, and S. Uchida, Phys. Rev. Lett. **74**, 3471 (1995).

[100] J. F. DiTusa, S-W. Cheong, J.-H. Park, G. Aeppli, C. Broholm, and C. T. Chen, Phys. Rev. Lett., **73**, 1857 (1994).

[101] G. Aeppli, T. E. Mason, S. M. Hayden, H. A. Mook, J. Kulda, Science **278**, 1432-1435 (1997).

[102] S. M. Hayden, G. Aeppli, H. Mook, D. Rytz, M. F. Hundley, and Z. Fisk, Phys. Rev. Lett. , **66**, p. 821 (1991).

[103] B. Keimer et al. , Phys. Rev. Lett. **67**, 1930 (1991).

[104] B. Sternlieb et al. , Phys. Rev. B **47**, 5320 (1993).

[105] J. Rossat-Mignod et al. , Physica B **169**, 58 (1991).

[106] S. Sachdev and J. Ye, Phys. Rev. Lett. **69**, 2411 (1992).

Chapter 3

ANGLE RESOLVED PHOTOEMISSION IN THE HIGH TEMPERATURE SUPERCONDUCTORS

Juan Carlos Campuzano

Department of Physics, University of Illinois at Chicago, 845 W. Taylor St., Chicago, IL 60607 and Argonne National Laboratory, 4700 S. Cass Ave., Argonne, IL 60439

jcc@uic.edu

Abstract This chapter present a review of the determination of the electronic structure of the high temperature superconductors by angle resolved photoemission spectroscopy, limited to our results on the quasi-two dimensional cuprate $Bi_2Sr_2CaCu_2O_8$. The review describes the methodology of data analysis of photoemission spectra in order to obtain information on the normal state electronic structure, the Fermi surface, the superconducting energy gap, and the pseudogap.

Keywords: photoemission, high temperature superconductors

1. Introduction

Shortly after the high temperature superconductors (HTSCs) were discovered, Anderson [1] suggested that they represent a new paradigm in condensed matter physics. Although these ideas have not been universally accepted, we will use angle resolved photoemission (ARPES) to show that these new materials indeed have an excitation spectrum quite unlike that of conventional metals. ARPES has played a major role in the study of the electronic excitations in the HTSCs, as it directly probes the single particle spectral function, and therefore offers a complete picture of the many body interactions. Here we will concentrate on discussing some of the strong interaction aspects of the ARPES results on the HTSCs. More extensive reviews are available [2, 3].

D. Baeriswyl and L. Degiorgi (eds.), Strong Interactions in Low Dimensions, 63–91.

2. Basics of Angle-Resolved Photoemission

We briefly discuss those aspects of the technique which will be useful in understanding the HTSCs, as there are more general treatments of ARPES in the literature [4]. The simplest model of ARPES is the three step model [5], which separates the process into photon absorption, electron transport through the sample, and emission through the surface. In the first step, the incident photon with energy $h\nu$ is absorbed by an electron in an occupied initial state, causing it to be promoted to an unoccupied final state, as shown in Fig. 3.1a. There is conservation of energy, such that

$$h\nu = BE + \Phi + E_{\text{kin}} \tag{3.1}$$

where Φ is the work function, and BE and E_{kin} are the binding and kinetic energies of the electron, respectively. We will not discuss the subsequent two steps of this model, as they only affect the number of emitted electrons, and thus the absolute intensity [4].

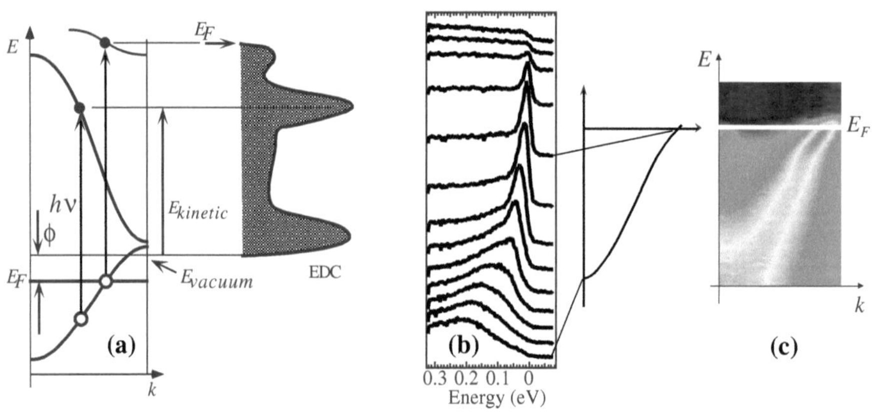

Figure 3.1. (a) Independent particle approximation view of ARPES; To the right is the resulting energy distribution curve (EDC); (b) Experimental EDCs along a path in momentum space in Bi2212; (c) Intensity map from these data.

The kinetic energy of the electrons is measured by an electron energy analyzer. If the number of emitted electrons is plotted as a function of their kinetic energy, as shown in Fig. 3.1a, peaks are found whenever an allowed transition takes place. Eq. (1) then yields the binding energy of the electron if the work function is known. Fortunately, for metallic samples one does not need to know Φ. By placing the sample in electrical contact with a good metal (e.g., polycrystalline gold) one can measure the binding energy of states with respect to the chemical potential (Fermi level E_F) of Au. The photoemission signal from Au will simply be a Fermi function convolved with the experimental energy resolution, and from the mid point of its leading edge one estimates E_F. Although the sample surface breaks (discrete) translational invariance in the direction normal to the surface, translational invariance in the directions parallel to the surface is still preserved, and thus $\mathbf{k}_{\|}$, the component of the electron momentum parallel to the surface, is conserved in photoemission. This allows us to obtain the in-plane momentum of the initial state by identifying it with the parallel momentum of the outgoing electron emitted along the direction (θ, ϕ) with kinetic energy E_{kin}, as shown in Fig. 3.2a.

$$k_{\|} = \sqrt{\frac{2mE_{\text{kin}}}{\hbar}} \sin\theta \cos\phi \tag{3.2}$$

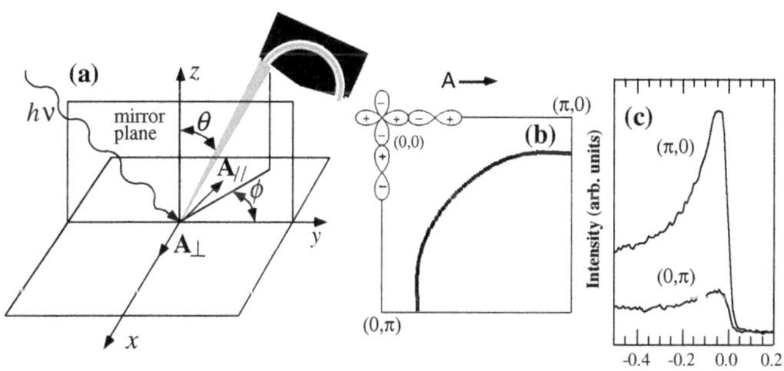

Figure 3.2. (a) Arrangement of the photon beam and detector. The angles refer to Eq. 3.2. (b) The Brillouin zone, showing the alignment of the Cu$3d_{x^2-y^2}$ orbitals hybridized with the O$2p$ orbitals relevant to superconductivity. The thick line is the Fermi surface. (c) EDC obtained at $\hbar\nu$ =22 eV at the $(\pi, 0)$ point of the zone. The top curve is with the electric vector along the mirror plane shown in (a), while the bottom curve has the electric vector perpendicular to that mirror plane. For details of these matrix element effects, see Ref. [6].

The momentum perpendicular to the sample surface k_z is not conserved, but this does not concern us since the HTSCs are quasi-two-dimensional (2D) materials with no observable k_z-dispersion. We will use the symbol $\mathbf{k}$ to simply denote the two-dimensional momentum parallel to the sample surface. In the independent particle approximation, the ARPES intensity as a function of momentum $\mathbf{k}$ (in the 2D Brillouin zone) and energy ω (measured with respect to the chemical potential) is given by Fermi's Golden Rule as

$$I(k,\omega) \propto |\langle\psi_f|\, \mathbf{A}\cdot\mathbf{p}\, |\psi_i\rangle|^2 f(\omega)\delta(\omega - \varepsilon_k) \tag{3.3}$$

where ψ_i and ψ_f are the initial and final states, $\mathbf{p}$ is the momentum operator, and $\mathbf{A}$ the vector potential of the incident photon. $f(\omega) = 1/[\exp(\omega/T) + 1]$ is the Fermi function at a temperature T in units where $\hbar = k_B = 1$, which ensures the physically obvious constraint that photoemission only probes *occupied* electronic states.

For noninteracting electrons, the emission at a given $\mathbf{k}$ is at a sharp energy $\epsilon_{\mathbf{k}}$ corresponding to the initial state dispersion, as shown in Fig. 3.1b. In going beyond the independent electron picture, the photoemission process quickly looses this simple interpretation (See for example Refs. [7, 8]), unless the experiment remains in the sudden approximation regime. This regime takes place when the photoelectron just created by the absorption of a photon leaves behind the hole at such high speed that the interaction between the two can be ignored. Then, provided that one can also ignore the final state lifetime and the background effects (described in detail in Ref [3]), equation 3.3 above becomes [9, 10]

$$I(\mathbf{k},\omega) = I_0(\mathbf{k})f(\omega)A(\mathbf{k},\omega) \tag{3.4}$$

where $I_0(\mathbf{k})$ includes all the kinematical factors and the square of the dipole matrix element (shown in Eq. 3.3), and $A(\mathbf{k},\omega)$ is the one-particle spectral function described in detail below. Randeria *et al.* [10] have shown how one can test for the applicability of the sudden approximation, which greatly strengthens the case for a simple $A(\mathbf{k},\omega)$ interpretation of ARPES data in the HTSCs.

The one-particle spectral function represents the probability of adding or removing a particle from the interacting many-body system, and is defined as $A(\mathbf{k},\omega) = -(1/\pi)\mathrm{Im}G(\mathbf{k},\omega + i0^+)$ in terms of the Green's function. It can be written as the sum of two pieces $A(\mathbf{k},\omega) = A_-(\mathbf{k},\omega) + A_+(\mathbf{k},\omega)$, where the spectral weight to add an electron to the system is given by $A_+(\mathbf{k},\omega) = Z^{-1}\sum_{m,n} e^{-\beta\mathcal{E}_m}|\langle n|c^\dagger_{\mathbf{k}}|m\rangle|^2\delta(\omega + \mathcal{E}_m - \mathcal{E}_n)$, and that to extract an electron is $A_-(\mathbf{k},\omega) = Z^{-1}\sum_{m,n} e^{-\beta\mathcal{E}_m}|\langle n|c_{\mathbf{k}}|m\rangle|^2\delta(\omega + \mathcal{E}_n - \mathcal{E}_m)$. Here $|m\rangle$ is an exact eigenstate of the many-body system with energy $\mathcal{E}_m$, Z is the partition function and $\beta = 1/T$. It follows from these definitions that $A_-(\mathbf{k},\omega) = f(\omega)A(\mathbf{k},\omega)$ and $A_+(\mathbf{k},\omega) = [1 - f(\omega)]\, A(\mathbf{k},\omega)$, where $f(\omega) = 1/[\exp(\beta\omega) + 1]$ is the Fermi function. Since an ARPES experiment involves removing an electron from the system, the simple golden rule Eq. 3.3 can be generalized to yield an intensity proportional to $A_-(\mathbf{k},\omega)$.

3. Analysis of ARPES Spectra: EDCs and MDCs

The one-electron Green's function can be generally written as $G^{-1}(\mathbf{k},\omega) = G_0^{-1}(\mathbf{k},\omega) - \Sigma(\mathbf{k},\omega)$ where $G_0(\mathbf{k},\omega) = 1/[\omega - \varepsilon_{\mathbf{k}}]$ is the Green's function of the noninteracting system, $\varepsilon_{\mathbf{k}}$ is the "bare" dispersion, and the (complex) self-energy $\Sigma(\mathbf{k},\omega) = \Sigma'(\mathbf{k},\omega) + i\Sigma''(\mathbf{k},\omega)$ encapsulates the effects of all the many-body interactions. Then using its definition in terms of ImG, we obtain the general result for a single state

$$A(\mathbf{k},\omega) = \frac{1}{\pi}\frac{\Sigma''(\mathbf{k},\omega)}{[\omega - \varepsilon_{\mathbf{k}} - \Sigma'(\mathbf{k},\omega)]^2 + [\Sigma''(\mathbf{k},\omega)]^2} \tag{3.5}$$

We emphasize that this expression is entirely general, and does not make any assumptions about the validity of perturbation theory or of Fermi liquid theory.

New electron energy analyzers, which measure the photoemitted intensity as a function of energy and momentum simultaneously, allow the direct visualization of the spectral function, as shown in Figs. 3.1c and 3.3a, and have also suggested new ways of plotting and analyzing ARPES data. In the traditional energy distribution curves (EDCs, panel c), the measured intensity $I(\mathbf{k},\omega)$ is plotted as a function of ω (binding energy) for a fixed value of $\mathbf{k}$. In the new [11, 12] momentum distribution curves (MDCs, panel b), $I(\mathbf{k},\omega)$ is plotted at fixed ω as a function of $\mathbf{k}$.

We now describe why in strongly correlated systems one should be very careful with the traditional EDC analysis. Note that the EDC lineshape is non-Lorentzian. This is due to (a) The asymmetry introduced by the Fermi function $f(\omega)$ which chops off the positive ω part of the spectral function, and (b) The self energy has non-trivial ω dependence. This makes even the full $A(\mathbf{k},\omega)$ non-Lorentzian in ω as seen from Eq.(3.5). Thus one is usually forced to model the self energy and make fits to the EDCs. At this point one is further hampered by the lack of detailed knowledge of the additive extrinsic background which itself has ω-dependence. (Although, as we shall see, the MDC analysis gives a new way of determining this background).

The MDCs have certain advantages in studying gapless excitations near the Fermi surface [12, 13, 14]. In an MDC, the intensity is plotted as a function of k varying normal to the Fermi surface in the vicinity of a fixed $\mathbf{k}_F(\theta)$, where θ is the angle parametrizing the Fermi surface. For k near k_F we may linearize the bare dispersion $\varepsilon_{\mathbf{k}} \simeq v_F^0(k - k_F)$, where $v_F^0(\theta)$ is the bare Fermi velocity. From Fig. 3.3b we find that the MDC can accurately fit to a Lorentzian lineshape, which together with Eq(3.5,

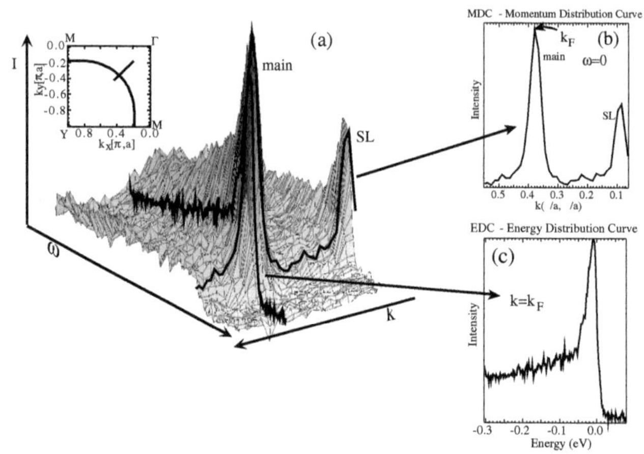

Figure 3.3. (a) The ARPES intensity as a function of **k** and ω at $h\nu$=22eV and T=40K. main is the main band, and SL a superlattice image. (b) A constant ω cut (MDC) from (a). (c) A constant **k** cut (EDC) from (a). The diagonal line in the zone inset shows the location of the **k** cut; the curved line is the Fermi surface.

implies that (i) the self-energy Σ is essentially independent of k normal to the Fermi surface, but can have arbitrary dependence on θ along the Fermi surface, and (ii) the pre-factor $I_0(\mathbf{k})$ does not have significant k dependence over the range of interest. The MDC is a Lorentzian centered at $k = k_F + [\omega - \Sigma'(\omega)]/v_F^0$, and has width (HWHM) $W_M = |\Sigma''(\omega)|/v_F^0$.

Thus the MDC peak position gives the renormalized dispersion, while its width is proportional to the imaginary self energy. Experimentally, excellent Lorentzian fits are invariably obtained (except when one is very near the bottom of the "band" or in a gapped state[15]).

Finally, note that the external background in the case of MDCs is also very simple. One can fit the MDC (at each ω) to a Lorentzian plus a constant (at worst Lorentzian plus linear in k) background. From this one obtains the value of the external background including its ω dependence. Now this ω-dependent background can be subtracted from the EDC also. Note that estimating this background was not possible from an analysis of the EDCs alone.

4. The Valence Band

The basic unit common to all cuprates is the copper-oxide plane, CuO_2. Some compounds have a tetragonal cell, $a = b$, such as the Tl compounds, but most have orthorhombic cells. There are two notations used in the literature for the reciprocal cell. The one used here, appropriate for Bi2212 and Bi2201, has $\Gamma - M$ along the $Cu - O$ bond direction, with $M \equiv (\pi, 0)$, and $\Gamma - X(Y)$ along the diagonal, with $Y \equiv (\pi, \pi)$. The main effect of the orthorhombicity in Bi2212 and Bi2201 is the superlattice modulation along the b axis, with Q_{SL} parallel to $\Gamma - Y$. Except when referring to this modulation, we will assume tetragonal symmetry in our discussions. For a complete review of the LDA-electronic structure of the cuprates, see Ref. [16].

The Cu ions are four fold coordinated to planar oxygens. Apical (out of plane oxygens) exist in some structures (LSCO), but not in others. Either way, the apical bond distance is considerably longer than the planar one, so in all cases, the cubic point group symmetry of the Cu ions is lowered, leading to the highest energy Cu state having $d_{x^2-y^2}$ symmetry. As the atomic $3d$ and $2p$ states are nearly degenerate, a characteristic which distinguishes cuprates from other 3d transition metal oxides, the net result is a strong bonding-antibonding splitting of the Cu $3d_{x^2-y^2}$ and O $2p$ σ states, with all other states lying in between. In the stoichiometric (undoped) material, Cu is in a d^9 configuration, leading to the upper (antibonding) state being half filled. According to band theory, the system should be a metal. But in the undoped case, integer occupation of atomic orbitals is possible, and correlations due to the strong on-site Coulomb repulsion on the Cu sites leads to an insulating state. That is, the antibonding band "Mott-Hubbardizes" and splits into two, one completely filled (lower Hubbard band), the other completely empty (upper Hubbard band) [17].

On the other hand, for dopings characteristic of the superconducting state, a large Fermi surface is observed by ARPES[18, 19]. Thus, to a first approximation, the basic electronic structure in this doping range can be understood from simple band theory considerations. The simplest approximation is to consider the single Cu $3d_{x^2-y^2}$ and two O $2p$ (x,y) orbitals. This dispersion is shown in Fig. 3.4(a)

In Fig. 3.4(b) we show an ARPES spectrum obtained at the $(\pi, 0)$ point of the Brillouin zone for Bi2212. Three distinct features can be observed: the bonding state at roughly -6eV, the antibonding state near the Fermi energy, and the rest of the states in between. This rest consists of the non-bonding state mentioned above, as well as the remainder of the Cu $3d$ and O $2p$ orbitals, plus states originating from the other (non Cu-O) planes. It is difficult to identify all of these "non-bonding" states, as their close proximity and broadness causes them to overlap in energy. The overall picture of the electronic structure of the valence band has the structure predicted by the simple chemical arguments given above, as shown in Fig. 3.4(c). The most important conclusion that one

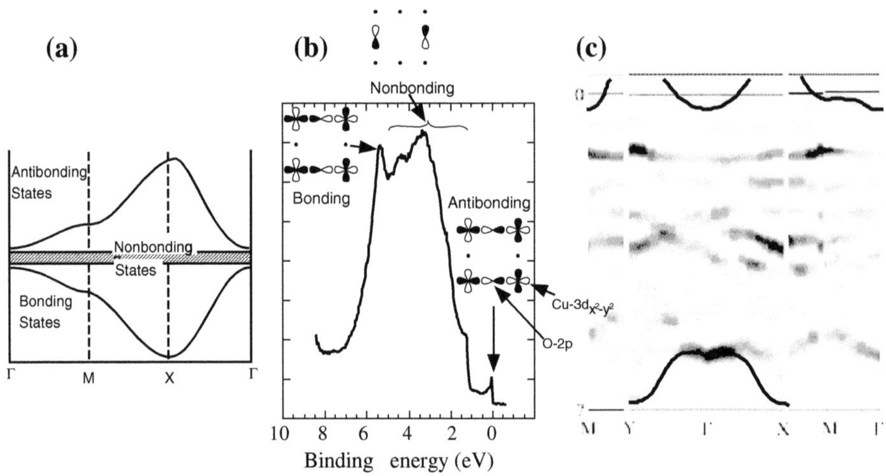

Figure 3.4. (a) Simple three band estimate of the electronic structure of the Cu-O plane states; (b) EDC showing the whole valence band at the $(\pi, 0)$ point; c) Intensity map of the whole valence band obtained by taking the second derivative of spectra such as the one in (b). The orbitals in (b) are based on the three band model, where black and white lobes correspond to positive and negative wavefunctions.

can derive from Fig 3.4 is the early prediction by Anderson [1], namely that there is a single state relevant to transport and superconducting properties. This antibonding state is well separated from the rest of the states, and therefore any reasonable theoretical description of the physical properties of these novel materials should arise from this single state.

Despite these simple considerations, correlation effects do play a major role, even in the doped state. The observed antibonding band width is about a factor of 2-3 narrower than that predicted by band theory [19].

5. Normal State Dispersion and the Fermi Surface

The first issue facing us is how do we define the Fermi surface in a system at high temperatures where there are no well-defined quasiparticles. Clearly, the traditional $T = 0$ definition of a Fermi surface defined by the jump discontinuity in $n(\mathbf{k})$ is not useful here. First, the HTSCs are superconducting at low temperatures. But even samples which have low T_c's have normal state peak widths at E_F which are an order of magnitude broader than the temperature[20, 14].

It is an experimental fact that in the cuprates ARPES sees broad peaks which disperse as a function of momentum and go through the

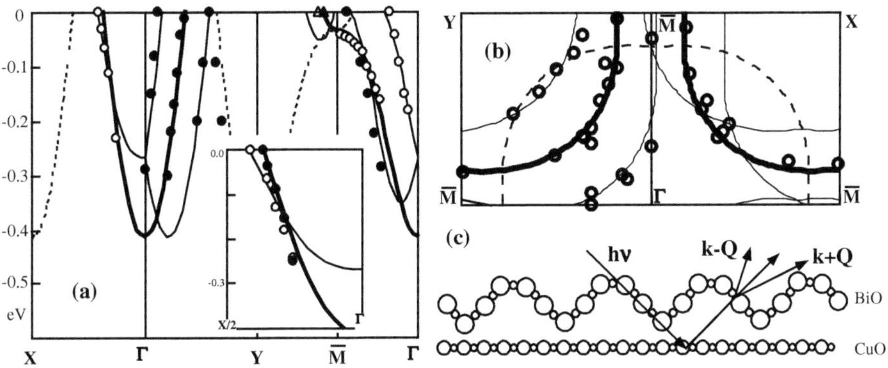

Figure 3.5. Dispersion (a) and Fermi surface (b) obtained from normal state measurements. The thick lines are obtained by a tight binding fit to the dispersion data of the main band with the thin lines $(0.21\pi, 0.21\pi)$ umklapps and the dashed lines (π, π) umklapps of the main band. Open circles in (a) are the data. In (b), filled circles are for odd initial states (relative to the corresponding mirror plane), open circles for even initial states, and triangles for data taken in a mixed geometry. The inset of (b) is a blowup of ΓX.

chemical potential at a reasonably well-defined momentum, as shown in Fig. 3.1c. We can thus adopt a practical definition of the "Fermi surface" in these materials as "the locus of gapless excitations". The first attempts to determine the Fermi surface in cuprates were made on YBCO [18], however, surface effects as well as the presence of chains appear to complicate the picture, so we will focus principally on Bi2212.

We begin with the results obtained by using the traditional method of deducing the dispersion and Fermi surface by studying the EDC peaks. This method was used for the cuprates by Campuzano *et al.*[18], Olson *et al.*[19], and Shen and Dessau [21], culminating in the very detailed study of Ding *et al.*[22]. Even though the use of EDC peak dispersion has some limitations, it has led to considerable understanding of the overall electronic structure, Fermi surface, and of superlattice effects in Bi2212. It is therefore worthwhile to review these results first, which are summarized in Fig. 3.5. Fig. 3.5a shows the dispersion of all the peaks observed in the EDCs. The Fermi surface crossings corresponding to these dispersing states are estimated from the **k**-point at which the EDC peak positions go through the chemical potential when extrapolated from the occupied side. The k_F estimates are plotted as open symbols in Fig. 3.5(b).

The thick curve in Fig. 3.5(a) is $\epsilon(\mathbf{k})$, a tight-binding fit [23] to the dispersion data in the Y-quadrant; this represents the main CuO_2 state. The thin curves in Fig. 3.5(a) are $\epsilon(\mathbf{k}\pm\mathbf{Q})$ umklapps, obtained by shifting the main band fit by $\pm\mathbf{Q}$ respectively, where $\mathbf{Q} = (0.21\pi, 0.21\pi)$ is the superlattice (SL) vector known from structural studies [24]. A lack of understanding of these SL effects has led to much confusion regarding such basic issues as the Fermi surface topology and the anisotropy of the SC gap (see Section 8). These SL bands arise from the diffraction of the outgoing photoelectron off the structural superlattice distortion (which lives primarily) on the Bi-O layer, thus leading to "ghost" images of the electronic structure at $\epsilon_{\mathbf{k}\pm\mathbf{Q}}$, as shown in Fig. 3.5c. We also have a few data points lying on a dashed curve $\epsilon(\mathbf{k} + \mathbf{K}_\pi)$ with $\mathbf{K}_\pi = (\pi, \pi)$; this is the "shadow band", first observed by Aebi et al. [11]. The physical origin of these "shadow bands" is not certain at the present time [3]. The thick curve in Fig. 3.5(b) is the Fermi surface contour obtained from the main band fit, while the Fermi surfaces corresponding to the SL bands are the thin lines and that for the shadow band is dashed. The main Fermi surface is a large hole-like barrel centered about the (π, π) point whose enclosed area corresponds to approximately 0.17 holes per planar Cu. One of the key questions is why *only one* CuO main band is found in Bi2212 which is a bilayer material with two CuO planes per unit cell. We discuss this important issue at the end of this Section 7. Other methodologies were subsequently developed for the determination of the Fermi surface, discussed in detail by Mesot, et al.[6] and in our recent review[3]. Of these, we describe the most straightforward determination of k_F and the near-E_F dispersion based on the MDCs. As shown in Section 3, the MDC peak position in the vicinity of the Fermi surface, i.e, near $(\mathbf{k} = \mathbf{k}_F, \omega = 0)$ is given by: $k = k_F + [\omega - \Sigma'(\omega)]/v_F^0$. Thus k_F is determined by the peak location of the MDC at $\omega = 0$, as shown in Fig. 3.3b. After plotting the MDC at $\omega = 0$, one simply reads k_F from the position of the peak. The fully renormalized Fermi velocity $v_F = v_F^0/[1 - \partial\Sigma'/\partial\omega]$ is given by the slope of the MDC peak dispersion. We note that the factor arising from the k-dependence of the self-energy is already included in v_f^0, so that $v_f^0 = v_f^{\text{bare}}[1 + \partial\Sigma'/\partial\varepsilon_{\mathbf{k}}]$. (To see this, note that the analysis of Section 3 can be easily generalized to retain the first order term $\partial\Sigma'/\partial\varepsilon_{\mathbf{k}}$ without spoiling the Lorentzian lineshape of the MDC provided this k-dependence does not enter Σ'').

The significance of this approach is that, as emphasized by Kaminski *et al.*[14], the dispersions of the EDC and MDC peak positions are actually different in the cuprates. This difference arises due to the non-Fermi liquid nature of the normal state, so that the EDC peak dispersion is *not* given by the condition $\omega - v_F^0(k - k_F) - \Sigma' = 0$ but also involves in general Σ''. In contrast the MDC peak dispersion is rigorously described by the expression described above, and is much simpler to interpret.

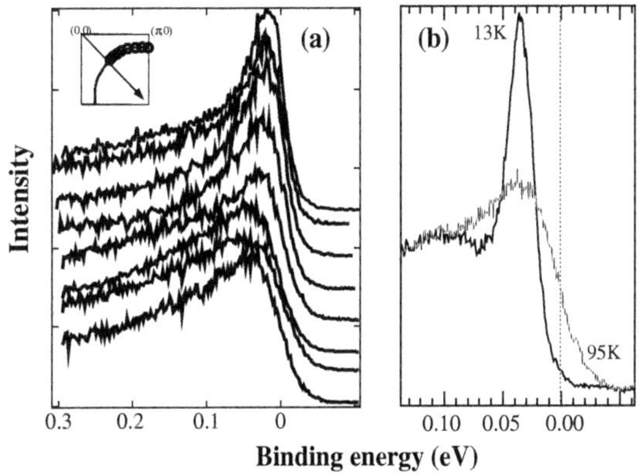

Figure 3.6. a) EDCs for an OD sample with $T_c = 87$K at T=100K, along the Fermi surface points indicated in the inset. The top curve corresponds to the nodal direction, while the bottom curve corresponds to the anti-nodal direction. b) Spectra at the $\mathbf{k}_F$ in the normal and superconducting states.

6. Absence of Quasiparticles in the Normal State

We now look in more detail at normal state spectra at the chemical potential for a near optimally doped sample with $T_c = 87$k. As can be seen in Fig. 3.6, these spectra are quite unusual, in that their widths are an order of magnitude larger than the temperature, and therefore controlled by many body interactions. Although, as discussed earlier, an MDC analysis is required to accurately determine the lifetime of the electron's initial state [14, 12], an order of magnitude estimate can be obtained from the uncertainty principle $\Delta E \Delta t \geq \hbar$ using the width of the EDC at the Fermi energy along the diagonal (the top curve in Fig. 3.6a), where there is no SL contamination. We find that the electron does not live much longer than a few femtoseconds. Although these spectra were obtained at 100K, where perhaps one would not expect long electron lifetimes, similar widths are obtained for the normal state of optimally doped samples with low T_c, such as Bi2201 [14]. One can then estimate the inelastic mean free path using the measured velocity $\partial k/\partial\varepsilon$ from Fig. 3.1 ($\approx 2 \times 10^7$ cm/s), to be of the order of 2-3 unit cells. Consequently, one should not think of quasiparticles as the elementary excitations in the normal state of optimally doped HTSCs, in sharp contrast to usual metals, where one would find some long-lived electrons at the Fermi surface. The short lifetime of excitations in the HTSCs is indicative of strong many body interactions, which cause the electrons to loose their coherence.

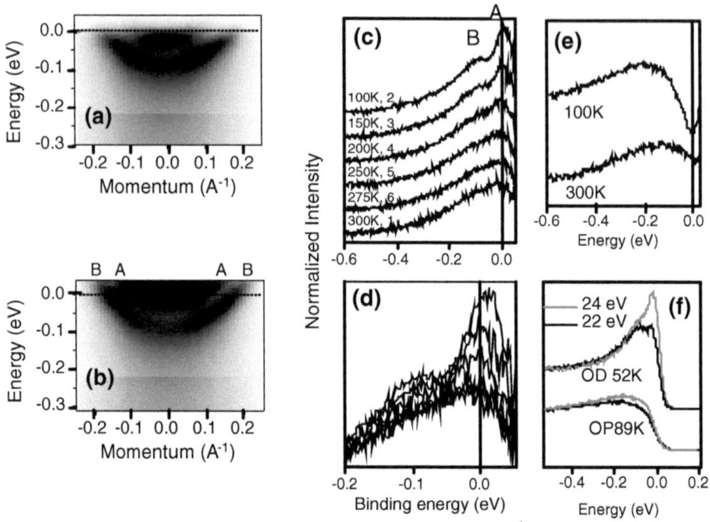

Figure 3.7. (a) Intensity versus momentum and energy for an OD sample with $T_c =$ 52K at T=100K, along $(\pi, 0) \rightarrow (\pi, pi)$, with plots centered at $(\pi, 0)$. (b) Data in (a) divided by the Fermi function. (c) EDCs at $(\pi, 0)$ (divided by the Fermi function) at various temperatures. All curves are overlapped in (d) to demonstrate lack of temperature dependence of the lineshape above 250K; (e) Spectrum at $(\pi, 0)$ (divided by the Fermi function) for an optimally doped ($T_c = 89$K) sample. (g) Raw data at $(\pi, 0)$ at two different photon energies for an overdoped ($T_c = 52$K sample) and an optimally doped ($T_c = 89$K) sample at $T = 100$K.

7. Bilayer Splitting

On very general grounds, one expects that the two CuO_2 layers in a unit cell of Bi2212 should hybridize to produce two electronic states, a bonding (B) and an antibonding (A) combination, which are even and odd, respectively, under reflection in a mirror plane mid-way between the layers. Electronic structure calculations [25] find that the intra-bilayer hopping as a function of the in-plane momentum $\mathbf{k}$ is of the form [26, 27] $t_{\perp}(\mathbf{k}) = -t_z(\cos k_x - \cos k_y)^2$. Thus the two bilayer states are degenerate along the zone diagonal. However they should have a maximum splitting at $\bar{M} = (\pi, 0)$ of order 0.25 eV, which may be somewhat reduced by many-body interactions. As described at great length above, and consistent with the fact that the electronic states at the chemical potential in normal state are not coherent, we did not find evidence for two states crossing the Fermi surface along $(\pi, 0) - (\pi, \pi)$ for the near optimal doped sample.

However, in the heavily overdoped Bi2212 samples, several authors [28, 29, 30] have recently found evidence of bilayer splitting, and we have characterized this splitting as a function of doping and temperature [31]. In panel (a) of Fig. 3.7 we plot raw ARPES data for an overdoped (OD) sample (T_C =52K) at T=100K, along a momentum cut centered at the $(\pi, 0)$ point of the Brillouin zone. In addition, in panel (b) we plot the data divided by the Fermi function, which approximates the true spectral function. The data reveal two dispersing bands due to the bilayer splitting, with the antibonding A band close to, and the bonding B band well below the chemical potential. In panel (c) we show the temperature dependence of the EDCs (raw data divided by the Fermi function). The bilayer splitting can clearly be seen at 100K, however above 250K the two bands are no longer observed. The sample was temperature-cycled when taking the data to ensure that the observed effect is intrinsic and not due to the sample aging (the numbers in the legend indicate the order of measurement). At 100K, one sees clearly the presence of two peaks, a sharp A peak near the chemical potential, and a broader B peak at about 100 meV below. As the temperature is increased, the peaks broaden and lose intensity, until only a single broad peak remains at 250K. In panel (d), we plot the curves for all temperatures without an offset to show that lineshape changes occur only up to 250K. Based on this, we argue that above 250K the system no longer exhibits coherent excitations, both in regards to inverse lifetime (spectral peak widths) and bilayer splitting (appearance of two separate spectral peaks).

We can contrast this behavior with that of an optimally doped sample (T_c = 89K) shown in panel (e), where the intensity plots do not indicate the presence of bilayer splitting, even at 100K. At 100K, only a single broad peak is seen, with no presence of bilayer splitting, indicating incoherent behavior. Instead, a pseudo-gap (to be discussed in detail below) is seen, centered at the chemical potential, which fills in as the temperature is increased. An important check can be made by analyzing the photon energy dependence of the data. It has been recently observed that the spectral lineshape changes as a function of photon energy for overdoped samples due to the relative weighting of the A and B peaks [28, 29]. This is clearly seen in panel (f), where data for the overdoped sample of (b) is shown for two different photon energies. In contrast, for the optimal doped sample, only a very small change with photon energy is observed, indicating the absence of bilayer splitting.

In summary, bilayer splitting is observed in the overdoped samples, indicating coherent electron behavior. However, this coherent behavior again disappears at higher temperatures.

8. The Superconducting State

Remarkable changes occur in the spectral lineshape as the HTSCs enter the superconducting (SC) state. In Fig. 3.6(b) we show ARPES spectra for near-optimal Bi2212 ($T_c = 87$ K) at $\mathbf{k}_F$ along $(\pi, 0)$ to (π, π) at two temperatures: $T = 13$ K, which is well below T_c, and $T = 95$ K, which is in the normal state. The T-dependent changes in the line shape may be understood as follows. At 95K one has a very broad $A(\mathbf{k}_F, \omega)$, with a maximum at $\omega = 0$, which is cutoff by the Fermi function. For $T < T_c$ a gap begins to open up and spectral weight shifts down to negative energies $\omega = -|\Delta(\mathbf{k})|$. See ref. [32] for a detailed analysis of the 13 K data along these lines. Another striking feature of the data is the sharpening of the peak with decreasing T. This indicates that the scattering rate Σ" of the quasiparticles, which determines the line width of $A(\mathbf{k}, \omega)$, drops sharply in the superconducting state in qualitative agreement with analysis of optical and microwave data [33]. Finally note that with decreasing linewidth there is a large increase in the peak intensity due to conservation of spectral weight. This then is a consequence of the strong T-dependence of Σ" as the gap opens up.

ARPES is the only available technique for measuring the momentum dependence of the energy gap, and complements phase-sensitive tests of the order parameter symmetry [34]. Thus ARPES has played an important role [35], [36] in establishing the d-wave order parameter in the high T_c superconductors [34].

We start by recalling particle-hole (p-h) mixing in the BCS framework (even though, as we have discussed, there are aspects of the data which are dominated by many body effects beyond weak coupling BCS theory). The BCS spectral function is given by

$$A(\mathbf{k}, \omega) = u_{\mathbf{k}}^2 \Gamma / \pi((\omega - E_{\mathbf{k}})^2 + \Gamma^2) + v_{\mathbf{k}}^2 \Gamma / \pi((\omega + E_{\mathbf{k}})^2 + \Gamma^2) \qquad (3.6)$$

where the coherence factors are $v_{\mathbf{k}}^2 = 1 - u_{\mathbf{k}}^2 = \frac{1}{2}(1 - \epsilon_{\mathbf{k}}/E_{\mathbf{k}})$ and Γ is a phenomenological linewidth. The normal state energy $\epsilon_{\mathbf{k}}$ is measured from E_F and the Bogoliubov quasiparticle energy is $E_{\mathbf{k}} = \sqrt{\epsilon_{\mathbf{k}}^2 + |\Delta(\mathbf{k})|^2}$, where $\Delta(\mathbf{k})$ is the gap function. Note that only the second term in Eq. 3.6, with the $v_{\mathbf{k}}$-coefficient, would be expected to make a significant contribution to the EDCs at low temperatures.

In the normal state above T_c, the peak of $A(\mathbf{k}, \omega)$ is at $\omega = \epsilon_{\mathbf{k}}$ as can be seen by setting $\Delta = 0$ in Eq. 3.6. In ARPES we would see a spectral peak which disperses through zero binding energy as $\mathbf{k}$ goes through $\mathbf{k}_F$. In the superconducting state, the spectrum changes from $\epsilon_{\mathbf{k}}$ to $E_{\mathbf{k}}$; see Fig. 3.8. As $\mathbf{k}$ approaches the Fermi surface the spectral peak shifts towards lower binding energy, but no longer crosses E_F. Precisely at $\mathbf{k}_F$

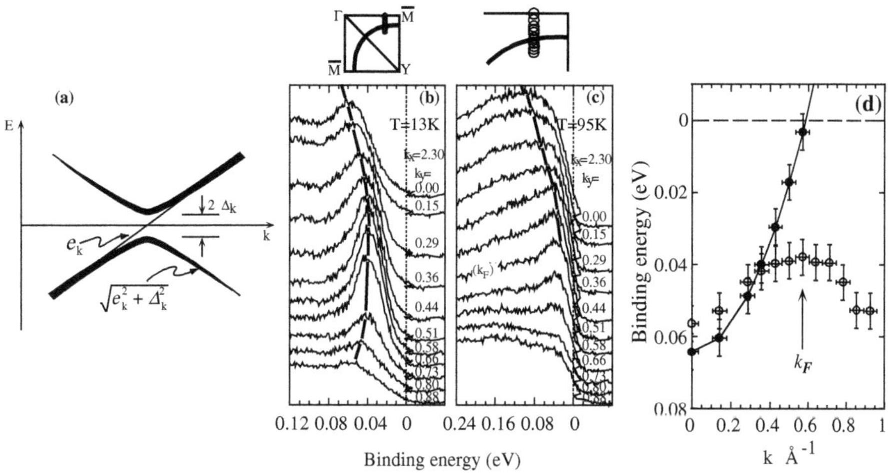

Figure 3.8. (a) Schematic dispersion in the normal (thin line) and superconducting (thick lines) states following BCS theory. The thickness of the superconducting state lines indicate the spectral weight given by the BCS coherence factors u and v. (b) Superconducting state and (c) normal state EDC's for a near optimal $T_c = 87$K Bi2212 sample for a set of **k** values (in units of $1/a$) shown in the Brillouin zone at the top. Note the different binding energy scales in panels (b) and (c). (d) Normal state dispersion (closed circles) and SC state dispersion (open circles) obtained from EDC's of panels (b) and (c). Note the back-bending of the SC state dispersion for **k** beyond $\mathbf{k}_F$ which is a clear indication of particle-hole mixing. The SC state EDC peak position at $\mathbf{k}_F$ is an estimate of the SC gap at that point on the Fermi surface.

the peak is at $\omega = |\Delta(\mathbf{k}_F)|$, which is the closest it gets to E_F. This is the manifestation of the gap in ARPES. As **k** goes beyond $\mathbf{k}_F$, in the region of states which were unoccupied above T_c, the spectral peak *disperses back*, receding away from E_F, although with a decreasing intensity (see Eq. 3.6). This is the signature of p-h mixing.

In Fig. 3.8b and c we show experimental evidence for particle-hole mixing [37]. Spectra for Bi2212 are shown for **k**'s along the cut in the inset. In the normal state in panel (c) we see the electronic state dispersing through E_F: the **k**'s go from occupied (top of panel) to unoccupied states (bottom of panel). The normal state dispersion is plotted as black dots in Fig. 3.8d. We see from Fig. 3.8b that the SC state spectral peaks do not disperse through the chemical potential, rather they first approach $\omega = 0$ and then recede away from it. The dispersion of the SC state is plotted as open circles in Fig. 3.8d, and the difference between the normal and SC state dispersions is just as in the cartoon in panel (a).

There are three important conclusions to be drawn from Fig. 3.8d. First, the bending back of the SC state spectrum for $\mathbf{k}$ beyond $\mathbf{k}_F$ is direct evidence for p-h mixing in the SC state. Second, the energy of closest approach to $\omega = 0$ is related to the SC gap that has opened up at the Fermi surface, and a quantitative estimate of this gap will be described below. Third, the location of closest approach to $\omega = 0$ ("minimum gap") coincides, within experimental uncertainties, with the $\mathbf{k}_F$ obtained from analysis of normal state data.

In fact by taking cuts in $\mathbf{k}$-space which are perpendicular to the normal state Fermi surface one can map out the "minimum gap locus" in the SC state, or for that matter in any gapped state (e.g., the pseudogap regime to be discussed later). We emphasize that particle-hole mixing leads to the appearance of the "minimum gap locus" and this locus in the gapped state gives information about the underlying Fermi surface. (By this we mean the Fermi surface on which the SC state gap appears below T_c). In fact, the observation of p-h mixing in the ARPES spectra is a clear way of asserting that the gap seen by ARPES is due to superconductivity rather than of some other origin, e.g., charge- or spin-density wave formation.

The quantitative extraction of the gap at low temperatures ($T \ll T_c$), which we now summarize, was discussed by Ding *et al.*[36]. In Fig. 3.9, we show the $T = 13$K EDCs for an 87K T_c sample for various points on the main band FS in the Y-quadrant. Each spectrum shown corresponds to the minimum observable gap along a set of $\mathbf{k}$ points normal to the FS, obtained from a dense sampling of $\mathbf{k}$-space. Details can be found in Ref. [32].

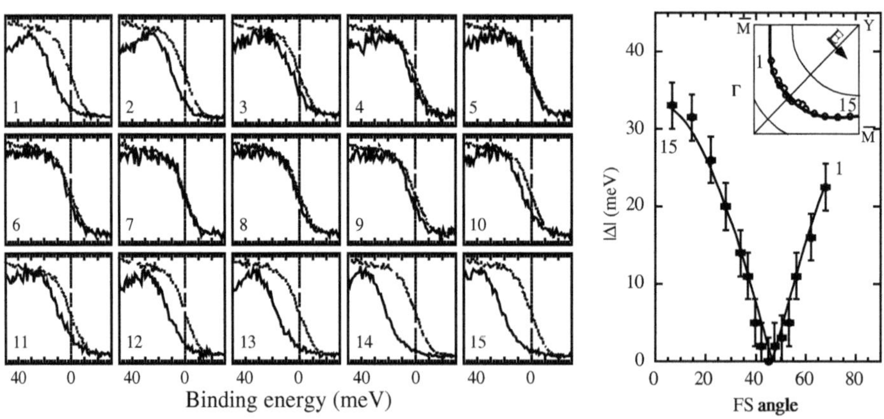

Figure 3.9. Bi2212 spectra (solid lines) for an 87K T_c sample at 13K and Pt spectra (dashed lines) versus binding energy (meV) along the Fermi surface in the Y quadrant. The photon polarization and BZ locations of the data points are shown in inset in the right panel.

We model the SC state data in terms of spectral functions [32, 36], avoiding the need to know the details of the self-energy and background by modeling only the leading edge of the spectra. We argue as follows: in the large gap region near $(\pi, 0)$, we see a linewidth collapse for frequencies smaller than $\sim 3\Delta$ upon cooling well below T_c. Thus for estimating the SC gap at the low temperature, it is sufficient to look at small frequencies, and to focus on the coherent piece of the spectral function with a resolution-limited leading edge. (Note this argument fails at higher temperatures, e.g., just below T_c). This coherent piece is modeled by the BCS spectral function, Eq. 3.6.

The other important question is the justification for using a coherent spectral function to model the rather broad EDC along and near the zone diagonal. As far as the early data being discussed here is concerned, such a description is self-consistent [32, 36], though perhaps not unique, with the entire width of the EDC accounted for by the large dispersion (of about 60 meV within our $\mathbf{k}$-window) along the zone diagonal. More recent data taken along $(0, 0)$ to (π, π) with a momentum resolution of $\delta\mathbf{k} \simeq 0.01\pi/a^*$ fully justifies this assumption by resolving coherent nodal quasiparticles in the SC state[20].

The gaps extracted from fits to the spectra of Fig. 3.9a are shown as filled symbols in Fig. 3.9b. For a detailed discussion of the error bars (both on the gap value and on the Fermi surface angle), and also of sample-to-sample variations in the gap estimates, we refer the reader to Ref. [36]. The angular variation of the gap obtained from the fits is in excellent agreement with $|\cos(k_x) - \cos(k_y)|$ form. The ARPES experiment cannot of course measure the phase of the order parameter, but this result is in strong support of $d_{x^2-y^2}$ pairing [34]. Moreover, the functional form of the anisotropy we find is consistent with electrons in the Cooper pair residing on neighboring Cu sites. That is, ARPES gives information on the spatial range of the pair interaction which is difficult to obtain from other techniques.

Upon varying the doping, the simple d-wave gap $\Delta = \Delta_0 \cos(2\phi)$ (Fig. 3.9b) is modified by the addition of the first harmonic $\Delta_{\mathbf{k}} = \Delta_{\max}[B\cos(2\phi) + (1-B)\cos(6\phi)]$, with $0 \leq B \leq 1$, as shown in Fig. 3.10[38]. Note that the $\cos(6\phi)$ term in the Fermi surface harmonics can be shown to be closely related to the tight binding function $\cos(2k_x) - \cos(2k_y)$, which represents next nearest neighbors interaction, just as $\cos(2\phi)$ is closely related to the near neighbor pairing function $\cos(k_x) - \cos(k_y)$. From Fig. 3.10 we find that while the overdoped data are consistent with $B \simeq 1$, the parameter B decreases as a function of underdoping.

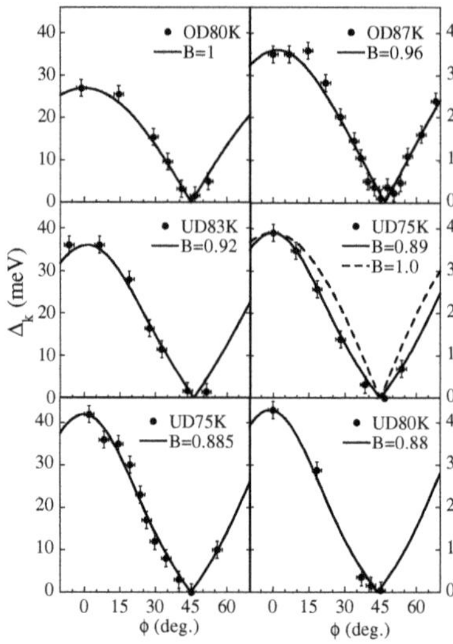

Figure 3.10. Values of the superconducting gap as a function of the Fermi surface angle ϕ obtained for a series of Bi2212 samples with varying doping[38]. Note two different UD75K samples were measured, and the UD83K sample has a larger doping due to sample aging[44]. The solid lines represent the best fit using the gap function: $\Delta_k = \Delta_{\max}[B\cos(2\phi) + (1-B)\cos(6\phi)]$ as explained in the text. The dashed line in the panel of an UD75K sample represents the gap function with B=1.

9. Pseudogap

We now describe one of the most fascinating developments in the study of high T_c superconductors: the appearance of a pseudogap above T_c in the underdoped side of the cuprate phase diagram. Briefly the "pseudogap" phenomenon is the loss of low energy spectral weight in a window of temperatures $T_c < T < T^*$; see Fig. 3.11. The pseudogap regime has been probed by many techniques like NMR, optics, transport, tunneling, μSR and specific heat; for reviews and references, see Refs. [39, 40]. ARPES, with its unique momentum-resolved capabilities, has played a central role in elucidating the pseudogap phenomenon [41, 42, 43, 44, 45].

9.1 Pseudogap near $(\pi, 0)$

In the underdoped materials, T_c is suppressed by lowering the carrier (hole) concentration as shown in Fig. 3.11. In the samples used by our group [43, 44, 45] underdoping was achieved by adjusting the oxygen

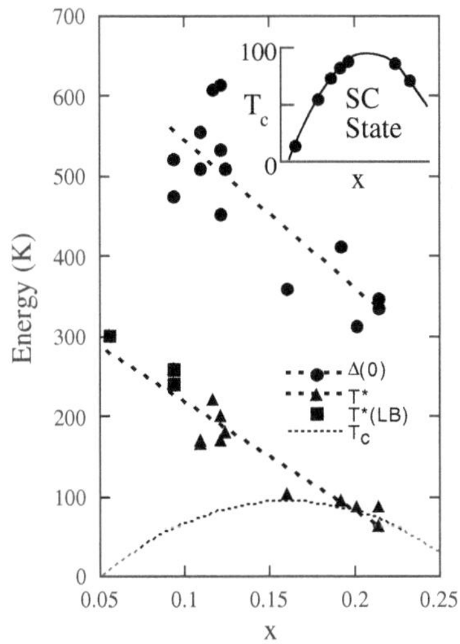

Figure 3.11. T^* (triangles for determined values and squares for lower bounds) and T_c (dashed line) as a function of hole doping x. The x values for a measured T_c were obtained by using the empirical relation $T_c/T_c^{max} = 1 - 82.6(x - 0.16)^2$ [46] with T_c^{max} =95 K. Also shown is the low temperature (maximum) superconducting gap $\Delta(0)$ (circles). Note the similar doping trends of $\Delta(0)$ and T^*.

partial pressure during annealing the float-zone grown crystals. These crystals also have structural coherence lengths of at least 1,250$\AA$ as seen from x-ray diffraction, and optically flat surfaces upon cleaving, similar to the slightly overdoped T_c samples discussed above. We denote the underdoped (UD) samples by their onset T_c: the 83K sample has a transition width of 2K and the highly underdoped 15K and 10K have transition widths > 5K. Other groups have also studied samples where underdoping was achieved by cation substitution [41, 42].

We now contrast the remarkable properties of the underdoped samples with the near-optimal Bi2212 samples which we have been mainly focusing on thus far. We will first focus on the behavior near the $(\pi, 0)$ point where the most dramatic effects occur, and come back to the very interesting **k**-dependence later. In Fig. 3.12 [47] we show the T-evolution of the ARPES spectrum at the $(\pi, 0) \to (\pi, \pi)$ Fermi crossing for an UD 83K sample. At sufficiently high temperature, the leading edge of the UD spectrum at $\mathbf{k}_F$ and the reference Pt spectrum coincide, but below a crossover temperature $T^* \simeq 180$K the leading edge midpoint of the spectrum shifts below the chemical potential. One can clearly see a loss of low energy spectral weight at 120K and 90K. It must be emphasized that this gap-like feature is seen in the normal (i.e., non-superconducting) state for $T_c = 83K < T < T^* = 180K$.

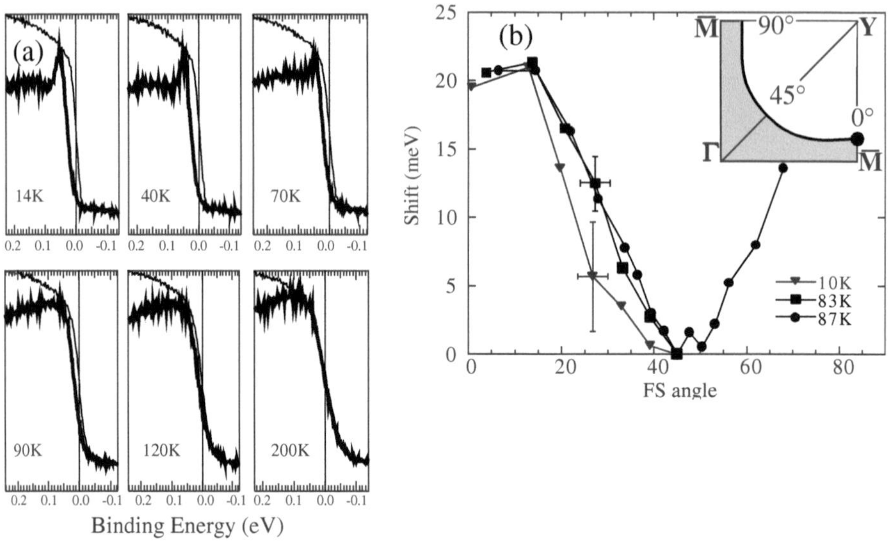

Figure 3.12. a) ARPES spectra at the dot in the inset of (b) for an 83 K underdoped sample at various temperatures (solid curves). The thin curves in each panel are reference spectra from polycrystalline Pt used to accurately determine the zero of binding energy at each temperature. b) Momentum dependence of the gap estimated from the leading-edge shift in samples with T_c's of 87K (slightly overdoped), 83K (UD) and 10K (UD), measured at 14K. For the sake of comparison between samples we made vertical offsets so that the shift at 45° is zero[43]. The inset shows the Brillouin zone with the large Fermi surface.

The doping dependence of the temperature T^*, below which a leading-edge pseudogap appears near $(\pi, 0)$, is shown in Fig. 3.11. Remarkably, T^* increases with underdoping, in sharp contrast with T_c, but very similar to the low temperature SC gap, a point we will return to at the end of the Section. The region of the phase diagram between T_c and T^* is called the pseudogap region. The trends of gap and T^* are in qualitative agreement with those obtained from other probes (see Ref. [39, 40]).

The T-dependence of the leading-edge midpoint shift appears to be completely smooth through the SC transition T_c. In other words, the normal state pseudogap evolves smoothly into the SC gap below T_c. Nevertheless, there *is* a characteristic change in the lineshape in passing through T_c associated with the appearance of a sharp feature below T_c in Fig. 3.12. This can be identified as the coherent quasiparticle peak for $T \ll T_c$. The existence of a SC state quasiparticle peak is quite remarkable given that the normal state spectra of UD materials are even broader than at optimality, and in fact become progressively broader with underdoping. In fact, the low temperature SC state spectra near $(\pi, 0)$ in the UD systems are in many ways quite similar to those at optimal doping, with the one crucial difference that the spectral weight in the coherent quasiparticle peak diminishes rapidly with underdoping [48, 49].

9.2 Anisotropy of the Pseudogap

We have already indicated that the pseudogap above T_c near the $(\pi, 0)$ point of the zone evolves smoothly through T_c into the SC gap below T_c, and thus the two also have the same magnitude. It is then interesting to know if the pseudogap above T_c has the same d-wave anisotropy as the SC gap below T_c.

The first ARPES studies [41, 42, 43] showed that the pseudogap is also highly anisotropic and has a $\mathbf{k}$-dependence which is very similar to that of the SC gap below T_c. Later work [45] further clarified the situation by showing that the anisotropy has a very interesting temperature dependence.

In Fig. 3.12b [43] we plot the leading edge shifts for three samples at 14K: the slightly overdoped 87K and UD 83K samples are in their SC states while the UD 10K sample is in the pseudogap regime. The gap estimate for each sample was made on the minimum gap locus. There is a flattening of the gap near the node, a feature that we discussed earlier for the SC gap in UD samples. The remarkable conclusion is that the normal state pseudogap has a very similar $\mathbf{k}$-dependence and magnitude as the SC gap below T_c.

9.3 Fermi Arcs

The T-dependence and anisotropy of the pseudogap was investigated in more detail in Ref. [45] motivated by the following question. Normal metallic systems are characterized by a Fermi surface, and optimally doped cuprates are no different despite the absence of sharp quasiparticles (see Section 5). On the underdoped side of the phase diagram, however, how does the opening of a pseudogap affect the locus of low lying excitations in $\mathbf{k}$-space?

In Fig. 3.13 we show ARPES spectra for an UD 83K sample at three $\mathbf{k}$ points on the Fermi surface for various temperatures. The superconducting gap, as estimated by the position of the sample leading edge midpoint at low T, is seen to decrease as one moves from point a near $(\pi, 0)$ to b to c, closer to the diagonal $(0, 0) \rightarrow (\pi, \pi)$ direction, consistent with a $d_{x^2-y^2}$ order parameter. At each $\mathbf{k}$ point the quasiparticle peak disappears above T_c as T increases, with the pseudogap persisting well above T_c, as noted earlier.

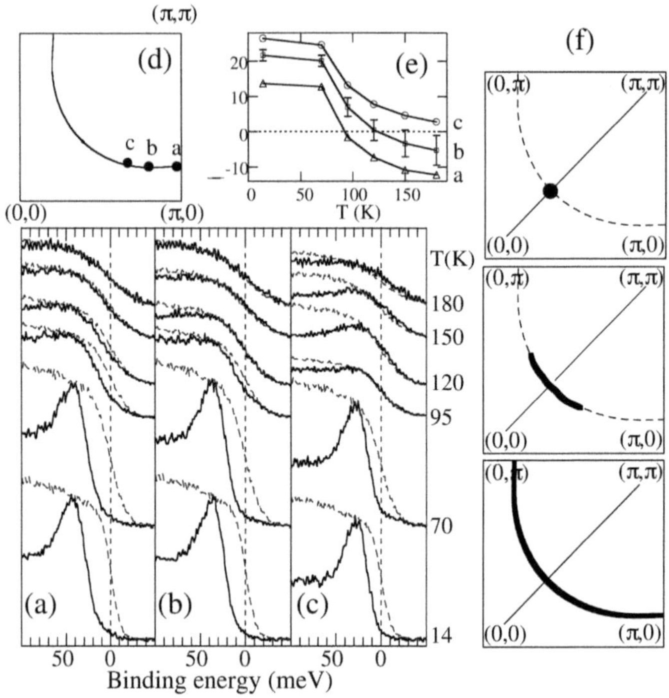

Figure 3.13. (a,b,c): Spectra taken at three **k** points in the Y quadrant of the zone (shown in (d)) for an 83K underdoped Bi2212 sample at various temperatures (solid curves). The dotted curves are reference spectra from polycrystalline Pt (in electrical contact with the sample) used to determine the chemical potential (zero binding energy). Note the closing of the spectral gap at different T for different **k**'s, which is also apparent in the plot (e) of the midpoint of the leading edge of the spectra as a function of T. Panels (f) show a schematic illustration of the temperature evolution of the Fermi surface in underdoped cuprates. The d-wave node below T_c (top panel) becomes a gapless arc above T_c (middle panel) which expands with increasing T to form the full Fermi surface at T^* (bottom panel).

The striking feature which is apparent from Fig. 3.13 is that the pseudogap at different **k** points closes at different temperatures, with larger gaps persisting to higher T's. At point a, near $(\pi, 0)$, there is a pseudogap at all T's below 180K, at which the Bi2212 leading edge matches that of Pt. As discussed above, this defines T^* above which the the largest pseudogap has vanished within the resolution of our experiment, and a closed contour of gapless excitations – a Fermi surface – is obtained. The surprise is that if we move along this Fermi surface to point b the sample leading edge matches Pt at 120K, which is smaller than T^*. Continuing to point c, about halfway to the diagonal direction, we find that the Bi2212 and Pt leading edges match at an even lower temperature of 95K. In addition, spectra measured on the same sample along the Fermi contour near the $(0,0) \to (\pi,\pi)$ line shows no gap at any T (even below T_c) consistent with $d_{x^2-y^2}$ anisotropy. One simple way to quantify the behavior of the gap is to plot the midpoint of the leading edge of the spectrum; see Fig. 3.13(e). We note that a leading edge midpoint at a negative binding energy, particularly for $\mathbf{k}$ point c, indicates the formation of a peak in the spectral function at $\omega = 0$ at high T. Further, we will say that the pseudogap has closed at a $\mathbf{k}$ point when the midpoint equals zero energy, in accordance with the discussion above. A clearer way of determining this will be presented below when we discuss the symmetrization method, but the results will be the same.

From Fig. 3.13, we find that the pseudogap closes at point a at a T above 180K, at point b at 120 K, and at point c just below 95 K. If we now view these data as a function of decreasing T, the picture of Fig. 3.13f clearly emerges. With decreasing T, the pseudogap first opens up near $(\pi, 0)$ and progressively gaps out larger portions of the Fermi contour. Thus one obtains gapless arcs which shrink as T is lowered, eventually leading to the four point nodes of the d-wave SC gap.

We next turn to a powerful visualization aid that makes these results very transparent. This is the symmetrization method introduced in Ref. [45], which effectively eliminates the Fermi function f from ARPES data and permits us to focus directly on the spectral function A. Given ARPES data described by[10] $I(\omega) = \sum_{\mathbf{k}} I_0 f(\omega) A(\mathbf{k}, \omega)$ with the sum over a small momentum window about the Fermi momentum $\mathbf{k}_F$, we can generate the symmetrized spectrum $I(\omega) + I(-\omega)$. Making the reasonable assumption of particle-hole (p-h) symmetry for a small range of ω and $\epsilon_{\mathbf{k}}$, we have $A(\epsilon_{\mathbf{k}}, \omega) = A(-\epsilon_{\mathbf{k}}, -\omega)$ for $|\omega|, |\epsilon|$ less than few tens of meV. It then follows, using the identity $f(-\omega) = 1 - f(\omega)$, that $I(\omega) + I(-\omega) = \sum_{\mathbf{k}} I_0 A(\mathbf{k}, \omega)$ which is true even after convolution with a (symmetric) energy resolution function; for details see the appendix of Ref. [6]. The symmetrized spectrum coincides with the raw data for $\omega \leq -2.2T_{eff}$, where $4.4T_{eff}$ is the 10%-90% width of the Pt leading edge, which includes the effects of both temperature and resolution. Non-trivial information is obtained for the range $|\omega| \leq 2.2T_{eff}$, which is then the scale on which p-h symmetry has to be valid. We have extensively checked this method, and studied in detail the errors introduced by incorrect determination of the chemical potential or of $\mathbf{k}_F$ (which lead to spurious narrow features in the symmetrized spectra), and the effect of the small (1° radius) $\mathbf{k}$-window of the experiment (which was found to be small).

In Fig. 3.14 we show symmetrized data for the 83K underdoped sample corresponding to the raw data of Fig. 3.13. To emphasize that the

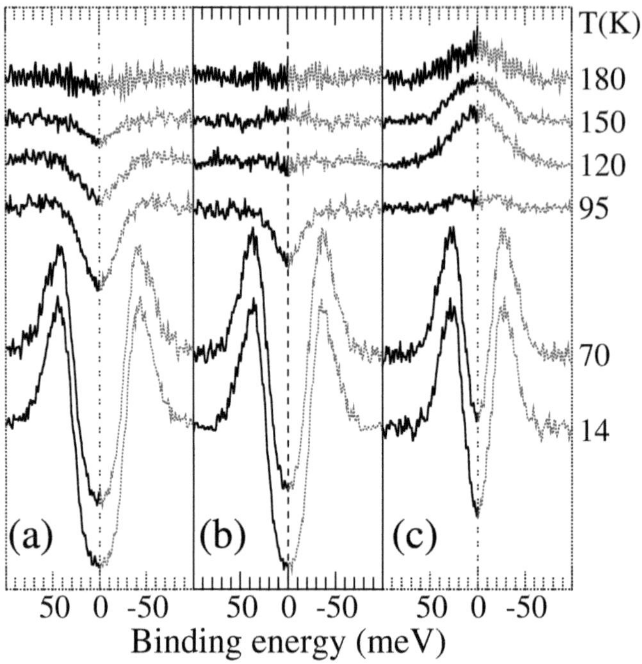

Figure 3.14. Symmetrized spectra corresponding to the raw spectra (a,b,c) of Fig. 3.13. The gap closing in the raw spectrum of Fig. 3.13 corresponds to when the pseudogap depression disappears in the symmetrized spectrum. Note the appearance of a spectral peak at higher temperatures in c.

symmetry is put in by hand, we show the $\omega > 0$ curve as a dotted line. At $\mathbf{k}$ point a near $(\pi, 0)$ the sharp quasiparticle peak disappears above T_c but a strong pseudogap suppression, on the same scale as the superconducting gap, persists all the way up to 180K (T^*). Moving to panels b and c in Fig. 3.14 we again see pseudogap depressions on the scale of the superconducting gaps at those points, however the pseudogap fills up at lower temperatures: 120K at b and 95K at c. In panel c, moreover, a spectral peak at zero energy emerges as T is raised. All of the conclusions drawn from the raw data in Fig. 3.13 are immediately obvious from the simple symmetrization analysis of Fig. 3.14. Near the $(\pi, 0)$ point the gap goes away with increasing temperature with the spectral weight filling-in, but with no perceptible change in the gap scale with T. On the other hand, at $\mathbf{k}_F$ points halfway to the node, one sees a suppression of the gap scale with increasing temperature.

We conclude this discussion with a brief mention of the implications of our results. We believe that the unusual T-dependence of the pseudogap anisotropy will be a very important input in reconciling the different crossovers seen in the pseudogap regime by different probes. The point here is that each experiment is measuring a **k**-sum weighted with a different set of **k**-dependent matrix elements or kinematical factors (e.g., Fermi velocity). For instance, quantities which involve the Fermi velocity, like dc resistivity above T_c and the penetration depth below T_c (superfluid density), should be sensitive to the region near the zone diagonal, and would thus be affected by the behavior we see at **k** point c. Other types of measurements (e.g. specific heat and tunneling) are more "zone-averaged" and will have significant contributions from **k** points a and b as well, thus they should see a more pronounced pseudogap effect. Interestingly, other data we have indicate that the region in the Brillouin zone where behavior like **k** point c is seen shrinks as the doping is reduced, and thus appears to be correlated with the loss of superfluid density[50]. Further, we speculate that the disconnected Fermi arcs should have a profound influence on magnetotransport given the lack of a continuous Fermi contour in momentum space.

9.4 Origin of the Pseudogap?

We conclude with a summary of ARPES results on the pseudogap and a brief discussion of its theoretical understanding.

As described above, the low-energy (leading edge) pseudogap has the following characteristics.

- The magnitude of the pseudogap near $(\pi, 0)$, i.e., the scale of which there is suppression of low energy spectral weight above T_c, is the same as the maximum SC gap at low temperatures. Further, both have the same doping dependence.

- There is a crossover temperature scale T^* above which the full Fermi surface of gapless excitations is recovered. The pseudogap near $(\pi, 0)$ appears below T^*.

- The normal state pseudogap evolves smoothly through T_c into the SC gap as a function of decreasing temperature.

- The pseudogap is strongly anisotropic with **k**-dependence which resembles that of the d-wave SC gap. The anisotropy of the pseudogap seems to be T-dependent leading to the formation of disconnected Fermi arcs below T^*.

- The pseudogap is "tied" to the Fermi surface, i.e., the minimum gap locus in the pseudogap regime coincides with the Fermi surface above T^* and the minimum gap locus deep in the SC state.

The simplest theoretical explanation of the pseudogap, qualitatively consistent with the ARPES observations, is that it arises due to pairing fluctuations above T_c [51, 40]. The SC gap increases with underdoping while T_c decreases. Thus in the underdoped regime T_c is not controlled by the destruction of the pairing amplitude, as in conventional BCS theory, but rather by fluctuations of the phase [51, 52] of the order parameter leading to the Uemura scaling $T_c \sim \rho_s$ [50]. Even though SC order is destroyed at T_c, the local pairing amplitude survives above T_c giving rise to the pseudogap features. A natural mechanism for such a pseudogap coming from spin pairing in a doped Mott insulator exists within the RVB framework [53], with the possibility of additional chiral current fluctuations [54].

More recently the pairing origin of the pseudogap has been challenged. Some experiments [55] have been argued to suggest a non-pairing explanation with a competition between the pseudogap and the SC gap. A specific realization of this scenario is the staggered flux or d-density wave mechanism [56] in which T^* is actually a phase transition below which both time-reversal and translational invariance are broken. A more subtle phase transition with only broken time-reversal has also been proposed [57] as the origin of the pseudogap.

Although a qualitative understanding of some of the characteristics of the pseudogap within the non-pairing scenarios is not clear at this time, these theories make sharp predictions about broken symmetries below T^* which can be tested. A very recent ARPES study [58] of circular dichroism finds evidence in favor of broken time reversal, thus casting some doubt on the pairing fluctuation ideas. The last word has clearly not been said on this subject, theoretically or experimentally, and the origin of the pseudogap remains one of the most important open questions in the field of high T_c superconductors.

Acknowledgments

Much of the experimental work described in this review was done in collaboration with Hong Ding, Adam Kaminski, Helen Fretwell, Kazimierz Gofron, Joel Mesot, Stephan Rosenkranz, Tsunehiro Takeuchi, and the group of Takashi Takahashi, including Takafumi Sato and Takayoshi Yokoya. We were very fortunate to have available to us the samples from Kazuo Kadowaki, T. Mochiku, David Hinks, Prasenjit Guptasarma, Boyd Veal, Z. Z. Li, and Helene Raffy. We have also benefited from many interactions over the years with Phil Anderson, Alex Abrikosov, Jim Allen, Cliff Olson, Ole Andersen, Al Arko, Bertram Batlogg, Arun Bansil, Matthias Eschrig, A. Fujimori, Peter Johnson, Bob Laughlin, Bob Schrieffer, Z. X. Shen, and Chandra Varma.

This work was supported by the National Science Foundation, Grant No. DMR 9974401 and the U.S. Department of Energy, Office of Science, under Contract No. W-31-109-ENG-38.

References

[1] P. W. Anderson, Science **235**, 1196 (1987).

[2] A. Damascelli, Z.-X. Shen, Z. Hussain, Rev. Mod. Phys. **75** (2002) 473.

[3] J.C. Campuzano, M.R. Norman, and M. Randeria, in *Superconductivity* (Springer-Verlag, Berlin, 2003), and references therein. Also, cond-mat/0209476.

[4] S. Hüfner, *Photoelectron Spectroscopy* (Springer-Verlag, Berlin, 1996), and references therein.

[5] C.N. Berglund and W.E. Spicer, Phys. Rev. **136**, A1030 and A1044 (1964).

[6] J. Mesot, *et al.*, Phys. Rev. B **63**, 224516 (2001).

[7] C. Caroli, *et al.*, Phys. Rev. B **8**, 4552 (1973).

[8] A. Bansil, M. Lindroos, and J.C. Campuzano, Proc. Mater. Res. Soc. **253**, 519 (1992); M. Lindroos, A. Bansil, K. Gofron, H. Ding, J.C. Campuzano, R. Liu, and B.W. Veal, Physica C **212**, 347 (1993).

[9] See, e.g., L. Hedin and S. Lundquist, *Solid State Physics* **23**, 1 (Academic, New York, 1969).

[10] M. Randeria, *et al.*, Phys. Rev. Lett. **74**, 4951 (1995).

[11] P. Aebi, *et al.*, Phys. Rev. Lett. **72**, 2757 (1994); J. Osterwalder, *et al.*, . Phys. A **60**, 247 (1995).

[12] A. V. Federov, *et al.*, Phys. Rev. Lett. **82**, 2179 (1999); T. Valla, *et al.*, Science **285**, 2110 (1999); T. Valla, *et al.*, Phys. Rev. Lett. **85**, 828 (2000).

[13] P.V. Bogdanov, *et al.*, Phys. Rev. Lett. **85**, 2581 (2000).

[14] A. Kaminski, *et al.*, Phys. Rev. Lett. **86**, 1070 (2001).

[15] M. R. Norman, *et al.*, Phys. Rev. B**64**, 184508 (2001).

[16] W. E. Pickett, Rev. Mod. Phys. **61**, 433 (1989).

[17] P. W. Anderson, *The Theory of Superconductivity in the High-T_c Cuprates* (Princeton Univ. Pr., Princeton, 1997).

[18] J. C. Campuzano, *et al.*, Phys. Rev. Lett. **64**, 2308 (1990).

[19] C. G. Olson, *et al.*, Phys. Rev. B **42**, 381 (1990).

[20] A. Kaminski, *et al.*, Phys. Rev. Lett. **84**, 1788 (2000).

[21] For a review of the first five years of work on the cuprates, see Sec. 4 and 5 of Z. X. Shen and D. S. Dessau, Phys. Repts. **253**, 1 (1995).

[22] H. Ding, *et al.*, Phys. Rev. Lett. **76**, 1533 (1996).

[23] M. R. Norman, *et al.*, Phys. Rev. B **52**, 615 (1995).

[24] R. L. Withers, *et al.*, J. Phys. **C 21**, 6067 (1988).

[25] S. Massida, J. Yu, and A.J. Freeman, Physica **C 152**, 251 (1988); O. K. Andersen, *et al.*, Phys. Rev. **B 49**, 4145 (1994).

[26] O. K. Andersen *et al.*, Phys. Chem. Solids **56**, 1573 (1995).

[27] S. Chakravarty, *et al.*, Science **261**, 337 (1994).

[28] D.L. Feng, *et al.*, Phys. Rev. Lett. **86**, 5550 (2001); D.L. Feng, *et al.*, Phys. Rev. B **65**, 220501 (2002).

[29] Y. D. Chuang, *et al.*, Phys. Rev. Lett. **87**, 117002 (2001); Y. D. Chuang, *et al.*, cond-mat/0107002.

[30] A. A. Kordyuk, *et al.*, Phys. Rev. Lett. **89**, 077003 (2002).

[31] A. Kaminski, et al., cond-mat/0210531.

[32] H. Ding, *et al.*, Phys. Rev. Lett. **74**, 2784 (1995) and **75**, 1425 (E) (1995).

[33] M.C. Nuss, *et al.*, Phys. Rev. Lett. **66**, 3305 (1991); D. A. Bonn, *et al.*, Phys. Rev. Lett. **68**, 2390 (1992).

[34] D. J. Van Harlingen, Rev. Mod. Phys **67**, 515 (1995); C. C. Tsuei and J. R. Kirtley, *ibid* **72**, 969 (2000).

[35] Z. X. Shen, *et al.*, Phys. Rev. Lett. **70**, 1553 (1993).

[36] H. Ding, *et al.*, Phys. Rev. B **54**, R9678 (1996).

[37] J. C. Campuzano, *et al.*, Phys. Rev. B **53**, R14737 (1996).

[38] J. Mesot, *et al.*, Phys. Rev. Lett. **83**, 840 (1999).

[39] T. Timusk and B. W. Statt, Rept. Prog. Phys. **62**, 61 (1999).

[40] M. Randeria, in *Proceedings of the International School of Physics "Enrico Fermi" Course CXXXVI on High Temperature Superconductors*, ed. G. Iadonisi, J. R. Schrieffer, and M. L. Chiafalo, (IOS Press, 1998), p. 53 - 75; [Varenna Lectures, 1997] cond-mat/9710223.

[41] D.S. Marshall, *et al.*, Phys. Rev. Lett. **76**, 4841 (1996).

[42] A. G. Loeser, *et al.*, Science **273**, 325 (1996).

[43] H. Ding, *et al.*, Nature **382**, 51 (1996).

[44] H. Ding, *et al.*, Phys. Rev. Lett. **78**, 2628 (1997).

[45] M. R. Norman, *et al.*, Nature **392**, 157 (1998).

[46] M.R. Presland, *et al.*, Physica C **176**, 95 (1991).

[47] H. Ding, *et al.*, J. Phys. Chem. of Solids **59**, 1888 (1998).

[48] D.L. Feng, *et al.*, Science **289**, 277 (2000).

[49] H. Ding, *et al.*, Phys. Rev. Lett. **87**, 227001 (2001).

[50] Y. J. Uemura, *et al.*, Phys. Rev. Lett. **62**, 2317 (1989).

[51] M. Randeria, *et al.*, Phys. Rev. Lett. **69**, 2001 (1992); N. Trivedi and M. Randeria, Phys. Rev. Lett. **75**, 312 (1995).

[52] V. Emery and S. Kivelson, Nature **374**, 434 (1995).

[53] G. Baskaran, *et al.*, Solid St. Comm. **63**, 973 (1987); G. Kotliar and J. Liu, Phys. Rev. B **38**, 5142 (1988); H. Fukuyama, Prog. Theor. Phys. Suppl. **108**, 287 (1992).

[54] P. A. Lee and X. G. Wen, Phys. Rev. Lett. **76**, 503 (1996) and Phys. Rev. B **63**, 224517 (2001).

[55] J. Tallon and J. Loram, Physica C **349**, 53 (2001).

[56] S. Chakravarty, *et al.*, Phys. Rev. B **63**, 094503 (2001).

[57] C. M. Varma, Phys. Rev. Lett. **83**, 3538 (1999).

[58] A. Kaminski, *et al.*, Nature **416**, 610 (2002).

Chapter 4

LUTTINGER LIQUIDS: THE BASIC CONCEPTS

K. Schönhammer

Institut für Theoretische Physik, Universität Göttingen, Bunsenstr. 9, D-37073 Göttingen, Germany

schoenh@theorie.physik.uni-Goettingen.de

Abstract This chapter reviews the theoretical description of interacting fermions in one dimension. The Luttinger liquid concept is elucidated using the Tomonaga-Luttinger model as well as integrable lattice models. Weakly coupled chains and the attempts to experimentally verify the theoretical predictions are discussed.

Keywords: Luttinger liquids, Tomonaga model, bosonization, anomalous power laws, breakdown of Fermi liquid theory, spin-charge separation, spectral functions, coupled chains, quasi-one-dimensional conductors

1. Introduction

In this chapter we attempt a simple selfcontained introduction to the main ideas and important computational tools for the description of interacting fermions in one spatial dimension. The reader is expected to have some knowledge of the method of second quantization. As in section 3 we describe a *constructive* approach to the important concept of bosonization, *no* quantum field-theoretical background is required. After mainly focusing on the Tomonaga-Luttinger model in sections 2 and 3 we present results for integrable lattice models in section 4. In order to make contact to more realistic systems the coupling of strictly $1d$ systems as well as to the surrounding is addressed in section 5. The attempts to experimentally verify typical Luttinger liquid features like anomalous power laws in various correlation functions are only shortly discussed as this is treated in other chapters of this book.

D. Baeriswyl and L. Degiorgi (eds.), Strong Interactions in Low Dimensions, 93–136.

2. Luttinger liquids - a short history of the ideas

As an introduction the basic steps towards the general concept of Luttinger liquids are presented in historical order. In this exposition the ideas are discussed without presenting all technical details. This is done in section 3 by disregarding the historical aspects aiming at a simple presentation of the important practical concepts like the " bosonization of field operators".

2.1 Bloch's method of "sound waves" (1934)

In a paper on incoherent x-ray diffraction Bloch [1] realized and used the fact that one-dimensional ($d = 1$) noninteracting fermions have the same type of low energy excitations as a harmonic chain. The following discussion of this connection is very different from Bloch's own presentation.

The low energy excitations determine e.g. the low temperature specific heat. Debye's famous T^3-law for the lattice contribution of three dimensional solids reads in $d = 1$

$$c_L^{\text{Debye}} = \frac{\pi}{3} k_B \left(\frac{k_B T}{\hbar c_s} \right), \tag{4.1}$$

where c_s is the sound velocity. At low temperatures the electronic contribution to the specific heat in the "Fermi gas" approximation of Pauli is also linear in T and involves the density of states of the non-interacting electrons at the Fermi energy. This yields for spinless fermions in $d = 1$

$$c_L^{\text{Pauli}} = \frac{\pi}{3} k_B \left(\frac{k_B T}{\hbar v_F} \right), \tag{4.2}$$

where v_F is the Fermi velocity. With the replacement $c_s \leftrightarrow v_F$ the results are *identical.* This suggests that apart from a scale factor the (low energy) excitation energies and the degeneracies in the two types of systems are identical. For the harmonic chain the excited states are classified by the numbers n_k of phonons in the modes ω_k whith n_k taking integer values from zero to infinity. The excitation energy is given by $E(\{n_{k_m}\}) - E_0 = \sum_{k_m} \hbar\omega_{k_m} n_{k_m}$. For small wave numbers k_m the dispersion is *linear* $\omega_{k_m} \approx c_s |k_m|$. Therefore the excitations energies are multiples of $\hbar c_s (2\pi/L)$ for periodic boundary conditions and multiples of $\Delta_B \equiv \hbar c_s \pi/L$ for fixed boundary conditions. The calculation of the partition function is standard textbook material. This is also true for noninteracting electrons but there the calculation involves fermionic occupation numbers n_k^F which take values zero and one. The two textbook calculations yield Eqs. (4.1) and (4.2), but through the "clever" use of the grand canonical ensemble in order to simplify the fermionic calculation the identity (apart from $c_s \leftrightarrow v_F$) remains mysterious. A deeper understanding involves two steps:

1) *Linearization* of the kinetic energy $\varepsilon_k = \hbar^2 k^2/(2m)$ of the free fermions around the Fermi point k_F for fixed boundary conditions or both Fermi points $\pm k_F$ for periodic boundary conditions. As the argument is simplest for fixed boundary conditions [2] which lead to $k_m = m\pi/L$ we discuss this case for the moment. Then the energies $\varepsilon_{k_n} - \varepsilon_F$ are integer multiples of $\Delta_F \equiv \hbar v_F \pi/L$ where v_F is the Fermi velocity.

2) *Classification* of any state of the Fermi system by the *number* n_j *of upward shifts by* j *units of* Δ_F with respect to the ground state. As the fermions are indistinguishable the construction of the $\{n_j\}$ shown in Fig. 4.1, where the highest occupied level in the excited state is connected with the highest occupied level in the ground state and so forth for the second, third ... highest levels, completely specifies the excited state. As the n_j can run from zero to infinity like bosonic quantum numbers and the excitation energy is given by $\sum_j (j\Delta_F)n_j$ the *canonical* partition function for the noninteracting fermions has the same form as the canonical partition function for the harmonic chain apart from $\Delta_F \leftrightarrow \Delta_B$ if one assumes the Fermi sea to be infinitely deep [3].
As we have linearized ω_k for small k as well as ε_k around k_F this equivalence only holds for the *low temperature* specific heats ($k_B T \ll \hbar\omega_{\max}, k_B T \ll \varepsilon_F$).

If we denote the creation (annihilation) operator of a fermion with $k_n = n\pi/L$ by $c_n^\dagger$ (c_n) and assume a *strictly linear* dispersion $\varepsilon_n^{\text{lin}} = \hbar v_F k_n$ for *all* $k_n > 0$ a more technical formulation of the discussed equivalence can be given by the *exact operator identity*

$$\begin{aligned} T &= \sum_{n=1}^{\infty} \hbar v_F k_n c_n^\dagger c_n \\ &= \frac{\hbar v_F \pi}{L}\left[\sum_{l=1}^{\infty} l\, b_l^\dagger\, b_l + \frac{1}{2}\mathcal{N}(\mathcal{N}+1)\right], \end{aligned} \tag{4.3}$$

where the operators b_l with $l \geq 1$ are defined as

$$b_l \equiv \frac{1}{\sqrt{l}} \sum_{m=1}^{\infty} c_m^\dagger c_{m+l} \tag{4.4}$$

and $\mathcal{N} \equiv \sum_{n=1}^{\infty} c_n^\dagger c_n$ is the fermionic particle number operator. The proof of the "Kronig identity" (4.3) is simple (see Ref. [4]) . The oper-

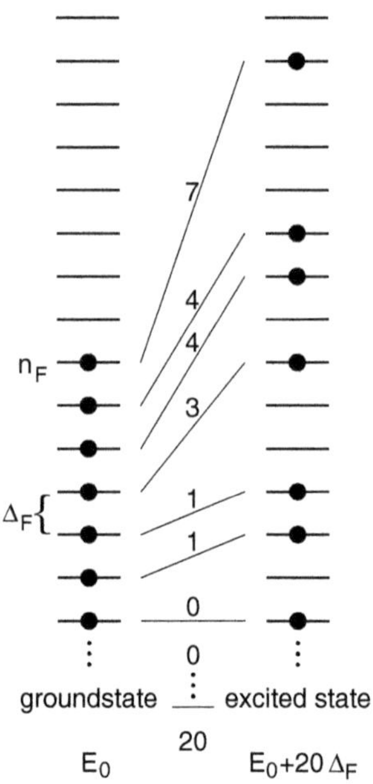

Figure 4.1. Example for the classification scheme for the excited states in terms of the numbers n_j of upward shifts by j units of Δ_F. In the example shown the nonzero n_j are $n_7 = 1, n_4 = 2, n_3 = 1$ and $n_1 = 2$.

ators b_l obey commutation relations $[b_l, b_{l'}] = 0$ and for $l \geq l'$

$$\left[b_l, b_{l'}^\dagger\right] = \frac{1}{\sqrt{ll'}} \sum_{m=1}^{l'} c_m^\dagger c_{m+l-l'}. \tag{4.5}$$

For all N-particle states $|\phi_N^{(M)}\rangle \equiv \prod_{n=1}^{N} c_{i_n}^\dagger |\mathrm{Vac}\rangle$ in which the $M(< N)$ lowest one-particle levels are *all occupied* one obtains

$$\left[b_l, b_{l'}^\dagger\right] |\phi_N^{(M)}\rangle = \delta_{ll'} |\phi_N^{(M)}\rangle \tag{4.6}$$

for $l, l' \leq M$, i.e. these operators obey *boson commutation relations* $[b_l, b_{l'}^\dagger] = \delta_{ll'} \hat{1}$ in this *subspace* of all possible N-particle states.
Later it turns out to be useful to work with $\tilde{T} \equiv T - \langle T \rangle_0 - \mu_0 \tilde{\mathcal{N}}$, where $\langle T \rangle_0 = \Delta_F n_F(n_F+1)/2$ is the ground-state energy, $\mu_0 = \Delta_F(n_F+1/2)$ is the chemical potential of the noninteracting fermions and $\tilde{\mathcal{N}} \equiv \mathcal{N} - n_F \hat{1}$. Then $\tilde{T}$ is of the form as the rhs of Eq. (4.3) with $\mathcal{N}(\mathcal{N}+1)$ replaced by $\tilde{\mathcal{N}}^2$.

2.2 Tomonaga (1950): Bloch's method of sound waves applied to interacting fermions

When a two-body interaction between the fermions is switched on, the ground state is no longer the filled Fermi sea but it has admixtures of (multiple) particle-hole pair excitations. In order to simplify the problem Tomonaga studied the *high density limit* where the range of the interaction is much larger than the interparticle distance, using *periodic* boundary conditions [5]. Then the Fourier transform $\tilde{v}(k)$ of the two-body interaction is nonzero only for values $|k| \leq k_c$ where the cut-off k_c is much smaller than the Fermi momentum $k_c \ll k_F$. This implies that for not too strong interaction the ground state and low energy excited states have negligible admixtures of holes deep in the Fermi sea and particles with momenta $|k| - k_F \gg k_c$. In the two intermediate regions around the two Fermi points $\pm k_F$, with particle-hole pairs present, the dispersion ε_k is *linearized* in order to apply Bloch's "sound wave method"

$$k \approx \pm k_F : \quad \varepsilon_k = \varepsilon_F \pm v_F(k \mp k_F). \tag{4.7}$$

Tomonaga realized that the Fourier components of the operator of the density

$$\begin{aligned}\hat{\rho}_n &= \int_{-L/2}^{L/2} \hat{\rho}(x) e^{-ik_n x} dx = \int_{-L/2}^{L/2} \psi^\dagger(x)\psi(x) e^{-ik_n x} dx \\ &= \sum_{n'} c_{n'}^\dagger c_{n'+n},\end{aligned} \tag{4.8}$$

where $c_n^\dagger(c_n)$ creates (annihilates) a fermion in the state with momentum $k_n = \frac{2\pi}{L} n$, plays a central role in the interaction term, as well as the kinetic energy. Apart from an additional term linear in the particle number operator [4], which is usually neglected, the two-body interaction is given by

$$\hat{V} = \frac{1}{2L} \sum_{n \neq 0} \tilde{v}(k_n) \hat{\rho}_n \hat{\rho}_{-n} + \frac{1}{2L} \mathcal{N}^2 \tilde{v}(0) \tag{4.9}$$

Tomonaga's important step was to split $\hat{\rho}_n$ for $|k_n| \ll k_F$ into two parts, one containing operators of *"right movers"* i.e. involving fermions near the right Fermi point k_F with velocity v_F and *"left movers"* involving fermions near $-k_F$ with velocity $-v_F$

$$\hat{\rho}_n = \sum_{n' \geq 0} c_{n'}^\dagger c_{n'+n} + \sum_{n' < 0} c_{n'}^\dagger c_{n'+n} \equiv \hat{\rho}_{n,+} + \hat{\rho}_{n,-} \tag{4.10}$$

where the details of the splitting for small $|n'|$ are irrelevant. Apart from the square root factor the $\hat{\rho}_{n,\alpha}$ are similar to the b_l defined in Eq. (4.4). Their commutation relations in the *low energy subspace* are

$$[\hat{\rho}_{m,\alpha}, \hat{\rho}_{n,\beta}] = \alpha m \delta_{\alpha\beta}\delta_{m,-n}\hat{1}. \tag{4.11}$$

If one defines the operators

$$b_n \equiv \frac{1}{\sqrt{|n|}} \begin{cases} \hat{\rho}_{n,+} & \text{for } n > 0 \\ \hat{\rho}_{n,-} & \text{for } n < 0 \end{cases} \tag{4.12}$$

and the corresponding adjoint operators $b_n^\dagger$ this leads using $\rho_{n,\alpha}^\dagger = \rho_{-n,\alpha}$ to the bosonic commutation relations

$$[b_n, b_m] = 0, \quad [b_n, b_m^\dagger] = \delta_{mn}\hat{1}. \tag{4.13}$$

Now the kinetic energy of the right movers as well as that of the left movers can be "bosonized" as in Eq. (4.3). The interaction $\hat{V}$ is bilinear in the $\hat{\rho}_n$ as well as the $\hat{\rho}_{n,\alpha}$. Therefore apart from an additional term containing particle number operators *the Hamiltonian for the interacting fermions is a quadratic form in the boson operators*

$$\begin{aligned}\tilde{H} &= \sum_{n>0} \hbar k_n \left\{ \left(v_F + \frac{\tilde{v}(k_n)}{2\pi\hbar} \right) \left(b_n^\dagger b_n + b_{-n}^\dagger b_{-n} \right) \right. \\ &\quad \left. + \frac{\tilde{v}(k_n)}{2\pi\hbar} \left(b_n^\dagger b_{-n}^\dagger + b_{-n} b_n \right) \right\} + \frac{\hbar\pi}{2L} \left[v_N \tilde{\mathcal{N}}^2 + v_J \mathcal{J}^2 \right] \\ &\equiv H_B + H_{\tilde{\mathcal{N}},\mathcal{J}}, \end{aligned} \tag{4.14}$$

where $\tilde{\mathcal{N}} \equiv \tilde{\mathcal{N}}_+ + \tilde{\mathcal{N}}_-$ is the total particle number operator relative to the Fermi sea, $\mathcal{J} \equiv \tilde{\mathcal{N}}_+ - \tilde{\mathcal{N}}_-$ the "current operator", and the velocities are given by $v_N = v_F + \tilde{v}(0)/\pi\hbar$ and $v_J = v_F$. Here v_N determines the energy change for adding particles without generating bosons while v_J enters the energy change when the difference in the number of right and left movers is changed. Similar to the discussion at the end of section 1.1 we have defined $\tilde{H} \equiv H - E_0^H - (\mu_0 + \tilde{v}(0)n_0)\tilde{\mathcal{N}}$, where E_0^H is the Hartree energy and n_0 the particle density. As the particle number operators $\tilde{\mathcal{N}}_\alpha$ commute with the boson operators $b_m(b_m^\dagger)$ the two terms H_B and $H_{\tilde{\mathcal{N}},\mathcal{J}}$ in the Hamiltonian *commute* and can be treated separately. Because of the translational invariance the two-body interaction only couples the modes described by $b_n^\dagger$ and b_{-n}. With the Bogoliubov transformation $\alpha_n^\dagger = c_n b_n^\dagger - s_n b_{-n}$ the Hamiltonian H_B can be brought into the form

$$H_B = \sum_{n\neq 0} \hbar\omega_n \alpha_n^\dagger \alpha_n + \text{ const.}, \tag{4.15}$$

where the $\omega_n = v_F|k_n|\sqrt{1+\tilde{v}(k_n)/\pi\hbar v_F}$ follow from 2×2 eigenvalue problems corresponding to the condition $[H_B, \alpha_n^\dagger] = \hbar\omega_n\alpha_n^\dagger$. For small k_n one obtains for smooth potentials $\tilde{v}(k)$ again a *linear* dispersion $\omega_n \approx v_c|k_n|$, with the *"charge velocity"* $v_c = \sqrt{v_N v_J}$, which is larger than v_F for $\tilde{v}(0) > 0$. The expression for the coefficients c_n and s_n with $c_n^2 - s_n^2 = 1$ will be presented later for the generalized model Eq. (4.17) . For fixed particle numbers N_+ and N_-, the excitation energies of the *interacting* Fermi system are given by $\sum_m \hbar\omega_m n_m$ with integer occupation numbers $0 \le n_m < \infty$. For small enough excitation energies the only difference of the excitation spectrum for fixed particle numbers with respect to the noninteracting case is the replacement $v_F \leftrightarrow v_c$.

In his seminal paper Tomonaga did not realize the anomalous decay of correlation functions of the model because in his discussion of the density correlation function he missed the $2k_F$-contribution discussed in section 3.

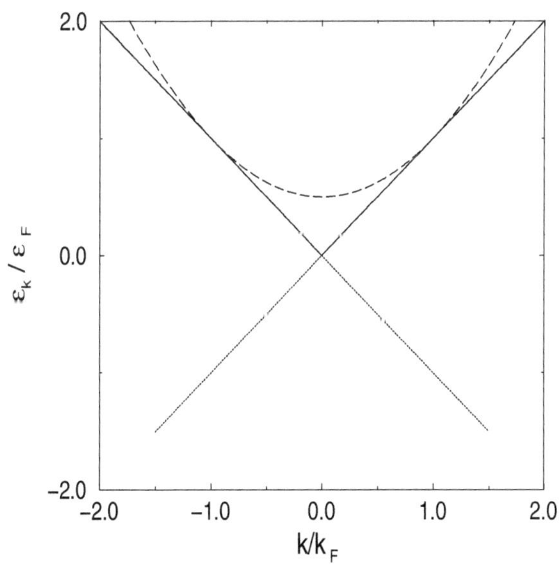

Figure 4.2. Energy dispersion as a function of momentum. The dashed curve shows the usual "nonrelativistic" dispersion and the full curve the linearized version used (apart from a constant shift) in Eq. (4.3) for $k > 0$ for fixed boundary conditions. The dot-dashed parts are the additional states for $k_0 = -1.5k_F$. The model discussed by Luttinger corresponds to $k_0 \to -\infty$.

2.3 Luttinger (1963): no discontinuity at the Fermi surface

Luttinger, apparently unaware of Tomonaga's work, treated spinless, massless fermions (in the relativistic sense, but $c \leftrightarrow v_F$) in one dimension, i.e. two *infinite* branches of right and left moving fermions with dispersion $\pm v_F k$ [6]. As Luttinger himself made an error with the fact that his Hamiltonian is not bounded from below, it is useful to switch from Tomonaga's to Luttinger's model keeping a *band cut-off* k_0 such that $k \geq k_0 = 2\pi m_0/L$ with $m_0 < 0$ for the right movers and correspondingly for the left movers (see Fig. 4.2). Fortunately Luttinger's error had no influence on his inquiry if a *sharp Fermi surface* exists in the exact ground state of the interacting model. After a rather complicated calculation using properties of so-called "Toeplitz determinants" Luttinger found that the average occupation $\langle n_{k,+}\rangle$ in the ground state for $k \approx k_F$ behaves as

$$\langle n_{k,+}\rangle - \frac{1}{2} \sim \left|\frac{k-k_F}{k_c}\right|^{\alpha_L} \text{sign}(k_F - k), \tag{4.16}$$

where α_L depends on the interaction strength (see below) [7]. "Thus, in this model, the smallest amount of interaction *always* destroys the discontinuity of $\langle n_k\rangle$ at the Fermi surface" [6]. This can be related to the fact that the equal time correlation functions $\langle \psi_\alpha^\dagger(x)\psi_\alpha(0)\rangle$ decay as $1/|x|^{1+\alpha_L}$ in the *interacting* system in contrast to $\langle \psi_\alpha^\dagger(x)\psi_\alpha(0)\rangle \sim 1/|x|^d$ (with $d = 1$) in the noninteracting case. Therefore α_L is called the *"anomalous dimension"*[8].

Apart from the different dispersion Luttinger also used a different interaction. In contrast to Tomonaga he only kept an interaction between the right and left movers but not the term $\sim \tilde{v}(k_n)(b_n^\dagger b_n + b_{-n}^\dagger b_{-n})$ in Eq. (4.14) . In the limit of a delta interaction of the right and left movers his model is identical to the *massless Thirring model* (1958) [9] at that time not well known in the solid state physics community.

2.4 Towards the "Luttinger liquid" concept

Luttinger's treatment of the Dirac sea was corrected in a paper by Mattis and Lieb (1965) [10] which also offered a simpler way to calculate $\langle n_{k,\alpha}\rangle$. The time dependent one-particle Green's function for the spinless Luttinger model was calculated by Theumann (1967) [11] by generalizing this method. She found *power law behaviour* in the corresponding spectral function $\rho(k,\omega)$, especially $\rho(k_F,\omega) \sim \alpha_L|\omega|^{\alpha_L-1}$, i.e. *no sharp quasiparticle for* $k = k_F$ consistent with Luttinger's result for the occupation numbers (Fig.4.3). For a delta interaction her res-

ults agreed with an earlier calculation for the massless Thirring model by Johnson (1961) [12]. Later the time dependent one-particle Green's function was calculated by various other methods, e.g. using Ward identities (Dzyaloshinski and Larkin (1974) [13]) as well as the important method of the *"bosonization of the field operator"* (Luther and Peschel (1974) [14]) which will be addressed in detail in section 3. It was first proposed in a different context by Schotte and Schotte (1969) [15].
What is now usually called the "Tomonaga-Luttinger (TL) model" is the following generalization of Eq. (4.14)

$$\begin{aligned}\tilde{H}_{TL} &= \frac{2\pi\hbar}{L}\sum_{n>0} n\left\{\left(v_F + \frac{g_4(k_n)}{2\pi\hbar}\right)\left(b_n^\dagger b_n + b_{-n}^\dagger b_{-n}\right)\right. \\ &+ \left.\frac{g_2(k_n)}{2\pi\hbar}\left(b_n^\dagger b_{-n}^\dagger + b_{-n} b_n\right)\right\} + \frac{\hbar\pi}{2L}\left\{v_N \tilde{\mathcal{N}}^2 + v_J \mathcal{J}^2\right\}, \quad (4.17)\end{aligned}$$

where $v_N = v_F + (g_4(0) + g_2(0))/2\pi\hbar$ and $v_J = v_F + (g_4(0) - g_2(0))/2\pi\hbar$. The interaction parameters $g_2(k_n)$ and $g_4(k_n)$ are allowed to be *different.* As Tomonaga's original model the TL model is exactly solvable, i.e. it can also be brought into the form of Eq. (4.15). The eigenvector components in $\alpha_n^\dagger = c_n b_n^\dagger - s_n b_{-n}$ are given by

$$c_n = \frac{1}{2}\left(\sqrt{K_n} + \frac{1}{\sqrt{K_n}}\right), \quad s_n = \frac{1}{2}\left(\sqrt{K_n} - \frac{1}{\sqrt{K_n}}\right) \quad (4.18)$$

with $K_n = \sqrt{v_J(k_n)/v_N(k_n)}$, where $v_{J(N)}(k_n) \equiv v_F + [g_4(k_n) \mp g_2(k_n)]/2\pi\hbar$. The frequencies are given by $\omega_n = |k_n|\sqrt{v_J(k_n)v_N(k_n)} \equiv |k_n|v_c(k_n)$. The TL-Hamiltonian corresponds to a fermionic Hamiltonian that *conserves* the number of right and left movers.

A more general model including spin and terms changing right movers into left movers and vice versa is usually called the "g-ology model". An important step towards the general Luttinger liquid concept came from the *renormalization group* (RG) study of this model. It was shown that for *repulsive* interactions (see section 3) the renormalized interactions flow towards a fixed point Hamiltonian of the TL-type unless in lattice models for commensurate electron fillings strong enough interactions (for the half filled Hubbard model discussed in section 4 this happens for arbitrarily small on-site Coulomb interaction U) destroy the metallic state by opening a Mott-Hubbard gap. The RG approach is described in detail in reviews by Sólyom (1979) [16] and Shankar (1994) [17]. These results as well as insight from models which allow an exact solution by the Bethe ansatz led Haldane [18, 19] to propose the concept of *Luttinger liquids* (LL) as a replacement of Fermi liquid theory in one dimension, which "fails because of the infrared divergence of certain vertices it assumes

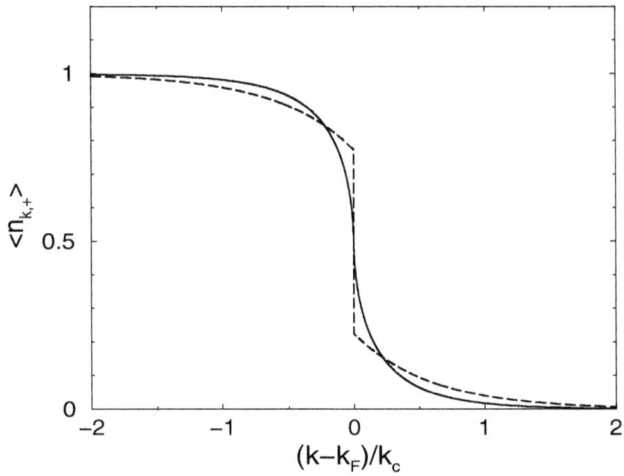

Figure 4.3. The full line shows the average occupation $\langle n_{k,+}\rangle$ for a TL model with $\alpha_L = 0.6$. The dashed line shows the expectation from Fermi liquid theory, where the discontinuity at k_F determines the quasi-particle weight Z_{k_F} in $\rho_+(k_F,\omega)$. As discussed following Eq. (4.48) this can also be realized in a TL model with $g_2(0) = 0$. There also the details of the interaction are specified.

to remain finite" [19] . At least for spinless fermions Haldane was able to show that "the Bogoliubov transformation technique that solves the Luttinger model provides a general method for resumming *all* infrared divergences present"[19]. Similar to Fermi liquid theory in higher dimensions this new LL phenomenology allows to describe the low energy physics in terms of a few constants, *two* for the spinless case: the "stiffness constant" $K \equiv K_0 = \sqrt{v_J/v_N}$ (also called g in various publications) and the *"charge velocity"* $v_c = \sqrt{v_J v_N}$. In his seminal paper Haldane showed explicitly that the LL relations survive in not exactly soluble generalizations of the TL model with a non-linear fermion dispersion. He also gave a clear presentation how to calculate general correlation functions and e.g. the occupancies shown in Fig. 4.3 for the TL model. The technical details are addressed in section 3.

Before we do this two additional important aspects of LL-behaviour should be mentioned. The first concerns the strong influence of impurities on the low energy physics [20, 21, 22, 23, 24, 25], especially the peculiar modification of the electronic properties of a LL when a *single* impurity with an arbitrarily weak backscattering potential is present. For a spinless LL with a *repulsive* two-body interaction, i.e. $K < 1$ a perturbative bosonic RG calculation [25] shows that the backscattering strength V_B is a *relevant* perturbation which grows as Λ^{K-1} when the flow parameter Λ is sent to zero. This leads to a breakdown of the perturbative analysis in V_B. On the other hand a weak hopping between the open ends of two semi-infinite chains is *irrelevant* and scales to zero as $\Lambda^{K^{-1}-1}$. Assuming that the open chain presents the only stable fixed point it was argued that at low energy scales even for a weak impurity physical observables behave as if the system is split into two semi-infinite chains. This leads to a conductance which vanishes with a power law in T at low temperatures [25]. A more technical discussion is presented in section 3.

Electrons are spin one-half particles and for their description it is necessary to include the spin degree of freedom in the model. For a fixed quantization axis the two spin states are denoted by $\sigma =\uparrow,\downarrow$. The fermionic creation (annihilation) operators $c^\dagger_{n,\pm,\sigma}(c_{n,\pm,\sigma})$ carry an additional spin label as well as the $\hat\rho_{n,\pm,\sigma}$ and the boson operators $b_{n,\sigma}$ which in a straightforward way generalize Eq. (4.12). The interactions $g_\nu(k)$ with $\nu = 2,4$ in Eq. (4.17) become matrices $g_\nu^{\sigma\sigma'}$ in the spin labels. If they have the form $g_\nu^{\sigma\sigma'}(k) = \delta_{\sigma,\sigma'}g_{\nu\|}(k) + \delta_{\sigma,-\sigma'}g_{\nu\perp}(k)$ it is useful to switch to new boson operators $b_{n,a}$ with $a = c, s$

$$\begin{aligned} b_{n,c} &\equiv \frac{1}{\sqrt{2}}\left(b_{n,\uparrow} + b_{n,\downarrow}\right) \\ b_{n,s} &\equiv \frac{1}{\sqrt{2}}\left(b_{n\uparrow} - b_{n,\downarrow}\right) \ , \end{aligned} \tag{4.19}$$

which obey $\left[b_{a,n}, b_{a',n'}\right] = 0$ and $\left[b_{a,n}, b^\dagger_{a',n'}\right] = \delta_{aa'}\delta_{nn'}\hat{1}$. The kinetic energy can be expressed in terms of "charge" (c) and "spin" (s) boson operators using $b^\dagger_{n,\uparrow}b_{n,\uparrow} + b^\dagger_{n\downarrow}b_{n\downarrow} = b^\dagger_{n,c}b_{n,c} + b^\dagger_{n,s}b_{n,s}$. If one defines the interaction matrix elements $g_{\nu,a}(q)$ via

$$\begin{aligned} g_{\nu,c}(q) &\equiv g_{\nu\|}(q) + g_{\nu\perp}(q) \\ g_{\nu,s}(q) &\equiv g_{\nu\|}(q) - g_{\nu\perp}(q) \ , \end{aligned} \tag{4.20}$$

and defines $\tilde{\mathcal{N}}_{\alpha,c(s)} \equiv (\tilde{\mathcal{N}}_{\alpha,\uparrow} \pm \tilde{\mathcal{N}}_{\alpha,\downarrow})/\sqrt{2}$ one can write the TL-Hamiltonian $\tilde H^{(1/2)}_{TL}$ for spin one-half fermions as

$$\tilde H^{(1/2)}_{TL} = \tilde H_{TL,c} + \tilde H_{TL,s} \ , \tag{4.21}$$

where the $\tilde H_{TL,a}$ are of the form Eq. (4.17) but the interaction matrix elements have the additional label a. The two terms on the rhs of Eq. (4.21) *commute*, i.e. the "charge" and "spin" excitation are completely independent. This is usually called "spin-charge separation". The "diagonalization" of the two separate parts proceeds exactly as before and the low energy excitations are "massless bosons" $\omega_{n,a} \approx v_a|k_n|$

with the *charge velocity* $v_c = (v_{J_c} v_{N_c})^{1/2}$ and the *spin velocity* $v_s = (v_{J_s} v_{N_s})^{1/2}$. The corresponding two stiffness constants are given by $K_c = (v_{J_c}/v_{N_c})^{1/2}$ and $K_s = (v_{J_s}/v_{N_s})^{1/2}$. Because of Eq. (4.21) the dependence of the velocities on the interaction strength (4.20) is obtained using the results for the spinless model following Eq. (4.18).

The low temperature thermodynamic properties of the TL model including spin, Eqs. (4.17) and (4.21), can be expressed in terms of the four velocities $v_{N_c}, v_{J_c}, v_{N_s}, v_{J_s}$ or the four quantities v_c, K_c, v_s, K_s. Due to spin-charge separation the specific heat has two additive contributions of the same form as in Eqs. (4.1) and (4.2). If we denote, as usual, the proportionality factor in the linear T-term by γ one obtains

$$\frac{\gamma}{\gamma_0} = \frac{1}{2}\left(\frac{v_F}{v_c} + \frac{v_F}{v_s}\right), \tag{4.22}$$

where γ_0 is the value in the noninteracting limit. To calculate the spin susceptibility χ_s one adds a term $-h\tilde{\mathcal{N}}_s$ to $\tilde{H}_{TL}^{(1/2)}$. Then by minimizing the ground state energy with respect to N_s one obtains $\langle \tilde{\mathcal{N}}_s \rangle \sim h/v_{N_s}$, i.e. χ_s is inversely proportional to v_{N_s}. If one denotes the spin susceptibility of the noninteracting system by $\chi_{s,0}$, this yields for the zero temperature susceptibility

$$\frac{\chi_s}{\chi_{s,0}} = \frac{v_F}{v_{N_s}} = K_s \frac{v_F}{v_s}. \tag{4.23}$$

For spin rotational invariant systems one has $K_s = 1$ [26]. The zero temperature compressibilty κ is proportional to $(\partial^2 E_0/\partial N^2)_L^{-1}$ which using Eqs. (4.17) and (4.21) leads to

$$\frac{\kappa}{\kappa_0} = \frac{v_F}{v_{N_c}} = K_c \frac{v_F}{v_c}. \tag{4.24}$$

A simple manifestation of spin-charge separation occurs in the time evolution of a localized perturbation of e.g. the the spin-up density. The time evolution $\alpha_{n,a}(t) = \alpha_{n,a} e^{-i\omega_{n,a} t}$ for $a = c, s$ implies

$$b_{n,a}(t) = b_{n,a}\left(c_{n,a}^2 e^{-i\omega_{n,a}t} - s_{n,a}^2 e^{i\omega_{n,a}t}\right) - b_{-n,a}^\dagger c_{n,a} s_{n,a}\left(e^{-i\omega_{n,a}t} - e^{i\omega_{n,a}t}\right) \tag{4.25}$$

If the initial state of the system involves a perturbation of *right movers* only, i.e. $\langle b_{n,a} \rangle = 0$ for $n < 0$ and the perturbation is sufficiently smooth ($\langle b_{n,a} \rangle \neq 0$ for $0 < n \ll n_c$ only) the initial perturbation is split into four parts which move with velocities $\pm v_c$ and $\pm v_s$ without changing the initial *shape*. If only the initial expectation values of the $b_{n,\uparrow}$ are different from zero one obtains for $\delta\langle\rho_\uparrow(x,0)\rangle \equiv F(x)$ using Eq. (4.12)

$$\delta\langle\rho_\uparrow(x,t)\rangle = \sum_a \left[\frac{1+K_a}{4}F(x-v_a t) + \frac{1-K_a}{4}F(x+v_a t)\right]. \qquad (4.26)$$

For the spin rotational invariant case $K_s = 1$ there is no contribution which moves to the left with the spin velocity. Already for the pure g_4-model with $K_c = 1$ but $v_c \neq v_s$ "spin-charge separation" of the distribution occurs. For the spinless model with $g_2 \neq 0$ the initial distribution splits into one right- and one left-moving part, which is often called "charge fractionalization" [27, 28]. Note that the splitting described in Eq. (4.26) is independent of the details of $F(x)$ like the corresponding total charge. An additional comment should be made: spin-charge separation is often described as the fact that when an electron is injected into the system *its* spin and charge move independently with different velocities. This is very misleading as it is a *collective* effect of the total system which produces expectation values like in Eq. (4.26).

The easiest way to understand the important manifestation of spin-charge separation in the momentum resolved one-particle spectral functions [29, 30] is to make use of the bosonization of the electronic field operators discussed in the next section.

3. Luttinger liquids - computational tools

In section 2 many of the important features of LL's like the absence of a discontinuity at the Fermi surface were presented without giving any details how these properties are actually determined. As the most important tool the bosonization of the field operators is presented in detail in this section. This method is then used to calculate correlation functions like the one-particle Green's function and the $2k_F$-density response function. In the second part of this section the TL model with additional interactions and (or) a one particle potential with a "backscattering" contribution is discussed. The model is no longer exactly solvable and RG arguments play an important role [16, 17, 25].

3.1 Bosonization of the field operator

In the following a selfcontained presentation of the bosonization of a *single* fermion operator including a straightforward construction of the particle number changing part ("Klein factor") is given. We present the bosonization of the field operator for the right movers described by the $c_{l,+}$ and just mention the corresponding result for the left movers.

The starting point are the commutation relations the $c_{l,+}$ obey for $m > 0$

$$[b_m, c_{l,+}] = -\frac{1}{\sqrt{m}} c_{l+m,+} \quad , \quad [b_m^\dagger, c_{l,+}] = -\frac{1}{\sqrt{m}} c_{l-m,+} \quad . \tag{4.27}$$

If (after taking the limit $m_0 \to -\infty$) one introduces the 2π-periodic auxiliary field operator $\tilde{\psi}_+(v)$, where v later will be taken as $2\pi x/L$

$$\tilde{\psi}_+(v) \equiv \sum_{l=-\infty}^{\infty} e^{ilv} c_{l,+} \ , \tag{4.28}$$

it obeys the simple commutation relations

$$[b_m, \tilde{\psi}_+(v)] = -\frac{1}{\sqrt{m}} e^{-imv} \tilde{\psi}_+(v) \ ; \ [b_m^\dagger, \tilde{\psi}_+(v)] = -\frac{1}{\sqrt{m}} e^{imv} \tilde{\psi}_+(v) \ . \tag{4.29}$$

Products of exponentials of operators *linear* in the boson operators

$$A_+ \equiv \sum_{n \neq 0} \lambda_n b_n^\dagger \quad ; \quad B_- \equiv \sum_{n \neq 0} \mu_n b_n \tag{4.30}$$

with arbitrary constants λ_n and μ_n obey similar commutation relations

$$[b_m, e^{A_+} e^{B_-}] = \lambda_m e^{A_+} e^{B_-} \quad ; \quad [b_m^\dagger, e^{A_+} e^{B_-}] = -\mu_m e^{A_+} e^{B_-} \ , \tag{4.31}$$

which follow from $[b_m, e^{\lambda b_m^\dagger}] = \lambda e^{\lambda b_m^\dagger}$. We therefore make the ansatz

$$\tilde{\psi}_+(v) = \hat{O}_+(v) e^{i\phi_+^\dagger(v)} e^{i\phi_+(v)}, \tag{4.32}$$

where the operator $i\phi_+(v)$ is given by [19]

$$i\phi_+(v) = \sum_{n=1}^{\infty} \frac{e^{inv}}{\sqrt{n}} b_n \ . \tag{4.33}$$

Then the operator $\hat{O}_+(v)$ *commutes* with all the b_m and $b_m^\dagger$. We next construct $\hat{O}_+(v)$ such that both sides of Eq. (4.32) yield identical matrix elements.

As $\tilde{\psi}_+(v)$ reduces the number of right movers by one, the operator $\hat{O}_+(v)$ also must have this property. In order to determine $\hat{O}_+(v)$ we work with the eigenstates of the noninteracting system (the limit $m_0 \to -\infty$ is implied and n_F is an arbitrary positive integer later related to k_F)

$$|\{m_l\}_b, \tilde{N}_+, \tilde{N}_-\rangle \equiv \prod_l \frac{(b_l^\dagger)^{m_l}}{\sqrt{m_l!}} \left(\prod_{n=m_0}^{n_F+\tilde{N}_-} c_{-n,-}^\dagger \right) \left(\prod_{r=m_0}^{n_F+\tilde{N}_+} c_{r,+}^\dagger \right) |\mathrm{Vac}\rangle. \tag{4.34}$$

It is easy to see that $\hat{O}_+(v)|\{0\}_b, \tilde{N}_+, \tilde{N}_-\rangle$ has no overlap to excited states

$$\begin{aligned}\langle\{m_l\}_b, \tilde{N}_+ - 1, \tilde{N}_-|\hat{O}_+(v)|\{0\}_b, \tilde{N}_+, \tilde{N}_-\rangle &= \\ \langle\{0\}_b, \tilde{N}_+ - 1, \tilde{N}_-|\prod_l \frac{(b_l)^{m_l}}{\sqrt{m_l!}}\hat{O}_+(v)|\{0\}_b, \tilde{N}_+, \tilde{N}_-\rangle \ . \quad (4.35)\end{aligned}$$

As $\hat{O}_+(v)$ commutes with the b_l the rhs of Eq. (4.35) vanishes unless all m_l are zero. This implies

$$\hat{O}_+(v)|\{0\}_b, \tilde{N}_+, \tilde{N}_-\rangle = c_+(\tilde{N}_+, \tilde{N}_-, v)|\{0\}_b, \tilde{N}_+ - 1, \tilde{N}_-\rangle \ , \qquad (4.36)$$

where $c_+(\tilde{N}_+, \tilde{N}_-, v)$ is a c-number. In order to determine $c_+(\tilde{N}_+, \tilde{N}_-, v)$ we calculate $\langle\{0\}_b, \tilde{N}_+ - 1, \tilde{N}_-|\tilde{\psi}_+(v)|\{0\}_b, \tilde{N}_+, \tilde{N}_-\rangle$ using Eq. (4.28) as well as Eq. (4.32). In the calculation of the matrix element with the fermionic form Eq. (4.28) we use Eq. (4.34) which yields

$$\langle\{0\}_b, \tilde{N}_+ - 1, \tilde{N}_-|c_{l,+}|\{0\}_b, \tilde{N}_+, \tilde{N}_-\rangle = (-1)^{\tilde{N}_-}\delta_{l,n_F+\tilde{N}_+} . \qquad (4.37)$$

The factor $(-1)^{\tilde{N}_-}$ occurs because we have to commute $c_{l,+}$ through the product of $N_- = -m_0 + 1 + n_F + \tilde{N}_-$ fermionic operators of the left movers if we assume $-m_0 + n_F$ to be odd. We note that *no* such factor occurs for the corresponding matrix element of the left movers. The calculation of the ground state to ground state matrix element of $\tilde{\psi}_+(v)$ using Eq. (4.32) is simple as both exponentials involving the boson operators can be replaced by the unit operator and the matrix element is just $c_+(\tilde{N}_+, \tilde{N}_-, v)$. The comparison therefore yields

$$c_+(\tilde{N}_+, \tilde{N}_-, v) = (-1)^{\tilde{N}_-}e^{iv(n_F+\tilde{N}_+)} \qquad (4.38)$$

and $c_-(\tilde{N}_+, \tilde{N}_-, v) = e^{-iv(n_F+\tilde{N}_-)}$. Together with Eq. (4.34) and the definition $\tilde{\mathcal{N}}_\alpha \equiv \mathcal{N}_\alpha - (-m_0 + 1 + n_F)\hat{1}$ this implies

$$\hat{O}_+(v)e^{-i(n_F+\tilde{\mathcal{N}}_+)v}(-1)^{\tilde{\mathcal{N}}_-}|\{0\}_b, \tilde{N}_+, \tilde{N}_-\rangle = |\{0\}_b, \tilde{N}_+ - 1, \tilde{N}_-\rangle \ . \quad (4.39)$$

If we apply the operator $\hat{O}_+(v)e^{-i(n_F+\tilde{\mathcal{N}}_+)v}(-1)^{\tilde{\mathcal{N}}_-}$ to the states in Eq. (4.34) and use again that $\hat{O}_+(v)$ commutes with the boson operators we obtain

$$\hat{O}_+(v)e^{-i(n_F+\tilde{\mathcal{N}}_+)v}(-1)^{\tilde{\mathcal{N}}_-}|\{m_l\}_b, \tilde{N}_+, \tilde{N}_-\rangle = |\{m_l\}_b, \tilde{N}_+ - 1, \tilde{N}_-\rangle \ . \qquad (4.40)$$

This shows that the operator $U_+ \equiv \hat{O}_+(v)e^{-i(n_F+\tilde{\mathcal{N}}_+)v}(-1)^{\tilde{\mathcal{N}}_-}$ is *independent* of v and given by

$$U_+ = \sum_{\tilde{N}_+, \tilde{N}_-}\sum_{\{m_l\}}|\{m_l\}_b, \tilde{N}_+ - 1, \tilde{N}_-\rangle\langle\{m_l\}_b, \tilde{N}_+, \tilde{N}_-| \ . \qquad (4.41)$$

It follows immediately that U_+ is *unitary*, i.e. $U_+U_+^\dagger = U_+^\dagger U_+ = \hat{1}$. From Eq. (4.41) one can infer that for arbitrary functions f of the number operator $\tilde{\mathcal{N}}_+$ one has $U_+ f(\tilde{\mathcal{N}}_+) = f(\tilde{\mathcal{N}}_+ + 1)U_+$.
To summarize we have shown that

$$\hat{O}_+(v) = U_+ e^{i(n_F+\tilde{\mathcal{N}}_+)v}(-1)^{\tilde{\mathcal{N}}_-} \; . \tag{4.42}$$

In $\hat{O}_-(u) = U_- e^{-i(n_F+\tilde{\mathcal{N}}_-)u}$ *no* factor $(-1)^{\tilde{\mathcal{N}}_+}$ appears and therefore $\hat{O}_+(v)$ and $\hat{O}_-(u)$ anticommute, which is necessary to enforce anticommutation relations between $\tilde{\psi}_+(v)$ and $\tilde{\psi}_-(u)$. It is an easy exercise to show that e.g. the anticommutation relations $[\tilde{\psi}_+(v), \tilde{\psi}_+(u)]_+ = 0$ are fulfilled. In the calculation the properties of $\hat{O}_+(v)$ as well as the factor in Eq. (4.32) involving the boson operators enter. If one replaces the operators $\hat{O}_\alpha(v)e^{-i\alpha v(\tilde{\mathcal{N}}_\alpha+n_F)}$ by "Majorana fermions" η_α which commute with the boson operators and obey the anticommutation relations $[\eta_\alpha, \eta_\beta]_+ = 2\delta_{\alpha\beta}\hat{1}$, as often done in the literature, this yields $[\tilde{\psi}_\alpha(v), \tilde{\psi}_\alpha(u)]_+ = [1 - \cos(u-v)]e^{i\alpha(u+v)(\tilde{\mathcal{N}}_\alpha+n_F)}$, i.e. a violation of the correct anticommutation relations. This implies that the U_α have to be properly treated. In many publications they are written as $U_\alpha = e^{i\hat{\theta}_\alpha}$, where the *phase operators* $\hat{\theta}_\alpha$ are assumed to obey the canonical commutation relations (CCR) $[\tilde{\mathcal{N}}_\alpha, \hat{\theta}_\alpha] = i\hat{1}$ [19]. We do not use this concept here because no phase operator can be constructed which obeys the CCR as an operator identity [4, 31, 32, 33, 34].

In the following we will always use the "normal ordered" form (all boson annihilation operators to the right of the creation operators) of the bosonization formula Eqs. (4.32, 4.33). Alternatively one introduces a convergence factor $e^{-n\lambda/2}$, whith $\lambda \to 0$ and works with the Hermitian Bose fields $\Phi_\alpha(v) \equiv \phi_\alpha(v) + \phi_\alpha^\dagger(v)$ as well as the fields $\Phi_+ \pm \Phi_-$. The derivatives of the latter fields are related to the total current and the deviation of the total charge density from its average value [35]. As we work with an interaction cut-off k_c the introduction of λ is not necessary and because of the space limitation this field-theoretic formulation is not used here.

3.2 Calculation of correlation functions for the TL model

In order to calculate correlation functions of the TL model with nonzero interactions it is necessary to express the field operator $\tilde{\psi}_+(v)$ Eq. (4.32) in terms of the $\alpha_n, \alpha_n^\dagger$ instead of the $b_n, b_n^\dagger$ because the former have a simple time dependence and for the temperature dependent expectation values one has $\langle \alpha_m^\dagger \alpha_n \rangle = \delta_{mn} n_B(\omega_n)$, where $n_B(\omega) = 1/(e^{\beta\omega} - 1)$ is the Bose function. For the ground state calculation all one needs is $\alpha_n|\Phi_0\rangle = 0$ without using the explicit form of the interacting ground state $|\Phi_0\rangle$.

For periodic boundary conditions one has $b_m = c_m\alpha_m + s_m\alpha_{-m}^\dagger$ where the operators α_m and $\alpha_{-m}^\dagger$ commute. Therefore $e^{i\phi_+(v)}$ (and $e^{i\phi_+^\dagger(v)}$) in Eq. (4.32) can be written as a product of two exponentials with the annihilation operators to the right. After once using the Baker-Hausdorff formula, $e^{A+B} = e^A e^B e^{-\frac{1}{2}[A,B]}$ if the operators A and B commute with $[A, B]$, in order to complete the process of normal ordering one obtains for the physical field operator $\psi_\alpha(x) = \tilde{\psi}_\alpha(2\pi x/L)/\sqrt{L}$ for a system of finite length L with periodic boundary conditions [36]

$$\psi_\alpha(x) = \frac{A(L)}{\sqrt{L}}\hat{O}_\alpha\left(\frac{2\pi x}{L}\right) e^{i\chi_\alpha^\dagger(x)} e^{i\chi_\alpha(x)} \tag{4.43}$$

with

$$i\chi_\alpha(x) = \sum_{m\neq 0} \frac{\theta(\alpha m)}{\sqrt{|m|}} \left(c_m e^{ik_m x}\alpha_m - s_m e^{-ik_m x}\alpha_{-m}\right) , \tag{4.44}$$

$A(L) \equiv e^{-\sum_{n=1}^\infty s_n^2/n}$ and $\theta(x)$ is the unit step function.

This is a very useful formula for the calculation of properties of one-dimensional interacting fermions. For the special choice $s_n = s(0)e^{-n/n_c}$ where $n_c = k_c L/2\pi$ is determined by the interaction cut-off, $A(L)$ can be calculated analytically using $\sum_{n-1}^\infty z^n/n = -\log(1-z)$. For $n_c \gg 1$ this yields $A(L) = (4\pi/k_c L)^{s^2(0)}$ which shows that the prefactor in Eq.(4.44) has an anomalous power law proportional to $(1/L)^{\frac{1}{2}+s^2(0)}$. This implies that the $c_{n,\alpha}$ scale like $(1/L)^{s^2(0)}$.

The time dependent operator $\psi_+(x,t)$ follows from Eq. (4.44) by replacing α_m and α_{-m} by $\alpha_m e^{-i\omega_m t}$ and $\alpha_{-m}e^{-i\omega_m t}$ and U_+ in $\hat{O}_+$ by $U_+(t)$. Various kinds of time dependent correlation functions can quite simply be calculated using this result. Here we begin with $iG_+^<(x,t) \equiv \langle\psi_+^\dagger(0,0)\psi_+(x,t)\rangle$.

As U_+ commutes with the bosonic operator the particle number changing operators lead to a simple time dependent factor

$$U_+^\dagger U_+(t)|\Phi_0(\tilde{N}_+,\tilde{N}_-)\rangle = e^{-i[E_0(\tilde{N}_+,\tilde{N}_-)-E_0(\tilde{N}_+-1,\tilde{N}_-)]t}|\Phi_0(\tilde{N}_+,\tilde{N}_-)\rangle. \tag{4.45}$$

As $\psi_+(x)$ in Eq.(4.43) is normal ordered in the α's one has to use the Baker-Hausdorff formula only once to normal order $\psi_+^\dagger(0,0)\psi_+(x,t)$.

This yields with $k_F = 2\pi n_F/L$

$$\begin{aligned} ie^{i\mu t}G_+^<(x,t) &= \frac{A^2(L)}{L}e^{ik_Fx}e^{[\chi_+(0,0),\chi_+^\dagger(x,t)]} \qquad (4.46) \\ &= \frac{e^{ik_Fx}}{L}e^{\sum_{n=1}^\infty \frac{1}{n}\left[e^{-i(k_nx-\omega_nt)}+2s_n^2\left(\cos(k_nx)e^{i\omega_nt}-1\right)\right]} \end{aligned}$$

where $\mu \equiv E_0(\tilde{N}_+,\tilde{N}_-) - E_0(\tilde{N}_+ - 1, \tilde{N}_-)$ is the chemical potential. The analytical evaluation of the sum (integral in the limit $L \to \infty$) in the exponent in Eq. (4.46) is not possible. An approximation which gives the correct large x and t behaviour [37] is to replace ω_n by v_ck_n. This yields for $L \to \infty$ with the exponential cut-off for the s_n used earlier [14]

$$ie^{i\mu t}G_+^<(x,t) = \frac{-i}{2\pi}\frac{e^{ik_Fx}}{x - v_ct - i0}\left[\frac{r^2}{(x - v_ct - ir)(x + v_ct + ir)}\right]^{s^2(0)}, \qquad (4.47)$$

where $r = 2/k_c$. As $\langle\psi_+^\dagger(0,0)\psi_+(x,0)\rangle$ for large x decays proportional to $(1/x)^{1+2s^2(0)}$ the *anomalous dimension* for the spinless model is given by

$$\alpha_L = 2s^2(0) = (K-1)^2/2K. \qquad (4.48)$$

Luttinger's result for $\langle n_{k,+}\rangle$ follows by performing the Fourier transform with respect to x. The full line in Fig.4.3 was calculated with $s_n^2 = 0.3e^{-2k_n/k_c}$, while the dashed curve corresponds to $s_n^2 = 0.6(k_n/k_c)e^{-2k_n/k_c}$. The latter example corresponds to an interaction with $g_2(k \to 0) \to 0$ which leads to a vanishing anomalous dimension α_L. In this case the occupancies $\langle n_{k,+}\rangle$ have a discontinuity at k_F as in a Fermi liquid [38]. An efficient numerical algorithm to calculate $\langle n_{k,+}\rangle$ for arbitrary forms of s_n^2 is described in the appendix of reference [2].

The spectral function $\rho^<(k,\omega)$ relevant for describing angular resolved photoemission is obtained from Eq. (4.47) by a *double* Fourier transform

$$\begin{aligned} \rho_+^<(k,\omega) &= \langle c_{k,+}^\dagger\delta[\omega + (H - E_0(\tilde{N}_+ - 1, \tilde{N}_-))]c_{k,+}\rangle \qquad (4.49) \\ &= \frac{1}{2\pi}\int_{-\infty}^{\infty} dte^{i\omega t}\int_{-\infty}^{\infty} dxe^{-ikx}ie^{i\mu t}G_+^<(x,t). \end{aligned}$$

As Eq. (4.47) is reliable in the large x and t limit its use in Eq. (4.49) correctly describes the spectral function for $k \approx k_F$ and $\omega \approx 0$ [39]. Using the variable substitutions $u_\mp = x \mp v_ct$ the double integral factorizes and with the additional approximation $i0 \to ir$ on the rhs of Eq. (4.47) one obtains [11, 14]

$$\rho_+^<(k_F+\tilde{k},\omega) \sim \theta(-\omega-v_c|\tilde{k}|)(-\omega+\tilde{k}v_c)^{\frac{\alpha_L}{2}-1}(-\omega-\tilde{k}v_c)^{\frac{\alpha_L}{2}}e^{r\omega/v_c}. \quad (4.50)$$

Without the additional approximation there is an additional weak dependence on $\omega + \tilde{k}v_c$ [29]. The complete spectral function $\rho_+(k,\omega) = \rho_+^<(k,\omega) + \rho_+^>(k,\omega)$, where $\rho_+^>(k,\omega)$ is defined via $iG_+^>(x,t) \equiv \langle\psi_+(x,t)\psi_+^\dagger(0,0)\rangle$ can be obtained using $\rho_+^>(k_F+\tilde{k},\omega) = \rho_+^<(k_F-\tilde{k},-\omega)$ which follows from the particle-hole symmetry of the model. The absence of a sharp quasi-particle peak is manifest from $\rho_+(k_F,\omega) \sim \alpha_L|\omega|^{\alpha_L-1}e^{-r|\omega|/v_c}$.

In order to calculate correlation functions of the spin one-half TL model the operators $b_{n,\sigma}$ which appear in the generalization of the bosonization formula Eqs. (4.32) and (4.33) have to be replaced by the spin and charge bosons $b_{n,\sigma} = (b_{n,c} + \sigma b_{n,s})/\sqrt{2}$. Because of the exponential occurence of the boson operators in Eq. (4.32) *and* spin-charge separation Eq.(4.21) the Green's function $G_{+,\sigma}^<(x,t)$ *factorizes* into a spin and a charge part, which both are of the form as the square root of the function on the rhs of Eq. (4.47) . This square root results from the factors $1/\sqrt{2}$ in the expression for the $b_{n,\sigma}$. In the spin factor the charge velocity v_c is replaced by the spin velocity v_s. For the average occupation numbers one again obtains Luttinger's result Eq. (4.16) with $\alpha_L = s_c^2(0) + s_s^2(0) \equiv \alpha_c + \alpha_s$. The individual contributions can be expressed in terms of the $K_a \equiv (v_{J,a}/v_{N,a})^{1/2}$ as $\alpha_a = (K_a-1)^2/(4K_a)$. As in the spinless model the fermionic (creation) annihilation operators $c_{n,\alpha,\sigma}^{(\dagger)}$ scale like $(1/L)^{\alpha_L/2}$. For spin rotational invariant systems one has $K_s = 1$, i.e. *no* contribution to the anomalous dimension α_L from the spin part of the Hamiltonian [26]. For the momentum integrated spectral functions one obtains $\rho_{\alpha,\sigma}(\omega) \sim |\omega|^{\alpha_L}$ as in the spinless model [40]. The k-resolved spectral functions $\rho_{\alpha,\sigma}(k,\omega)$ on the other hand show a drastic difference to the model without spin. The delta peaks of the noninteracting model are broadened into *one* power law threshold Eq. (4.47) in the model without spin and *two* power law singularities (see Fig. 4.4) in the model including spin [29, 30, 37] (for $\alpha_L < 1/2$ in the case of a spin independent interaction). The "peaks" disperse *linearly* with $k - k_F$.

It is also straightforward to calculate various response functions for the TL model. We discuss the density response function $R(q,z) \equiv -\langle\langle\hat{\rho}_q;\hat{\rho}_{-q}\rangle\rangle_z/L$ of the spinless model for $q \approx 0$ and $q \approx \pm 2k_F$, where

$$\langle\langle\hat{A};\hat{B}\rangle\rangle_z \equiv -\frac{i}{\hbar}\int_0^\infty \langle[A(t),B]\rangle e^{izt}dt \quad (4.51)$$

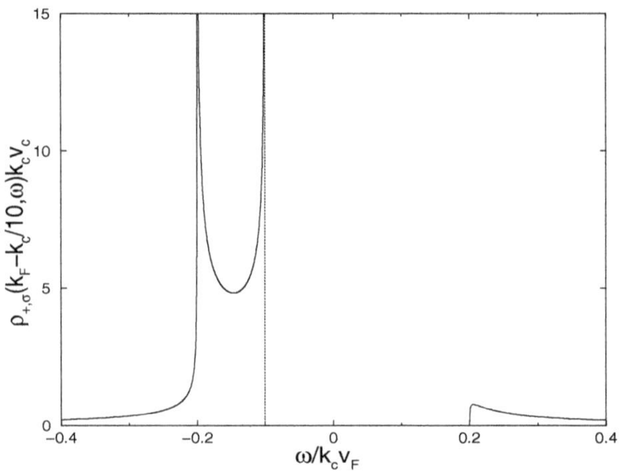

Figure 4.4. Spectral function $\rho_{+,\sigma}(k_F+\tilde{k},\omega)$ as a function of normalized frequency for $\tilde{k}=-k_c/10$ for the TL- model with a spin independent interaction. The parameters are chosen such that $v_c=2v_F$ and α_L=1/8.

involves the retarded commutator [41] and z is a frequency above the real axis. From the decomposition [42] $\psi(x)\approx\psi_+(x)+\psi_-(x)$ of the field operator $\psi(x)$ in the original Tomonaga model it is obvious that the operator $\hat{\rho}(x)=\psi^\dagger(x)\psi(x)$ of the density (see Eq. (4.8)) has two very different contributions

$$\begin{aligned}\hat{\rho}(x) &\approx \hat{\rho}_+(x)+\hat{\rho}_-(x)+\left(\psi_+^\dagger(x)\psi_-(x)+h.c.\right) \\ &\equiv \hat{\rho}_0(x)+\hat{\rho}_{2k_F}(x).\end{aligned} \qquad (4.52)$$

The spatial Fourier transform of $\hat{\rho}_0$ is *linear* in the boson operators Eq. (4.12) and the $q\approx 0$ contribution of the density response function $[R(q,z)]_0$ defined with the operators $(\hat{\rho}_0)_q$ follows using the (linear) equations of motion for the $b_n(t)$ as

$$[R(q,z)]_0=\frac{1}{\pi\hbar}\frac{q^2v_J(q)}{[qv_c(q)]^2-z^2} \qquad (4.53)$$

This exact result for the $q\approx 0$ contribution agrees with the RPA result for the original Tomonaga model. This fact, not mentioned in Tomonaga's paper [5] as the RPA paper by Bohm and Pines [43] was not yet published, was "discovered" many times in the literature. For the spin 1/2-model $[R(q,z)]_0$ has an additional factor 2 and one has to replace v_J by v_{J_c}.

The real part of the $(q=0)$ frequency dependent conductivity $\sigma(\omega+i0)$ follows from $[R(q,\omega+i0)]_0$ by multiplication with $ie^2\omega/q^2$ and taking the limit $q \to 0$. This yields for the spinless model

$$(\hbar/e^2)\mathrm{Re}\sigma(\omega + i0) = v_J\delta(\omega) = Kv_c\delta(\omega) \tag{4.54}$$

For the Galilei invariant Tomonaga model Eq. (4.14) one has $v_J = v_F$, i.e. the weight D of the zero frequency "Drude peak" is independent of the interaction, as expected. As D apart from a constant is given by the second derivative of $E_0(\Phi)/L$ with respect to the magnetic flux through the 1d ring [44], K (or K_c) can be obtained from a *ground state calculation* for microscopic lattice models using $K_{(c)} = (D\kappa/D_0\kappa_0)^{1/2}$, where κ is the compressibility discussed in Eq. (4.24). The anomalous decay of the correlation functions for these models, which are more difficult to calculate directly, can then be quantitatively predicted if Haldane's LL concept is taken for granted. For a *weak* two-body interaction the result for $K_{(c)} - 1$ *linear* in the interaction follows from first order perturbation theory for the ground-state energy, which involves the (non-selfconsistent) Hartree and Fock terms. As they are independent of the magnetic flux, D/D_0 has *no* term linear in $\tilde{v}$, i.e. $K_c \approx (\kappa/\kappa_0)^{1/2} = (v_F/v_{N_c})^{1/2}$, which holds exactly for Galilei invariant continuum models [45]. Performing the second derivative of $E_0^{(1)}(N)$ with respect to N yields [46]

$$K_c = 1 - \frac{2\tilde{v}(0) - \tilde{v}(2k_F)}{2\pi\hbar v_F} + O(\tilde{v}^2). \tag{4.55}$$

In the spinless case the factor 2 in front of $\tilde{v}(0)$ is missing in the result for K. Instead of D as the second input besides κ one can obtain v_c directly by calculating the lowest charge excitation energy (see section 4).

The easiest way to calculate the $q \approx \pm 2k_F$ contribution to the density response is to use the bosonization of the field operators [14]. The first step is to normal order $\psi_+^\dagger(x)\psi_-(x)$ using Eq. (4.43) . This gives a factor $e^{[\chi_+(x),\chi_-^\dagger(x)]}$ which using $[\chi_+(x),\chi_-^\dagger(x)] = -2\sum_{m>0} c_m s_m/m$ together with the factor $A^2(L)$ leads to

$$\psi_+^\dagger(x)\psi_-(x) = \frac{a_0}{L}\left(\frac{4\pi}{k_cL}\right)^{K-1}\hat{O}_+^\dagger(\frac{2\pi x}{L})\hat{O}_-(\frac{2\pi x}{L})e^{-i\Delta\chi^\dagger(x)}e^{-i\Delta\chi(x)} \tag{4.56}$$

with

$$\Delta\chi(x) \equiv \chi_+(x) - \chi_-(x) = -i\sum_{m>0}\sqrt{\frac{K_m}{m}}\left(e^{ik_mx}\alpha_m - e^{-ik_mx}\alpha_{-m}\right).$$

Here a_0 is a dimensionless constant of order unity and the exponent $K-1$ of the second factor on the rhs follows using $2s_m^2 + 2c_m s_m = K_m - 1$. The importance of this factor for impurity scattering in Luttinger liquids was first pointed out by Mattis (1974) [21] and will be discussed later. The calculation of the two terms of the commutator $\langle[\psi_+^\dagger(x,t)\psi_-(x,t), \psi_-^\dagger(0,0)\psi_+(0,0)]\rangle$ is then straightforward and one obtains for the spectral function of the $q \approx \pm 2k_F$ response function the power law behaviour [14]

$$\mathrm{Im}[R(\pm 2k_F + Q, \omega)]_{2k_F} \sim \mathrm{sign}(\omega)\theta(\omega^2 - v_c^2 Q^2)\left(\frac{\omega^2 - v_c^2 Q^2}{v_c^2 k_c^2}\right)^{K-1} \tag{4.57}$$

The *static* $\pm 2k_F + Q$ response diverges proportionally to $|Q|^{2(K-1)}$ which has to be contrasted with the logarithmic singularity in the noninteracting case. In the model including spin the exponent $2K-2$ is replaced by $K_c + K_s - 2$.

The pair propagator $P(q,\omega)$ resulting from the response function for $\hat{A} = \psi_+^\dagger(x)\psi_-^\dagger(x)$ and $\hat{B} = \psi_-(0)\psi_+(0)$ was found by Luther and Peschel to be the same as the $2k_F$-density response, provided the *sign* of the interaction is reversed [14]. An *attractive* interaction leads to a power law divergence in $P(q=0,\omega=0)$ as the temperature is lowered, indicative of large *pairing fluctuations.*

3.3 The TL model with additional interactions and perturbations

The exact solution of the TL model essentially depends on the fact that the numbers of right and left movers are *conserved.* This symmetry can be destroyed by a one-particle potential with $\pm 2k_F$-Fourier components or by interaction terms which change the individual particle numbers, like $2k_F$-"backscattering" terms or Umklapp-terms for a half-filled band. With such additional terms the model is in general no longer exactly solvable. Important insights about the influence of such terms have come from a (perturbational) RG analysis [16, 25].

3.3.1 Impurity in a spinless TL model . We begin with the spinless model with an additional impurity which is described by

$$\hat{V}_I = \int [V_F(x)\hat{\rho}_0(x) + V_B(x)\hat{\rho}_{2k_F}(x)]\, dx \equiv \hat{V}_F + \hat{V}_B, \tag{4.58}$$

where $\hat{V}_F$ describes the forward and $\hat{V}_B$ the backward scattering due to the impurity and the two different operators for the densities are defined in Eq. (4.52). As the forward scattering term is *linear* in the boson operators it can be treated in an exact way. The backscattering term has the property $[\hat{V}_B, \hat{\mathcal{N}}_\alpha] \neq 0$ and the model can no longer be solved exactly (except for $K = 1/2$ and a special assumption about V_B, as discussed below). For a zero range impurity it follows directly from Eq. (4.56) that $\hat{V}_B$ scales as $(1/L)^K$ while $\tilde{H}_{TL}$ in Eq. (4.17) scales as $1/L$. Therefore the influence of $\hat{V}_B$ depends crucially on the *sign* of the two-body interaction [21, 22]. For *repulsive* interactions one has $K < 1$ which shows that $\hat{V}_B$ is a *relevant* perturbation. For $K > 1$, i.e. an attractive interaction, $\hat{V}_B$ is irrelevant. A detailed RG analysis of the problem was presented in a seminal paper by Kane and Fisher [25]. For a zero range backscattering potential and two-body interaction they mapped the problem to a *local* bosonic sine-Gordon model [25, 35, 47]. The subsequent RG analysis shows that the backscattering amplitude scales as Λ^{K-1} when the flow parameter Λ is sent to zero [48], as can be anticipated from Eq. (4.56) . This leads to the breakdown of the perturbational analysis in V_B for repulsive interactions. As already mentioned in section 2 this analysis was supplemented by a RG analysis of a weak hopping between two semi-infinite chains. The weak hopping scales to zero like Λ^{α_B} for repulsive interactions, where $\alpha_B = K^{-1} - 1$ is the *boundary exponent.* It describes e.g. the different scaling $\rho(x, \omega) \sim |\omega|^{\alpha_B}$ of the local spectral function near a hard wall boundary of a LL [25, 49, 50]. These scaling results together with the asumption mentioned in section 2 leads to the "split chain scenario" in which even for a *weak* impurity the observables at low energies behave as if the system is split into two chains with fixed boundaries at the ends. Within the bosonic field theory this assumption was verified by quantum Monte Carlo calculations [51] and the thermodynamic Bethe ansatz [52].

This implies e.g. for the local density of states $\rho(x, \omega) \sim |\omega|^{\alpha_B}$ for small $|\omega|$ and x *near the impurity* like in a LL near a hard wall. The transmission through the impurity vanishes near k_F proportional to $\sim |k - k_F|^{2\alpha_B}$ which leads to a conductance $G(T)$ which vanishes with temperature T in power law fashion $G(T) \sim T^{2\alpha_B}$ [25].

Additional insight comes from the analysis for the special value $K = 1/2$ [25, 35, 32] . For $V_B(x) = V_B\delta(x)$ the expression for $\Delta\chi(0)$ in Eq. (4.56) can be written in terms of new boson operators $\tilde{\alpha}_m \equiv (\alpha_m - \alpha_{-m})/\sqrt{2}$. If one neglects the momentum dependence of K_m in Eq. (4.56) and puts $K_m = 1/2$ one obtains $i\Delta\chi(0) = \sum_{m\geq 1} \tilde{\alpha}_m/\sqrt{m}$ as in the bosonization of a *single* field operator Eqs. (4.32) and (4.33) . It is then possible to *refermionize* the $K = 1/2$-TL model with a zero range impurity. Even the Klein factors can properly be handled [32] and one obtains a model of "shifted noninteracting Fermi oscillators" which can be solved exactly introducing an auxiliary *Majorana fermion* [35, 32]. Unfortunately the local densities of states *cannot* be calculated exactly because of the complicated nonlinear relationship between the original fermion operators and the fermion operators which diagonalize the shifted Fermi oscillator problem [32]. Additional results for the transport through a spinless LL containing one impurity were obtained by mapping the problem onto the boundary sine-Gordon model and using its integrability [53].

In order to bridge the two regimes treated by Kane and Fisher one can use a fermionic RG description bearing in mind that it is perturbational in the two-body interaction [54, 55]. It shows that the *long range* oscillatory effective impurity potential is responsible for the "splitting", for site impurities as well as for hopping impurities of arbitrary strength. For realistic parameters *very large* systems are needed to reach the asymptotic open chain regime [55]. Hence only special mesoscopic systems, such as very long carbon nanotubes, are suitable for experimentally observing the impurity induced open boundary physics.

For a discussion of the impurity problem in the TL model including spin see also reference [56].

3.3.2 The TL- model with additional two-body interactions. Tomonaga was well aware of the limitations of his approach for more generic two-body interactions ("In the case of force of too short range this method fails" [5]). We therefore first discuss Tomonaga's continuum model in this short range limit $k_c \gg k_F$ opposite to the one considered in section 2. Then low energy scattering processes with momentum transfer $\approx \pm 2k_F$ are possible and have to be included in the theoretical description of the low energy physics.

In the "g-ology" approach one linearizes the nonrelativistic dispersion around the two Fermi points and goes over to right- and left-movers as in section 2. Then the "$2k_F$"-processes are described by the additional interaction term

$$H_{\text{int}}^{(1)} = \sum_{\sigma,\sigma'} \int \left(g_{1\|}\delta_{\sigma,\sigma'} + g_{1\perp}\delta_{\sigma,-\sigma'} \right) \psi^\dagger_{+,\sigma}(x)\psi^\dagger_{-,\sigma'}(x)\psi_{+,\sigma'}(x)\psi_{-,\sigma}(x)dx. \tag{4.59}$$

For a spin-independent two particle interaction one has $g_{1\|} = g_{1\perp} = g_1$. For the zero range interaction assumed in Eq. (4.59) one has to introduce band cut-offs to regularize the interaction term. The RG flow equations for the cut-off dependent interactions on the one-loop level are quite simple [16]. If s runs from zero to infinity in the process of integrating out degrees of freedom one obtains for spin-independent interactions

$$\begin{aligned}\frac{dg_1(s)}{ds} &= -\frac{1}{\pi\hbar v_F}g_1^2(s) \\ \frac{dg_2(s)}{ds} &= -\frac{1}{2\pi\hbar v_F}g_1^2(s)\end{aligned} \tag{4.60}$$

and g_4 is not renormalized. Obviously $g_1(s)$ can be obtained from the first equation only

$$g_1(s) = \frac{g_1}{1 + s\frac{g_1}{\pi\hbar v_F}}, \tag{4.61}$$

where g_1 is the starting value. It is easy to see that $g_1(s) - 2g_2(s) = g_1 - 2g_2$ holds by subtracting twice the second equation from the first in Eq. (4.60) . In the following we use the notation $g_\nu^* \equiv g_\nu(s \to \infty)$. Now one has to distinguish two cases:

For $g_1 \geq 0$ one renormalizes to the *fixed line* $g_1^* = 0$, $g_2^* = g_2 - g_1/2$ and the *fixed point Hamiltonian is a TL model* which shows the generic importance of the TL model for repulsive interactions. In this case the g_1-interaction is called *marginally irrelevant.* For the nonrelativistic continuum model with a spin independent interaction one has $g_{2c}^* = 2\tilde{v}(0) - \tilde{v}(2k_F)$ and $g_{2s}^* = 0$ and for the stiffness constant $K_c = [(2\pi v_F + g_{4c}^* - g_{2c}^*)/(2\pi v_F + g_{4c}^* + g_{2c}^*)]^{1/2} \approx 1 - [2\tilde{v}(0) - \tilde{v}(2k_F)]/(2\pi\hbar v_F)$ and $K_s = 1$. Due to the approximations made, also here only the result for $K_c - 1$ *linear* in $\tilde{v}$ is reliable. The agreement with the direct calculation Eq. (4.55) shows explicitly to leading order in the interaction that Haldane's Luttinger liquid concept is consistent.

For $g_1 < 0$ the solution (4.61) shows that $g_1(s)$ *diverges* at a finite value of s. Long before reaching this point the perturbational analysis breaks down and all one can say is that the flow is towards *strong coupling.* In this case the g_1- interaction is called *marginally relevant.* In order to obtain an understanding of the strong coupling regime it is useful to bosonize the additional interaction $H_{\text{int}}^{(1)}$ in Eq. (4.59) [57]. The term proportional to $g_{1\parallel}$ is of the form of a $g_{2\parallel}$-interaction and therefore bilinear in the boson operators Eq. (4.19) . For the $g_{1\perp}$-term one uses the bosonization of the field operators Eqs. (4.32) and (4.33) with additional spin labels. As the $g_{1\perp}$-term contains field operators $\psi_{\alpha\uparrow}^\dagger(x)\psi_{\alpha\downarrow}(x)$ of *opposite* spin it only involves "spin bosons" Eq. (4.19), which implies "spin-charge separation" also for this model [58]. The charge part stays trivial with massless charge bosons as the elementary interactions. Luther and Emery showed that for a particular value of $g_{1\parallel}$ the $g_{1\perp}$-term can be written as a product of spinless fermion field operators and the exact solution for the spin part of the Hamiltonian is possible using *refermionization* [57], discussed earlier in connection with the backscattering impurity. The diagonalization of the resulting problem of noninteracting fermions is simple and shows that the spectrum for the spin excitations is *gapped.* It is generally believed that these properties of *Luther-Emery phases* are not restricted to the solvable parameter values.

Strong coupling phenomena which lead to deviations from LL-properties with gapped phases are discussed in detail in section 4 for *lattice models.* There in case of *commensurate* filling *Umklapp processes* can become important, e.g. for half filling where two left movers from the vicinity of the left Fermi point are scattered into two right movers near the right Fermi point or vice versa. As $G = 4k_F$ is a reciprocal lattice vector such a scattering process is a low energy process conserving quasi-momentum. In the g-ology model such processes are described by an additional term

$$H_{\text{int}}^{(3)} = \frac{1}{2}\sum_{\sigma,\sigma'}\int g_3^{\sigma,\sigma'}(x-y)\left[\psi^\dagger_{+,\sigma}(x)\psi^\dagger_{+,\sigma'}(y)\psi_{-,\sigma'}(y)\psi_{-,\sigma}(x)e^{2ik_F(x+y)}\right. \\ \left. +H.c.\right]dxdy \tag{4.62}$$

Umklapp processes for $\sigma = \sigma'$ are only possible for nonzero interaction range.

4. Results for integrable lattice models

As mentioned in subsection 2.4, results for integrable models which can be solved exactly by the Bethe ansatz played a central role in the emergence of the general "Luttinger liquid" concept [18]. It is therefore appropriate to shortly present results for the two most important lattice models of this type, the model of spinless fermions with nearest neighbour interaction and the $1d$-Hubbard model. (We put $\hbar = 1$ in this section.)

4.1 Spinless fermions with nearest neighbour interaction

The one-dimensional single band lattice model of spinless fermions with nearest neighbour hopping matrix element $t(> 0)$, and nearest neighbour interaction U (often called V in the literature) is given by

$$H = -t\sum_j\left(c_j^\dagger c_{j+1} + H.c.\right) + U\sum_j \hat{n}_j\hat{n}_{j+1} \equiv \hat{T} + \hat{U}, \tag{4.63}$$

where j denotes the sites and the $\hat{n}_j = c_j^\dagger c_j$ are the local occupation number operators. In the noninteracting limit $U = 0$ one obtains for lattice constant $a = 1$ the well known dispersion $\epsilon_k = -2t\cos k$. For the following discussion of the interacting model ($U \neq 0$) we mainly focus on the *half filled* band case $k_F = \pi/2$ with $v_F = 2t$. In contrast to the (continuum) Tomonaga model Umklapp terms appear when the interaction term in Eq. (4.63) is written in the k-representation. As discussed below they are *irrelevant* at the noninteracting ($U = 0$) fixed point [17]. Therefore the system is a *Luttinger liquid* for small enough values of $|U|$. The large U limit of the model is easy to understand: For $U \gg t$ charge density wave (CDW) order develops in which only every other site is occupied thereby avoiding the "Coulomb penalty". For large but negative U the fermions want to be as close as possible and phase separation occurs. For the quantitative analysis it is useful that the model in Eq. (4.63) can be exactly mapped to a $S = 1/2$-Heisenberg chain with uniaxially anisotropic nearest neighbour exchange ("XXZ" model) in a magnetic field by use of the Jordan-Wigner transformation [59]. For $U > 0$ this model is also called the antiferromagnetic Heisenberg-Ising model. The point $U \equiv U_c = 2t$ corresponds to the *isotropic* Heisenberg model. For $U > 2t$ the Ising term dominates and the ground state is a well defined doublet separated by a gap from the continuum and long range antiferromagnetic order exists. For $-2t < U \leq 2t$ there is no long range magnetic order and the spin-excitation spectrum is a gapless continuum. The mapping to the XXZ-model therefore suggests that the spinlesss fermion model Eq. (4.63) in the half filled band case is a Luttinger liquid for $|U| < 2t$.

Before we present the exact results for the Luttinger liquid parameters K and v_c from the Bethe ansatz solution [18, 60], we shortly discuss the RG approach to the model. A perturbative RG calculation around the free fermion fixed point is discussed in detail in Shankar's review article [17]. The first step is to write the four fermion matrix elements of the interaction $\hat{U}$ in Eq. (4.63) in the k-representation. This yields for a chain of N sites with periodic boundary condition and values $k_j = 2\pi j/N$ in the first Brillouin zone

$$\langle k_1, k_2|\hat{U}|k_3, k_4\rangle = \frac{2U\cos(k_1 - k_3)}{N} \sum_{m=0,\pm 1} \delta_{k_1+k_2,k_3+k_4+2\pi m} \tag{4.64}$$

The $m = 0$ term on the rhs of Eq. (4.64) represents the direct scattering terms and the $m = \pm 1$ terms the Umklapp processes. The matrix element antisymmetrized in k_3 and k_4 is proportional to $\sin[(k_1 - k_2)/2]\sin[(k_3 - k_4)/2]$. Therefore the low energy Umklapp Hamiltonian scales like $(1/L)^3$ which shows that it is strongly irrelevant at the free field fixed point [17]. This analysis confirms the Luttinger liquid behaviour for small values of U, but gives no hint about the critical value U_c for the CDW transition. With the separation $\hat{U} \equiv \hat{U}_0 + \hat{U}_{\text{Umklapp}}$ implied by Eq. (4.64) one can do better by first treating $\hat{T} + \hat{U}_0$ by bosonization and then perform the RG analysis around the corresponding TL fixed point to get information for which value of U the Umklapp term starts to be a relevant perturbation. For this analysis it is easier to work directly with the unsymmetrized matrix elements in Eq. (4.64). As $k_1 - k_3 \approx \pm\pi$ for the low energy Umklapp processes this leads after extending the (linearized) dispersion of the right and left movers from $-\infty$ to ∞ to a g_3-interaction with a range of order $r = a$. The scaling dimension of the corresponding $H_{\text{int}}^{(3)}$ follows using bosonic normal ordering as in Eq. (4.56). For $x - y$ of order r or smaller one obtains

$$\psi_+^\dagger(x)\psi_+^\dagger(y)\psi_-(y)\psi_-(x)L^2 \sim \left(\frac{x-y}{L}\right)^2 \left(\frac{r}{L}\right)^{4(K-1)} \times (U_+^\dagger)^2 U_-^2 e^{2ik_F(x+y)} e^{iB^\dagger(x,y)} e^{iB(x,y)} , \tag{4.65}$$

where $B(x,y) = \chi_-(x) + \chi_-(y) - \chi_+(x) - \chi_+(y)$ with $\chi_\alpha(x)$ defined in Eq. (4.44). The first factor on the rhs is due to the Pauli principle and describes the same physics as the two sin-factors mentioned above for small arguments. Therefore the second factor has to provide more than two powers of L to make the Umklapp term a relevant perturbation, which happens for $K < 1/2$. As discussed below, the exact Bethe ansatz result for K yields $U_c = 2t$. If one uses the simple *linear* approximation for $K - 1$ in Eq. (4.55) one obtains with Eq. (4.64) $K^{\text{lin}} = 1 - U/(\pi t)$ for the critical value $U_c^{\text{lin}}/t = \pi/2$, not too bad an approximation.

Exact analytical results for the Luttinger liquid parameters for the half filled model can be obtained from the Bethe Ansatz solution [18, 60, 61]. It is *not* necessary to address the anomalous decay of the correlation functions directly, but one can use a ground state property and the lowest charge excitation to extract the parameters, as was dicussed in connection with Eq. (4.55). This yields for the stiffness constant $K = \pi/[2\arccos(-U/2t)]$ and for the charge velocity $v_c = \pi t\sqrt{1-(U/2t)^2}/[\pi - \arccos(-U/2t)]$. For repulsive interactions $U > 0$ the value of K decreases monotonously from the noninteracting value $K = 1$ to $K = 1/2$ for $U = 2t$, which corresponds to an anomalous dimension $\alpha_L = (K + 1/K)/2 - 1 = 1/4$. For attractive interactions K diverges when U approaches $-2t$, and the charge velocity v_c goes to zero. Results for the Luttinger liquid parameter K for less than half filled bands are shown in Fig. 4.5 [62]. The limit $a \to 0$ and $n \to 0$ corresponds to the continuum limit. As the interaction goes over to a contact interaction its effect vanishes because of the Pauli principle and K goes to 1. For small enough values of U the linear approximation

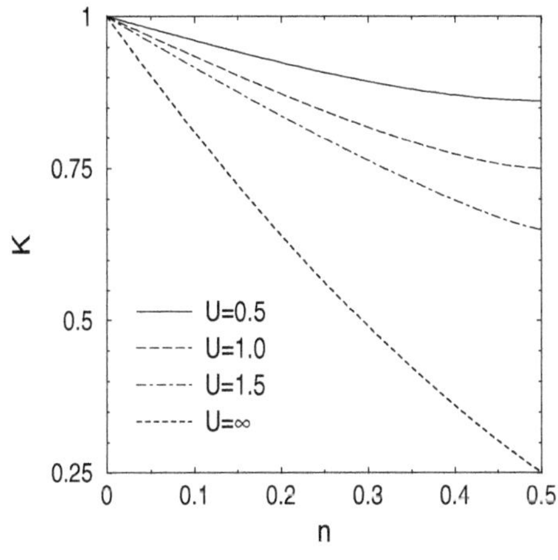

Figure 4.5. Luttinger liquid parameter K from the Bethe ansatz solution as a function of the band filling n for different values of U $(t = 1)$. The short dashed curve shows the infinite U result $(1/2 + |n - 1/2|)^2$.

Eq. (4.55) $K^{\text{lin}} = 1 - U \sin(n\pi)/\pi t$ provides a good approximation for *all* values of n, in contrast to the Hubbard model discussed below. In the infinite U limit the Bethe Ansatz equations simplify considerably and the ground-state energy as well as low lying excited states can be calculated analytically [61]. With these results it is easy to show that $K = (1-n)^2$ holds for $0 < n < 1/2$, i.e. $K = 1/4$, is the lower bound for K in the LL regime of the model [18]. The corresponding upper bound of the anomalous dimension is $\alpha_L = 9/8$. In order to achieve larger values of α_L the model in Eq. (4.63) has to be generalized to include longer range interactions [63].

4.2 The Hubbard model

As there exists an excellent review on the LL behaviour in the $1d$-Hubbard model [64], the following discussion will be rather short. As the model includes spin the on-site interaction between electrons of opposite spins is not forbidden by the Pauli principle. This is taken as the only interaction in the model. The $1d$ Hubbard Hamiltonian reads

$$H = -t \sum_{j,\sigma} \left(c_{j,\sigma}^{\dagger} c_{j+1,\sigma} + H.c. \right) + U \sum_{j} \hat{n}_{j,\uparrow} \hat{n}_{j,\downarrow}. \tag{4.66}$$

In the *extended* Hubbard model a next nearest interaction term $V \sum_j \hat{n}_j \hat{n}_{j+1}$ with $\hat{n}_j \equiv \hat{n}_{j,\uparrow} + \hat{n}_{j,\downarrow}$ is added [65]. In order to show the important difference to the spinless model Eq. (4.63) we again first discuss the *half-filled* band case, which is metallic for $U = 0$. For $U \gg t$ the "Coulomb penalty" is avoided when each site is *singly* occupied. Then only the spin degrees of freedom matter. In this limit the Hubbard model can be mapped to a spin-1/2 Heisenberg antiferromagnet with an exchange coupling $J = 4t^2/U$. In the charge sector there is a large gap $\Delta_c \sim U$ while the spin excitations are gapless. The $1d$ Hubbard model can also be solved exactly using the Bethe ansatz [66] and properties like the charge gap or the ground-state energy can be obtained by solving Lieb and Wu's integral equation. In contrast to the spinless model described in the previous subsection the charge gap in the Hubbard model is finite for *all* $U > 0$. While for $U \gg t$ it is asymptotically given by U it is exponentially small, $\Delta_c \approx (8t/\pi)\sqrt{U/t}\exp(-2\pi t/U)$, for $0 < U \ll t$. This shows that the Umklapp term is no longer irrelevant at the free field fixed point. The Pauli principle factor of Eq. (4.65) is missing as the interaction is between electrons of *opposite* spin. The Umklapp term is therefore a marginal perturbation. The RG analysis [16] shows that the Umklapp term is marginally *relevant* while the $2k_F$-backscattering ("g_1") interaction is marginally *irrelevant* for $U > 0$ as discussed following Eq. (4.60).

When the band is *not* half filled Umklapp is not a low energy process and the Hubbard model is a Luttinger liquid with $K_s = 1$. The LL parameters K_c and v_a can be obtained by (numerically) solving Lieb an Wu's integral equation [67]. Even for $0 < U \ll t$ the perturbative result Eq. (4.55) works well only for intermediate filling $n \equiv N_{el}/N \approx 0.5$, where N_{el} is the number of electrons (half filling corresponds to $n = 1$) . In the limit $n \to 0$ the Fermi velocity $v_F = 2t\sin(\pi n/2)$ goes to zero but $2\tilde{v}(0) - \tilde{v}(2k_F) = U$ stays finite and the correction term increases with decreasing n in contrast to the spinless model. The Bethe ansatz results show that $K_c \to 1/2$ for $n \to 0$ as well as $n \to 1$ for *all* $U > 0$. For $U \to \infty$ it leads to $K_c \to 1/2$ for *all fillings* n different from 1. In this limit the velocities are given by $v_c = 2t\sin(\pi n)$ and $v_s = (2\pi t^2/U)[1 - \sin(2\pi n)/(2\pi n)]$, i.e. the spin velocity goes to zero [64, 67]. The $U = \infty$ results for v_c and K_c can be understood *without* the Bethe ansatz solution. Double occupancies of the lattice sites are forbidden and the system behaves like a system of *noninteracting spinless fermions* with k_F replaced by $2k_F$ [64]. The spin degrees of freedom play no role and any configuration of the spins gives an eigenfunction of the same energy. This immediately explains the result for v_c mentioned above. For a TL model with spin one obtains (for fixed $N_\uparrow - N_\downarrow$) from Eqs. (4.17) and (4.21) $L(\partial^2 E_0/\partial N^2)_L = \pi v_{N_c}/2$, while the factor 1/2 is missing in the spinless case. The formula for the spinless case can be used to calculate $L(\partial^2 E_0/\partial N^2)_L$ for $U = \infty$ with v_N replaced by

$v_F(2k_F)$, using the spinless fermion analogy. This yields $v_{N_c} = 2v_c$ i.e. $K_c = 1/2$.

As the calculation of correlation functions not only requires excitation energies but also many electron matrix elements which are difficult to evaluate using the Bethe ansatz, various numerical methods have been used to study e.g. the manifestation of spin-charge separation in the one-particle spectral function [68, 69]. The Bethe ansatz approach simplifies in the infinite U limit [70]. After earlier work [71, 72] the frequency dependent optical conductivity of the $1d$ Hubbard model was also studied using Bethe ansatz methods [73, 74], as well as the dynamical density-matrix renormalization group [74].

5. Weakly coupled chains: the Luttinger to Fermi liquid transition

Strictly one-dimensional systems are a theoretical idealization. Apart from this even the coupling to an experimental probe presents a non-trivial disturbance of a Luttinger liquid. Unfortunately the weak coupling of a $1d$ system to such a probe as well as the coupling between several LLs is theoretically not completely understood [26]. The coupling between the chains in a very anisotropic $3d$ compound generally, at low enough temperatures, leads to true *long-range order*. The order develops in the phase for which the algebraic decay of the correponding correlation function of the single chain LL is slowest [64]. This can lead e.g. to charge-density wave (CDW), spin-density wave (SDW) order or superconductivity.

In the following we shortly address some important issues of the coupled chain problem, which are a prerequisite for the theoretical descriptions of the attempts to experimentally verify LL behaviour. In the first part of this section theoretical aspects of the problem of an infinite number of coupled chains are addressed. This is followed by a short discussion of the (approximate) experimental realizations of LLs. As there are other chapters in this book addressing this question the discussion will be rather short.

5.1 Theoretical models

We consider a system of $N_\perp$ coupled chains described by the Hamiltonian

$$H = \sum_{i=1}^{N_\perp} H_i + \sum_{i \neq j} H_{ij}^{(ee)} + \sum_{n,(\sigma)} \sum_{i,j} t_{\perp,ij} c^\dagger_{n,(\sigma),i}, c_{n,(\sigma),j} \tag{4.67}$$

where the H_i are the Hamiltonians of the individual chains, the $H_{ij}^{(ee)}$ represent the two-body (Coulomb) interaction of electrons on different chains and the last term $H^{(t_\perp)}$ describes the hopping between the chains with $t_{\perp,ij}$ the transverse hopping matrix elements and the $c_{n,(\sigma),i}^{(\dagger)}$ the (creation) annihilation operators of one-particle states with quasi-momentum k_n along the chain i and spin σ (if spin is included in the model). The individual H_i can be TL-Hamiltonians Eq. (4.17) or lattice Hamiltonians like in Eqs. (4.63) or (4.66).

We address the question if LL physics survives in such a model. The second and the third term on the rhs of Eq. (4.67) describe different types of couplings between the chains. If the transverse hopping is neglected ($t_\perp \equiv 0$) the model can be solved exactly for special assumptions about the two-body interaction and the H_i. If the individual chains are described by TL-Hamiltonians Eq. (4.17) and the interaction $H_{ij}^{(ee)}$ can be expressed in terms of the densities $\hat{\rho}_{n,(a),\alpha,i}$ the exact solution is possible by bosonization [75, 76]. This is important when the long range Coulomb interaction is taken into account. For a *single* chain the corresponding one-dimensional Fourier transform $\tilde{v}(q)$ (which has to be regularized at short distances) has a logarithmic singularity for $q \to 0$. This leads to $K_{(c)} = 0$ and the divergence of the anomalous dimension, i.e. the system is *not* a LL. The $4k_F$ harmonic of the density-density correlation function shows a very slow decay almost like in a Wigner crystal [77]. The Coulomb coupling *between* the chains removes this singularity and a three-dimensional extended system of coupled chains *is* a LL [75]. The corresponding anomalous dimension can be calculated and leads to values of order unity for realistic values of the coupling constant $e^2/(\pi\hbar v_F)$ [76]. If $2k_F$-scattering terms of the interaction are kept the model can no longer be solved exactly and a more complicated scenario emerges in the parquet approximation [78].

The inclusion of the transverse hopping presents a difficult problem even if the inter-chain two-body interactions are *neglected.* This is related to the fact that the transverse hopping is a *relevant* perturbation for $\alpha_L < 1$ [79, 80, 81]. This can easily be seen if the individual chains are described by TL-Hamiltonions Eq. (4.17), scaling like $1/L$. As discussed in section 3 the $c_{n,(\sigma),i}^{(\dagger)}$ scale like $(1/L)^{\alpha_L/2}$. As $H^{(t_\perp)}$ involves products of creation and annihilation operators on *different* chains no further boson normal ordering is necessary and $H^{(t_\perp)}$ scales as $(1/L)^{\alpha_L}$. This suggests "confinement" for $\alpha_L > 1$: if an extra electron is put on the j-th chain it stays there with probability close to 1 even in the long time limit. This conclusion can be questioned as RG calculations perturbative in $t_\perp$ demonstrate that the hopping term generates new and relevant interchain two-particle hoppings. These calculations show that the system flows to a strong-coupling fixed point which cannot be determined within the approach [81, 82].

If inter-chain two-body interactions are *included* the relevance of hopping terms can be different. When only density-density and current-current interactions between the wires are included, as discussed above [75, 76], the possible relevance around this Gaussian model, recently called *sliding* LL [83, 84, 85], can be different. If the single chains are in the spin-gapped Luther-Emery regime [57] single-particle hopping between the chains is irrelevant and the coupled system can show power-law correlations characteristic of a $1d$-LL [83, 85]. For the spinless model single particle and pair hoppings can be irrelevant for strong enough forward interactions [84].

In the following we concentrate on the Luttinger to Fermi liquid crossover. In order to get a *quantitative* picture it is desirable to study models which allow *controlled approximations.* The simple perturbative calculation in $t_\perp$ for the calculation of the one-particle Green's function by Wen [80] discussed below is unfortunately only controlled in the rather unphysical limit when the transverse hopping is independent of the distances of the chains ($t_{\perp,ij} \equiv t_\perp$)[86]. The (retarded) one-particle Green's function G is expressed in terms of the selfenergy Σ

$$G(k_\|, \vec{k}_\perp, z; t_\perp) = \frac{1}{z - \epsilon_{k_\|,\vec{k}_\perp} - \Sigma(k_\|, \vec{k}_\perp, z; t_\perp)}, \tag{4.68}$$

where $\epsilon_{k_\|,\vec{k}_\perp}$ denotes the energy dispersion for the noninteracting model and $z = \omega + i0$ is the frequency above the real axis. For small $t_\perp$ the dispersion can be linearized around $k_\| = \pm k_F$ near the *open* noninteracting Fermi surface. This yields $\epsilon_{k_\|,\vec{k}_\perp} \approx \pm v_F(k_\| \mp k_F) + t_\perp(\vec{k}_\perp)$. In the context of Fermi liquid theory the selfenergy is studied in (all orders) perturbation theory in the two-body interaction v around the noninteracting limit. This can be done using standard Feynman diagrams. In the present context one wants to study how the LL behaviour for *finite* two body interaction and finite anomalous dimension is modified by the transverse hopping. Similar to perturbation theory for the Hubbard model around the atomic limit nonstandard techniques have to be used [87]. The simplest approximation, which corresponds to the "Hubbard I" approximation for the Hubbard model, is to replace Σ in Eq. (4.68) in zeroth order in $t_\perp$ by the selfenergy $\Sigma^{(\mathrm{chain})}(k_\|, z)$ of a single chain

[80]. This approximation first used by Wen reads for $k_{\|} \approx k_F$

$$G(k_{\|}, \vec{k}_{\perp}, z; t_{\perp})_{\text{Wen}} = \frac{1}{\left[G_{+}(k_{\|}, z)\right]^{-1} - t_{\perp}(\vec{k}_{\perp})}, \tag{4.69}$$

where G_+ is determined by the spectral function ρ_+ discussed following Eq. (4.50) via a Hilbert transform. In the asymptotic low-energy regime this yields $G_+(k_F + \tilde{k}_{\|}, z) = A_0[(\tilde{k}_{\|}/k_c)^2 - (z/\omega_c)^2]^{\alpha_L/2}/(z - v_c\tilde{k}_{\|})$ for spinless fermions, with $\omega_c \equiv k_c v_c$ and $A_0 = \pi\alpha_L/[2\sin(\pi\alpha_L/2)]$. Wen's approximate Green's function leads to a spectral function with the *same* range of continua as $\rho_+(k_{\|}, \omega)$. In *addition* there can be *poles* at $\omega_{k_{\|},\vec{k}_{\perp}}$, determined by setting the denominator in Eq. (4.69) equal to zero. The poles located at $\omega_{k_{\|},\vec{k}_{\perp}} = 0$ determine the Fermi surface $\tilde{k}_{\|}(\vec{k}_{\perp})$ of the *interacting* coupled system. From Eq. (4.69) and the simple form of G_+ one obtains $A_0(\tilde{k}_{\|}/k_c)^{(1-\alpha_L)} = t_{\perp}(\vec{k}_{\perp})$, which shows that the reduction of warping of the Fermi surface (FS) by the interaction is proportional to $[t_{\perp}(\vec{k}_{\perp})/\omega_c]^{\alpha_L/(1-\alpha_L)}$. This is shown in Fig. 4.6 for a two dimensional system of coupled chains. If one writes $t_{\perp}(\vec{k}_{\perp}) \equiv t_{\perp}c(\vec{k}_{\perp})$, with $c(\vec{k}_{\perp})$

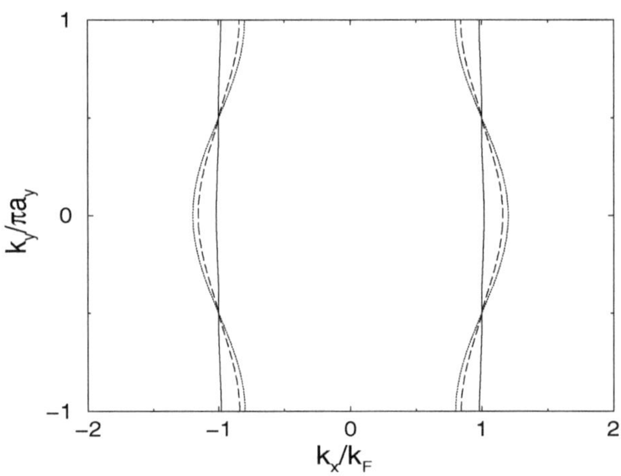

Figure 4.6. Fermi surface "flattening" in Wen's approximation for coupled chains for different values of the the anomalous dimension α_L for a single chain. The dotted lines show the noninteracting FS, the long dashed curves correspond to $\alpha_L = 0.125$ and the full ones to $\alpha_L = 0.6$. At $\alpha_L = 1$ the FS degenerates to two parallel lines as without interchain coupling, called the "confinement transition".

a dimensionless function, the new effective low energy scale is given by $t_{\text{eff}} = \omega_c(t_{\perp}/\omega_c)^{1/(1-\alpha_L)}$. The weights $Z_{\vec{k}_{\perp}}$ of the poles for $\vec{k}$ values on the Fermi surface are also proportinal to $[t_{\perp}(\vec{k}_{\perp})/\omega_c]^{\alpha_L/(1-\alpha_L)}$. Wen's approximate solution has the Fermi liquid type property of quasi-particle poles with nonzero weight on the Fermi surface, except at the special points where $t_\perp(\vec{k}_\perp)$ vanishes. The improved treatment by Arrigoni [88] shows that this peculiar vanishing of the quasi-particle weights is an artefact of Wen's approximation. The new idea involved is to let the number of "perpendicular" dimensions "$d-1$" go to infinity. This extends the original idea of the "dynamical mean field theory" (DMFT) [89], where one treats the Hubbard model in infinite dimensions as an effective *impurity* problem to the case of a *chain* embedded in an effective medium. Results are obtained by carrying out a resummation of *all* diagrams in the $t_\perp$-expansion which contribute in this large dimension limit [88]. This approach shows explicitly how the leading order Wen approximation is uncontrolled at low energies. For the case of weakly coupled one-dimensional Mott insulators one expects the approximation to be better controlled [90].

Despite the Fermi liquid like properties at energy scales much smaller than $t_{\rm eff}$ the coupled chain system can nevertheless show LL like properties for energy scales larger than $t_{\rm eff}$ if there is a large enough energy window to the high energy cutoff $\tilde{\omega}_c$ which describes the regime where the asymptotic LL power laws hold for a single chain. Then for temperatures lower than $\tilde{\omega}_c$ but higher than $t_{\rm eff}$ the system behaves like a LL. The integrated spectral functions $\rho^<_{\alpha,(\sigma)}(\omega)$ probed by angular integrated photoemission, for example, show approximate power law behaviour $\sim(-\omega)^{\alpha_L}$ for temperatures larger than $t_{\rm eff}$ in the energy window $k_BT<-\omega<\tilde{\omega}_c$. Unfortunately little is known about the value of $\tilde{\omega}_c$ for microscopic models. An exception is the Tomonaga model Eq. (4.14) with a constant $\tilde{v}(k)$ up to the cutoff k_c, where the high energy cutoff $\tilde{\omega}_c$ equals $\omega_c=\min(v_c,v_s)k_c$ [4]. This implies for the integrated spectral function for the very large U Hubbard model with periodic boundary conditions that the power law $|\omega|^{\alpha_L}$ only holds in a narrow energy window $\sim v_s$, which vanishes proportional to $1/U$ in the $U\to\infty$ limit [65]. Another example is the Hubbard model at boundaries where $\tilde{\omega}_c$ can be very small for small U [50].

As an alternative way to treat the "anisotropic large dimension model" [88] one can try to solve the resulting chain-DMFT equations numerically, using e.g. a quantum Monte Carlo algorithm [91]. In this reference the H_i were chosen as Hubbard Hamiltonians (4.66) with chain lengths of 16 and 32 sites. The results for a partly filled band as a function of temperature indicate in fact a crossover from a LL to a FL at the estimated crossover scale as the temperature is lowered. In agreement with Arrigoni [88] the authors find that the quasi-particle weight is more uniform along the Fermi surface than suggested by Wen's approximation

Eq. (4.69). At half filling and low but finite temperatures the crossover from the Mott insulator to FL was examined (the intermediate LL regime was too narrow to be visible). In the future it is to be expected that this method applied to longer chains and additional nearest neighbour interaction will provide important results which allow a more realistic comparison with experimental work.

Because of space limitations the interesting field of a *finite* number of coupled chains cannot be discussed here [92].

5.2 On the experimental verification of LL behaviour

There exist several types of experimental systems where a predominantly $1d$ character can be hoped to lead to an (approximate) verification of the physics of Luttinger liquids. In the following we present a short list of the most promising systems and discuss some of the experimental techniques which have been used. As these topics are also discussed in other chapters of this book we do not attempt a complete list of references but only refer to most recent papers or to review articles on the subject.

The following systems look promising:

- Highly anisotropic "quasi-one-dimensional" conductors

  There has been extensive work on organic conductors, like the Bechgaard salts [93, 94], as well as inorganic materials [95, 96].

- Artificial quantum wires

  Two important types of realizations are quantum wires in *semiconductor heterostructures* [97, 98] or quantum wires on *surface substrates* [99, 100].

- Carbon nanotubes

  The long cylindrical fullerenes called quantum nanotubes are also quantum wires but have been listed separately because of their special importance in future applications like "molecular electronics" [101, 102]. Using the peculiar band structure of the π-electrons of a single graphite plane it was shown that single wall "armchair" nanotubes should show LL behaviour with $K_c \sim 0.2 - 0.3$ down to very low temperatures [103, 104], despite the fact that *two* low energy channels are present.

- Fractional quantum Hall fluids

  Electrons at the edges of a two-dimensional fractional quantum Hall system can be described as a *chiral Luttinger liquid* [105]. The power law tunneling density of states observable in the tunneling current-voltage characteristics shows power laws of extraordinary quality [106]. The theoretical predictions for general filling factors between the Laughlin states $\nu = 1$ and $\nu = 1/3$ [107, 108] are not borne out by experiment [109]. As in these chiral LLs the right- and left-movers are *spatially separated* the edge state transport is quite different from the case of quantum wires and FQH fluids are not further discussed in the following.

Promising experimental techniques to verify LL behaviour are:

- High resolution photoemission

  One of the earliest claims of possible verification of Luttinger liquid behaviour was from angular integrated photoemission of the Bechgaard salt $(TMTSF)_2PF_6$, which showed a power law supression at the chemical potential with an exponent of order 1 over an energy range of almost one eV [110]. There are serious doubts that this suppression can be simply explained by the LL power law behaviour [94]. Therefore a large number of other quasi-one-dimensional conductors were examined [94, 95, 96, 111]. In addition periodic arrays of quantum wires on surface substrates were studied by angular resolved photoemisssion (ARPES), but the interpretation of a two peak structure as spin-charge separation [99] was questioned [100]. Spin-charge separation was shown to occur in the $1d$ Hubbard model also at higher energies on the scale of the conduction band width [69, 70, 73]. Recent ARPES spectra of TTF-TCNQ were interpreted with the $1d$ Hubbard model at finite doping to show signatures of spin-charge separation over an energy scale of the conduction band width. As for the Hubbard model $K_c > 1/2$ for $n \neq 1$ which implies $\alpha_L < 1/8$ for the anomalous dimension the experimentally found nearly linear spectral onset at low energies cannot be explained within the same model. ARPES data for the "Li purple bronze" seem to compare favorably to the LL lineshape [96]. For the quasi-one-dimensional antiferromagnetic insulators $SrCuO_2$ and Sr_2CuO_3 ARPES spectra have been interpreted to show evidence of spin-charge separation [112]. For a more in depth discussion see the chapter by Grioni in this book.

- Transport

  As discussed in section 3 even a single impurity has a drastic effect on the conductance of a LL, which vanishes as a power law with temperature. Another issue is the "conductance puzzle" of a clean

LL. There has been an extended discussion whether the quantized value e^2/h for noninteracting electrons in a single channel is modified by the interaction to $K_c e^2/h$ [113, 114]. Apparently the answer depends sensitively on the assumptions made about the contacts, a very delicate theoretical as well as experimental problem [115]. Experimental results are available for cleaved edge overgrowth quantum wires [97] as well as carbon nanotubes [116, 117, 118]. In the nanotubes the authors observe approximate power laws of the conductance which seem to be consistent with LL behaviour. A detailed dicussion of transport through quantum wires is presented in the chapter by Yacoby. For a recent theoretical discussion of experimental results on the interchain transport in the Bechgaard salts see references [119, 120]. There the question of energy scales and the importance of the proximity of the incipient Mott insulator are addressed.

- Optical properties

  Optical properties have long been used to investigate electronic properties of quasi-one-dimensional systems [121]. The optical behaviour of different Bechgaard salts was analyzed recently using LL concepts [122]. At low energies, smaller than about ten times the Mott gap, the importance of dimerization and interchain hopping was pointed out [123]. As there is a separate chapter about the optical response in chains and ladders it will not be discussed further here.

Obviously neither the list of systems nor that of methods is coming close to being complete. They were presented to show that there are intensive experimental activities in the attempt to verify the elegant LL concept put forward by theoreticians. Further work on both sides is necessary to come to unambiguous conclusions.

Acknowledgments

For useful comments on the manuscript the author would like to thank J. Allen, E. Arrigoni, D. Baeriswyl, L. Bartosch, J. von Delft, R. Egger, F. Essler, F. Gebhard, A. Georges, T. Giamarchi, M. Grayson, P. Kopietz, V. Meden, W. Metzner, and J. Sólyom.

References

[1] F. Bloch, Helv. Phys. Acta **7**, 385 (1934)

[2] K. Schönhammer and V. Meden, Am. J. Physics **64**, 1168-1176 (1996)

[3] Obviously there is the restriction $\sum_{j>0} n_j \leq n_F$ which shows in agreement with Eq. (4.6) that the mapping is straightforward only for the low energy excitations which leave all levels deep in the Fermi sea occupied. For noninteracting fermions the canonical partition function can also be calculated analytically when the energy levels have a lower bound (see K. Schönhammer, Am. J. Phys. **68**, 1033 (2000)).

[4] K.Schönhammer, "Interacting fermions in one dimension: The Tomonaga-Luttinger model", cond-mat/9710330.

[5] S. Tomonaga, Prog. Theor. Phys. **5**, 544-569 (1950)

[6] J.M. Luttinger, J. Math. Phys. **4**, 1154-1162 (1963)

[7] Eq. (4.16) describes the leading deviations of $\langle n_{k,+} \rangle$ from 1/2 for $0 \leq \alpha_L < 1$. For $\alpha_L > 1$ there is an additional contribution *linear* in $k - k_F$ while in the special case $\alpha_L = 1$ logarithmic corrections are present.

[8] As the greek letter α is used as the index to distinguish the right and left movers we write α_L ("Luttinger") for the anomalous dimension.

[9] W. Thirring, Ann. Phys. (N.Y.), **3**, 91 (1958)

[10] D.C. Mattis and E.H. Lieb, J. Math. Phys. **6**, 304 (1965)

[11] A. Theumann, J. Math. Phys. **8**, 2460 (1967)

[12] K. Johnson, Nuovo Cimento **20**, 733 (1961)

[13] I.E. Dzyaloshinski and A. I. Larkin, Sov. Phys.-JETP, **38**, 202 (1974)

[14] A. Luther and I. Peschel, Phys. Rev.B **9**, 2911-2919 (1974)

[15] D. Schotte and U. Schotte, Phys. Rev. **182**, 479 (1969)

[16] J. Sólyom, Adv. Phys. **28** , 201 (1979)

[17] R. Shankar, Rev. Mod. Phys. **66**, 129 (1994)

[18] F.D.M. Haldane, Phys. Rev. Lett. **45**, 1358 (1980)

[19] F.D.M. Haldane, J. Phys. C **14**, 2585-2919 (1981)

[20] L.P. Gorkov and I.E. Dzyaloshinski, JETP Letters, **18**, 401 (1973)

[21] D. C. Mattis, J. Math. Phys. **15**, 609 (1974)

[22] A. Luther and I. Peschel, Phys. Rev. Lett. **32**, 992 (1974)

[23] W. Apel and T.M. Rice, Phys. Rev. B **26**, 7063 (1982)

[24] T. Giamarchi and H.J. Schulz, Phys. Rev. B **37**, 325 (1988)

[25] C.L. Kane and M.P.A. Fisher, Phys.Rev.Lett. **68**, 1220, (1992); Phys. Rev. B **46**, 15233 (1992)

[26] For a theoretical review see: J. Voit, Rep. Prog. Phys. **58**, 977-1116 (1995)

[27] M.P.A. Fisher and L.I. Glazman, "Transport in a one-dimensional Luttinger liquid", in *Mesoscopic Electron Transport*, NATO series E, Vol. **345**, 331 (Kluwer Academic Publishing, Dordrecht, 1997)

[28] K.-V. Pham, M. Gabay and P. Lederer, Phys.Rev. B **61**, 16397 (2000)

[29] V. Meden and K. Schönhammer, Phys. Rev. B **46**, 15753 (1992)

[30] J. Voit, Phys. Rev. B **47**, 6740 (1993)

[31] K. Schönhammer, Phys. Rev. B **63**, 245102 (2001)

[32] J.v. Delft and H. Schoeller, Ann. Phys. (Leipzig) **7**, 225 (1998).

[33] The CCR $[\tilde{\mathcal{N}}, \hat{\theta}] = i\hat{1}$ cannot hold as an operator identity. This commutation relation would imply $[\sin \alpha\tilde{\mathcal{N}}, \hat{\theta}] = i\alpha \cos \alpha\tilde{\mathcal{N}}$. If one puts $\alpha = \pi$ this leads to a *contradiction* as $\sin \pi\tilde{\mathcal{N}}$ is zero because $\tilde{\mathcal{N}}$ has integer eigenvalues, but $\cos \pi\tilde{\mathcal{N}}$ is nonzero (see P. Jordan, Z. Phys. **44**, 1 (1927)).

[34] K. Schönhammer, Phys. Rev. A **66**, 014101 (2002)

[35] A. O. Gogolin, A. A. Nersesyan, and A. M. Tsvelik, *Bosonization and Strongly Correlated Systems* (Cambridge University Press, Cambridge, 1998)

[36] Some authors factor out $e^{i\alpha k_F x}$ in the definition of the $\psi_\alpha(x)$, which is contained in our $\hat{O}_\alpha(2\pi x/L)$.

[37] The subtleties of the large x *and* t behaviour are discussed in V. Meden, Phys. Rev. B **60**, 4571 (1999)

[38] V. Meden, "Spektrale Eigenschaften niedrigdimensionaler korrelierter Elekronensysteme", PhD thesis, Gttingen (1996), http://www.theorie.physik.uni-goettingen.de /~ meden

[39] The naive expectation that the k-resolved spectral function reduces to the noninteracting delta-peak for $|k - k_F| \gg k_c$ is false. This is discussed in: K. Schönhammer and V. Meden, Phys. Rev. B **47**, 16205 (1993)

[40] In contrast to the result for the occupancies (see [7]) the $|\omega|^{\alpha_L}$ result holds without a restriction on the value of α_L.

[41] D.N. Zubarev, Soviet Physics Uspekhi **3**, 320 (1960)

[42] This decomposition is valid only for calculating *low energy properties.*

[43] D. Bohm and D. Pines, Phys. Rev. **85**, 338 (1952)

[44] W. Kohn, Phys. Rev. **133** A, 171 (1964)

[45] F.D.M. Haldane, Phys. Rev. Lett. **47**, 1840 (1981)

[46] It is assumed that the (quasi-) momentum matrix elements $v_{k_1,k_2;k_3,k_4}$ of the two-body interaction have the property $v_{k_1,k_2;k_1,k_2} = \tilde{v}(0)$ and $v_{k_1,k_2;k_2,k_1} = \tilde{v}(|k_1 - k_2|)$. Then it is also easy to show directly that $v_J = v_F + O(\tilde{v}^2)$ for non Galilei invariant models.

[47] In this mapping the Klein factors are *not* taken into account in a rigorous way.

[48] The corresponding weak coupling RG for the disorder problem with *many* impurities was in fact performed earlier [24].

[49] M. Fabrizio and A. Gogolin, Phys. Rev. B **51**, 17827 (1995)

[50] V. Meden, W. Metzner, U. Schollwöck, O. Schneider, T. Stauber, and K. Schönhammer, Eur. Phys. J. **16**, 631-646 (2000)

[51] K. Moon, H. Yi, C.L. Kane, S.M. Girvin, and M.P.A. Fisher, Phys.Rev. Lett. **71**, 4381 (1993)

[52] P. Fendley, A.W.W. Ludwig, and H. Saleur, Phys. Rev. Lett. **74**, 3005 (1995)

[53] R. Egger, H. Grabert, A. Koutouza, H. Saleur, and F. Siano, Phys. Rev. Lett. **84**, 3682 (2000)

[54] D. Yue, L.I. Glazman, and K.A. Matveev, Phys. Rev. B **49**, 1966 (1994)

[55] V. Meden, W. Metzner, U. Schollwöck, and K. Schönhammer, Phys.Rev.B **65**, 045318 (2002); J. Low Temp. Phys. **126**, 1147 (2002)

[56] A. Furusaki and N. Nagaosa, Phys. Rev. B **47**, 4631 (1993)

[57] A. Luther and V. Emery, Phys. Rev. Lett. **33**, 589 (1974)

[58] Here again the Klein factors are treated in a sloppy way. Strictly speaking even the fact that the Hamiltonian is a sum of commuting boson and particle number parts is lost.

[59] see e.g. E. Lieb, T. Schultz and D. Mattis, Annals of Physics **16**, 407 (1961)

[60] C.N. Yang and C.P. Yang, Phys. Rev. **150**, 321, 327 (1966)

[61] M. Fowler and M.W. Puga, Phys.Rev. B **18**, 421 (1978)

[62] As the numerical results by Haldane [18] are not easily extracted from his figure, the results shown were obtained by an independent numerical solution of the Bethe Ansatz equations (V. Meden, private communication).

[63] S. Capponi, D. Poilblanc, and E. Arrigoni, Phys. Rev. B **57**, 6360 (1998)

[64] H.J. Schulz, "Fermi liquids and non-Fermi liquids" in *Mesoscopic Quantum Physics, Les Houches, Session LXI, 1994* (Elsevier, Amsterdam, 1995)

[65] for a review see e.g. : F. Mila and K. Penc, J. of Electron Spectr. and Rel. Phen. **117-118**, 451 (2001), and references therein

[66] E. H. Lieb and F. Y. Wu, Phys. Rev. Lett. **20**, 1445 (1968)

[67] H.J. Schulz, Phys. Rev. Lett. **64**, 2831 (1990); there is a factor 2 missing in the denominator on the rhs of Eq. (8).

[68] J.Favand, S. Haas, K. Penc, and E. Dagotto, Phys. Rev. B **55**, R4859 (1997)

[69] M.G. Zacher, E. Arrigoni, W. Hanke, and J.R. Schrieffer, Phys. Rev. B **57**, 6370 (1998)

[70] K. Penc, K. Hallberg, F. Mila and H.Shiba, Phys.Rev. B **55**, 15475 (1997)

[71] F. Gebhard, K. Bott, M. Scheidler, P. Thomas, and S. W. Koch, Phil. Mag. B **75**, 13 (1997)

[72] F. B. Gallagher and S. Mazumdar, Phys. Rev. B **56**, 15025 (1997)

[73] J. M. P. Carmelo, N. M. R. Peres, and P.D. Sacramento, Phys. Rev. Lett. **84**, 4673 (2000)

[74] E. Jeckelmann, F. Gebhard, and F.H.L. Essler Phys. Rev. Lett. **85**, 3910 (2000); F.H.L. Essler, F. Gebhard, and E. Jeckelmann, Phys. Rev. B **64**, 125119 (2001)

[75] H.J. Schulz, J. Phys. C **16**, 6769 (1983)

[76] P. Kopietz, V. Meden, and K. Schönhammer, Phys. Rev. Lett. **74**, 2997 (1995); Phys. Rev. B **56**, 7232 (1997). The effective low energy scale t_{eff} discussed after Eq. (4.69) is *not* recovered in the higher dimensional bosonization scheme.

[77] H.J. Schulz, Phys. Rev. Lett. **71**, 1864 (1993)

[78] L.P. Gorkov and I.E. Dzyaloshinski, Sov. Phys. JETP **40**, 198 (1975)

[79] C. Bourbonnais, F. Creuzet, D. Jerome, K. Bechgaard, and A. Moradpour, J. Phys. (Paris) Lett. **45**, L755 (1984)

[80] X. G. Wen, Phys. Rev. B **42**, 6623 (1990)

[81] C. Bourbonnais and L.G. Caron, Int. J. Mod. Phys. B **5**, (1991)

[82] S.A. Brazowski and V.M. Yakovenko, Sov. Phys. JETP **62**, 1340 (1985); A.A. Nerseyan, A. Luther, and F.V. Kusmartsev, Phys. Lett. A**176**, 363 (1993)

[83] V.J. Emery, E. Fradkin, S.A. Kivelson, and T.C. Lubensky, Phys. Rev. Lett. **85**, 2160 (2000)

[84] A. Vishwanath and D. Carpentier, Phys. Rev. Lett. **86**, 674 (2001)

[85] R. Mukhopadyhay, C. L. Kane, and T. C. Lubensky, Phys. Rev. B **64**, 045120 (2001)

[86] D. Boies, C. Bourbonnais, and A.M.S. Tremblay, Phys. Rev. Lett. **74**, 68 (1995)

[87] W. Metzner, Phys. Rev. B **43**, 8549 (1991)

[88] E. Arrigoni, Phys. Rev. Lett. **83**, 128 (1999); Phys. Rev. B **61**, 7909 (2000)

[89] W. Metzner and D. Vollhardt, Phys. Rev. Lett. **62**, 324 (1989); A. Georges, G. Kotliar, W. Krauth, and M.J. Rozenberg, Rev. Mod. Phys. **68**, 13 (1996)

[90] F.H.L. Essler and A. Tsvelik, Phys. Rev. B **65**, 115117 (2002)

[91] S. Biermann, A. Georges, A. Liechtenstein, and T. Giamarchi, Phys. Rev. Lett. **87**, 276405 (2001)

[92] see e.g. M. Tsuchiizu, P. Donohue, Y. Suzumura, and T. Giamarchi, Eur. Phys. J. B **19**, 185 (2001)

[93] D. Jerome and H.J. Schulz, Adv. Phys. **31**, 299 (1982)

[94] M. Grioni and J. Voit, "High resolution photoemission studies of low dimensional system", in: H. Stanberg, H. Hughes (Eds.), *Electron Spectroscopies Applied to Low-Dimensional Materials*, Kluwer Academic Publishers, (2000) p. 501

[95] G.H. Gweon, J.D.Denlinger, J.W. Allen, R. Claessen, C.G.Olson, H.Höchst, J. Marcus, C. Schlenker, and L.F. Schneemeyer, J. of Electron Spectr. and Rel. Phen. **117-118**, 481 (2001)

[96] J. Allen, Solid State Commun. **123**, 496 (2002)

[97] R. de Picciotto, H.L. Störmer, L.N. Pfeiffer, K.W. Baldwin, and K.W. West, Nature **397**, 598 (1999)

[98] O.M. Auslaender, A. Yacobi, R. de Piciotto, K.W. Baldwin, L.N.Pfeiffer, and K.W. West, Science **295**, 825 (2002)

[99] P. Segovia, D. Purdie, M. Hengsberger, and Y. Baer, Nature **402**, 504 (1999)

[100] R. Losio, K.N. Altmann, A. Kirakosian, J.-L. Lin, D. Y. Petrovykh, and F.J. Himpsel, Phys. Rev. Lett. **86**, 4632 (2001)

[101] R. Saitoh, G. Dresselhaus, and M.S. Dresselhaus, *Physical Properties of Carbon Nanotubes*, Imperial College Press (1998)

[102] C. Dekker, Physics Today **52**(5), 22 (1999)

[103] R. Egger and A. Gogolin, Phys. Rev. Lett. **79**, 5082 (1997); Eur. Phys. J. B **3**, 281 (1998)

[104] C. Kane, L. Balents, and M.P.A. Fisher, Phys. Rev. Lett. **79**, 5086 (1997)

[105] X.G. Wen, Phys.Rev. B **43**, 11025 (1991)

[106] A. M. Chang, L.N. Pfeiffer, and K.W. West, Phys. Rev. Lett. **77**, 2538 (1998)

[107] C.L. Kane, M.P.A. Fisher, and J. Polchinski, Phys. Rev. Lett. **72**, 4129 (1994)

[108] L.S. Levitov and B.I. Halperin, Phys. Rev. Lett. **80**, 141 (1997)

[109] M. Grayson, D.C. Tsui, L.N. Pfeiffer, K.W. West, and A.M. Chang, Phys. Rev. Lett. **80**, 1062 (1998)

[110] B. Dardel, D. Malterre, M. Grioni, P.Weibel, Y.Baer, J. Voit, and D. Jerome, Europhys. Lett.**24**, 687 (1993)

[111] R. Claessen, M.Singh, U. Schwingschlögl, P. Blaha, M. Dressel, and C. S. Jacobsen, Phys. Rev. Lett. **88**, 096402 (2002)

[112] C. Kim, J. of Electron Spectr. and Rel. Phen. **117-118**, 503 (2001)

[113] D. L. Maslow and M. Stone, Phys. Rev B **52**, R5539 (1995)

[114] I. Safi and H.J. Schulz, Phys. Rev B **52**, R17040 (1995)

[115] R. Egger and H. Grabert, Phys. Rev. B **58**, 10761 (1998)

[116] M. Bockrath, D.H. Cobden, J. Lu, A.G. Rinzler, R.E. Smalley, L. Balents, and P.L. McEuen, Nature **397**, 598 (1999)

[117] Z. Yao, H. Postma, L. Balents and C. Dekker, Nature **402**, 273 (1999)

[118] H. Postma, T. Teepen, Z. Yao, M. Grifoni and C. Dekker, Science **293**, 76 (2000)

[119] A. Georges, T. Giamarchi, and N. Sandler, Phys. Rev. B **61**, 16393 (2000)

[120] S. Biermann, A. Georges, T. Giamarchi, and A. Liechtenstein, cond-mat/0201542, to appear in Proc. of the Windsor NATO summer school (editors: I. Lerner and B. Altshuler)

[121] G. Grüner, *Density Waves in Solids*, Addison-Wesley Publishing Company (1994)

[122] A. Schwarz, M. Dressel, G. Grüner, V. Vescoli, L. Degiorgi, and T. Giamarchi, Phys. Rev. B **58**, 1261 (1998)

[123] D. Controzzi, F.H.L. Essler, and A.M. Tsvelik, Phys. Rev. Lett. **86**, 680 (2001)

Chapter 5

PHOTOEMISSION IN QUASI-ONE-DIMENSIONAL MATERIALS

M. Grioni
Institut de Physique des Nanostructures, Ecole Polytechnique Fédérale, CH-1015 Lausanne, Switzerland
marco.grioni@epfl.ch

Abstract Photoelectron spectroscopy, both in the momentum-integrated (PES) and momentum-resolved (ARPES) modes, is a direct probe of single particle excitations in solids. This Chapter presents a brief review of high-resolution PES and ARPES data on quasi-one dimensional (1D) materials. On a broad (~eV) energy scale, band mapping experiments reveal the expected features of translational periodicity and of coexisting potentials. On a smaller scale, the spectral properties of 1D systems are incompatible with those of conventional (3D) metals. They suggest that strong interactions, and possibly the peculiar correlations predicted by theory, shape their electronic structure. Characteristic signatures of the transitions to non-metallic broken-symmetry states develop from the unusual spectral lineshapes.

Keywords: Photoemission, quasi one-dimensional systems, Peierls transition

1. Introduction

There is more than one good reason to be interested in the electronic structure of one-dimensional solids. At a very fundamental level, these systems provide the simplest playground to test elementary ideas on the formation of extended states and periodic bands which are at the heart of solid state physics [1], without the unnecessary geometrical complications of a three-dimensional material. And yet, as discussed throughout this book, life in 1D is actually much more interesting than a simple independent-particle band picture might suggest. Due to the unique Fermi surface (FS) topology in 1D - two points - and the consequent

D. Baeriswyl and L. Degiorgi (eds.), Strong Interactions in Low Dimensions, 137–163.

perfect nesting, several electronic instabilities oppose and ultimately destroy the metallic state, giving rise to an extremely rich variety of broken symmetry phases [2]. The peculiarities of the 1D behavior are of course not limited to the non-metallic ordered phases. The 'normal' state is shaped by singular correlations, which reflect the dramatically reduced phase space. The nature of this correlated state, and its properties, are qualitatively different from those of conventional '3D' metal [3, 4, 5]. One dimensional systems therefore give us one of the simplest examples of *non-Fermi liquid* behavior.

Photoelectron spectroscopy - photoemission, in short - is one of the best experimental techniques available to address the various aspects of the 1D phenomenology [6]. It is traditionally associated with the study of electronic states in solids, and ARPES band mapping experiments provide us with striking demonstrations of the reality of the concept of bands. ARPES is also a direct probe of the FS, complementary to traditional magnetotransport techniques. Compared to other probes of the electronic structure of solids, it presents the crucial advantage of a remarkable selectivity and freedom in the choice of the energy *and* momentum of the electronic state under investigation. Finally, photoemission is sensitive in various ways to correlations. Here again, its most appealing feature is the immediate connection between experiment and the fundamental single particle properties which can be calculated by theoretical models of strongly correlated materials.

The last decade has witnessed a general surge of interest for the spectral properties of 1D systems, and the beginning of a consistent research activity based on photoemission. The availability of high-quality single crystals of *real* quasi-one-dimensional compounds has played a major role. These anisotropic materials built from 1D structural building blocks - infinite linear chains of transition metal atoms or stacks of flat organic molecules [7, 8], exhibit genuine one-dimensional properties, namely open Fermi surfaces. In the following, with a common abuse of language, we will simply refer to them as '1D'. Several 1D compounds, both inorganic and organic, have already been synthesized in sizes ($\sim$ mm) compatible with the stringent requirements of photoemission or optical experiment. More recently, ARPES experiments have also been performed on ordered artificial 1D structures grown at the interface with a suitable substrate. The rapid development of the whole field has also been fostered by parallel rapid improvements in the energy and momentum resolution of ARPES, and by the inflationary growth of 'gap spectroscopy' [9].

We give here a synthetic overview of the main results and of the open questions raised by these experiments. Some of these issues have been previously discussed in a broader perspective in a recent review [10]. The Chapter is organized as follows. Section 2 discusses some specific aspects of an ARPES experiment in 1D, and the spectral signatures of translational periodicity. Section 3 briefly describes the spectral features of normal interacting systems (Fermi liquids). The unusual properties of quasi-1D materials are compared with the 1D spectral properties predicted by theory in Section 4. Section 5 describes the changes in the electronic structure associated with the transitions to broken symmetry states. Finally, the growing evidence for strong electron-phonon interactions in 1D Peierls materials is illustrated in Section 6.

2. Tracking the 1D bands

The experimental set-up, interpretation, and limitations of a PES or ARPES have been presented in many excellent reviews, e.g. in [6, 11, 12] and some specific aspects are discussed in other Chapters of this book [9], and will not be discussed here. For the present purposes it is sufficient to remember here few basic notions.

In a photoemission experiment photoelectrons are generated by the interaction of a solid with a monochromatic photon beam. Photoelectrons are collected either within a narrow emission angle (ARPES) or after integration over a broad angular range (PES), and their kinetic energy is then measured. Since energy and momentum are conserved in the interaction, a measurement of these quantities for the photoelectron contains information on the energy and momentum of the electron inside the solid. Systematic measurements as a function of the emission angles in principle allow one to reconstruct the electron dispersion relation $E(k)$. In reality, mapping the bands of a 3D solids is not straightforward, because the component of the electron wavevector perpendicular to the surface ($k_\perp$) is undetermined, unless the discontinuity of the crystal potential at the surface is known, or special measurement strategies are used. Luckily, the uncertainty on $k_\perp$ is not a limitation for D<3, e.g. in two-dimensional systems, like surfaces or the 2D high-T_c cuprates [9, 13]. For the same reason, band mapping is particularly easy in 1D systems, provided that the 1D 'chain' direction lies within surface. Simple inspection shows that the in-chain momentum $k_\parallel$ is uniquely determined by measuring the photoelectron emission angle off the surface normal θ and the kinetic energy ε_K by $k_\parallel = (2m\varepsilon_K/\hbar^2)^{1/2}\,\sin\theta$. The initial state energy $\varepsilon(k)$ and photoelectron kinetic energy are related by $\varepsilon_K = \varepsilon(k) + \hbar\omega - \phi$, where $\hbar\omega$ is the photon energy, ϕ is the spectrometer's work function, and the energy zero coincides with the Fermi level E_F.

The expected output of an ideal ARPES experiment on a 1D metal with a single conduction band has a simple structure. The spectrum consists of a single peak, which disperses in energy and emission angle - i.e. wavevector - reproducing the band dispersion up to the Fermi wavevector k_F, where the peak position coincides with E_F. Such a *Fermi level crossing*, which identifies a point of the FS, is the typical ARPES fingerprint of a metal. Integration of the photoelectron signal over the whole Brillouin zone (BZ) as in a PES experiment, yields a quantity which is closely related to the momentum-integrated energy dependent density of states (DOS). In the PES spectrum, the DOS is modulated by dipole transition matrix elements and possibly distorted due to the strong surface sensitivity - few atomic layers - of the measuement. It is also weighted by the Fermi-Dirac function, which generates a *metallic Fermi edge*. The resolution- or temperature-limited edge centered at E_F, is the characteristic PES signature of a metal.

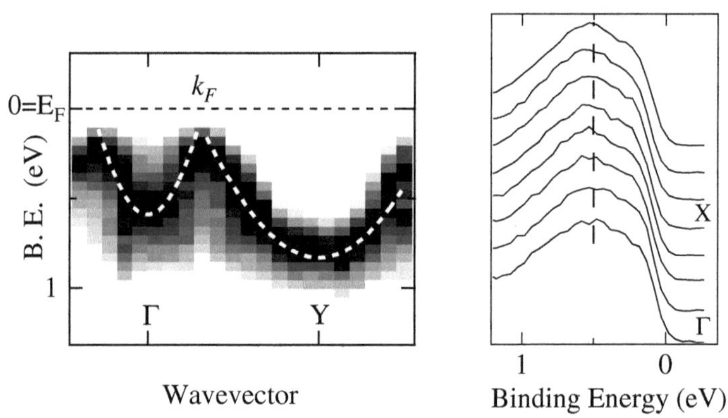

Figure 5.1. ARPES spectra of TTF-TCNQ from Ref. 15 showing dispersion along the 1D 'chain' direction (*left*) but not in a perpendicular direction (*right*).

For an experimentalist this simple picture provides a convenient guideline, supported by ARPES experiments on various metallic quasi-1D systems, like the organic compound TTF-TCNQ (Fig. 5.1) [14]. TTF-TCNQ is a 1D metal due to charge transfer between the weakly-coupled segregated stacks of flat TTF and TCNQ molecules [8]. The ARPES intensity map reproduces the expected 1D dispersion, with the donor (TTF) and acceptor (TCNQ) bands crossing at the Fermi wavevector k_F, as required by overall charge neutrality. The absence of dispersion in a direction perpendicular to the chains (Fig. 5.1, right) confirms the essentially 1D character of the electronic structure. Any effects of transverse coupling must be quite small and masked by the intrinsic linewidth.

The peculiar FS topology of a 1D material can be revealed by a more thorough sampling of the reciprocal space [15, 16], as in Fig. 5.2 for another typical 1D material, $(TaSe_4)_2I$[17]. The angular intensity map shows nearly flat and parallel FS sheets, in the first and higher BZs. The data, which demonstrate the existence of an open FS were actually recorded *below* (0.25 eV) E_F. Contrary to intuition, and unlike the case of a normal metal, the definition of the FS would be much worse in a map acquired *exactly* at E_F. This is a puzzling and crucial aspect of the spectroscopy of 1D metals, which we will reconsider below.

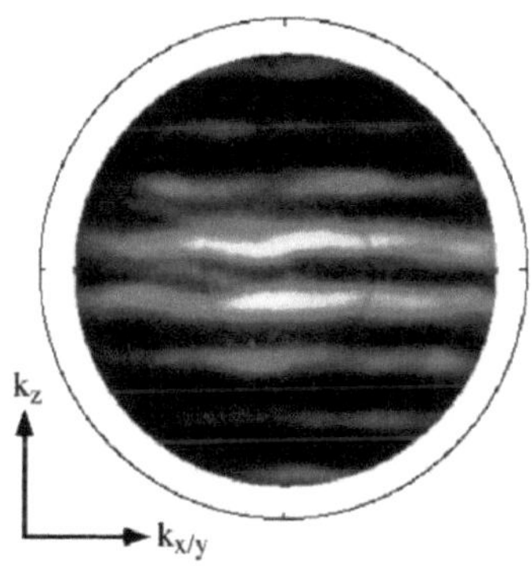

Figure 5.2. ARPES intensity map of $(TaSe_4)_2I$ from Ref. 18. The map, acquired 250 meV below E_F, shows the almost flat sheets of the open Fermi surface, perpendicular to the chain direction 'z', and their replicas in higher Brillouin zones

.

Due to their *geometrical* simplicity 1D compounds are ideal model systems to study periodic bands in solids. Elementary considerations [1] show that bands can be equivalently described in the *extended*, *repeated*, or *reduced-zone* pictures, taking advantage of the equivalence of Bloch states whose wavevectors differ by a reciprocal lattice vector G. It is legitimate to ask whether this ambiguity would carry over to an ARPES band mapping experiment, leading to the observation of replicas (or *shadows*) of the main band. The multiple images of the FS of Fig.5.2 suggest in a very pictorial way that this must be the case. The issue becomes more subtle when electrons are simultaneously subject to coexisting periodic potentials, as in charge-density-wave (CDW) materials. There, electrons feel both the periodic potential of the unperturbed lattice, and an additional potential with wavevector Q_{CDW} due to the modulation of the atomic positions. This problem has a simple solution when the two periodicities are commensurate, i.e. when the ratio

(G/Q_{CDW}) is a rational number, because in that case the whole band structure can be folded into a common reduced BZ.

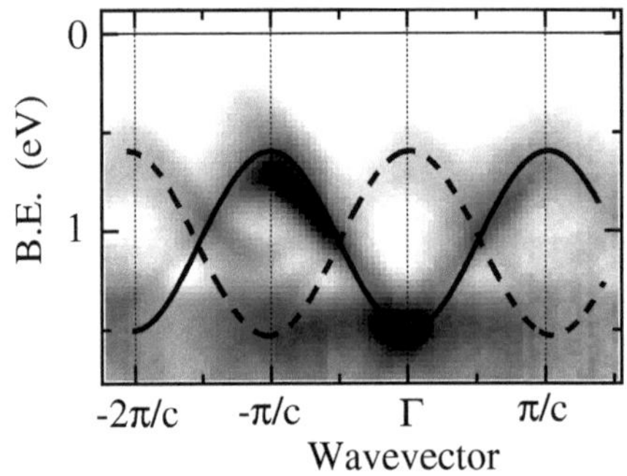

Figure 5.3. ARPES band dispersion for the 1D insulator $(NbSe_4)_3I$ from Ref. 20, showing the effect of a doubled periodicity.

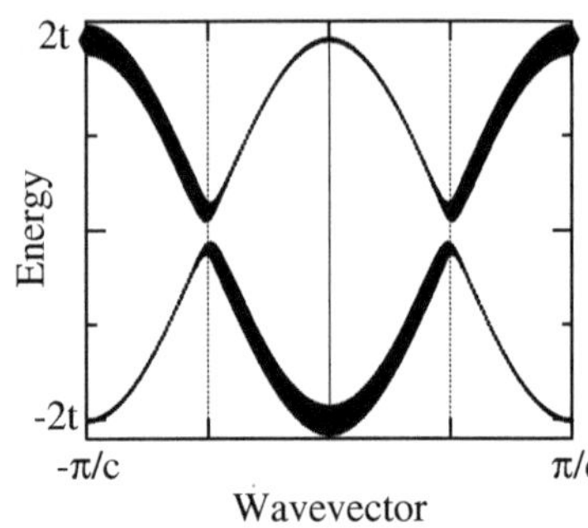

Figure 5.4. Calculated bands for a situation analogous to that of Fig. 3. The line thickness is proportional to the spectral weight.

The insulating 1D compound $(NbSe_4)_3I$ offers a practical realization of an electronic stystem subject to two commensurate potentials. The valence band maxima coincide with the boundaries ($k = \pm(\pi/c)$) of the BZ determined by the periodicity along the 1D Nb chains, but a structural motif *around* the chains effectively doubles the spatial periodicity, and halves the BZ [18]. The ARPES map of Fig.5.3 shows a dispersing band, with maxima at $k = \pm(\pi/c)$, which is repeated in the second and higher BZs, but with a reduced intensity. It also shows a second weaker shadow band, offset by $G' = (\pi/c)$, consistent with a folding of the main band at $k = \pm(\pi/2c)$, the boundaries of the reduced BZ. The experimental band dispersion clearly shows signatures of both periodicities, but the ambiguity built in the definition of the wavevectors is lifted, because the ARPES intensities do not exhibit the full periodicity of the total potential. The data of Fig. 5.3[19] show that it is necessary to distinguish between the band energies, i.e. the eigenvalues of the periodic Hamiltonian, which are not directly accessible, and the experimentally observable spectral weight. A simple model clarifies this point. Figure 5.4 illustrates the calculated dispersion of a tight-binding band $\varepsilon(k)$= -$2t$ cos(kx), with periodicity $G = (2\pi/c)$, subject to an additional potential of periodicity $G' = (\pi/c)$ and strength $V' = 0.2\ t$. The thickness of the lines is proportional to the spectral weight, i.e. for each k and band index, to the squared modulus of the corresponding eigenvector. As in the experiment, the spectral weight is piled up along the 'main band', and only a small fraction goes to the shadow band. Increasing V' has two effects: i) the energy gaps at $k = \pm(\pi/2c)$ widen, and ii) spectral weight is progressively transferred to the shadow bands. Clearly, periodicity alone cannot determine the solution of the problem, which critically depends on the strength of the perturbing potential.

A more intriguing case is that of two incommensurate potentials, where the ratio G/G' is not a rational number. This is actually the rule in CDW systems, where the periodicity of the CDW is dictated by the size of the FS ($Q_{CDW} = 2k_F$), and is therefore independent of the lattice periodicity [2]. Now the problem has no obvious solution, because the two potentials do not admit a common reduced BZ, and it is not possible to fold bands in a consistent way at zone boundar ies. Strictly speaking the solid is not periodic: the band structure could present a fractal character, with an infinite hierarchy of exponentially small gaps [20]. The issue has a fundamental interest even if the case of two incommensurate potentials would be indistinguishable from that of suitable commensurate approximants in real materials and at finite temperature, and even at T=0 if the zero point motion of the ions were taken into account.

The ARPES data of Fig. 5.5 for the Peierls system $(TaSe_4)_2I$ indicate the solution of this puzzle. This material is similar to $(NbSe_4)_3I$ considered above, but k_F is slightly bigger than the zone boundary ($k_F \sim 1.085\pi/c$), and the CDW is incommensurate with the lattice [21]. Even here, the main band and and its shadow in the higher BZs, are readily identified and the dispersion appears to be periodic. Actually, a closer inspection of the top of the band reveals traces of both underlying incommensurate periodicities [22], again reflecting the uneven distribution of spectral weight over the - strictly nonperiodic - eigenvalues. This problem is harder to model than the commensurate case, and cannot be solved analytically. Numerical solutions however deliver a clear physical message: the weights of higher order umklapp processes decrease exponentially, so that only the first-order shadow bands and gaps - and not the full underlying fractal bandstructure - are experimentally observable s[22].

In summary, this Section has shown that bands and Fermi surfaces with strong one-dimensional character have been observed by ARPES in real 1D compounds. Notice however that the observation of spectral features dispersing with the expected periodicity is not sufficient to conclude that a simple band picture would adequately describe these materials. Actually, the spectral properties of 1D systems are complex and suggest strong, and possibly peculiar interactions, which distinguish 1D systems from conventional 3D materials, as discussed in the next section.

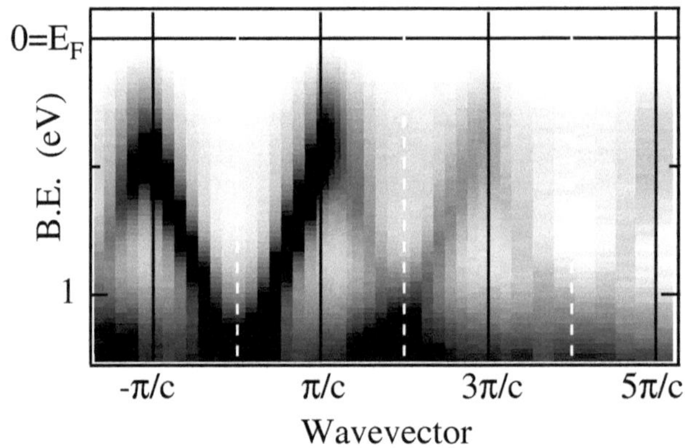

Figure 5.5. ARPES band dispersion for the incommensurate CDW system $(TaSe_4)_2I$ from Ref. 23.

.

3. The spectral function

The previous discussion implicitely assumed the validity of an independent particle picture. Such an assumption is questionable in the presence of strong interactions, and is certainly inadequate in 1D, where correlations deeply modify the nature of the electronic states. It is therefore necessary to resort to a more general view of ARPES in interacting systems.

Landau's Fermi liquid theory [23] is the paradigm for electrons in solids. It is based on the one-to-one correspondence between noninteracting electrons and weakly interacting *quasiparticles* (QPs), the elementary excitations of the correlated system. Residual interactions between the QPs limit their lifetime (τ), but their fermionic character imposes that the scattering rate τ^{-1} tends asymptotically to 0 at E_F: $\tau^{-1} \sim (E - E_F)^2$. The QPs are therefore well defined excitations near the Fermi surface.

The one-to-one correspondence between noninteracting particles and QPs is reflected in the spectral properties. QP bands can be mapped by ARPES, even if the dispersion $\varepsilon(k)$, the spectral width ΔE and the spectral weight of the QPs are renormalized by the interactions. In particular, the typical fingerprints of a metal - Fermi level crossings and a metallic Fermi edge - survive in the interacting system..

Formally, the ARPES spectrum $I(k, \omega)$ is proportional - via transition matrix elements - to the spectral density function $A(k, \omega)$ [24], where $\omega = (E - E_F)$. $A(k, \omega)$ is a fundamental theoretical quantity derived from the one-particle Green's function of the interacting system: $A(k, \omega) = -(1/\pi) Im[G(k, \omega)]$, and:

$$G(k, \omega) = \frac{1}{\omega - \varepsilon(k) - \Sigma(k, \omega)}. \tag{5.1}$$

Here $\varepsilon(k)$ represents the ideal dispersion of the independent electrons, and all effects of correlation are contained in the self energy Σ. Such a close relation between the photoemission spectrum and the propagator is not surprising, since $G(k, \omega)$ is precisely defined in terms of addition and removal of one particle. At T=0 ARPES strictly probes the $\omega < 0$ part of $A(k, \omega)$, while at finite temperatures $I(k, \omega) \sim A(k, \omega) f(T, \omega)$, and the high-energy cutoff is provided by the Fermi-Dirac distribution $f(T, \omega)$.

In the non-interacting limit Σ=0 and G has poles at $E_0(k) = \varepsilon(k)$, which yields for the spectral function the simple form $A(k, \omega) = \delta(k, \omega)$. In an interacting system, one obtains:

$$A(k, \omega) = \frac{1}{\pi} \frac{|Im\Sigma(k, \omega)|}{|\omega - \varepsilon(k) - Re\Sigma(k, \omega)|^2 + |Im\Sigma(k, \omega)|^2}. \tag{5.2}$$

Near the Fermi surface, this expression can be recast in the phenomenological form:

$$A(k, \omega) = \frac{Z_k}{\pi} \frac{\Gamma}{(\omega - E(k))^2 + \Gamma^2} + A_{inc}. \tag{5.3}$$

The first term - the QP peak or *coherent* part of the spectral function - is centered at the renormalized energy $E(k) = \varepsilon(k) + Re\Sigma$. Its energy width $\Gamma = 2\ Im\Sigma \sim \omega^2$ is related to the QP lifetime by $\tau = (\hbar/\Gamma)$, and its momentum width to the inverse of the QP coherence length: $\Delta k = (l)^{-1}$. The QP weight $Z_k < 1$ measures the overlap of the QP with the corresponding free-electron state. $Z(k_F)$ in particular is the same factor which renormalizes quantities like the electron effective mass, and the linear term of the specific heath. Qualitatively, strong correlations yield small Z's and therefore small QP weights. The spectral weight (1-Z) removed from the QP peak is redistributed, possibly over a much larger energy scale, to the *incoherent* part of $A(k, \omega)$, and represents the dressing of the QP.

Some crucial predictions of FL theory, namely that of a quadratic energy dependence of the QP lifetime near the FS, have been verified by accurate ARPES measurements in model metals, like the layered compound $TiTe_2$ [25]. The specific contributions to the QP scattering rate

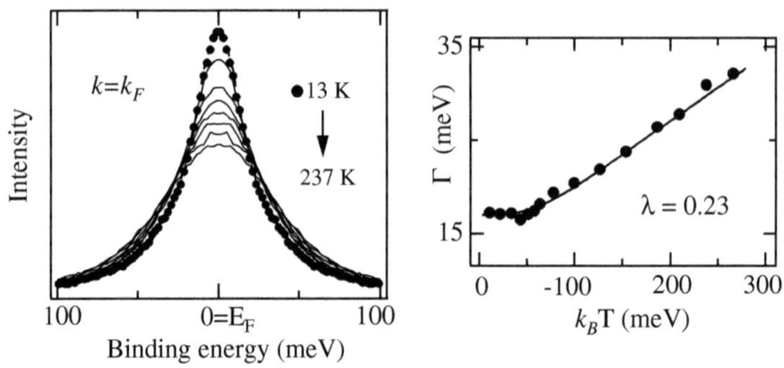

Figure 5.6. QP spectra of the 2D metal $TiTe_2$ measured at the Fermi surface. The increasing linewidth (*right*) reflects the temperature-dependent phonon scattering. The spectra have been symmetrized around E_F. From Ref. 34.

can also be determined from the spectra [26, 27, 28, 29, 30, 31, 32, 33]. At T=0 and on the FS the only lifetime-limiting mechanism is scattering on impurities, but at finite temperature QP scattering by phonons rapidly becomes dominant. This is illustrated by the $TiTe_2$ data of Fig. 5.6. The spectra measured at k_F have been symmetrized around E_F to remove the perturbing temperature dependence of the Fermi function, as customary in work on the high T_C cuprates. Notice that symmetrized curves peak at E_F, as expected for a QP at the Fermi surface. The linewidth increases with temperature reflecting the increasing phonon scattering. The temperature dependence is linear at sufficiently high temperature ($\sim$ 60 K for $TiTe_2$), as predicted by theory [34], and the slope yields the value of the *e-ph* coupling parameter λ which measures the strength of the interaction.

The good qualitative and even quantitative agreement between theory and experiment observed in normal metals sharply contrasts with the results on quasi-1D materials. The first measurements of organic 1D systems like TTF-TCNQ in the '70s already revealed an apparent lack of intensity in the vicinity of the Fermi level [35]. However, only two decades later high-resolution data could definitely establish this surprising anomaly [36]. This is illustrated by the momentum-integrated spectra of three typical 1D compounds, in Fig. 5.7. The spectra do not exhibit the characteristic metallic Fermi step, which is well visible in the spectrum of $TaSe_2$, a (2D) metallic reference. The intensity is vanishingly small at E_F, and strongly reduced over a broad (0.1 - 0.5 eV) energy range. Such deep *pseudogaps* are in contrast with the metallic properties of the materials, and are also much broader than the real gaps of the respective low-temperature insulating phases (see below).

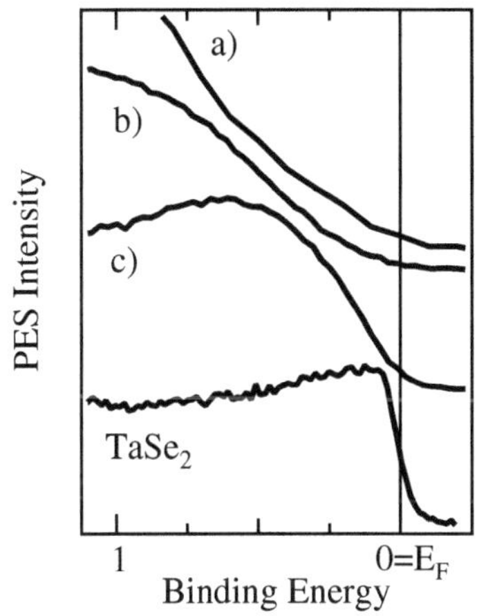

Figure 5.7. The PES spectra of the 1D compounds $(TMTSF)_2PF_6$ (a), $(TaSe_4)_2I$ (b) and $K_{0.3}MoO_3$ (c) do not exhibit the typical Fermi edge, unlike a 2D reference.

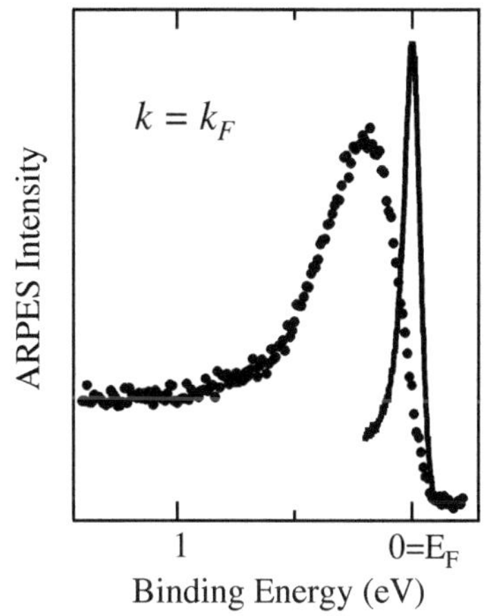

Figure 5.8. The ARPES spectrum of $K_{0.3}MoO_3$ is compared with the QP spectrum of the Fermi liquid reference $TiTe_2$ (solid line). Both are measured at k_F and T=200 K.

The ARPES spectra of 1D compounds are even more striking. Figure 5.8 shows the data for the Peierls compound $K_{0.3}MoO_3$ (the 'blue bronze'), measured at $k = k_F$ in the normal metallic state. A normal metal would exhibit a QP peak at E$_F$, indicative of a Fermi level crossing, as in the spectrum of $TiTe_2$ shown here for comparison. By contrast, the spectrum of the 1D compound is peaked well below E_F, and is much broader than the normal metal reference. If the QP energy is identified, as usual, with the peak position, the data of Fig. 5.8 again indicate a broad pseudogap (this issue will be reconsidered in Section 6) . Furthermore, while the FL reference becomes progressively sharper at low temperature, as in Fig. 5.6, the 1D spectrum is essentially temperature-independent.

The absence of a metallic Fermi step and of Fermi surface crossing, and the broad lineshapes are common spectral features of the 1D materials studied so far. The spectra of TTF-TCNQ, shown in Fig. 5.1 are no exception. The TTF and TCNQ bands seem to cross the Fermi level at k_F, but when the spectra are collected with high resolution in the vicinity of the Fermi surface, the 'pseudogapped' lineshape appears clearly [14]. Xue et al. [37] have reported a possible exception to this 'empirical rule' in $Li_{0.9}Mo_6O_{17}$, but the evidence for a Fermi level crossing was not confirmed by two independent sets of data [38, 39, 40].

4. Spectral evidence for the Luttinger liquid?

It is perhaps not surprising that 1D systems do not exhibit the spectral fingerprints of normal metals. The Fermi liquid concept is not valid in 1D, where the leading terms of the FL expansion and the corrections due to coupling of the QP to the collective modes are of the same order. Haldane's Luttinger liquid (LL) conjecture [3] based on the solution [41] of the Tomonaga-Luttinger model [42, 43], maintains that a 1D system with gapless charge and spin degrees of freedom, have only collective excitations, described as charge (*holons*) and spin (*spinons*) fluctuations. Individual electrons or holes are not stable excitations, and rapidly separate into holons and spinons.

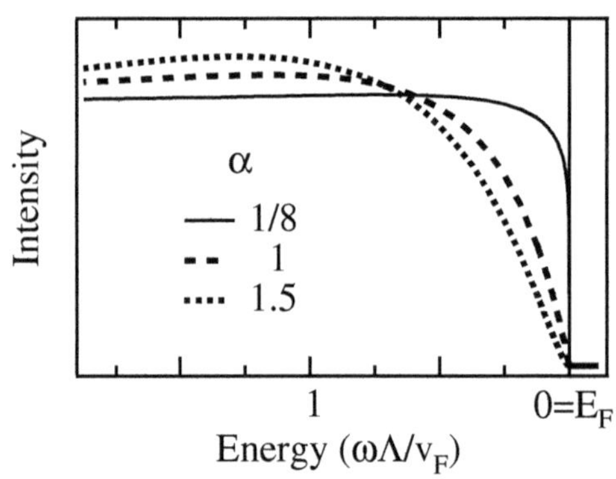

Figure 5.9. Calculated PES spectra of the Luttinger model for three values of the characteristic exponent α. Λ is an energy cutoff of the model. From Ref. 47.

The lack of QPs influences the electronic properties of 1D systems, and determines the structure of the spectral function. The calculated momentum-integrated (PES) LL spectrum $\rho(\omega)$, shown in Figure 5.9 for T=0, exhibits a power-law dependence $\rho(\omega) \sim \omega^{\alpha}$ [44, 45, 5]. The exponent α depends on the strength and range of the interactions, and is related to the parameter K_ρ of the Luttinger model by $\alpha = (1/8)(K_\rho + K_\rho^{-1}\text{-}2)$. Unlike a normal metal, the LL has no spectral intensity at the Fermi level (ω=0). It is therefore tempting to draw an analogy with the experimental spectra of Fig. 5.7 [46], even if some caution is necessary because the power-law dependence is only an asymptotic expression valid near E_F, and its range of validity is ill-defined.

The actual lineshape strongly depends on the value of the exponent. When α is small, the spectrum has a steep leading edge, which would be essentially indistinguishable from a conventional metallic Fermi edge. This is still the case for α=1/8 (K_ρ=1/2), the largest exponent compatible with the standard Hubbard model [47]. A comparison with experiment, however, would suggest larger α values, of order 1, for which the lineshape has a positive concavity over an extended energy region below E_F. Such large α values indicate strong and long-range correlations. They could still be compatible with extended versions of the Hubbard model, or perhaps with more elaborate theoretical schemes [48]. It is not clear whether they would also be compatible with metallic behavior, or whether they would lead to an ordered insulating ground state [49]. From the experimental point of view, there are some independent indications of similarly large Luttinger exponents in optical [50, 51], transport [52] and NMR [53] data on the organic Bechgaard salts.

A possible justification for the large exponents of the PES spectra has been proposed independently by Voit [54] and by Eggert *et al.* [55]. They pointed out that for a given interaction strength represented by K_ρ, considerably larger α values are obtained if the 1D system is bounded by a defect or an impurity. Such defects must be common in real materials, namely at surfaces. The LL exponents measured by surface sensitive techniques like photoemission could then be different - typically larger - from the 'bulk' exponents describing optical, transport or thermodynamic properties.

The momentum-resolved (ARPES) spectral function (Fig. 5.10) bears the most revealing fingerprints of the Luttinger liquid. Remarkably, the dispersing QP peak of the FL is replaced by two distinct singularities representing the spinon and holon excitations. These peaks are degenerate at k_F, but disperse with different velocities, reflecting spin-charge separation. In spite of intense efforts, the experimental evidence for the LL scenario of Fig. 5.10 remains scarce. The observation of separately dispersing features for ordered Au lines grown on stepped Si(111) (Fig. 5.11) [56] was hailed as the possible smoking gun for the LL. However, the case for holons and spinons has been weakened by the subsequent observation by ARPES of two distinct Fermi level crossings, supporting two separate bands [57].

Spin-charge separation has also been claimed in $Li_{0.9}Mo_6O_{17}$[58]. This 1D material remains metallic to an unusually low temperature (T=24 K) and therefore lends itself to a high-resolution ARPES study. The spectra are compatible with the calculated LL lineshapes if a large exponent ($\alpha \sim 0.9$) is assumed. Unfortunately, for such large values of α the holon divergence is attenuated and the spinon feature is entirely

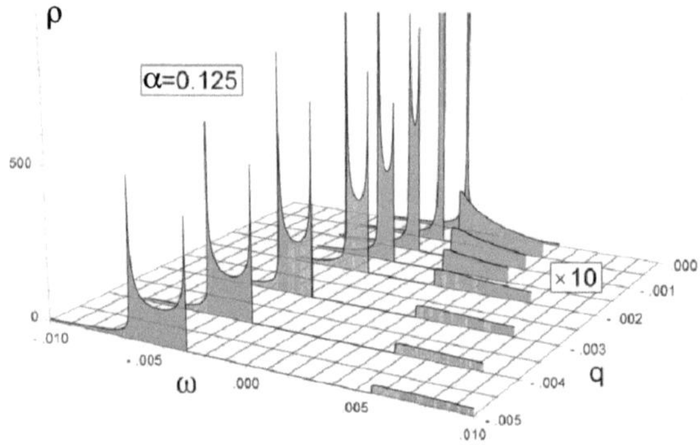

Figure 5.10. Calculated momentum-resolved spectral function of the Luttinger model for α=1/8, showing separate spinon and holon branches. From Ref.45.

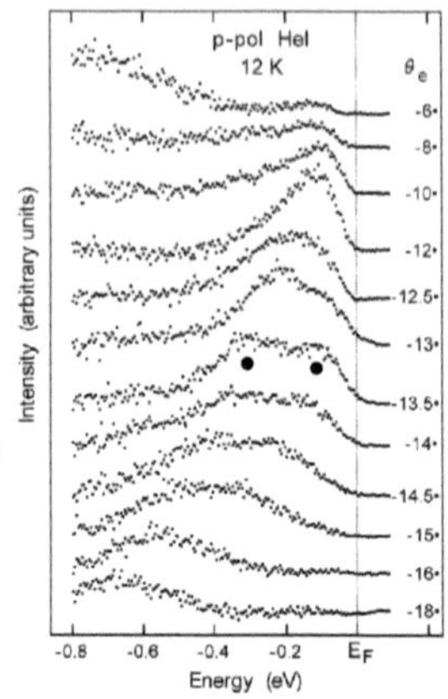

Figure 5.11. ARPES spectra of ordered metallic lines at the Au-Si(111) interface. The markers highlight separately dispersing features showing similarities with the spinon-holon branches of the LL. Adapted from Ref. 57.

suppressed [44], so that the most distinctive signature of spin-charge separation - two separate peaks - is lost. In this limit the LL lineshape exhibits a broad leading edge which extrapolates to zero at E_F. Similarly broad onsets have been observed in the organic Bechgaard salts [59], TTF-TCNQ [14] and in inorganic Peierls systems [36, 16, 60, 61].

The anomalous spectral lineshape is not the only critical element in favor or against a LL interpretation of the ARPES spectra. In 1D the spectral weight distribution exhibits peculiar features which reflect the possibility of sharing the energy and momentum of the photohole between the independent spinon and holon. Calculations within the Hubbard [62] and t-J[63, 64] models predict a continuous intensity distribution between the low-energy spinon band - of bandwidt $\sim J$ - and the higher energy holon band - of bandwidth $\sim t$ -, as well as shadows of the holon band crossing E_F at $k_F \pm 2k_F$ (Fig. 5.12). Similar features have been observed in 1D insulators like $SrCuO_2$ [64] Sr_2CuO_3 [65] and NaV_2O_5 [66]. Those materials are certainly not Luttinger liquids, due to the charge gap, but do exhibit spin-charge separation. More recently, it has been suggested that the spectra of the normal state of TTF-TCNQ could be compatible with the scenario of Fig. 5.12 [67].

An interpretation of the photoemission data based on the LL hypothesis must explain the puzzling similarity of the spectra of materials

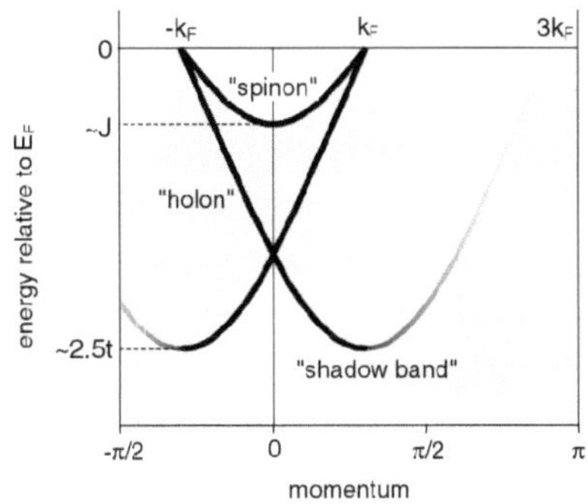

Figure 5.12. Calculated spectral weight distribution for the 1D $t-J$ model. From Ref. 68.

with rather disparate physical properties. This concern is eased by the realization that the spectral functions of 1D systems with charge gaps (Mott insulators) or spin gaps (Peierls systems) present similarities with that of the LL [68]. Nevertheless, the need to assume similarly strong and long-range interactions to describe the spectra of all 1D materials studied so far, is somewhat disturbing.

Finally, the possibility of observing the spectral signatures of the LL in real materials, as opposed to ideal 1D systems, is still an open theoretical question. Strictly speaking, any transverse coupling between 1D systems would immediately destroy the LL and yield a normal Fermi liquid [69]. The LL signatures however could reappear at sufficiently large energies and high temperatures. For instance, when $k_BT > t_\perp$ the warping of the Fermi surface due to a transverse coupling $t_\perp$ may be considered irrelevant, and the systems would be effectively one-dimensional. Of course things are not so simple, and a full analysis of the coupled LLs, is required to make firm predictions [70, 71, 72]. This much debated issue has also clear implications on theoretical models of the (quasi-2D) high-T_c cuprate superconductors, where Luttinger-like properties have been invoked to explain the unusual properties of the normal state [73].

5. Observing the spectral consequences of the electronic instabilities

The unique nesting properties of the Fermi surface makes the metallic state particularly vulnerable in 1D. Superconductivity, CDW, SDW, spin-Peierls, Mott insulator instabilities compete in the 'normal' state and the most rapidly divergent response function drives the system into the corresponding broken-symmetry state [74]. In a mean-field (MF) approach, the metal-non metal transition occurs at a finite temperature T_{MF}, and is accompanied by the removal of the Fermi surface and by the opening of an energy gap in the density of states (Fig. 5.13). At this level, the treatment of the CDW, SDW and SC instabilities is analogous [2]. The DOS of all the ordered phases is described by the BCS function, with a peak at the temperature-dependent gap energy $\Delta(T/T_{MF})$ which saturates to the (weak coupling) low-temperature limit $(\Delta_0/k_BT_{MF}) = 3.5$. Characteristic changes are also expected in the momentum-resolved spectra, similar to those observed in the high-T_C cuprates [9]. The energies of the QP peaks within an energy Δ of E_F is affected, and shadow bands appear, corresponding to the new periodicity.

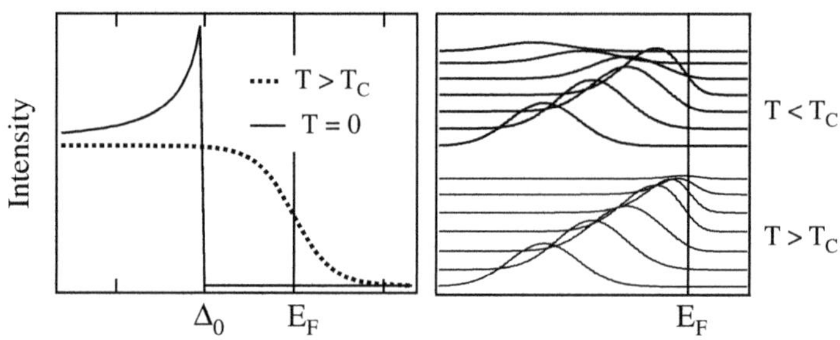

Figure 5.13. Schematic picture of the mean-field metal-non metal transition. The opening of an energy gap below T_C is visible both in the PES (*left*) and in the ARPES (*right*) spectra

The mean-field analysis ignores the disruptive effect of fluctuations which oppose long-range order in one dimension. Phase transitions are observed in real materials, but only as a consequence of transverse coupling. They occur at a temperature $T_C < T_{MF}$ and are three-dimensional in character. Between T_C and T_{MF} most physical properties are affected by fluctuations in space and time of the order parameter. The simple picture of Fig. 5.13 is modified by the appearence of a pseudogap above the real transition temperature [75, 76, 77] Nevertheless, the T=0 and high temperature ($T > T_{MF}$) limits do not change. In particular, all spectral changes induced by the transition should be confined within an energy Δ_0 of the Fermi level.

From the results presented above, it should be clear that the mean-field scenario is not a good starting point to interpret the ARPES data of 1D materials, which do not exhibit the expected signatures of the normal metallic state (Fermi edge, Fermi level crossing). Characteristic spectral changes associated with the phase transitions have been observed,

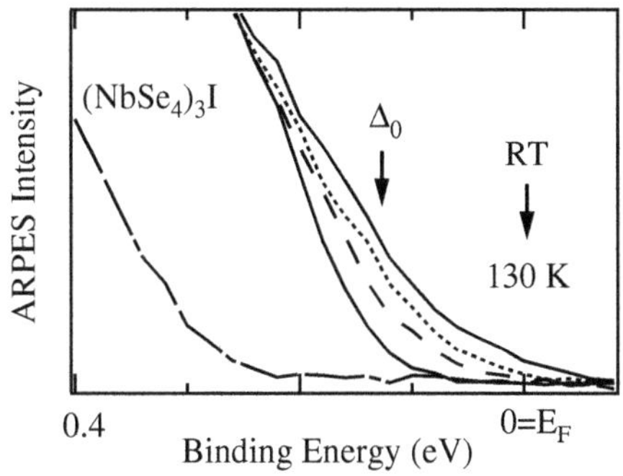

Figure 5.14. Spectra of $(TaSe_4)_2I$ measured at k_F showing a temperature-dependent shift of the leading edge below the Peierls transition, and of the temperature-independent insulating reference $(NbSe_4)_3I$

.

namely in materials with a CDW ground state. These changes occur over the characteristic energy scale of the transition (the gap) *above* the unconventional lineshape. A strong correlation has been observed between the temperature-dependent gap and the position of the spectral leading edge. Figure 5.14 shows the high-energy end of the ARPES spectrum of $(TaSe_4)_2I$[60] measured at k=k_F between RT and 100 K, in the CDW phase. The peak is located at 0.4 eV, rather than at the estimated gap edge (50-75 meV). The extrapolation to the baseline of the almost linear onset defines an energy E* which varies between E*~ 0 above the Peierls temperature T_P=273 K, and E*~ 0.1 eV below 150 K. The insulating reference $(NbSe_4)_3I$ exhibits a similar lineshape [19], but the extrapolated leading edge yields a temperature independent E* ~ 0.3 eV, again close to (half) the gap energy.

A more objective [60] analysis is possible, based on the temperature changes of the ARPES intensity distribution (Fig. 5.15).The momentum distribution curve (MDC) is obtained by 'cutting' the intensity map at a constant energy. A cut performed at T=300 K at E_F yields a lorentzian MDC centered at k_F. This is the lineshape expected for a QP at the FS. At lower temperatures the leading edge of the spectrum recedes from E_F, as shown in Fig. 5.14, but the intensity and linewidth of the MDC can be recovered at a higher binding energy $\Delta^*(T)$. This energy shift follows the expected temperature dependence of the CDW gap below the 2nd order Peierls transition, and saturates to the gap energy Δ_0 ~ 120 meV. Therefore it behaves as a phenomenological order parameter for the transition.

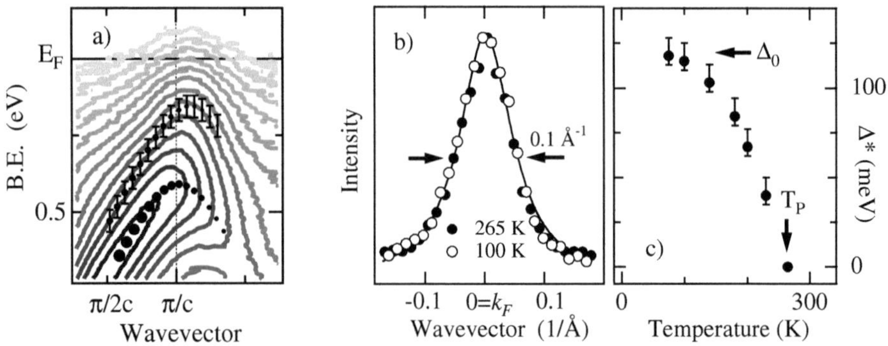

Figure 5.15. (a) Intensity map of $(TaSe_4)_2I$ near k_F at T=300 K, showing constant-intensity lines; (b) momentum distribution curves (MDCs) measured at E=0 (265 K) and E=110 meV (100 K); (c) temperature-dependent shift extracted from the MDCs below T_P. From Ref. 62.

The blue bronze $K_{0.3}MoO_3$ exhibits an analogous temperature-dependent shift of the spectral leading edge below the Peierls transition (T_P=180 K) [36, 16, 78, 61]. The ARPES data for this material also show the influence of fluctuations above T_P. The intensity measured at the Fermi surface decreases continuously from the metallic phase through the transition and into the CDW state. The temperature dependence of the ARPES signal parallels the evolution of the spin susceptibility, which reflects the progressive opening of a fluctuation-induced pseudogap in the metallic phase [61, 79]

Examples of spectral changes induced by transitions to ordered ground states have been observed in several 1D systems, including Peierls compounds like $NbSe_3$ [80], the Mott-Hubbard system $BaVS_3$ [81], and the (possibly SDW) compound $Li_{0.9}Mo_6O_{17}$ [37]. Organic materials like the $(DCNQI) - Cu$ salts [82, 83], $TTF - TCNQ$ [14, 67] and the Bechgaard salt $(TMTSF)ClO_4$ [84] also exhibit characteristic temperature-dependent spectral changes. More recently, spectral signatures of a Peierls transition, were also observed in an artificial 1D system consisting of self-assembled In chains at the Si(111) surface [85, 86]. The ARPES spectra show typical 1D dispersions and lineshapes, and a temperature-dependent shift of the intensity at the Fermi surface analogous to that observed in single crystal samples.

These clear spectral signatures of metal-non metal transitions in 1D systems confirm that the electronic structure is modified in the broken-symmetry states. They also prove that photoemission is sensitive to these changes, and is therefore potentially a powerful probe of the interactions that lead to the instabilities, and of the nature of the ground state. However, one should not forget that the spectra cannot be described by a simple mean-field approach, neither in the metallic nor in the insulating phases, and that the agreement between theory and experiment is not substantially improved even by the inclusion of fluctuations. The following section describes a possible alternative scenario which overcomes these difficulties.

6. A polaronic scenario

The results presented in the previous sections have shown that in 1D systems there is no direct relation between the QP energy and the position of the ARPES peak. Temperature-dependent data like those of Fig. 5.15 show that the extrapolated leading edge of the spectrum at $k{=}k_F$ coincides with E_F in the metallic phase, and with the edge of the energy gap below T_P. It is therefore tempting to identify, for all wavevectors, the QP energy with the ARPES leading edge. Clearly the properties of such QPs must be peculiar. The unusual spectral lineshape, the vanishingly small coherent intensity, and the broad lorentzian MDCs, all point to strong interactions and rapid QP decay through scattering. In particular, the QP coherence length is quite short ($l = (\Delta k)^{-1} \sim 0.1$ Å) for both $(TaSe_4)_2I$ and $K_{0.3}MoO_3$. As discussed in Section 2, interations progressively move spectral weight from the coherent QP peak, to the incoherent part of the spectrum, at energies which can be much larger than the QP energy.

The properties of Peierls systems are shaped by the interaction between electrons and the lattice, which leads to the ordered CDW ground state. The problem of determining the spectral properties of a dense electron system under the effect of strong electron-phonon scattering is notoriously difficult [87]. Several aspects and limiting cases of this problem have been discussed in the literature [88, 89, 90, 91] but the theoretical results cannot be directly compared with an ARPES experiment. Physical insight is gained by considering the much simpler and exactly soluble case of a single electron coupled to a harmonic oscillator of frequency Ω [34]. The spectrum (Fig. 16) exhibits a progression of peaks equally spaced by the energy $\hbar\Omega$. The highest energy (the 'zero phonon' or adiabatic peak) corresponds to a transition between the lowest energy configurations of the initial and final states. The remaining peaks are energy losses, i.e. transitions where the oscillator is left in an excited state. The whole spectrum has a poissonian envelope, with a maximum at E $\sim \langle n \rangle \hbar\Omega$, where $\langle n \rangle$ is the average number of vibrations present in the ground state. This lineshape describes quite well the spectra of diatomic molecules [92].

In a solid, the 'zero-phonon' peak becomes the coherent QP peak, and the coupling to the continuous phonon spectrum forbids the observation of the discrete satellite progression. As the strength of the coupling to the oscillator increases, more and more spectral weight is moved from the coherent to the incoherent part of the spectrum and the poissonian envelope evolves into a gaussian lineshape. The situation illustrated in the figure corresponds to this strong coupling limit: the very weak QP peak is hidden at the leading edge of the gaussian spectrum. In this limit the QP is quite different from a bare electron. It can be described as a small polaron, i.e. an electron heavily dressed by phonons, which moves coherently with the local lattice deformation.

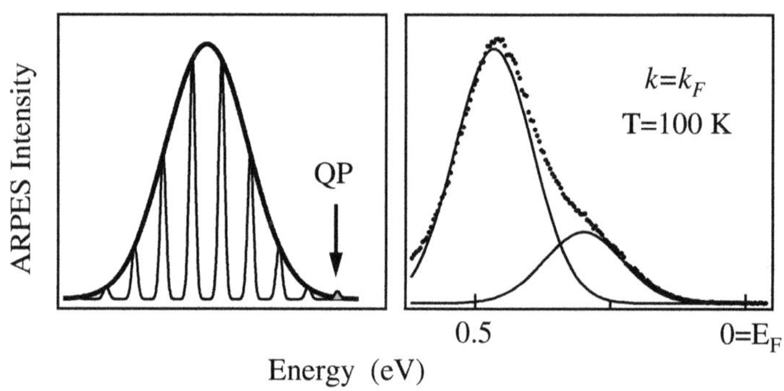

Figure 5.16. (*Left*) ARPES spectral function of an electron coupled to a harmonic oscillator in the strong coupling limit. The envelope is gaussian. (*Right*) The spectrum of $(TaSe_4)_2I$ is well described by the sum of two 'polaronic' gaussian lineshapes. From Ref. 62.

.

Figure 5.16 shows a fit to the spectrum of $(TaSe_4)_2I$ which uses two gaussian 'polaronic' lineashapes [60], representing the bonding and antibonding band states predicted by band structure calculations [21] . In the polaronic scenario, the peak of each gaussian component simply follows the QP dispersion, and is therefore only indirectly related to the QP energy. The peak-onset energy separation contains information on the structure of the phonon cloud via the relation $\Delta E = \langle n \rangle \hbar\Omega$. For both $(TaSe_4)_2I$ and $K_{0.3}MoO_3$ this analysis yields $\langle n \rangle \sim 10$, with a rather large uncertainty reflecting the poor knowledge of which phonon modes are the most relevant. The electrons are therefore heavily dressed, and their masses are strongly renormalized. This is consistent with an analysis of the optical conductivity which yields effective masses 30-40 times larger than the band values [21]. The vanishing coherent ARPES intensity is the most direct manifestation of the strong renormalization.

Such heavy carriers may seem to be incompatible with the broad bands ($\sim$ 1 eV, see Fig. 5.5) seen in the ARPES experiments, but the contradiction is only apparent. Again, to interpret the ARPES spectra it is necessary to distinguish between eigenvalues and spectral weight, as discussed in section 2. The experimental dispersion represents the distribution of spectral weight over a dense family of extremely narrow polaronic bands. The same physical situation can be described from a different but equivalent point of view which emphasizes the photoemission final state [93]. The istantaneous creation of a photoelectron introduces a hole in a frozen lattice. This hole could disperse in the unrelaxed lattice with a bandwidth corresponding to that of a band structure calculation, but it is rapidly slowed down as the lattice relaxes around it and the polaron is formed. Experiments which probe the response of the hole in the time domain measure a QP group velocity two orders of magnitude smaller than the velocity derived from the ARPES bands [94], supporting the polaronic picture. A similar polaronic scenario may describe the properties of the colossal magnetoresistance manganites [95], but the effect of strong *e-ph* interactions seems to be especially dramatic and pervasive in 1D.

The polaronic scenario has an interesting corollary with possible consequences on the search for non-conventional correlated states in 1D. The typical spectral signatures of spin-charge separation (Fig. 5.10) are confined to energies close to the ARPES leading edge, and it can be assumed that the combined spectral weights of the spinon and holon branches would be commensurate with the (weak) coherent QP weight of the polaron model. In the strong coupling limit, the spectrum essentially reflects the *e-ph* interaction, and the Luttinger features would be exceedingly weak and hard to observe. Spectroscopic investigations of spin-charge separation should therefore primarily target materials with dominant electronic correlations and weak *e-ph* coupling, with SDW or Mott-insulator ground states, rather than on Peierls systems.

7. Conclusions

One decade of intense spectroscopic investigations of 1D systems has produced many exciting results, and raised several still open questions. The surprising results of the first pioneering experiments have been confirmed with greater accuracy, and considerably extended. Some of these developments have been reviewed in this Chapter.

We have verified the existence of bands with strong 1D character and directly observed the predicted open Fermi surfaces. The data on one hand illustrate the effects of translational invariance on the electronic states, postulated in elementary textbooks. At the same time, they reveal the fundamental difference between the dispersion of a band structure calculation and the real observable, the momentum distribution of spectral weight, which critically depends on the strength of the periodic potentials.

High energy and momentum resolution data, and technical advances like the capability of collecting full energy-wavevector intensity maps, have allowed us to observe the spectral signatures of the transitions from the normal metallic states to the low-temperature broken symmetry phases. In particular, it has been possible to correlate the temperature-dependent removal of spectral weight in the vicinity of the Fermi surface with the onset of long-range order.

The most peculiar and stimulating aspect of these results remains the observation, consistently verified on a variety of 1D materials, of unusual spectral lineshapes. The deep pseudogaps at the Fermi level and the absence of Fermi surface crossings, are clearly incompatible with the properties of conventional metals. These lineshapes suggest that strong and/or peculiar interactions are at work in these materials, although the spectral evidence for the Luttinger liquid predicted by theory, and discussed in other Chapters of this book, remains elusive. Signatures of spin-charge separation have been observed in insulating materials. The spectra of specific 1D metallic compounds exhibit hints of the Luttinger phenomenology, but the interpretation of the parameters extracted from the spectra remains problematic. There is a growing consensus that the interpretation of the complex experimental data will require more realistic models, which should take into account important elements like the real band structure, the influence of the underlying instabilities, and also extrinsic factors like surface-specific properties (relaxation, reconstruction, non-stoichiometry) and defects. The strong spectroscopic evidence for strongly renormalized polaronic carriers demonstrates that an appropriate treatment of the electron-phonon interaction is mandatory, at least in Peierls systems.

From the experimental point of view, the pace is limited by the available single crystals of sufficient quality and size. In this respect the recently demonstrated fabrication of one-dimensional ordered structures could considerably broaden the scope of this research. Several laboratories are currently exploring surface and nanoscience technologies which could be exploited to build novel artificial 1D systems. One can only dream of the new opportunities offered by macroscopic arrays of identical nanotubes, or of self-assembled metal wires where the nature and strength of the intra-chain and of the substrate-mediated transverse interactions could be under the control of the investigator. These dreams define some of the exciting scientific goals and of the major technical challenges for the next decade.

Acknowledgments

The ARPES experiments performed in Lausanne were part of the PhD work of L. Perfetti and F. Zwick. It is a pleasure to acknowledge the fruitful and stimulating collaborations with L. Degiorgi (ETHZ), G. Grüner (UCLA), D. Jérome (Orsay), H. Höchst (SRC - Wisconsin), F. Del Dongo (Parma), and the invaluable theoretical support by J. Voit (Bayreuth).

References

[1] C. Kittel, *Quantum Theory of Solids* (J. Wiley and Sons, New York, 1963).

[2] G. Grüner, in "Density Waves in Solids", (Addison-Wesley, Reading, MA, 1994).

[3] F. D. M. Haldane, J. Phys. C **14**, 2585 (1981).

[4] J. Voit, Rep. Prog. Phys. **58**, 977 (1995).

[5] K. Schönhammer, this book.

[6] S. Hüfner,*Photoelectron Spectroscopy,* Springer Verlag, Berlin, (1995).

[7] P. Monceau (ed.), *Electronic Properties of Inorganic Quasi-One-Dimensional Compounds* , (Reidel, Dordrecht, 1985).

[8] D. Jérome and H.J. Schulz, Adv. Phys. **31**, 299 (1982).

[9] J.C. Campuzano, this book.

[10] M. Grioni and J. Voit, in *Electron Spectroscopies Applied to Low-dimensional Materials* , H. Stanberg and H. Hughes editors, (Kluwer, Dordrecht, 2000).

[11] S.D. Kevan ed., *Angle-resolved Photoemission - Theory and current application* (Elsevier, Amsterdam 1992).

[12] C.-O. Almbladth and L. Hedin, in *Handbook on Synchrotron Radiation,* Vol. I, Koch E.-E. editor, (North-Holland, 1983).

[13] A. Damascelli, Z-X. Shen, and Z. Hussain, Rev. Mod. Phys. **xx**, xx (2002).

[14] F. Zwick *et al.*, Phys. Rev. Lett. **81**, 1058 (1998).

[15] K.E. Smith, K. Breuer, M. Greenblatt, and W. McCarrol, Phys. Rev. Lett. **70**, 3772 (1993).

[16] G.-H. Gweon *et al.*, J. Phys. : Condens. Matter **8**, 9923 (1996); G.-H. Gweon *et al.*, J. Electron Spectrosc. Relat. Phenom. **117-118**, 481 (2001).

[17] S. Hüfner *et al.*, J. Electron Spectrosc. Relat. Phenom. **100**, 191 (1999).

[18] P. Gressier *et al.*, Inorg. Chem. **23**, 1221 (1984).

[19] V. Vescoli *et al.*, Phys. Rev. Lett. **84**, 1272 (2000).

[20] H. Hiramoto and M. Kohmoto, Int. J. Mod. Phys. B **6**, 281 (1992).

[21] M.-H. Whangbo and L.F. Schneemeyer, Inorg. Chem. **25**, 2424 (1986); J.-L. Mozos, P. Ordejón and E. Canadell, Phys. Rev. B **65**, 233105 (2000).

[22] J. Voit *et al.*, Science **290**, 501 (2000).

[23] L. D. Landau, Sov. Phys. JETP **3**, 920 (1957); **5**, 101 (1957); **8**, 70 (1959). P. Nozières, *Interacting Fermi systems,* W. A. Benjamin, Inc., New York, (1964).

[24] L. Hedin and S. Lundqvist, in Solid State Physics, Vol. 23, H. Ehrenreich, D. Turnbull, and Seitz F. eds (Academic, New York 1969) p. 1.

[25] R. Claessen *et al.*,Phys. Rev. Lett. **69**, 808 (1992); Phys. Rev. B **54**, 2453 (1996).

[26] B.A. McDougall, T. Balasubramanian, and E. Jensen, Phys. Rev. B **51**, 13891 (1995).

[27] M. Hengsberger, D. Purdie, P. Segovia, M. Garnier, and Y. Baer, Phys. Rev. Lett. 83**83**, 592 (1999); M. Hengsberger, R. Frésard, D. Purdie, P. Segovia, and Y. Baer, Phys. Rev. B. **60**, 10796 (1999).

[28] Ph. Hofmann, Y.Q. Cai, Ch. Grütter, and J.H. Bilgram, Phys. Rev. Lett. **81**, 1670 (1998).

[29] T. Valla, A.V. Fedorov, P. D. Johnson, S. L. Hulber, Phys. Rev. Lett. **83**, 2085 (1999).

[30] E. Rotenberg, J. Schaefer, and S.D. Kevan, Phys. Rev. Lett. **84**, 2925 (2000).

[31] G. Nicolay, F. Reinert, S. Schmidt, D. Ehm, P. Steiner, and S. Hüfner, Phys. Rev. B **62**, 1631 (2000).

[32] J.J. Paggel, T. Miller, and T.-C. Chiang, Phys. Rev. Lett. **83**, 1415 (1999)

[33] L. Perfetti *et al.*, Phys. Rev. B **64**, 115102 (2001).

[34] G.D. Mahan, *Many-Particle Physics* 3rd ed. (Plenum, New York 1986).

[35] W.D. Grobman *et al.*, Phys. Rev. Lett. **32**, 534 (1974); P. Nielsen, A.J. Epstein, and D.J. Sandman, Solid St. Commun. **15**, 53 (1974); E.E. Koch and W.D. Grobman, Solid St. Commun. **23**, 49 (1977).

[36] B. Dardel *et al.*, Phys. Rev. Lett. **67**, 3144 (1991).

[37] J.X. Xue, Phys. Rev. Lett. **83**, 1235 (1999).

[38] G.-H. Gweon, Phys. Rev. Lett. **85**, 3985 (2000).

[39] M. Grioni *et al.*, Physica Scripta T **66**, 172 (1996).

[40] The origin of this discrepancy, possibly related to different surface conditions, is not clear at present.

[41] D. C. Mattis and E. H. Lieb, J. Math. Phys. **6**, 304 (1963)

[42] J. Luttinger, J. Math. Phys. **4**, 1154 (1963).

[43] S. Tomonaga, Prog. Theor. Phys. **5**, 554 (1950).

[44] J. Voit, Eur. Phys. J. B **5**, 505 (1998) and Proceedings of the NATO Advanced Research Workshop on The Physics and Mathematical Physics of the Hubbard Model, San Sebastian, October 3-8, 1993, Edited by D. Baeriswyl, D. K. Campbell, J. M. P. Carmelo, F. Guinea, and E. Louis; Plenum Press, New York (1995); J. Voit, Phys. Rev. B **47**, 6740 (1993), J. Phys.: Condens. Matter **5**, 8305 (1993).

[45] V. Meden and K. Schönhammer, Phys. Rev. B **46**, 15753 (1992); K. Schönhammer and V. Meden, Phys. Rev. B **47**, 16205 (1993) and (E) **48**, 11521 (1993).

[46] B. Dardel *et al.*, Europhys. Lett. **24**, 687 (1993).

[47] H. J. Schulz, Phys. Rev. Lett. **64**, 2831 (1990).

[48] E. Papa and A. Tsvelik, cond-mat/0004007.

[49] F. Mila and X. Zotos, Europhys. Lett. **24**, 133 (1993).

[50] V. Vescoli *et al.*, Science **281**, 1181 (1998).

[51] L. Degiorgi,this book.

[52] J. Moser *et al.*, Eur. Phys. J. B **1**, 39 (1998).

[53] P. Wzietek *et al.*, J. Phys. I **3**, 171 (1993).

[54] J. Voit, Y. Wang, and M. Grioni, Phys. Rev. B **61**, 7930 (2000).

[55] S. Eggert, H. Johannesson, and A. Mattsson, Phys. Rev. Lett. **76**, 1505 (1996); A.E. Mattsson, S. Eggert, and H. Johannesson, Phys. Rev. B **56**, 15615 (1997).

[56] P. Segovia *et al.*, Nature (London) **402**, 504 (1999).

[57] R. Losio *et al.*, Phys. Rev. Lett. **86**, 4632 (2001).

[58] J.D. Denlinger *et al.*, Phys. Rev. Lett. **82**, 2540 (1999).

[59] F. Zwick *et al.*, Phys. Rev. Lett. **79**, 3982 (1997).
[60] L. Perfetti *et al.*, Phys. Rev. Lett. **87**, 216404 (2001).
[61] L. Perfetti *et al.*, Phys. Rev. B **66**, 075107 (2002).
[62] K. Penc, K. Hallberg, F. Mila, and H. Shiba, Phys. Rev. Lett. **77**, 1390 (1996) and Phys. Rev. B **55**, 15475 (1997).
[63] H. Suzuura and N. Nagaosa, Phys. Rev. B **56**, 3548 (1997).
[64] C. Kim *et al.*, Phys. Rev. Lett. **77**, 4054 (1996) and Phys. Rev. B **56**, 15589 (1997).
[65] H. Fujisawa *et al.*, Solid State Commun. **106**, 543 (1998).
[66] K. Kobayashi*et al.*, Phys. Rev. Lett. **80**, 3121 (1998).
[67] R. Claessen *et al.*, Phys. Rev. Lett. **88**, 096402 (2002).
[68] J. Voit, J. Phys.: Condens. Mat. **8**, L-779 (1996) and Eur. Phys. J. B **5**, 505 (1998).
[69] C. Castellani, C. di Castro, and W. Metzner, Phys. Rev. Lett. **72**, 316 (1994).
[70] H. Schulz, Int. J. Mod. Phys. B **5**, 57 (1991).
[71] C. Bourbonnais and L.G. Caron, Int. J. Mod. Phys. B **5**, 1033 (1991).
[72] D.G. Clarke, S.P. Strong and P.W. Anderson, Phys. Rev. Lett. **72**, 3218 (1994).
[73] P.W. Anderson, Phys. Rev. Lett. **64**, 1839 (1990).
[74] L. Solyom, Adv. Phys. **28**, 201 (1979).
[75] P. A. Lee, T. M. Rice, and P. W. Anderson, Phys. Rev. Lett. **31**, 462 (1973).
[76] R. H. McKenzie, Phys. Rev. B **52**, 16428 (1995).
[77] N. Shannon and R. Joynt, J. Phys.: Condens. Matter **8**, 10493 (1996)
[78] A.V. Fedorov *et al.*, J. Phys.: Condens. Matt. **12**, L191 (2000).
[79] D.C. Johnston, Phys. Rev. Lett. **52** 2049 (1984).
[80] J. Schaefer *et al.*, Phys. Rev. Lett. **87** 196403 (2001).
[81] M. Nakamura *et al.*, Phys. Rev. B **49**, 16191 (1994).
[82] I.H. Inoue et al., Phys. Rev. B **47**, 12917 (1993).
[83] T. Takahashi et al., Phys. Rev. B **53**, 1790 (1996).
[84] F. Zwick *et al.*, Solid State Commun. **113**, 179 (2000).
[85] H.W. Yeom, Phys. Rev. Lett. **82** 4898 (1999).
[86] O. Gallus *et al.*, Eur. Phys. J. B **20**, 313 (2001).

[87] J.T. Devreese, Encyclopedia of Applied Physics, Vol. 14, 383 (1996).

[88] E.V.L. de Mello and J. Ranninger, Phys. Rev. B **55**, 14872 (1997).

[89] J.M. Robin, Phys. Rev. B **56**, 13634 (1997).

[90] S. Ciuchi, F. de Pasquale, S. Fratini, and D. Feinberg, Phys. Rev. B **56**, 4494 (1997).

[91] A.S. Mishchenko, N.V. Prokof'ev, A. Sakamoto, and B.V. Svistunov, Phys. Rev. B **62**, 6317 (1997).

[92] L. Asbrink, Chem. Phys. Lett. **7**, 549 (1970).

[93] V. Perebeinos and P.B. Allen, Phys. Rev. Lett. **85**, 5178 (1973).

[94] J. Demsar, K. Biljakovic, D. Mihailovic, Phys. Rev. Lett. **83**, 800 (1999).

[95] D.S. Dessau *et al.*, Phys. Rev. Lett. **81**, 192 (1998).

Chapter 6

ELECTRODYNAMIC RESPONSE IN "ONE-DIMENSIONAL" CHAINS

L. Degiorgi

Laboratorium für Festkörperphysik, ETH Zürich, CH-8093 Zürich, Switzerland

degiorgi@solid.phys.ethz.ch

Abstract This Chapter reviews selected optical results on low dimensional systems. It mainly focuses the attention on the linear chain Bechgaard salts, as representative systems of "real" one-dimensional materials. These results are discussed in connection with findings on other prototype one-dimensional compounds, like the ladders, and novel one-dimensional quantum wire, like the carbon nanotubes. We address a variety of relevant problems and concepts associated with the physics of an interacting electron gas in low dimensions, as non-Fermi liquid behaviour, dimensionality crossover, and specifically the appearance of the Luttinger liquid state.

Keywords: optical properties, quasi one-dimensional systems

1. Introduction

The Fermi liquid (FL) theory is extremely general and robust, and has been one of the cornerstones of the theory of interacting electrons in metals for the last half century. The theory is based on the recognition that the low-lying excitations of an interacting electron system are in a one-to-one correspondence with those of the dilute gas of quasiparticles, only with renormalized energies [1]. FL theory has been thoroughly tested on a variety of materials and is usually valid in higher than one dimension (1D). For instance some Kondo systems (like the Heavy Electron materials) seems to generally fit into the Fermi liquid scenario [2]. One possible notable exception is the normal state of the two-dimensional (2D) copper oxide-based high-temperature superconductors (HTSC) [3, 4]. Recently, a great deal of interest

D. Baeriswyl and L. Degiorgi (eds.), Strong Interactions in Low Dimensions, 165–193.

has been devoted to the possible breakdown of the FL framework in quasi-one dimensional materials.

In a strictly 1D interacting electron system, the FL state is replaced by a state where interactions play a crucial role, and which is generally referred to as a Luttinger liquid (LL). Various Chapters of this book deal with issues associated with the Physics in one dimension. Nevertheless, a short repetition of some essential features seems us in order. The 1D state predicted by the LL theory [5, 6, 7] is characterized by features such as spin-charge separation and the absence of a sharp edge in the momentum distribution function $n(k)$ at the Fermi wave vector k_F (i.e., of the fact that, in the Fermi liquid language, the renormalization factor $Z \to 0$ at k_F). The first immediate consequence of the absence of the discontinuity at k_F in the momentum distribution function is the powerlaw behaviour of the density of states (DOS) $\rho(\omega) \sim\mid \omega \mid^{\alpha}(\omega{=}E_F - E)$. The exponent α in this expression reflects the nature and strength of the interaction. The non-Fermi liquid nature of the LL is also manifested by the absence of single-electron-like quasiparticles and by the non-universal decay of the various correlation functions. Fortunately, well-defined techniques to treat such interactions exist in one dimension, and the physical properties of LL are theoretically well characterized and understood [8, 9, 10, 11, 12, 13].

Spectroscopic techniques are in general very useful in order to shed light on the microscopic electronic properties of a variety of materials, including obviously strongly correlated systems [2, 3, 4, 14]. Optical studies, in particular, gave a rather thorough insight into the ground-state properties of the quasi one-dimensional materials [15, 16]. Broken symmetry ground states, like charge (CDW) or spin (SDW) density wave as well as spin-Peierls, are characterized by clear optical signatures revealing the characteristic energy scales of the systems. Among the most important features we remind the collective excitations and the single particle gaps [15, 16]. Several review articles have been devoted to these issues, and in this respect Refs. [15, 16] should give a broad perspective. However, the direct experimental observation of the novel LL quantum state in real "quasi" one-dimensional materials is still very much debated. Because of the concepts mentioned above, we should limit here our discussion to the normal-state properties only, as emerged from the dynamics of the charge transport.

This Chapter will be organized as follows: first we will address more specifically the motivation for applying optical techniques in the context of the physics in low dimensions. Second, we will present selected results on the Bechgaard salts. The discussion will be devoted to the comparison between the experimental data and the theoretical scenarios, exploring alternative points of view for the physics of low dimensional systems. The conclusion will summarize the state-of-the-art as far as the optical response in low dimensions is concerned, flashing furthermore on the remaining open problems and future perspectives.

It is hopeless to account in full length for the huge amount of work, both experimental and theoretical, recently performed in this vast field of research. Hard choices will be done, neglecting unfortunately several relevant contributions. This should not be considered as undermining those ideas and results but it is aimed to focus the attention on paramount examples of 1D materials and on a few selected issues as starting point for a debate, which is also partially covered by other Chapters of this book.

1.1 Optical response in quasi-one dimensional systems

On a general ground the LL, which describes gapless 1D-fermion systems, may be unstable towards the formation of a spin or a charge gap [12]. Spin gaps are obtained in microscopic 1D models including electron-phonon coupling, and are relevant to the description of the normal state of superconductors and Peierls (CDW) insulators [12]. The second instability, which concerns the one-dimensional chain-like systems discussed here, is a more typical consequence of electronic correlations and leads to charge gaps. In passing, we shall mention that, strictly speaking, 1D systems with one gapped channel, either charge or spin, do not belong to the universality class of the Luttinger model, but rather to that of the related Luther-Emery (LE) model [12]. Nevertheless, the two models exhibit several common, typically 1D features, like spin-charge separation [11]. At half filling, and more generally at commensurate values of band filling $n = p/q$ (with p and q integers), the electron-electron interaction together with Umklapp scattering (which arises when the lattice periodicity is also involved [17]) drive the system to a Mott insulating state with a charge correlation gap E_{gap}.

Of course, real materials are only quasi-one-dimensional, and the interchain hopping integral (i.e., the charge transfer integral $t_\perp$) is finite in the two transverse directions. In a very crude way, $t_\perp$ drives the delocalization of the charge carriers between parallel chains. It has been theoretically conjectured that such a delocalization can be viewed as an effective doping into the upper Hubbard band [18, 19], leading to deviations from the commensurate filling (which is insulating).

Therefore, by increasing $t_\perp$ there is a dimensionality crossover, which is accompanied by the evolution from a one-dimensional Mott insulator to a "doped" semiconductor. When $t_\perp$ is large enough so that $t_\perp > E_{gap}$, then it is energetically favorable for the charge carrier to hop between parallel chains [17, 20].

The measurement of the absorption spectrum is a suitable method in order to reveal the unusual nature of correlations in 1D. Particularly, the optical properties collected at different polarizations of light allow disentangling the physics for anisotropic materials. There is an ample theoretical literature, which is hopeless to review here in great detail. We refer to another contribution in this book, which is a good starting point for a broad theoretical perspective [21]. As introduction to the theoretical expectation for the electrodynamic response in one dimension, we briefly sketch the optical response as suggested by Ref. [17], which catches the essential features. Later in the discussion we will quote a few more interesting approaches, which are not necessarily sharing common points of view.

For a strict one-dimensional Mott insulator, the charge correlation gap E_{gap}, corresponding to the excitation between the lower and upper Hubbard bands, appears as a standard squared root singularity in the charge excitation spectrum (i.e., in the real part $\sigma_1(\omega)$ of the optical conductivity (Fig. 6.1)). The scenario calculated for a doped one-dimensional Mott semiconductor consists of a Mott gap (as reminder of the original 1D limit with Umklapp scattering process) and a zero-energy mode (i.e., theoretically a $D\delta(\omega)$ function at $\omega = 0$, representing the Drude resonance of the effective metallic contribution with scattering rate $\Gamma = 0$) for small doping levels (Fig. 6.1). The spectral weight D encountered in the "Drude resonance" is proportional to the effective charge doping, induced by the interchain coupling $t_\perp$, in the upper Hubbard band. Of course for the low-energy mode, this is an oversimplified view, since the interchain hopping makes the system two dimensional, and the low-energy feature is unlikely to be described by a simple one-dimensional theory [17, 20]. Because of the zero energy mode and the finite dc conductivity, the transition across the charge correlation gap is now a pseudogap excitation. Therefore, due to the deviation from the strict one-dimensional limit, such an excitation is no longer a singularity at E_{gap} but rather a broad feature (Fig. 6.1). $t_\perp$ becomes ineffective at high enough temperatures or frequencies. Consequently for $T, \omega \gg t_\perp$ and $T, \omega \gg E_{gap}$ we expect the 1D physics (i.e., the Mott insulating state) to be dominant. At such large energy scales for both ω and T, the warping of the Fermi surface, induced by $t_\perp$, is neither relevant nor ob-

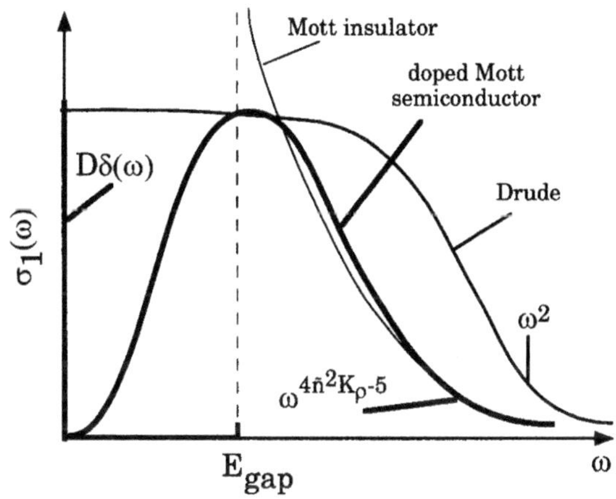

Figure 6.1. Optical conductivity of the Mott insulator and of the doped Mott semiconductor [17]. The simple Drude behaviour is shown for comparison.

servable. The theory predicts a powerlaw behaviour $\sigma_1(\omega){\sim}\omega^{-\gamma}$(Fig. 6.1) [12, 17, 20]. The exponent γ is characteristic of the correlations dominating in the system. Consequences and implications of such a powerlaw behaviour on the intrinsic physical properties of quasi one-dimensional systems will be treated and further developed in the discussion.

1.2 Low dimensional materials

Nature has recently provided us with remarkable materials, which have considerably boosted the experimental and theoretical understanding of electrons with a dimensionality less than three. Low dimensional organic conductors (like, e.g., the quasi one-dimensional Bechgaard salt chain or the so-called quasi two-dimensional $BEDT$ systems) can be described as having units of strongly linked molecules, which are chains and planes for one-dimensional (1D) and two-dimensional (2D) conductors, respectively, with weak interactions (i.e., $t_\perp$) between the units. They thus form a three-dimensional lattice solid, but the electrons are confined either along one direction or within a plane. Nevertheless, at low temperatures, when k_BT is smaller than $t_\perp$, one should rather consider the low-dimensional conductor as a three-dimensional anisotropic conductor [22, 23]. One can readily see that dimensionality crossovers should exist in real solids.

The Bechgaard salts are linear chain compounds based on the organic molecule tetramethyltetraselanofulvalene ($TMTSF$) and its sulfur analog tetramethyltetrathiafulvalene ($TMTTF$). Their composition is $(TMTSF)_2X$ or $(TMTTF)_2X$, where X is a counterion such as Br, ReO_4, ClO_4, PF_6 and AsF_6. There is a charge transfer of one electron to each of the counterions. The bandwidth is anisotropic and band structure calculations lead to transfer integrals $t_a \sim 250\ meV$, $t_b \sim 25\ meV$ and $t_c \sim 1\ meV$ along the three crystallographic directions [22, 24]. The anisotropy of the dc resistivities is in good agreement with these values [22, 25, 26, 27].

Figure 6.2 displays our own resistivity results $\rho(T)$ of the prototype Bechgaard salt compounds along the chain axis [25] whose optical properties will be primarily discussed here. There is a good agreement with the literature data [22, 26, 27]. The $(TMTSF)_2X$ salts have a clear metallic behaviour down to T_{SDW} of 12 and 6 K for $X = PF_6$ and ClO_4, respectively, where the SDW phase transition takes place. The $TMTTF$ salts, on the other hand, have a metal-insulator phase transition at $T_\rho \sim 100 - 200\ K$. Such an insulating phase was ascribed to charge localization [22, 28]. Below temperatures of about 10 K, there is furthermore a spin-Peierls and a SDW phase transition for $(TMTTF)_2PF_6$ and $(TMTTF)_2Br$, respectively [23, 29].

As stated in the introduction, the optical properties of the Bechgaard salts will be the main topic of the present review. Other low-dimensional systems attracted a lot of attention, as well. For instance, the discovery of high-temperature superconductivity raised renewed interest in low-dimensional quantum magnets [30]. In this respect, the antiferromagnetic $S = 1/2$ two-leg ladder systems have recently got much attention, because their properties are reminiscent of some encountered in the high-temperature superconducting cuprates [30]. Particularly, the layered cuprate $Sr_{14-x}Ca_xCu_{24}O_{41}$ has been studied vigorously. Dagotto *et al.* first proposed that even-leg ladders have a spin liquid ground state with a spin gap and that doping (e.g., by Ca-substitution) into the spin ladders results in pairing of the doped holes [31]. Through hole-doping, the ground state should be either a charge density wave or superconducting [30, 31].

Recently novel prototype 1D quantum wires, like carbon nanotubes and DNA, raised a lot of interest, as well. In the discussion we will

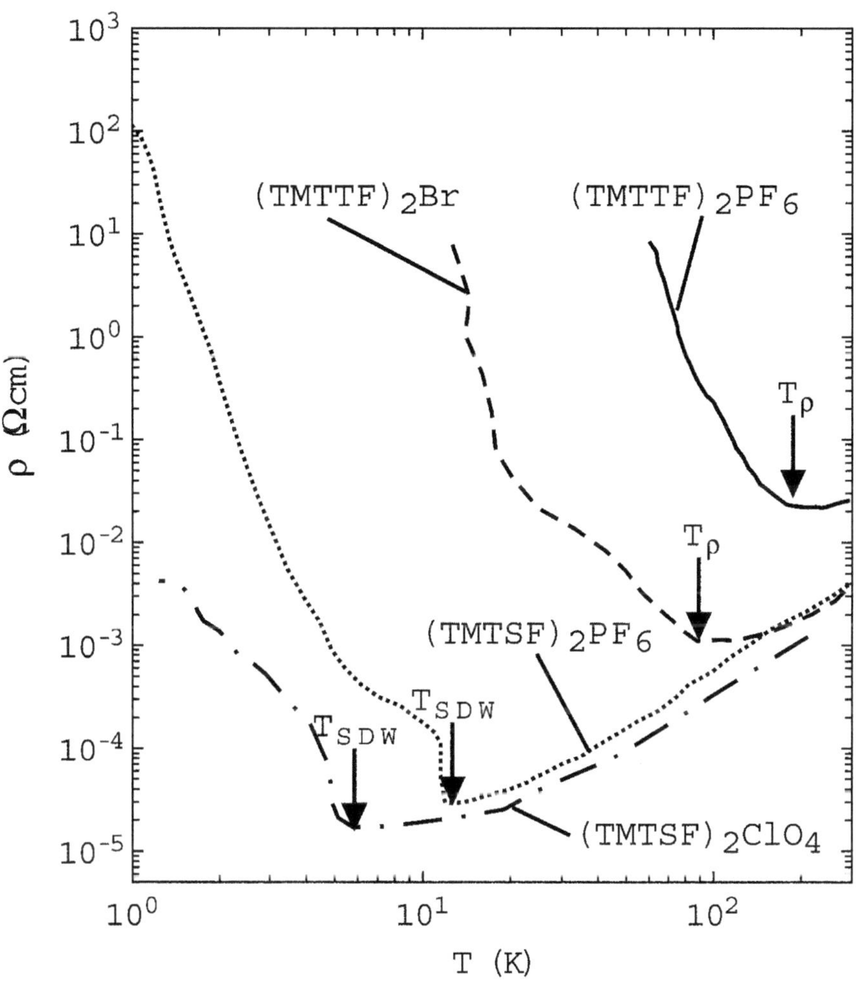

Figure 6.2. Temperature dependence of the dc resistivity of Bechgaard salts measured along the chain axis [25]. The data are taken on our samples and are in agreement with previously published results [22, 26, 27].

return to these latter materials for well-targeted and focused comparison with the linear chain organic Bechgaard salts.

2. Optical experiment and results

By combining the results from different spectrometers in the microwave, millimeter, submillimeter, infrared, visible and ultraviolet ranges, we have achieved the electrodynamic response over an extremely broad range ($10^{-5} - 10\ eV$) [25, 32, 33]. At all frequencies up to and including the midinfrared, we placed the samples in an optical cryostat and measured the reflectivity as a function of temperature between 5 and 300 K. From $2x10^{-3}$ to 5 eV, the polarized reflectance measurements were performed, employing four spectrometers with overlapping frequency ranges; while in the microwave and millimeter wave spectral range, the spectra were obtained by the use of a resonant cavity perturbation technique [32, 33, 34]. The full set of the optical functions, like the real part σ_1 of the optical conductivity, was achieved by performing the Kramers-Kronig transformation of the reflectivity. Appropriate extrapolations of the spectra beyond the measured spectral range were performed [35].

Figure 6.3 summarizes the temperature dependence of the optical reflectivity $R(\omega)$ along the different crystallographic directions, obtained on large single crystals of Bechgaard salts. As illustration, we present here data at selected temperatures for the $(TMTTF)_2Br$ and $(TMTSF)_2PF_6$ salts. For both $TMTTF$ and $TMTSF$ salts there is a clear anisotropy of $R(\omega)$ between different polarization directions, which is also well represented by the different energy position of the plasma edge-like feature in $R(\omega)$, the plasma edge being the more or less sharp rise of $R(\omega)$ so that $R(\omega)\rightarrow$ 100% for $\omega \rightarrow 0$. Generally speaking, for both $TMTTF$ and $TMTSF$ salts at 300 K there is a metallic behaviour with a sharp plasma edge at about 1 eV along the a-axis (Fig. 6.3a), while along the b-axis (Fig. 6.3b) $R(\omega)$ corresponds to an overdamped metallic trend with the onset of the broad plasma edge at 0.2 eV. The two families of Bechgaard salts differentiate more at low temperatures. On the one hand, the plasma edge of $(TMTSF)_2PF_6$ along the chain a-axis at $T \sim 20\ K$ is more steeper than at 300 K and $R(\omega)$ rises at low energies (Fig. 6.3a). Along the b-axis for the $TMTSF$-based salt, there is a clear crossover from an overdamped plasma edge at 300 K into a sharp one at 20 K (Fig. 6.3b). On the other hand, for the $TMTTF$ salts, $R(\omega)$ at 20 K is insulating like along the a-axis (i.e., $R(\omega) \simeq const.$ for $\omega \rightarrow 0$ (Fig. 6.3a)) and is temperature independent along the b-axis (i.e., the plasma edge keeps also at 20 K its overdamped-like shape (Fig. 6.3b), as already seen at 300 K). Finally, along the c-axis $R(\omega)$ of the $TMTSF$ compound (Fig. 6.3a) is insulating-like and temperature independent [25, 36].

Figure 6.4 displays the temperature dependence of the real part $\sigma_1(\omega)$ of the optical conductivity in different Bechgaard salts for $E\|a$, the chain direction. Part (a) and (b) show $\sigma_1(\omega)$ for the PF_6 and Br compounds of the $TMTTF$ family, while part (c) displays $\sigma_1(\omega)$ for $(TMTSF)_2PF_6$. We can immediately remark that the temperature

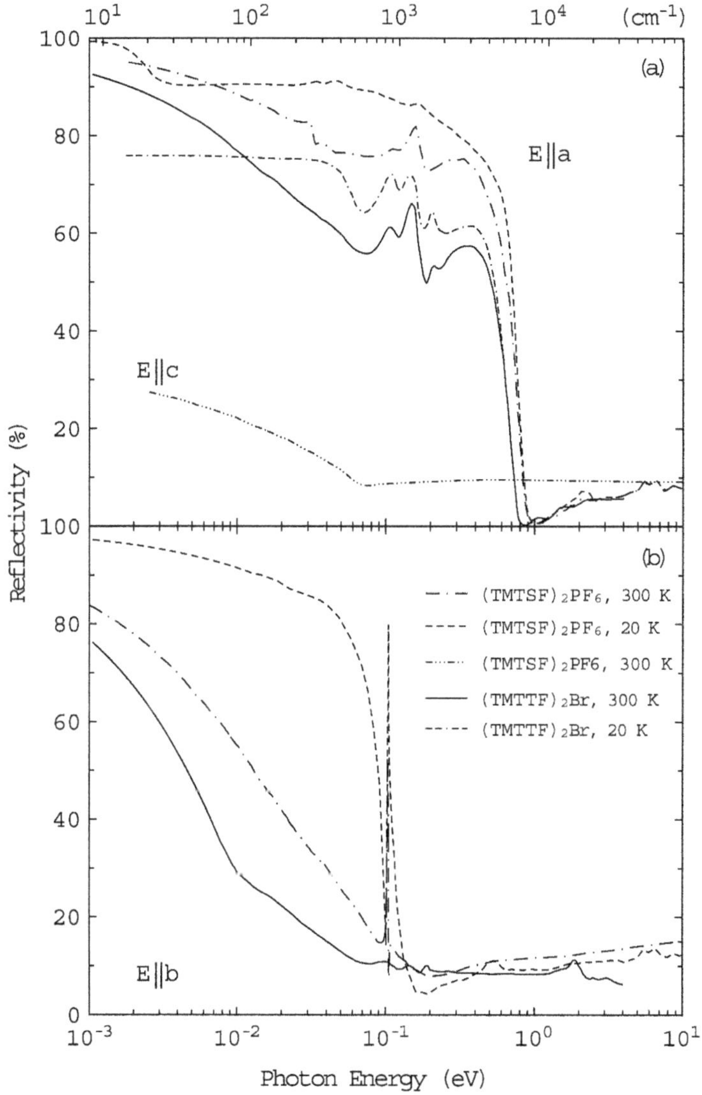

Figure 6.3. Optical reflectivity (a) along the a- and c-direction and (b) along the b-direction, for $(TMTTF)_2Br$ and $(TMTSF)_2PF_6$ at $T = 300\ K$ and $T = 20\ K$. Note the logarithmic energy scale [33, 34, 36].

dependence of $\sigma_1(\omega)$ is quite important in all compounds, except in $(TMTTF)_2PF_6$ (shown in Fig. 6.4 at $10\,K$ only), which is a very poor conductor at 300 K and is basically in an insulating state at any other temperature [36]. The data presented here are in broad agreement

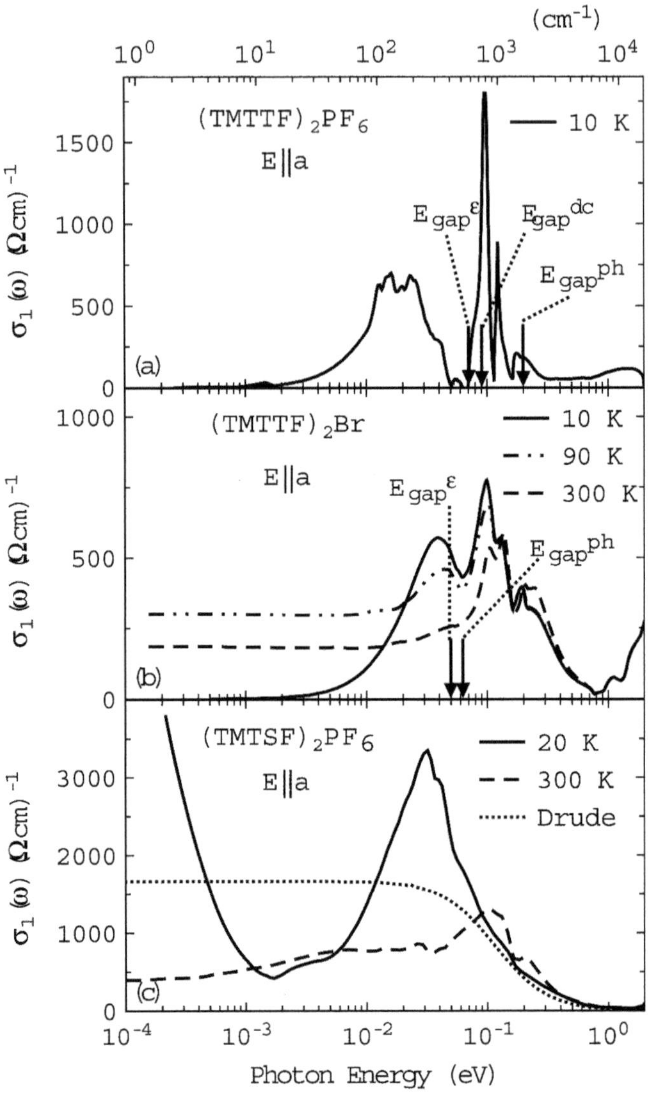

Figure 6.4. On chain optical conductivity of (a) $(TMTTF)_2PF_6$, (b) $(TMTTF)_2Br$, and (c) $(TMTSF)_2PF_6$ at temperatures above the transitions to the broken symmetry ground states. The arrows indicate the gaps observed by dielectric (ϵ) response, dc resistivity and photoemission (ph). A simple Drude component is also shown in part (c). Note that photoemission measures the quantity $E_{gap}/2$, assuming that the Fermi level is in the middle of the gap [14, 25].

with previous, less detailed studies [37, 38, 39, 40], and they include frequencies below the conventional optical spectra.

The optical conductivity of the $TMTTF$ salts displays several absorption features mostly ascribed to lattice vibrations (phonon modes), mainly due to the inter and intramolecular vibrations of the $TMTTF$ unit. These vibrations can be even enhanced by the electron-phonon coupling [40]. The phonon modes are particularly observed in the $TMTTF$ salts, because the screening by free electrons is less effective due to the insulating state at low temperatures. The identification of the charge gap (E_{gap}, Fig. 6.1), based on the optical spectra, is not straightforward, because of the large phonon activity. Thus, the results of other measurements were used. The gap values obtained for $(TMTTF)_2X$ with different techniques are displayed in Fig. 6.4. The dc conductivity as well as the low-temperature dielectric constant (ϵ) measured at 100 GHz are consistent with $E_{gap} = 87$ and 50 meV for $(TMTTF)_2PF_6$ and $(TMTTF)_2Br$, respectively [34, 41]. Therefore, the absorption in $\sigma_1(\omega)$ near 99 meV for $(TMTTF)_2PF_6$ is associated with E_{gap}; this is also supported by the fact that this feature has the correct spectral weight [34], defined as:

$$\int_0^{\omega_c} \sigma_1(\omega) d\omega \simeq n/m$$

with ω_c a cut-off frequency, n the charge carrier concentration and m the effective mass [35]. Similarly, for $(TMTTF)_2Br$ below 90 K (i.e., at $T < T_\rho$) there is the progressive disappearance of the spectral weight in FIR (i.e., $\omega_c < 40$ meV) with decreasing temperatures. The missing spectral weight mainly piles up at the gap feature around 50 meV. Moreover, the photoemission experiments suggest a gap of 0.2 eV for $(TMTTF)_2PF_6$ and 60 meV for $(TMTTF)_2Br$. Because the band is partially filled, the charge gap in both of the $(TMTTF)_2X$ salts is a correlation rather than a single-particle gap.

The optical properties of the $(TMTSF)_2X$ analogs (Fig. 6.4c), for which the dc conductivity gives evidence for metallic behaviour down to low temperatures (just above the SDW phase transition at T_{SDW}), are markedly different from those of a simple metal. A well-defined absorption feature around 25 meV (later ascribed to the correlation pseudogap) and a zero-frequency mode [32, 33, 34, 42] are observed at low temperatures. The latter mode at low temperatures is narrower in $(TMTSF)_2ClO_4$ (not shown here) than in $(TMTSF)_2PF_6$ [34]. The combined spectral weight of the two modes is in full agreement with the known carrier concentration of $1.4x10^{21}$ cm^{-3} and a band mass that is very close to the free electron mass [22]. For both $(TMTSF)_2X$ salts, the zero-frequency mode has small spectral weight on the order of 1%

of the total weight, obtained by integrating $\sigma_1(\omega)$ up to $\omega_c \sim 1\ eV$. Nevertheless, this mode is responsible for the large metallic conductivity. Figure 4c also indicates that the two components of $\sigma_1(\omega)$ clearly develop at low temperatures. In fact, there is a progressive narrowing of the effective metallic contribution to $\sigma_1(\omega)$ as well as a piling up of spectral weight at the pseudogap feature with decreasing temperatures (Fig. 6.4c). Furthermore, the dc-limit of $\sigma_1(\omega)$ is in fair agreement with σ_{dc} values from the transport data, as also confirmed by the so-called Hagen-Rubens extrapolation of our original absorption or reflectivity measurements (Fig. 6.3) [34, 35].

3. Discussion

The optical response of the organic Bechgaard salts (Fig.6.3 and 4) is mainly characterized by the gap-like feature (i.e., a finite energy mode) in all compounds and by the narrow zero frequency mode, which only appears in the $(TMTSF)$ salts. In both cases, due to full charge transfer from the organic molecule to the counter ions, the $TMTTF$ or $TMTSF$ stacks have a quarter-filled hole band. There is also a moderate dimerization, which is somewhat more significant for the $TMTTF$ family [22]. Therefore, depending on the importance of this dimerization, the band can be described as either half-filled (for a strong dimerization effect) or quarter-filled (for weak dimerization). Due to the commensurate filling, a strictly one-dimensional Luttinger liquid with Umklapp scattering effects transforms into a Mott insulating state (Fig. 6.1), which is dominated by the charge correlation gap excitation. Indeed, the $(TMTTF)_2X$ salts, with $X = PF_6$ or Br, are insulators at low temperatures [22] with a substantial (Mott) charge gap (Fig. 6.4a-b). For these compounds the correlation gap E_{gap} is so large that the interchain hopping ($t_\perp$) is not relevant. Charge carrier hopping on parallel chains is here strongly suppressed leading to a truly 1D insulating phase [17, 20]. In analogy to the $TMTTF$ salts, the strong FIR excitation of the $TMTSF$ salts (Fig. 6.4c) is ascribed, within the scenario depicted in Fig. 6.1, to the so-called pseudo charge correlation gap. The existence of a gap feature in the metallic state, containing nearly all of the spectral weight, is at first sight similar to what is expected for a band-crossing transition for simple semiconductors, which would result in a semimetallic state.

The nearly temperature-independent magnetic susceptibility, which gives strong evidence for a gapless spin excitation spectrum (this has often been interpreted as a Pauli susceptibility or as the susceptibility due to a large exchange interaction) [22], demonstrates that the state is not a simple semimetal. The existence of a gap, or pseudogap in the charge excitations (Fig. 6.4) with the absence of a gap for spin excitations, indicates spin-charge separation in the metallic state. This spin-charge separation is, however, distinct from that of a 1D LL, in which both excitations are gapless but have different dispersion velocities. Here, it is the Umklapp scattering, which leads to gapped charge excitations.

The narrow zero-energy mode seen in the spectra of the $TMTSF$ family at low frequencies (Fig. 6.4c) is the experimental optical fingerprint of the theoretically predicted Drude peak (Fig. 6.1) for the quasi two-dimensional Mott semiconductor [17, 20]. Such a peak contains only 1% of the carriers (i.e., 1% of the total spectral weight). Although no real doping exists from a chemistry point of view, the narrow Drude peak originates from deviations of commensurability due to the interchain coupling ($t_{\perp}$). If $t_{\perp}(\sim t_b)$ between chains is relevant (i.e., $t_{\perp} > E_{gap}$), small deviations from commensurate filling due to the warping of the Fermi surface exist, and should lead to effects equivalent to real doping on a single chain.

Hall effect measurements should give in principle an alternative experimental point of view on the issue of the small spectral weight encountered in the zero energy mode. There are two sets of data, which appeared simultaneously [43, 44]. However, the Hall constant was not measured in the same magnetic field configuration, making a direct comparison between the two experiments less straightforward. On the one hand, Moser *et al.* found characteristic power-law behaviours in agreement with transport and optical finding (see below), which were interpreted as a possible manifestation of LL state [44]. On the other hand, Mihaly *et al.* does not invoke the LL framework and suggest a rather conventional Fermi liquid scenario [43]. This is based on the finding that the Hall constant is not enhanced as it would be expected by the anomalously small spectral weight in the narrow zero energy mode of the optical conductivity (Fig. 6.4c). Their charge carrier concentration is consistent with the total spectral weight estimation up to 1 eV in the optical conductivity and agrees with the chemical counting. Nevertheless, they do not exclude a LL state at large temperature or energy scale (i.e., $T > 400\ K$), where indeed the 1D limit is fully recovered [43]. It remains to be seen how one can reconcile the Hall effect conclusion with the fact that from optics only a 1% fraction of the total spectral weight is effectively involved in the charge dynamics at low energies and the remaining amount of spectral weight is associated with finite energy excitations. This is a puzzle which awaits new theoretical thoughts about the Hall effect in one-dimension [13, 20].

The zero energy mode and the characteristic charge correlation (pseudo) gap in the optical conductivity raised quite a bit of interest among the theorists. For instance, the expression "infrared puzzle" was specifically coined in order to address the small amount of spectral weight in the zero-energy mode. This latter feature as well as the other experimental findings, presented here, led to a variety of approaches, mainly based on the Mott-Hubbard model [17, 20, 45, 46, 47, 48, 49, 50, 51]. The fundamental problem of the conductivity in one-dimensional systems and quantum wires has been also addressed from a general perspective in Refs. [52, 53, 54, 55]. Space limitation does not allow a thorough discussion of those theoretical ideas. We shall just mention that various experimental features can be reproduced reasonably well. For complementary theoretical discussions we refer to other contributions in the book [11, 13, 21].

We now look more carefully to the high frequency tail of the charge correlation (pseudo) gap (Fig.6.4). Figure 6.5 shows the frequency dependence of the optical conductivity of the finite energy mode (charge gap) in the $(TMTSF)_2X$ salts in a log-log representation where the optical conductivity and photon energy were normalized by the maximum value σ_{peak} and energy of the gap resonance ω_{peak}, respectively. We clearly observe a well distinguishable and characteristic power-law behaviour. There is a direct relation to the theoretical expectation. Indeed, at frequencies greater than $t_\perp$, the interchain electron transfer is irrelevant and calculations based on the 1D Hubbard model should be appropriate [17, 20]. The theoretical expectation (Fig. 6.1) consists in a powerlaw of the frequency-dependent optical conductivity $\sigma_1(\omega){\sim}\omega^{-\gamma}$ for frequencies greater than $t_\perp$ and E_{gap} but less than the on-chain bandwidth $4t_a$. The theory also predicts that the exponent $\gamma = 5 - 4\tilde{n}^2 K_\rho$, K_ρ being the so-called Luttinger liquid parameter and $\tilde{n}$ the degree of commensurability [12, 17, 20]. Our results for $X = PF_6$, AsF_6 and ClO_4 are consistent with an exponent γ=1.3 [25, 33].

The results on the powerlaw behaviour are very robust and allow us to discriminate among different regimes and type of correlations. Optical data collected on $(TMTSF)_2ReO_4$ are, in this respect, quite compelling [56]. A Peierls system with dominant CDW correlations, like $(TMTSF)_2ReO_4$, belongs to the universality class of the LE model [12]. The high frequency tail of the gap feature at 200 meV, displayed in Fig. 5 at 10 K, below the Peierls transition at 180 K, deserves a special attention. At photon energies much larger than the energy scale set by

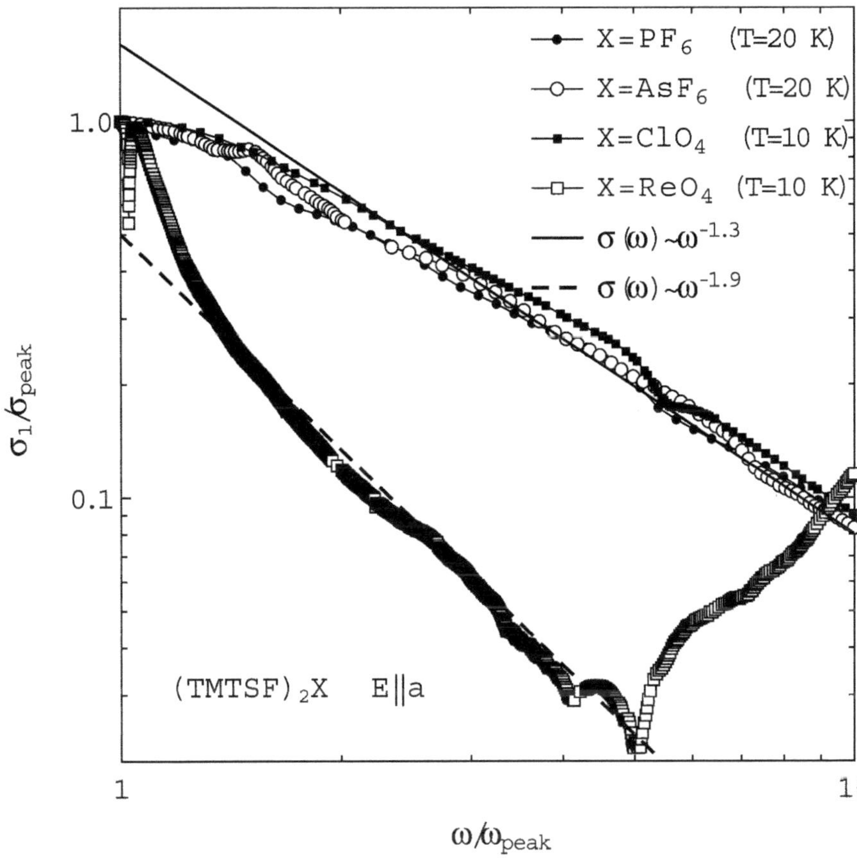

Figure 6.5. The frequency dependence of the optical conductivity in the spectral range of the finite energy mode (charge gap) in the $(TMTSF)_2X$ salts. The maximum value of $\sigma_1(\omega)$ and the frequency, where the maximum occurs, of the charge gap are represented by σ_{peak} and ω_{peak}, respectively. The solid line is the powerlaw $\sigma_1(\omega)\sim\omega^{-\gamma}$ with γ=1.3 (for $X = PF_6$, AsF_6 and ClO_4) and γ=1.9 (for $X = ReO_4$) [33, 34, 56].

the transverse charge transfer integral $t_\perp$ ($\sim$ 10 meV for the ReO_4 salt), $\gamma \sim 3$ for a rigid lattice with only Umklapp scattering off the single periodic potential [12, 57], as appropriate for a 1D band insulator. When the coupling to phonons is also included, as in CDW systems, the theory predicts $\gamma \sim 2$ [57]. The optical measurements on the ReO_4 salt up to 12 eV allow us to carefully search for such a power law behaviour, since the high frequency tail of the gap feature (between 200 meV and 1 eV) is not at all affected by the high frequency extrapolation necessary for the

Kramers-Kronig analysis. A fit of the optical conductivity at $\omega > E_{gap}$ (Fig. 6.5) yields the exponent $\gamma = 1.9$ at $10K$ ($\gamma = 1.7$ at 300 K, not shown here) [56]. This value is in good agreement with the predictions ($\gamma \sim 2$) of the LE scenario. Furthermore, $\gamma \sim 2$ is considerably different from the value $\gamma = 1.3$ found for other $(TMTSF)_2X$ (X=PF_6, AsF_6 and ClO_4) salts (Fig. 6.5). γ=1.3 corresponds in fact to the regime of the LL (or LE liquid), where strong and long-range electronic correlations dominate.

Due to the apparent contradiction of having a rather large (Mott) correlation gap ($\sim$ 12 meV) and a good metallic dc conductivity in the $TMTSF$ family (Fig. 4c), it was proposed that the finite energy mode in $\sigma_1(\omega)$ is due to the dimerization gap Δ_{dim} [40, 58]. This would be the case for an extremely strong (nearly infinite) repulsion, with the quarter-filled band being transformed into a half-filled band of (nearly noninteracting) spinless fermions. It was then argued that the real charge gap E_{gap} should be smaller, of the order of 50 K [58]. Attributing the finite energy mode in $\sigma_1(\omega)$ to the dimerization gap, it would imply that $\sigma_1(\omega) \sim \omega^{-3}$ for $\omega \gg \Delta_{dim}$, as in a simple semiconductor. The observed power law with $\gamma \sim 1.3$ (Fig. 6.5) differs significantly from this prediction [58], making such an interpretation of the data quite unlikely.

The characteristic exponent γ can in principle be used to obtain the Luttinger liquid parameter K_ρ, which controls the decay of all correlation functions [12, 17, 20, 33]. In our case and making the assumption that quarter-filled band Umklapp scattering (i.e., $\tilde{n} = 2$) is dominant in the $TMTSF$ family, it then follows γ=5-16K_ρ and $K_\rho \sim 0.23$ [33], which is in reasonable agreement with photoemission [14, 25, 59, 60] and transport data [26, 61].

Angle integrated photoemission spectra of the $TMTSF$ salts family were interpreted as fingerprint of the expected power law behaviour $\rho(\omega) \sim |\omega|^\alpha$ in the density of states within the LL scenario [14, 25]. The exponent α reflects again the strength of the interactions, and is related to the fundamental charge correlation parameter K_ρ of the Luttinger model by the expression $\alpha=(K_\rho+K_\rho^{-1}-2)/4$. An analysis of the experimental data yields $K_\rho \sim 0.2$ [14, 25, 59, 60, 62]. Nevertheless, a Luttinger (or Luther-Emery)-like interpretation for photoemission faces two main objections: the unexpectedly large value of α, and the absence of any k-dependence in the ARPES spectra [14].

As far as the transport results are concerned, the quarter filled Umklapp scattering scenario has been also confirmed by recent results of the dc resistivity collected along the chain a-axis and the least conducting c-axis [26]. At high temperatures (i.e., in the 1D limit) characteristic powerlaw behaviours $\rho_c \sim T^{-1.12}$ and $\rho_a \sim T$ are clearly identified, implying $K_\rho \sim 0.2$ [26]. These powerlaw behaviours are in agreement with the theory [17, 20, 61], as well as with the optical experiment.

It is worth mentioning that Controzzi *et al.* [50] also work out the optical conductivity of one-dimensional Mott insulators on the basis of the exact solution of the Sine-Gordon model. At least at very large energies, where 3D effects are unimportant, they found a powerlaw behaviour of $\sigma_1(\omega)$ with $K_\rho \sim 0.2$ [50]. Such a K_ρ would be consistent with the experimental data. However, Controzzi *et al.* also claim that the leading asymptotic behaviour of $\sigma_1(\omega)$ obtained in the perturbation theory is a good approximation only at extremely large frequencies [50]. The powerlaw in $\sigma_1(\omega)$, found experimentally (Fig. 6.5), remains then a rather astonishing feature, since it occurs in the spectral range just above E_{gap}.

There was also some discussion about the behaviour of $\sigma_1(\omega)$ at $\omega < E_{gap}$. Giamarchi conjectured a ω^3-behaviour [17], which however has never been found experimentally [33, 34] and is not even substantiated by rigorous theoretical arguments. At frequencies smaller than the correlation gap there is the progressive crossover into a two- or even three-dimensional scenario, so that the estimated behaviour of $\sigma_1(\omega)$ based on purely 1D model is no longer practicable.

Even though optical investigations shed light on various types of correlations and in principle can discriminate among different power-law behaviours (Fig. 6.5) as fingerprint of the nature of the one-dimensional state, some results are still puzzling and not fully understood. For instance, the high frequency tail of the optical gap of the supposed truly one-dimensional insulator $(NbSe_4)_3I$ displays a power-law with exponent $\gamma \sim 4.25$, as shown in Fig. 6.6 [63]. This exponent is distinctly different from what has been measured and predicted in other one-dimensional systems (Fig. 6.5). Importantly, the exponent here is also larger than the values $\gamma \sim 3$, predicted for a rigid 1D band insulator, or $\gamma \sim 2$ when phonons are included [57], as observed in typical CDW systems like $(TMTSF)_2ReO_4$ (Fig. 6.5). Such a discrepancy to standard models for $\sigma_1(\omega)$ is outside experimental uncertainties. The large exponent suggests that current relaxation involves a less efficient mechanism in $(NbSe_4)_3I$ than those treated theoretically hitherto. Apparently, 1D band insulators are not so well understood as might be assumed. In order to shed light on this

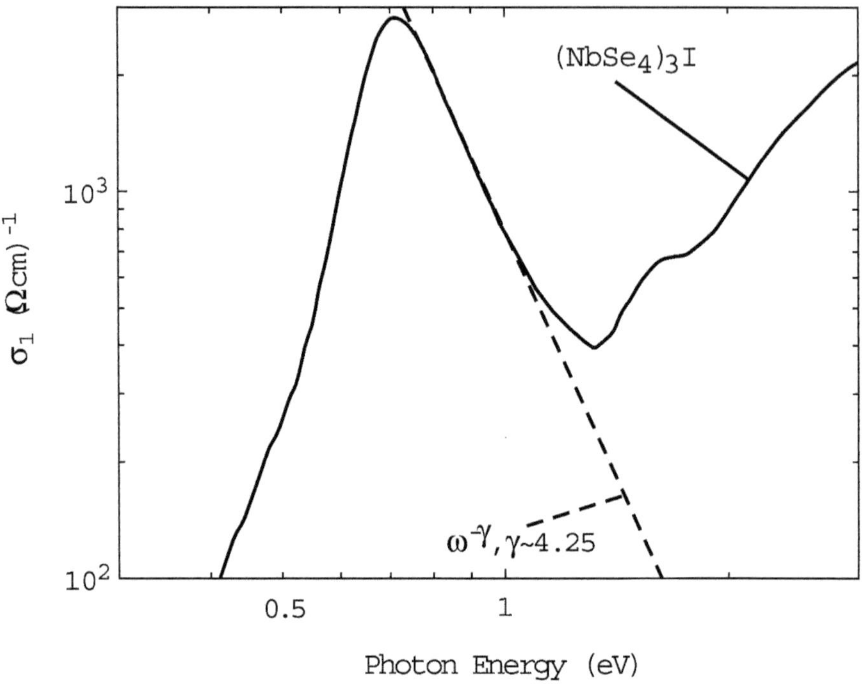

Figure 6.6. The optical conductivity of $(NbSe_4)_3I$ at 300 K in the mid-infrared spectral range, showing the power-law behaviour at the high frequency tail of the optical gap [63]. Note the logarithmic scales.

puzzle, it could be interesting to perform spectroscopic experiments on other well-known 1D insulators. Obvious candidates are the insulating phases of Peierls systems, but also polymers, e.g., polyacetylene.

The dimensionality crossover, induced by the increasing $t_\perp(\sim t_b)$ upon pressure or by changing the chemistry (i.e., going from the $TMTTF$ to the $TMTSF$ family), is one of the central issues, when discussing the physics of low dimensional systems. There is a well-established order for $t_\perp$ among the four salts investigated. The $(TMTTF)_2X$ analogs are more anisotropic than the $(TMTSF)_2X$ ones [22]. The values of the transfer integrals along the crystallographic directions (i.e., t_a, t_b and t_c) for both groups of salts are in broad agreement with tight-binding model calculations and with the trend indicated by the plasma frequency [23, 32, 33]. It is particularly instructive to compare E_{gap} with t_b. This is shown in Fig. 6.7 for the four measured Bechgaard salts. To arrive at a scale for t_b, we took the calculated values as averages for the $(TMTSF)_2X$ and $(TMTTF)_2X$ salts, respectively, and assumed that pressure changes t_b in a linear fashion. The positions of the various salts along the horizontal axis of Fig. 6.7 reflect this choice [34], with pressure values taken from the literature [22, 23]. Such a scale has been widely used when discussing the broken symmetry ground states of these materials.

The solid line in Fig. 6.7 represents the overall behaviour of the correlation gap (see also Fig. 6.4). Various experiments give slightly different values of the gap. This is probably due to the differences in the curvature of the band, which is scanned differently by different experiments, or due to the different spectral response functions involved. The decrease going from the $(TMTTF)_2X$ to the $(TMTSF)_2X$ analogs may represent various factors [45], such as the decreasing degree of dimerization and the slight increase in the bandwidth along the chain direction, as evidenced by the greater value of the plasma frequency measured along the chain direction in the $(TMTSF)_2X$ salts [34, 40]. The dotted line representing $2t_b$ (this is half the bandwidth perpendicular to the chains in the tight-binding approximation) crosses the full line displaying the behaviour of E_{gap} between the salts exhibiting insulating and metallic behaviour, whereas the dotted line representing t_b crosses the solid line between the two metallic salts. Therefore, the experiments strongly suggest that a crossover from a non-conducting to a conducting state occurs when the correlation gap exceeds the unrenormalized single particle transfer integral t_b between the chains by a factor A (i.e., $E_{gap} = At_b$), which is on the order of but somewhat greater than 1.

Additional evidence for a pronounced qualitative difference between states with $At_b < E_{gap}$ and $At_b > E_{gap}$ is given by plasma frequency studies along the b direction (i.e., perpendicular to the chain). As shown in Fig. 6.3, there is no well-defined plasma frequency for the insulating state of the $TMTTF$ salts, and we regard this as evidence for the confinement [64] of electrons on individual chains. In fact, the reflectivity has a temperature independent overdamped-like behaviour along the b-axis. Conversely, the electrons become deconfined as soon as $E_{gap} \sim At_b$ (Fig. 6.7). Such a deconfinement is manifested by the onset of a sharp plasma edge in the low-temperature reflectivity spectra of the $TMTSF$ salts along the b-axis (Fig. 6.3) [34]. This conclusion is not entirely unexpected: a simple argument (the same as one would advance for a band-crossing transition for an uncorrelated band semiconductor) would suggest that to create an electron hole pair with the electron and hole residing on neighbouring chains, an energy comparable to the gap would be required. The confinment-deconfinment crossover by tuning

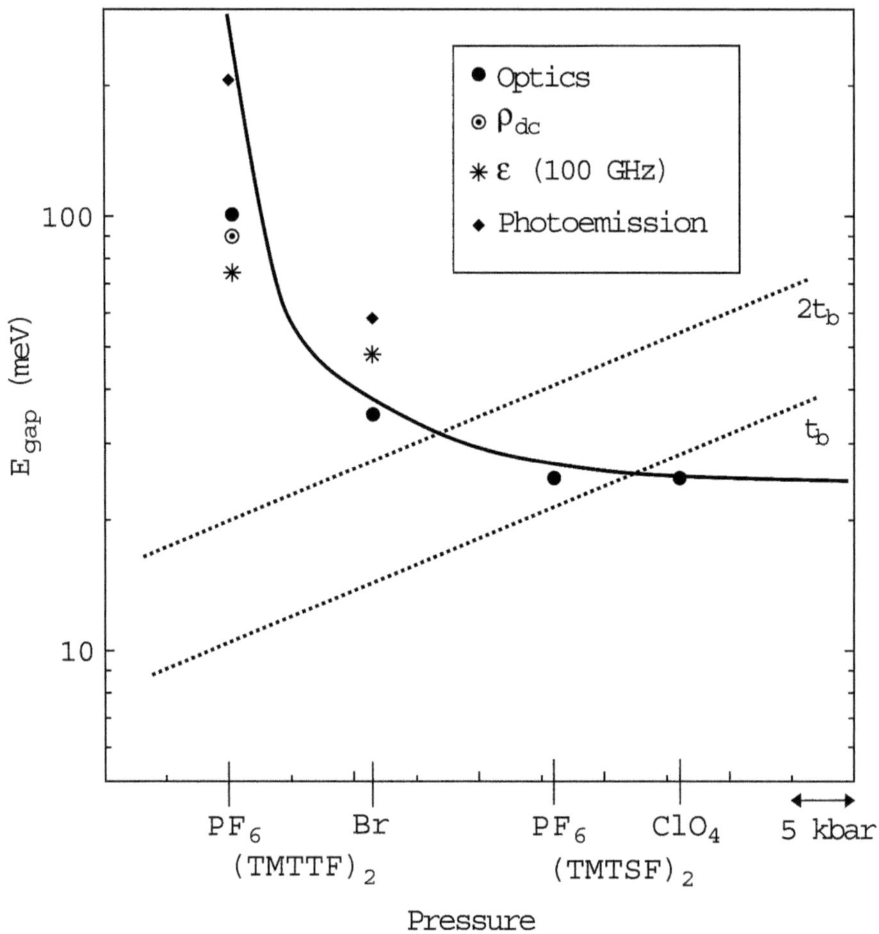

Figure 6.7. The pressure dependence of the (Mott) correlation gap, as established by different experimental methods, and of the transfer integral, perpendicular to the chains, t_b, for the Bechgaard salts [34]. The horizontal scale was derived with the results of pressure studies [16, 22, 23, 24].

$t_\perp$ is indeed a dimensionality driven insulator-metal transition [34].

There are several fundamental unsolved problems about the dimensionality crossover, though. The major discrepancy between theory and experiment concerns the relevant energy scale governing such a dimensionality crossover. Indeed, various theories [64, 65, 66, 67, 68] suggest a strong renormalization of the relevant interchain transfer integral

$t_\perp^{eff} \sim t_\perp(t_\perp/t_a)^{\alpha/1-\alpha}$ with $\alpha = (K_\rho + K_\rho^{-1})/4 - 1/2$ [66], for coupled Luttinger liquids. K_ρ extracted from the γ exponent, discussed previously, would lead to unreasonably low values of $t_\perp^{eff}$, substantially smaller than the bare $t_\perp \sim t_b$ estimated from the experiment and incompatible with the observed metallic behaviour in the $TMTSF$ salts at low frequency [33, 34]. Some of these studies [64, 65, 66, 67], however, do not take into account the periodicity of the underlying lattice and the resulting Umklapp scattering. Such a scattering has a marked influence on the effect of interchain transfer. A renormalization group treatment [67] of two coupled Hubbard chains predicts a crossover between confinement (that is, no interchain single-electron charge transfer) and deconfinement, at $At_b = E_{gap}$, with the value of A estimated to be between 1.8 and 2.3, quite in agreement with the optical findings. A transition or crossover from an insulator to a metal has also been conjectured by Bourbonnais [68], on the basis of studies of arrays of coupled chains, where also Umklapp scattering has been taken into account. Discrepancies between theoretical predictions and experimental estimates are not totally surprising, if one considers that the experiment probes the transverse optical response over an energy scale of about 0.1 eV. At these energies, self-energy effects at the origin of the renormalization of $t_\perp$ may be irrelevant. It follows that from optics $t_\perp^{eff}$ can be closer to the bare $t_\perp$. In that sense, the onset of the transverse plasma edge as a function of (chemical) pressure may not coincide with the one found from low-energy probes like dc transport [26] and NMR [69].

We also note that, in the absence of pressure-dependent optical studies, it remains to be determined whether the onset of the transverse plasma edge [34] which was observed going from the $(TMTTF)_2X$ to $(TMTSF)_2X$ salts (Fig. 3), coincides with the insulator-to-metal transition found in transport [26] and nuclear magnetic resonance (NMR) measurements [69]. Although optical experiments under pressure are difficult to conduct, studies of the pressure dependence of the dielectric constant, combined with dc transport data, could clarify this issue.

In order to broaden the perspectives on the dimensionality-crossover issue it is rather compelling to establish a comparison with the ladder cuprates. We want to argue that a similar behaviour occurs in the ladder systems, as well. The analogy between the $Sr_{14-x}Ca_xCu_{24}O_{41}$ ladder systems and the Bechgaard salts has been first pointed out by Mayaffre *et al.* in the course of transport and NMR investigations under pressure for the compound with $x = 12$ [70]. The degree of confinement of the

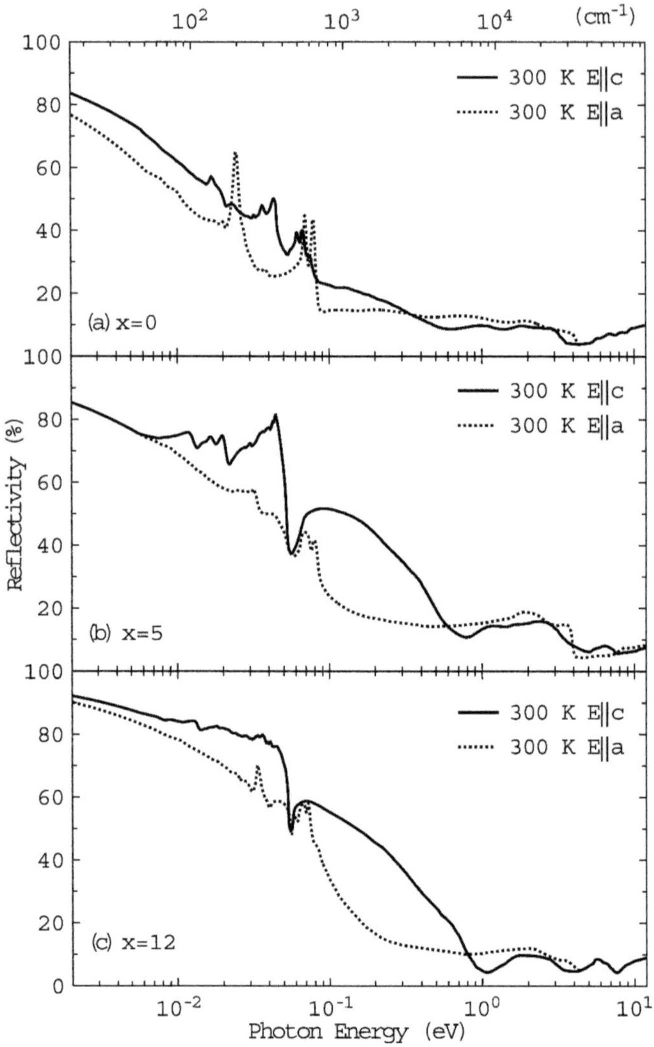

Figure 6.8. Optical reflectivity of $Sr_{14-x}Ca_xCu_{24}O_{41}$ single crystals at $T = 300\ K$ along the c- and a-axes for $x = 0$ (a), $x = 5$ (b) and $x = 12$ (c). Note the logarithmic energy scale [73].

carriers along the ladders could correlate with the size of the spin gap [31]: hole pairs are responsible for the conduction within the ladders as long as the magnetic forces can provide the binding of two holes on the same rung. The vanishing of the spin gap upon application of pressure [70] or its reduction upon Ca-substitution could thus be responsible for the dissociation of the pairs making in turn the hopping (deconfinement) of the transverse (a-axis) single particle easier [71]. Therefore, pressure is believed to have an effect similar to that of the Ca-substitution and, similarly to the Bechgaard salts, might induce a change in the intrinsic dimensionality of the system. Ca-substitution can be regarded as a chemical pressure because of its smaller ionic radius, leading to the lattice contraction, and its major effect on the ladders is to increase the hole density [72]. Nagata *et al.* also suggest that the application of pressure on ladder systems triggers the dimensionality crossover from one to two due to the enhancement of interladder interactions [72]. Therefore, pressure is more important than doping since the increased pressure induces not only superconductivity but also coherent charge dynamics perpendicular to the ladders.

The optical data on ladder cuprates (Fig. 6.8) support the above arguments about the analogy with the Bechgaard salts [73]. We observe that the increasing Ca-substitution induces the formation of a well developed plasma edge feature in $R(\omega)$ along the c-axis. This is actually to be expected when the material undergoes small to large doping. From the spectral weight arguments similar to Ref. [74], the development of the plasma edge upon doping along the c-axis is the consequence of the transfer of holes from the CuO_2 chains' reservoir to the ladder conduction paths. The total spectral weight for $x = 12$, encountered in $\sigma_1(\omega)$ for $E\|c$ up to approximately $\omega_c \sim 1\ eV$ but before the onset of interband transitions, corresponds to a plasma frequency of about $1\ eV$. By assuming the charge carrier mass equal to the free electron mass m_e as in Ref. [74], we also obtain a hole density per ladder-Cu of about $n \sim 0.2$ [73]. However, the Drude weight ascribed to the effective metallic contribution in $\sigma_1(\omega)$ at low frequencies ($\omega < 60\ meV$) corresponds to a plasma frequency of about $0.2\ eV$ (for $x = 12$ and $E\|c$). This would suggest that either the free charge carriers have an effective mass of about $25m_e$ or alternatively that only a small fraction of holes $n_{eff} < n$ is effectively involved in the metallic contribution. Moreover, along the transverse direction (a-axis, Fig. 6.8) one can observe an increased metallicity upon doping, manifested by an incipient plasma edge development. This bears a remarkable similarity with the behaviour in Bechgaard salts (Fig. 6.3) and indicates a similar confinement-deconfinement crossover upon Ca-substitution [73].

4. Conclusion

Optical data on the prototype linear chain Bechgaard salts reveal the peculiarity of the one-dimensional interacting electron gas response. The unusual spectral features of the Bechgaard salts prove that these materials are certainly not simple anisotropic band metals. Clear deviations from the Fermi liquid behaviour have been identified and several aspects hint to a possible manifestation of a Luttinger or Luther-Emery liquid in the normal phases [25, 26, 27]. Interestingly, different experiments and particularly different type of spectroscopies reveal such deviations on different energy scales. The characteristic energy scale in the optical conductivity data is the Mott-Hubbard gap, of about $\sim 25\ meV$ in $(TMTSF)_2PF_6$. The salient feature of the photoemission spectra is the much larger ($\sim 10^2\ meV$) pseudogap. Both techniques, on the other hand, point to a characteristic Luttinger parameter $K_\rho \sim 0.2$, and therefore to strong, long-range 1D correlations. Also, both the photoemission and optical data discriminate between the conducting $TMTSF$ and the insulating $TMTTF$ salts, with a reasonable agreement on the gap size of the latter [14].

Several issues concerning the electronic structure of these materials are still open, and it is not yet clear whether a comprehensive description is possible within the existing theoretical scenarios. The optical response in $TMTSF$ salts has been interpreted in terms of a dimensionality crossover induced by the interchain coupling $t_\perp$. These results (Fig. 6.4c) account for a so-called incipient 2D Fermi liquid scenario [26] where a low energy 3D Fermi liquid behaviour (i.e., narrow zero energy mode) coexists with the high-energy (1D) LL state (i.e., powerlaw on the high frequency tail of the charge correlation pseudogap). More theoretical work is certainly necessary to put the analysis of the data on firmer ground. Also, it remains to be explained why theory predicts the confinement-deconfinement crossover when E_{gap} is of the order of the renormalized transfer integral $t_\perp^{eff}$, while experimentally the bare $t_\perp$ turns out to be the relevant energy scale. It seems that a theory of the dimensionality crossover in coupled Luttinger liquid that would be completely consistent with the present data is still lacking [50].

We would like to conclude this Chapter with an outlook on future perspectives. Very interesting in the field of low-dimensional systems are graphitic carbon needles, which have been discovered in carbon rods after an arc discharge by high-resolution transmission electron microscopy [75]. A needle typically consists of a few microtubules centered coaxially about the needle axis, and is hollow. A microtubule has the form of a rolled graphitic sheet with a diameter of a few nanometers. A lot of interest in such a new form of carbon is also associated to the possible and potentially interesting technical applications and uses. A partial list includes superstrong cables, wires for nanosized electronic devices, charge-storage devices in batteries, and tiny electron guns for flat-screen television [76]. It is expected that the graphitic microtubules exhibit a variety of properties in electronic conduction, from a typical semiconductor to a good metal, depending on the tubule structure, i.e., chirality [77]. The tubule morphology suggests that these systems should represent a fascinating class of novel quasi-one-dimensional structures and can be considered as the ultimate realization of a one-dimensional quantum wire. The LL behaviour was also identified in these metallic single walled nanotubes (SWNT) [78]. Optical investigations, besides revealing the anisotropy of the charge dynamics [79], also display characteristic optical fingerprints of a quasi-one-dimensional system [80].

We also would like to mention another class of tubular structures consisting in the tungsten and molybdenum disulphide (WS_2 and MoS_2) materials [81]. Synthesis of such nanotubes made of atoms other than carbon may be possible and tubes as small as 15 nm have been found. The efficient synthesis of identical single wall MoS_2 nanotubes is expected to lead the way to the synthesis of the other related dichalgogenide systems, even in the sub-nanometer range. This will open new perspectives and will facilitate the investigation of truly single-tube properties and related quantum effects. Optical investigations will allow us understanding the still puzzling and in many respect controversial results on 1D quantum wire in general.

Finally, electronic excitations and motion of electric charges are well known to play a significant role in a wide range of macromolecules of biological interest, and long-range electron transfer may be possible for bio-polymers in general. In particular, electron transfer involving the DNA double helix is thought to be important in shaping its (biological) properties. In DNA transport or propagation of information, radiation damage and repair as well as biosynthesis seem to be governed and driven by electron transport. While DNA crystals are transparent insulators with a bandgap exceeding the visible energy range (corresponding to 2 eV), various experiments suggest that for DNA strands in a biological environment or in a solution the electron states are in general fundamentally different. Recent measurements suggest long-range and extremely rapid electron mediated interaction between donors and acceptors placed at various positions along the DNA helix with the implication that DNA can be viewed as a one-dimensional molecular wire. This notion is supported by recent measurements [82] of the dc current-voltage characteristics across individual DNA segments. An electrical conductivity of the order of 1000 $(\Omega cm)^{-1}$ was inferred from the measurements - comparable or larger than the conductivity of many so-called linear chain metals where the motion of electronic charges proceeds along one direction. These experiments were conducted at room temperature only and there are fundamental unresolved questions concerning contact effects and charge injection into the DNA helix in the course of the measurements [83]. To make the picture, if possible, even more confusing one just needs to quote recent controversial and contradictory reports in the literature, claiming a variety of transport phenomena [84, 85], even proximity induced superconductivity [86]. Optics, which is a well-known contacts-free experimental tool, is expected to play a leading role in addressing the issue of the relevant energy scales and excitations, as well as scattering effects influencing the transport properties of DNA. This is left for the future.

Acknowledgments

This Chapter is based on the original work by V. Vescoli and B. Ruzicka. The author is very grateful to D. Baeriswyl, C. Bourbonnais, P.M. Chaikin, M. Dressel, T. Giamarchi, M. Grioni, G. Gruner, D. Jerome, F. Mila and J. Voit for many illuminating discussions. Research at ETH Zurich was supported by the Swiss National Foundation for the Scientific Research.

References

[1] C. Kittel, in "Quantum Theory of Solids" (J. Wiley and Sons, New York, 1963).

[2] L. Degiorgi, Rev. Mod. Phys. **71**, 687 (1999).

[3] J. Campuzano, this book.

[4] D. van der Marel, this book.

[5] J. Luttinger, J. Math. Phys. **4**, 1154 (1963).

[6] S. Tomonaga, Prog. Theor. Phys. **5**, 554 (1950).

[7] for a review, see H.J. Schulz, Int. J. Mod. Phys. B **5**, 57 (1991).

[8] V.J. Emery, in "Highly conducting one dimensional solids", Eds. J.T. Devreese, R.P. Evrard and V.E. van Doren (Plenum, New York, 1979), p. 327.

[9] J. Solyom, Adv. Phys. **28**, 209 (1979).

[10] F. Haldane, J. Phys. C **14**, 2585 (1981).

[11] K. Schönhammer, this book.

[12] J. Voit, Eur. Phys. J. B **5**, 505 (1998) and Proceedings of the NATO Advanced Research Workshop on The Physics and Mathematical

Physics of the Hubbard Model, San Sebastian, October 3-8, 1993, Edited by D. Baeriswyl, D. K. Campbell, J. M. P. Carmelo, F. Guinea, and E. Louis; Plenum Press, New York (1995).

[13] X. Zotos and P. Prelovšek, this book.

[14] M. Grioni, this book.

[15] G. Gruner, Rev. Mod. Phys. **60**, 1129 (1988).

[16] G. Gruner, in "Density Waves in Solids", (Addison-Wesley, Reading, MA, 1994).

[17] T. Giamarchi, Physica B **230-232**, 975 (1997).

[18] H. Mori *et al.*, J. Phys. Soc. Jap. **63**, 1639 (1994).

[19] N. Bulut *et al.*, Phys. Rev. Lett. **72**, 705 (1994).

[20] S. Biermann *et al.*, cond-mat/0201542.

[21] A.J. Millis, this book.

[22] D. Jerome and H.J. Schulz, Adv. Phys. **31**, 299 (1982).

[23] D. Jerome, in "Strongly interacting fermions and high T_C superconductivity", Eds. B. Doucot and J. Zinn-Justin (Elsevier Science, 1995), p. 249.

[24] L. Ducasse *et al.*, J. Phys. C **19**, 3805 (1986).

[25] V. Vescoli *et al.*, Eur. Phys. J. B **13**, 503 (2000).

[26] J. Moser *et al.*, Eur. Phys. J. B **1**, 39 (1998).

[27] K. Bechgaard *et al.*, Solid State Commun. **33**, 1119 (1980).

[28] C. Coulon *et al.*, J. Physique (Paris) **43**, 1059 (1982).

[29] J.P. Pouget *et al.*, Mol. Cryst. Liq. Cryst. **79**, 129 (1982).

[30] E. Dagotto and T.M. Rice, Science **271**, 618 (1996).

[31] E. Dagotto *et al.*, Phys. Rev. B **45**, 5744 (1992).

[32] M. Dressel *et al.*, Phys. Rev. Lett. **77**, 398 (1996).

[33] A. Schwartz *et al.*, Phys. Rev. B **58**, 1261 (1998).

[34] V. Vescoli *et al.*, Science **281**, 1181 (1998).

[35] F. Wooten, in "Optical Properties of Solids", (Academic Press, New York, 1972).

[36] V. Vescoli *et al.*, Solid State Commun. **111**, 507 (1999).

[37] C.S. Jacobsen *et al.*, Phys. Rev. Lett. **46**, 1142 (1981).

[38] C.S. Jacobsen *et al.*, Solid State Commun. **38**, 423 (1981).

[39] C.S. Jacobsen *et al.*, Phys. Rev. B **28**, 7019 (1983).

[40] D. Pedron *et al.*, Phys. Rev. B **49**, 10893 (1994).

[41] S.E. Brown, private communication.

[42] N. Cao *et al.*, J. Phys. I (France) **6**, 1719 (1996).
[43] G. Mihaly *et al.*, Phys. Rev. Lett. **84**, 2670 (2000).
[44] J. Moser *et al.*, Phys. Rev. Lett. **84**, 2674 (2000).
[45] K. Penc and F. Mila, Phys. Rev. B **50**, 11429 (1996).
[46] M. Mori *et al.*, J. Phys. Soc. Jap. **65**, 3604 (1996).
[47] M. Tsuchiizu *et al.*, J. Phys. Soc. Jap. **68**, 1809 (1999).
[48] A. Georges *et al.*, Phys. Rev. B **61**, 16393 (2000).
[49] E. Jeckelmann *et al.*, Phys. Rev. Lett. **85**, 3910 (2000).
[50] D. Controzzi *et al.*, Phys. Rev. Lett. **86**, 680 (2001).
[51] J.M.P. Carmelo *et al.*, Phys. Rev. Lett. **84**, 4673 (2000).
[52] S.S. Kancharla *et al.*, Phys. Rev. B **64**, 085119 (2001).
[53] A. Rosch and N. Andrei, Phys. Rev. Lett. **85**, 1092 (2000).
[54] X. Zotos *et al.*, Phys. Rev. B **53**, 983 (1996).
[55] X. Zotos *et al.*, in "Open problems in strongly correlated electron sytems", Eds. J. Bonca *et al.*, Kluwer (2001), p. 273.
[56] F. Zwick *et al.*, Solid State Commun. **113**, 179 (2000).
[57] P.A. Lee *et al.*, Solid State Commun. **14**, 703 (1974).
[58] F. Favand and F. Mila, Phys. Rev. B **54**, 10435 (1996).
[59] B. Dardel *et al.*, Europhys. Lett. **24**, 687 (1993).
[60] F. Zwick *et al.*, Phys. Rev. Lett. **79**, 3982 (1997).
[61] C. Bourbonnais and D. Jerome, in "Advances in Synthetic Metals, Twenty years of Progress in Science and Technology", edited by P. Bernier, S. Lefrant, and G. Bidan (Elsevier, New York, 1999), p. 206.
[62] B. Dardel *et al.*, Phys. Rev. Lett. **67**, 3144 (1991).
[63] V. Vescoli *et al.*, Phys. Rev. Lett. **84**, 1272 (2000).
[64] P.W. Anderson, Proc. Natl. Acad. Sci. U.S.A. **92**, 6668 (1995).
[65] F. Mila and D. Poilblanc, Phys. Rev. Lett. **76**, 287 (1996).
[66] C. Bourbonnais and L.G. Caron, Int. J. Mod. Phys. B **5**, 1033 (1998).
[67] Y. Suzumura *et al.*, Phys. Rev. B **57**, R15040 (1998).
[68] C. Bourbonnais, in "Strongly interacting fermions and high-T_c superconductivity", Eds. B. Doucot and J. Zinn-Justin (Amsterdam, 1995), p. 307.
[69] P. Wzietek *et al.*, J. Phys. I **3**, 171 (1993).
[70] H. Mayaffre *et al.*, Science **279**, 345 (1998).

[71] For the ladders the "chain" axis is the c-axis, the transverse direction is the a-axis and the least conducting axis is the b-axis.

[72] T. Nagata *et al.*, Phys. Rev. Lett. **81**, 1090 (1998).

[73] B. Ruzicka *et al.*, Eur. Phys. J. B **6**, 301 (1998).

[74] T. Osafune *et al.*, Phys. Rev. Lett. **78**, 1980 (1997).

[75] S. Iijima, Nature **354**, 56 (1991).

[76] R.F. Service, Science **281**, 940 (1998).

[77] X. Blase *et al.*, Phys. Rev. Lett. **72**, 1878 (1994).

[78] M. Bockrath *et al.*, Nature **397**, 598 (1999).

[79] F. Bommeli *et al.*, Solid State Commun. **99**, 513 (1996).

[80] B. Ruzicka *et al.*, Phys. Rev. B **61**, R2468 (2000).

[81] R. Tenne *et al.*, Nature **360**, 444 (1992).

[82] H.W. Fink and C. Schönenberger, Nature **398**, 407 (1999).

[83] P. Tran *et al.*, Phys. Rev. Lett. **85**, 1564 (2000).

[84] D. Porath *et al.*, Nature **403**, 635 (2000).

[85] F.D. Lewis *et al.*, Nature **406**, 51 (2000).

[86] A.Yu. Kasumov *et al.*, Science **291**, 280 (2001).

Chapter 7

OPTICAL CONDUCTIVITY AND CORRELATED ELECTRON PHYSICS

A. J. Millis
Department of Physics, Columbia University, 538 West 120th St., N.Y., N.Y. 10027 USA *
millis@phys.columbia.edu

Abstract This article reviews the uses of optical conductivity, and in particular optical spectral weight in elucidating the physics of correlated electron physics. An introduction to the fundamental theoretical concepts is given, followed by a summary of results obtained on specific models and a discussion of available data.

Keywords: Optical Conductivity; Correlated Electron Systems; High Temperature Superconductivity; Colossal Magnetoresistance; Heavy Fermion Compounds

1. Introduction

The *optical conductivity* $\sigma_{ab}(q,\omega)$ is the linear response function relating the current, j, in the a direction to an applied transverse electric field, E, in the b direction:

$$j_a(q,\omega) = \sigma_{ab}(q,\omega)E_b(q,\omega) \tag{7.1}$$

One may approximately distinguish two kinds of optical processes: promotion of an electron from one orbital to another on the same ion (as in the $1s \rightarrow 2p$ hydrogenic transition familiar from elementary atomic physics) or motion of an electron from one place in the sample to another. The latter contribution means that the conductivity contains information on the ease with which electrons may move through the solid, and

*Partial support provided by the US National Science Foundation under the MRSEC program and DMR-00081075 and also by the Rutgers University Center for Materials Theory

D. Baeriswyl and L. Degiorgi (eds.), Strong Interactions in Low Dimensions, 195–235.

is of interest because the essence of the 'strong correlations' problem is the interplay between the localizing effect of replusive electron-electron interactions and the delocalizing effect of wave function hybridization.

To measure the conductivity one must apply a transverse electric field to the sample. For most frequencies of interest this may be accomplished by exposing the sample to electromagnetic radiation; indeed the conductivity is usually inferred by exposing a material sample to electromagnetic radiation and measuring the resulting reflection and transmission properties. The frequency scales of interest for 'correlated electron' physics are typically several electron volts or less. At these frequencies the magnetic field component of the incoming radiation is negligible and one may think of the applied radiation as corresponding to a transverse electric field only. Although electromagnetic fields are transverse in free space, a longitudinal component may be generated inside a solid. In most situations of present interest, the symmetry is such that this does not occur, and the possibility will be mostly be neglected here (an important exception of current interest concerns 'bilayer plasmons' in the c-axis conductivity of high-T_c materials; for a discussion and other references see e.g. [1]). Finally, the radiation wavelength is in almost all cases very large relative to any relevant length scale in the solid; thus one usually sets $q \to 0$ and discusses $\sigma(\omega)$. (Note that the crucial length scale is not the free space wavelength $\lambda = c/(2\pi\omega)$ but rather the penetration depth $\delta \sim \sqrt{1/\omega\sigma}$ [2] characterizing the decay of incident radiation inside the solid. In the 'correlated electron' materials of interest the strong correlations themselves constrain the motion of electrons, leading normally to a small conductivity and thus to a sufficiently long δ that one may neglect the spatial variation of the field inside the solid).

We henceforth consider the $q \to 0$ limit, introducing the vector potential A by writing $\overrightarrow{E}(\omega) = \frac{i\omega}{c}\overrightarrow{A}(\omega)$:

$$j_a(\omega) = \sigma_{ab}(\omega)\frac{A_b(\omega)}{i\omega/c} \tag{7.2}$$

This paper will review the theory of $\sigma(\omega)$ (with particular emphasis on sum rules), show how measurements of σ have been used to elucidate some aspects of the physics of presently interesting materials including high-T_c cuprates and the 'colossal' magnetoresistance manganites, outline some open issues and mention desirable improvements in experimental technique. Almost all of the specific results presented may be found in the literature somewhere, but I hope a unified presentation will be useful. The reader's attention is also drawn to Chapter 8, which treats the optical conductivity of high-T_c superconductors and touches on many of issues considered here.

2. Theory

2.1 Fundamental definitions

2.1.1 Basics. We are interested in the current induced in a system to a transverse applied electric field. To compute this one needs a Hamiltonian and a coupling to the electric field. The most fundamental Hamiltonian and coupling one would write is

$$H = \int d^3r \frac{1}{2m}\left((\nabla - \frac{ie}{c}A)\psi(r)\right)^2 + H_{el-ion} + H_{el-el} + H_{ion} \quad (7.3)$$

where ψ is the electron field operator, m is the electron mass, and we include electron-ion and electron-electron interations. We have not written the coupling of the electric field to the ions explicitly because the very large (relative to m) ion mass means that at any reasonable electron density the ion contribution to σ is negligible (exceptions can occur for materials with a very strong electronic anisotropy: the c-axis conductivity of high-T_c materials is an important example). The way the vector potential enters is dictated by gauge invariance.

The electrical current density operator $\overrightarrow{j}$ is defined in general as

$$\overrightarrow{j} = \frac{c}{V}\frac{\delta H}{\delta \overrightarrow{A}} \quad (7.4)$$

where V is the volume of the sample. From Eqs (7.3), (7.4) one finds thc familiar expression

$$\overrightarrow{j} = \int d^3r \left(\frac{ie}{m}\left(\psi^*(r)\overrightarrow{\nabla}\psi(r) - (\overrightarrow{\nabla}\psi^*(r))\psi(r)\right) - \frac{ie}{mc}\overrightarrow{A}\psi^*(r)\psi(r)\right) \quad (7.5)$$

and then using the usual Kubo formulae for linear response along with the relation between E and A gives [3]

$$\sigma_{ab}(\omega) = \frac{ine^2}{m(\omega + i\varepsilon)}\delta_{ab} + \frac{i\chi_{ab}(\omega)}{\omega + i\varepsilon} \quad (7.6)$$

where the current-current correlation function χ may be written in terms of the exact eigenstates $|n\rangle$, energies E_n and partition function Z as

$$\chi_{ab}(\omega) = \sum_{n,m} \frac{e^{-\beta E_n}}{Z}\left[\frac{\langle n|j_a|m\rangle\langle m|j_b|n\rangle}{\omega - (E_m - E_n) + i\varepsilon} - \frac{\langle n|j_b|m\rangle\langle m|j_a|n\rangle}{\omega + (E_m - E_n) + i\varepsilon}\right] \quad (7.7)$$

The first term in Eq (7.7) comes from the term proportional to A in Eq (7.6) and the relation $n = \frac{1}{V}\int d^3r\langle\psi^*(r)\psi(r)\rangle$ (the $\langle\rangle$ denotes thermal and quantal expectation value) has been used.

The conductivity is a causal function with real (σ') and imaginary (σ'') parts. The real part corresponds to transitions in which energy is absorbed or removed. The real and imaginary parts are related by a Kramers-Kronig relation which may be written ($\mathcal{P}$ denotes principal part)

$$\sigma''(\omega) = \int_{-\infty}^{\infty} \frac{dx}{\pi} \mathcal{P} \frac{\sigma'(x)}{\omega - x} \tag{7.8}$$

2.1.2 Charge Stiffness. The position, $\vec{R}$, of a charged particle which is freely accelerated by a spatially uniform electric field $\vec{E}$ obeys the equation of motion

$$m \frac{\partial^2 \vec{R}}{\partial t^2} = e\vec{E} \tag{7.9}$$

so after Fourier transformation on time the current $\vec{j} = e\partial\vec{R}/\partial t$ carried by a density n of such particles becomes

$$\vec{j} = \frac{ine^2}{m(\omega + i\varepsilon)} \vec{E} = ne^2 \left(\mathcal{P} \left(\frac{i}{\omega m} \right) + \frac{\pi}{m} \delta(\omega) \right) \vec{E} \tag{7.10}$$

In other words, particles which may be freely accelerated by an applied field lead to a delta function contribution to the real part of the conductivity.

One may write in general

$$\sigma'_{ab}(\omega) = \frac{ie^2 D_{ab}}{(\omega + i\varepsilon)} + \sigma'_{reg,ab}(\omega) \tag{7.11}$$

where $\lim_{\omega \to 0} \omega \sigma'_{reg}(\omega) = 0$. The quantity D is the **Drude weight** or **charge stiffness** and is defined by this equation. Comparison to Eq (7.10) shows that D measures the ability of the electrons in a given system to freely accelerate in an applied electric field.

In a system without translation invariance (e.g. a disordered system), $D = 0$. In a system with at least a discrete translation invariance (e.g. electrons in an ideal periodic lattice) at $T = 0$ (so there is no thermally induced disorder), one may have $D \neq 0$ (in which case one has a metallic ground state); or $D = 0$ (in which case one has a non-metallic ground state). This classification of metals and insulators in perfectly translation invariant environments is due to Kohn [4]. The quantity D is the order parameter characterizing the metallic state (in a clean system) and one may think of a Mott transition [5] as driving D to 0.

From the Kubo formula, Eq (7.6) one obtains the fundamental expression

$$D_{ab} = \frac{n}{m} - \sum_{n,m} \frac{e^{-\beta E_n}}{Z} \left[\frac{\langle n|j_a|m\rangle\langle m|j_b|n\rangle}{(E_m - E_n)} - \frac{\langle n|j_b|m\rangle\langle m|j_a|n\rangle}{(E_m - E_n)} \right] \quad (7.12)$$

Kohn [4] introduced a very useful alternative expression. Diagonalize the D tensor and consider a system which is periodic, with period L, in a direction corresponding to a principal axis of D. The system then has the topology of a torus. Introduce a flux ϕ through this torus and compute the change in ground state energy E_0. Because the flux may be represented by a vector potential $A_\phi = \phi/L$ one finds

$$D = L \frac{d^2 E_0}{d\phi^2}|_{\phi=0} \quad (7.13)$$

This expression is useful for numerical computations on finite lattices, and also for certain formal develoments given below. It may also be extended to non-zero temperatures by consideration of the change in free energy with respect to ϕ [6]. A comment however is in order: the derivative in Eq (7.13) is defined in terms of the difference in ground state energies $E_0(\Delta\phi) - E_0(0)$. In order for this to make sense the change $\Delta\phi$ must be small enough that no level crossings occur. In a typical $d-$dimensional non-superfluid system, the spacing between low-lying levels is of order $\Delta E \sim L^{-d}$ so one requires a $\Delta\phi\langle 1/(DL^{d-1})$. In $d = 1$ a $\Delta\phi$ of order unity may be used to obtain energy differences numerically; in $d > 1$ the maximum permissible $\Delta\phi$ vanishes rapidly as system size increases. Remarkably the D defined in this manner is still relevant to physical properties. For further discussion, see [7].

Of course, a real system never has perfect translation invariance. In a non-superfluid system the presence of weak breaking of translation invariance (for example from a low density of imperfections in the crystal), leads to a broadening of the delta function; typically into a Lorentzian, with area given approximately by D and width given by the residual scattering rate. If the residual scattering is small, the Lorentzian is reasonably well defined (although there is always some ambiguity about continuing it into σ'_{reg}) and its area may be found approximately. As the residual scattering increases, the ambiguities increase. A schematic representation of the conductivity is shown in Fig. (7.1).

2.1.3 Extended Drude Parametrization. We have seen that scattering processes (due for example to electron-impurity, electron-electron and electron-phonon scattering) may broaden the delta function of a non-superconducting metal. For example, the familiar 'Drude' approximation [8] leads for noninteracting electrons in the presence of static, impurity scattering to

$$\sigma_{drude}(\omega) = \frac{e^2 D}{-i\omega + \Gamma} \tag{7.14}$$

with Γ a scattering rate and D equal to that of the pure system, up to corrections of order $(\Gamma/D)^2$. This familiar form motivates the widely used 'extended drude' parametrization of the conductivity, namely

$$\sigma(\omega) = \frac{e^2 D}{-i(1+\lambda(\omega))\omega + \Gamma(\omega)} \tag{7.15}$$

with frequency dependent scattering rate $\Gamma(\omega)$ and 'mass enhancement' $\lambda(\omega)$ *defined* via

$$\Gamma(\omega) = e^2 D \,\mathrm{Re}\left[\sigma(\omega)^{-1}\right] \tag{7.16}$$

$$1 + \lambda(\omega) = e^2 D \,\mathrm{Im}\left[\sigma(\omega)^{-1}\right] \tag{7.17}$$

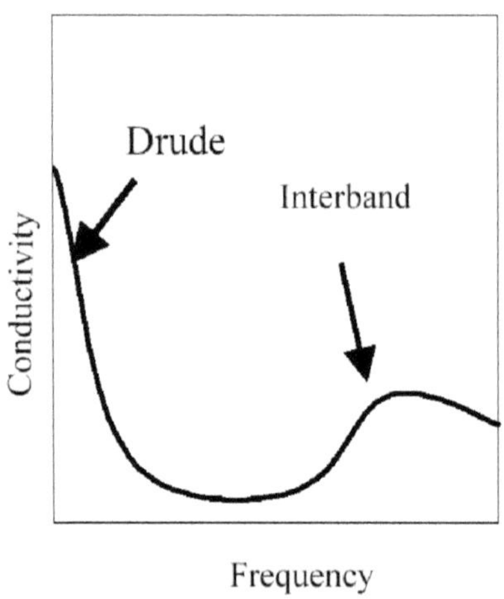

Figure 7.1. Qualitative representation of conductivity of 'typical' solid, showing in-band ('Drude') and interband transitions.

Note that because σ^{-1} is the response function yielding the field induced by an applied current, it is causal, implying that Γ and λ are related by a Kramers-Kronig relation. Extraction of the magnitude of Γ and λ from data requires a choice for D but the functional form is independent of D. Finally, it is important to be aware that although Γ and λ are always well defined, their interpretation in terms of scattering and mass renormalization depends crucially on physical assumptions, in particular that interband transitions do not contribute appreciably to the conductivity in the frequency range of interest. For a detailed discussion of the application of 'extended drude' ideas to high-temperature cuprate superconductors, see [9]

2.2 Spectral Weights and Sum Rules

2.2.1 Definition and f-sum rule. **Spectral weight** K is $(2/\pi)$ times the integral of the real part of the conductivity over some frequency range:

$$K(\Omega) = \int_0^{\Omega} \frac{2d\omega}{\pi}\sigma'(\omega) \tag{7.18}$$

Taking the $\omega \to \infty$ limit of Eq (7.8) and using Eqs (7.6), (7.7) yields the familiar f-sum rule for the total spectral weight in the conductivity:

$$\frac{ne^2}{m} = K(\infty) = \int_0^{\infty} \frac{2d\omega}{\pi}\sigma'(\omega) \tag{7.19}$$

This derivation, due to Kohn [4], shows that the f-sum rule follows directly from causality and gauge invariance, and does not depend on boundary conditions or anything else. It is thus completely general, but in the solid state physics context not particularly useful, because the integral must be taken over all frequencies, so the conductivity integral on the right hand side of Eq. (7.19) includes transitions from deep core levels to very high lying continuum levels while the quantitiy n is the total density of all electrons in the solid. Further, the generality of the sum rule means that it contains no information about the specifics of the system (for example, about what makes high-T_c materials different from silicon). Important system-specific questions include 'how is the spectral weight distributed in frequency?' and 'how does the distribution of spectral weight change when temperature or other parameters vary?'.

2.2.2 Down-folded models and restricted sum rules. In condensed matter physics one usually does not wish to deal with the full complexity of a solid. Rather, one deals with an "effective" or "down-folded" model which focusses on a (typically low energy) subset of the degrees of freedom. One obtains this model (at least notionally) by integrating out the other degrees of freedom in the solid. One widely used example (especially common in studies of transition metal oxides) is the 'tight binding + interactions model'. In many transition metal oxides, band theory suggests that the electronic states of interest lie in relatively narrow bands reasonably well separated from other bands. In this case it is reasonable to model the bands of interest via

$$H = -\sum_{i,\delta}(t^{ab}(\delta)c^{+}_{i,a}c_{i+\delta,b} + H.c.) + H_{int} \tag{7.20}$$

Here i denotes sites on a lattice, δ denotes a displacement connecting two lattice sites and a, b denote orbitals of electrons on a given site. The hopping matrix elements $t^{ab}(\delta)$ are often estimated from fits to a band theory calculation or, more correctly, from a down-folding procedure applied to a band calculation [10] and H_{int} denotes interactions of various sorts. Such models are more amenable to theoretical analysis than is the full Hamiltonian, Eq. (7.3). In most cases of physical interest t is negligible if δ is greater than one or two lattice constants. It is important to remember that interactions not included in conventional band calculations may in principle affect the values and indeed the form of the hopping part of the Hamiltonian. This issue has not been thoroughly studied, and in practice the procedure of obtaining hopping parameters from band theory and adding interactions is almost universally adopted.

To determine the optical conductivity of such models one requires a means of coupling an electromagnetic field to the model. The coupling usually adopted is the *Peierls substitution* $t(\delta) \to t(\delta)e^{i\frac{e}{c}\mathbf{A}\cdot\delta}$. This follows from the combined assumptions of gauge invariance and reasonably spatially localized wave functions, so that in the presence of a slowly spatially varying electromagnetic field one has $c_i^+ \to c_i^+ e^{i\int^{R_i}\mathbf{A}\cdot\mathbf{dl}}$. One must also assume that H_{int} involves only density-density (charge or spin) interactions, so that it is trivially gauge invariant. This assumption has been questioned by Hirsch [11] who argues that 'occupation modulated hopping' terms of the form $c_i^+ c_j(n_i + n_j)$ are important. These terms have not so far been much studied by other workers.

The Peierls assumption implies that the only important optical processes are those which move an electron from one site in the crystal to another. If there are optically allowed transitions between orbitals on the same site (e.g. the familiar hydrogenic $1s \to 2p$ transition), these must be added separately.

Once an A dependence is determined it seems plausible that the current may be computed from the definition $j_r = \frac{c}{NV_{cell}}\delta H/\delta A_r$ (here V_{cell} is the volume of a unit cell, N is the number of unit cells in the crystal and r is a Cartesian coordinate). The Peierls coupling, for example, yields

$$j_r = -\sum_{i,\delta}\left[it^{ab}(\delta)(e^{i\mathbf{A}\cdot\delta}c_{i,a}^+ c_{i+\delta,b} - e^{-i\mathbf{A}\cdot\delta}c_{i,b}^+ c_{i+\delta,a})\delta_r\right] \tag{7.21}$$

We emphasize the advantage of this procedure: if, as occurs for example in the Peierls substitution case, the A dependence of the effective model may be found from general arguments, one obtains optical matrix elements *without explicit computation of wave function overlaps* (of course these are implicitly included in H; for example in the Peierls case through the overlap matrix $t^{ab}(\delta)$).

Application to effective models of the arguments justifying the f-sum-rule Eq (7.19) yields a *restricted sum rule,* which relates the oscillator strength in the conductivity of the effective model to the expectation value of an operator given by the part of j linear in A. The form of the operator depends on the model; in the particular case of the Peierls coupled tight binding model the operator is the hopping amplitude weighted by the distance hopped and one finds

$$\frac{2}{\pi e^2}\int_0^\infty d\omega\sigma_{r,s}^{tb}(\omega) = \frac{1}{V_{cell}}\sum_{i,\delta}(t^{ab}(\delta)\delta_r\delta_s\langle c_{i,a}^+ c_{i+\delta,b}\rangle + H.c.) \qquad (7.22)$$

Here r, s denote cartesian directions, σ^{tb} denotes that portion of the conductivity arising from transitions among orbitals described by Eq (7.20) and again the sum rule applies both to infinite and finite lattices and to arbitrary boundary conditions. The restricted sum rule was apparently first noted by Maldague [12] and was rediscovered and its importance stressed by Baeriswyl et al [13]; the present derivation comes from [14].

The restricted sum rule relates the spectral weight in a subset of all optical transitions (those described by the low energy effective model) to an expectation value, which may depend on many parameters including temperature and interaction strength. This should be contrasted to the full f-sum-rule which relates the integral over all optical transitions to the total carrier density and bare mass, which are temperature and interaction independent. The difference between the full and restricted sum rules is made up by transitions involving orbitals not included in the effective model. In particular, temperature and interaction strength dependence of the restricted sum rule is compensated by transitions between orbitals included in the effective model and orbitals not included in the effective model. *At present there is no understanding of the relevant orbitals or energy range over which the full f-sum-rule is restored.*

We now specialize to tight binding models and make the further assumptions that the lattice has orthorhombic symmetry with three lattice constants $a_{1,2,3}$ and that the only appreciable hopping is between nearest neighbor sites. Then $\sigma_{r,s} \sim \sigma_r\delta_{r,s}$ and

$$\frac{2}{\pi e^2}\int_0^\infty d\omega\sigma_r^{tb}(\omega) = \frac{a_1^2}{a_2a_3}\sum_{i,\hat{r}}(t^{ab}(\hat{r})\langle c_{i,a}^+ c_{i+\hat{r},b}\rangle + H.c.) \equiv \frac{a_1^2}{a_2a_3}K_r \qquad (7.23)$$

In other words, if the hopping is only nearest neighbor a measurement of the optical conductivity gives the expectation value of the hopping amplitude, i.e. the electron **kinetic energy K.** (Note that an optical experiment moves electrons in one direction only; thus it yields the kinetic energy of motion in that direction). Thus if the assumptions leading to Eq (7.23) apply, then an optical measurement yields a fundamental parameter of a many-body Hamiltonian, namely the expectation value of the kinetic energy. This is important because in the non-interacting limit, $E = -K$ so the ground state wave function is the one which extremizes K. If interactions are fundamentally important (i.e. if the ground state wave function is fundamentally different from the band theory one) then K will be substantially *reduced* from its non-interacting value. An explicit example of this phenomenon will be found below in the section concerning the Hubbard model.

On the other hand, for electrons interacting with phonons the situation is different. The standard theory of electron-phonon interactions involves two parameters: a dimensionless coupling, conventionally denoted λ and the ratio of a typical phonon frequency ω_D to a typical electronic energy t. In most cases $\omega_D/t << 1$, and the expansion parameter describing fluctuations about the ground state is $\lambda\omega_D/t$. Provided λ is less than a (model dependent) critical value λ_c, the ground state is essentially that given by band theory, the physics is described by the familiar Migdal-Eliashberg theory and one finds among other things that $K = K_{band} - \mathcal{O}(\lambda\omega_D/t)$. For $\lambda > \lambda_c$ the ground state is fundamentally reconstructed (typically to an insulating 'polaronic' state) and K drops rapidly. For an explicit example see [15].

2.2.3 Down-folding and optical matrix elements. The procedure of obtaining optical matrix elements via knowledge of the A dependence of an effective Hamiltonian is appealing and is important in practical terms. A more careful derivation is therefore desirable. Consider a formal 'down-folding' procedure: separate the Hilbert space into a low energy sector (L) involving orbitals of interest and a high energy (H) sector which we do not wish to explicitly consider. The Schroedinger equation $H\psi = E\psi$ may be written in obvious notation as

$$\begin{pmatrix} H_H & H_M \\ H_M^* & H_L \end{pmatrix} \begin{pmatrix} \psi_H \\ \psi_L \end{pmatrix} = E \begin{pmatrix} \psi_H \\ \psi_L \end{pmatrix} \tag{7.24}$$

The high energy subspace may be formally eliminated, leading to an equation involving only the low energy subspace which is

$$H_{eff}(E)\psi_L \equiv \left[H_L + H_M^*(E - H_H)^{-1}H_M\right]\psi_L = E\psi_L \tag{7.25}$$

If we choose $\langle\psi_L|\psi_L\rangle = 1$ the exact (normalized) eigenfunction is

$$|\psi\rangle = \frac{\left[(E - H_H)^{-1}H_M|\psi_L\rangle + |\psi_L\rangle\right]}{\sqrt{1 + \langle\psi_L|H_M^*(E - H_H)^{-2}H_M|\psi_L\rangle}} \tag{7.26}$$

Using the same splitting one can write the exact current operator in a form which depends explicitly only on the components of the wave function in the L sector. If N_1 and N_2 are the (energy-dependent) normalization factors for the exact wave-functions corresponding to $\psi_{L1,2}$ then one finds

$$\begin{aligned} j_{12} &= [\langle \psi_{l1}|H_M^*(E_1 - H_H)^{-1} j_H (E_2 - H_H)^{-1} H_M |\psi_{L2}\rangle \\ &+ \langle \psi_{l1}|H_M^*(E_1 - H_H)^{-1} j_M |\psi_{L2}\rangle \\ &+ \langle \psi_{l1}|j_M^*(E_2 - H_H)^{-1} H_M |\psi_{L2}\rangle \\ &+ \langle \psi_{l1}|j_L|\psi_{L2}\rangle]/N_1 N_2 \end{aligned} \tag{7.27}$$

Comparison of Eqs (7.25), (7.27) shows that the diagonal matrix elements of j are correctly given by the $\langle\psi_L|j|\psi_L\rangle = \langle\psi_L|\delta H_{eff}(E)/\delta A|\psi_L\rangle$ (note the presence of E in the normalization!). *Off diagonal* matrix elements within the L subspace are not in general simply related to functional derivatives of H_{eff} because of the two energies occurring in the matrix element. If in the energy range of interest one may neglect the variation of the operators above with E then the matrix elements are correctly given by functional derivatives of H_{eff} with respect to A. In the same way neglect of the variation of H_{eff} with E allows one to derive a restricted sum rule as above.

2.2.4 Accuracy of Peierls Phase Approximation. As mentioned above, a particularly convenient and widely-used effective model involves a tight-binding parametrization of band theory, along with some interactions. It is therefore important to consider how well the $A-$dependence of H_{eff} is approximated by the Peierls phase ansatz. One issue concerns the importance of on-site ('1S-2P' like) transitions. This obviously depends on the system in question and cannot be discussed in general. Another concerns the possible relevance of the occupation modulated hopping terms in the interaction. Too little is known to allow discussion here. Even if these issues are neglected, a crucial question arises, related to the fact (seen e.g. in Eq (7.26)) that the physical wave function represented by the operator c_i^+ (which we like to think of as creating an electron in an 'atomic-like' state of wave function ϕ_i) in fact has a non-negligible fraction of its charge density coming from other orbitals, and may not be particularly well localized, so that the Peierls ansatz does not accurately describe the change in the wave function in the presence of a vector potential. There is also a nontrivial choice of basis aspect. To understand this, suppose the Hamiltonian, Eq. (7.20) is accepted (including the Peierls phase coupling to A). One may then change the basis from the original one ϕ_i to $\psi_i = \sum_j D(i-j)\phi_j$. In the new basis t changes to $t_{i-j}e^{iA\cdot(R_i-R_j)} \rightarrow t'_{i-j}(A) = \sum_{mn} D^*(i-m)t_{m-n}e^{iA\cdot(R_m-R_n)}D(n-j) \neq t'_{i-j}e^{iA\cdot(R_i-R_j)}$. Thus,the Peierls phase ansatz can at most be correct in one basis, and does not have to be correct in any basis.

There is presently substantial interest in transition metal oxides in which the important electronic states are relatively narrow bands derived from transition metal d-states (hybridized with oxygen p-states) and are relatively isolated from other bands. For such systems I believe that the 'correct' basis choice for t_{i-j} is the one in which the Peierls phase ansatz most nearly approximates diagonal matrix elements (in particular those giving rise to the charge stiffness) computed via other techniques. For example, it will be seen below that within band theory there is a standard expression for the charge stiffness, which one may compare to that obtained from the Peierls ansatz in a given tight binding basis.

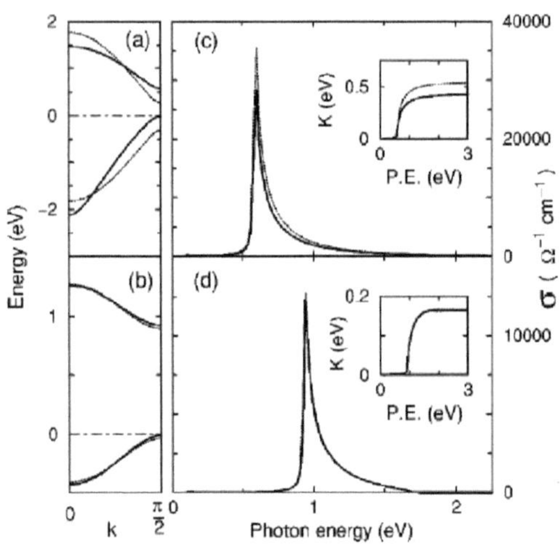

Figure 7.2. Comparison of exact and Peierls-ansatz computations of the conductivity and spectral weight for 'Doubled Kronig Penney Model' of electrons in one dimension, from [16].

The accuracy of the Peierls-phase matrix elements has not been the subject of systematic study in realistic situations. Ahn and Millis [16] have investigated the Kronig-Penney model of electrons in one spatial dimension in the presence of a periodic array of delta functions. They compared the exact conductivity to the result obtained by making a nearest neighbor tight-binding fit to the lowest-lying bands and found that in all reasonable situations even the oversimplified nearest neighbor approximation yielded spectral weights accurate to within 10%. Some of their results are reproduced as Fig. (7.2).

2.3 Simple Examples

2.3.1 Galilean Invariant Models. A situation frequently encountered in textbooks and occasionally in practice is the Galilean-invariant limit. If no ions are present (or the spatially varying part of the ionic potential can be neglected, as is the case for a low density of electrons in a clean semiconductor), and if (as is usually the case) H_{el-el} is a function of relative positions of electrons only, then the current is proportional to the momentum and is conserved: $\left[H, \overrightarrow{j}\right] = 0$. In this circumstance $\langle n|j_a|m\rangle = 0$ if $n \neq m$. Further, states $|n\rangle$ excited with non-negligible thermal probability $e^{-\beta E_n}/Z$ have a current (expectation value of j) which vanishes in the infinite system size limit. Thus at least in the infinite system size limit the current-current correlation term χ can usually be neglected and one has (GI stands for Galilean-invariant)

$$\sigma_{ab}^{GI}(\omega) = \left(\frac{ne^2}{im(\omega - i\varepsilon)}\right)\delta_{ab} \tag{7.28}$$

In summary, in a Galilean-invariant model (with arbitrary but Galilean-invariant interactions) the real part of the conductivity is entirely concentrated at $\omega = 0$: the response to a non-zero-frequency field is entirely reactive; the carriers are simply freely accelerated by the electric field.

2.3.2 Band Theory. Modern band theory is not a non-interacting theory. Electron-electron interactions are taken into account by different approximations to density functional theory [17]. The implementation most widely used is based on an effective single particle Schroedinger equation involving a non-local 'exchange-correlation potential' which contains a significant contribution from electron-electron interactions and is determined by a self consistency condition. Solving this equation yields a set of effective one-electron energy levels $\varepsilon_n^{eff}(p)$ and wavefunctions $\psi_{n,p}(r)$ such that (μ is the chemical potential) $E_{DFT} = \sum_n \int \frac{d^3p}{(2\pi)^3}\varepsilon_n^{eff}(p)\Theta(\mu - \varepsilon_n^{eff}(p))$ is a good approximation to the ground state energy (the approximation would be exact if the exact non-local potential were used and if the equation could be solved exactly, and the energetics obtained from standard approximations are often remarkably good in practice). The $\varepsilon_n^{eff}(p)$ and $\psi_{n,p}(r)$ themselves have in principle no rigorous meaning, but are often interpreted as actual electron energy levels and wave functions. Within band theory for a static, perfectly ordered lattice one expects that the real part of the conductivity has a delta function contribution from states at the fermi surface (if the material is predicted to be a metal) and interband contributions generically separated from $\omega = 0$ by an energy gap, so $\sigma^{reg}(\omega) = 0$ for a non-zero range of frequencies around $\omega = 0$ [8].

The arguments leading to Eq (7.13) may be applied to the density functional formalism to obtain an expression for D. This expression would be exact if the flux dependence of the exact exchange-correlation potential were known. In the band theory literature, the flux dependence of the exchange correlation potential is neglected, so the flux only enters in the derivative term of the Schroedinger equation. D may then be computed easily and is [8]:

$$D_{band,ab} = \sum_n \int \frac{d^3p}{(2\pi)^3} \frac{\partial \varepsilon_n^{eff}(p)}{\partial p_a} \frac{\partial \varepsilon_n^{eff}(p)}{\partial p_b} \delta(\varepsilon_n^{eff}(p) - \mu) \qquad (7.29)$$

In other words, if the flux dependence of the exchange-correlation potential is neglected then D is given by the average over the fermi surface of the appropriate components of the fermi velocity multiplied by the density of states. For many correlated materials, this is a very poor approximation to D, and the discussion of fermi liquid theory in the next subsection shows that neglect of the flux dependence of the exchange correlation potential is in principle incorrect.

An extensive literature exists on optical properties computed using Eq (7.6) with band theory wave functions used to compute the matrix elements, and with additional interaction corrections added using various roughly speaking perturbative extensions of band theory (most notably the 'GW' approximation); for a review see, e.g. [18]. Recent important work has addressed electron-hole correlation effects [19]. This approach involves extremely heavy computations, and has not yet been widely applied to 'strongly correlated' transition metal oxides (but see [20]).

2.3.3 Integrable Systems. One expects on general grounds that at $T > 0$, $D = 0$: even in the absence of disorder, many-body interactions will broaden the delta function in some manner. However, theoretical work over the last decade [6, 21] strongly sugggests that in many *integrable* one dimensional models, $D > 0$ at all T, so the systems are infinitely conducting at all temperatures. This peculiar result is apparently a consequence of the infinite number of conservation laws characteristic of integrable systems. The issue is discussed in more detail in Chapter 11.

2.3.4 Fermi Liquid Theory. The low energy properties of many interacting electron systems are believed to be well described by L. D. Landau's 'fermi liquid theory' (for references, see e.g. [3, 22]).

One crucial property of a fermi liquid is the existence of electronic 'quasiparticles' which behave in many ways as conventional electrons, but with a renormalized dispersion $v^*(p) = \frac{m_{band}(p)}{m^*(p)} \frac{\partial \varepsilon_n^{eff}(p)}{\partial p_a}$ where the momentum label indicates position on the fermi surface and we have defined a 'mass enhancement' m^*/m_{band}. Response functions of fermi liquids are determined by a combination of quasiparticle dispersion and 'Landau parameters' representing the feedback of the system on a given electronic state. Fermi liquid theory leads to

$$D_{observed} = D_{qp}(1 + F_{1s}/d) \tag{7.30}$$

where D_{qp} is obtained by using the quasiparticle dispersion in Eq (7.29), d is the dimensionality and F_{1s} is a Landau parameter. In a Galilean-invariant system the mass renormalization is independent of position on the (spherical) fermi surface and the relation $m^*/m = 1 + F_{1s}/d$ which follows from Galilean invariance ensures that D is unrenormalized. In a non-galilean-invariant system there is no such relation. If the many-body renormalizations involve a strictly momentum independent self energy then $F_{1s} = 0$ [23] and in 'heavy fermion' materials the effects due to velocity renormalizations are much larger (factor ~ 100) than those due to F, although the Landau parameter effects have been measured in a few cases and are important at the factor of two level [24]. On the other hand, in a fermi liquid near a ferromagnetic transition the effective mass diverges as the transition is approached, but because the critical modes involve long wavelength fluctuations (i.e. mainly forward scattering) one does not expect the conductivity to be strongly affected. Thus in this case the Landau parameter must diverge along with the effective mass.

2.3.5 Superconductivity and Density Waves. As the temperature is decreased a fermi liquid may become unstable to various forms of long ranged order. Two particularly instructive special cases are the superconducting and density wave instabilities.

A *superconductor* is characterized by a non-vanishing superfluid stiffness $\rho_S = \frac{1}{4e^2} \lim_{q\to 0, \omega\to 0} (\omega\sigma(q,\omega))$ where the 1/4 is conventional and refers to the charge $2e$ of a Cooper pair. The magnetic field penetration depth may be inferred from the $q \to 0$ $\omega = 0$ limit of the response, and in superconductors the order of limits does not matter, so that $\sigma(\omega)$ has a term proportional to $\delta(\omega)$ whose magnitude may be inferred from the penetration depth [25]. A superconductor is also characterized by an energy gap Δ which in conventional superconductors is much less than the characteristic conduction band energy scale E_F. In the conventional theory, up to terms of relative order $(\Delta/E_F)^2$ the total spectral weight in a superconductor is the same as that of the corresponding normal system, so the weight in the superfluid stiffness is mainly transferred down from higher frequencies (within the conventional theory it comes from ω of the order of a few times Δ).

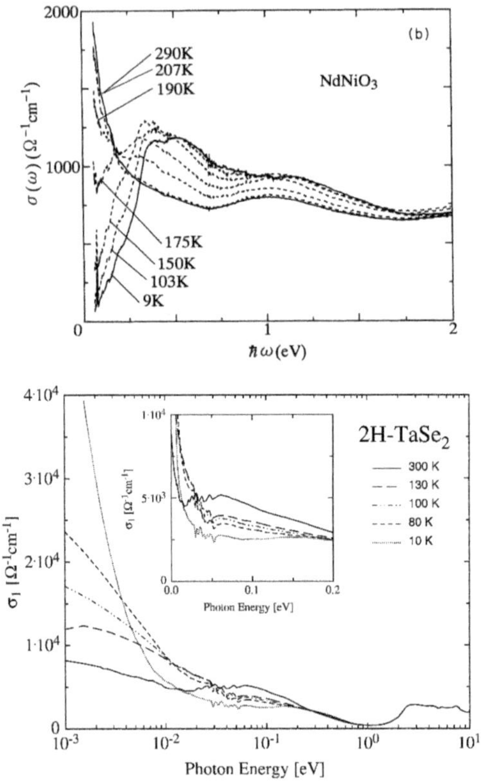

Figure 7.3. Upper panel: Measured conductivity of charge density wave system $NdNiO_3$ [28] showing temperature evolution consistent with conventional considerations. Lower panel: Measured conductivity of charge density wave system $TaSe_2$ [29] showing temperature evolution inconsistent with conventional expectations.

These arguments were introduced and experimentally verified by Tinkham in the late 1950s and are beautifully explained in [25]. In high temperature superconductors, conservation of spectral weight as temperature is varied across the superconducting transition was verified at the 10% level in the early 1990s [26]. Of course, small changes in conduction band spectral weight are expected as the temperature is varied through the superconducting transition temperature T_c and recent improvements in experimental technique have allowed these changes (which seem to be at the 1% level) to be observed [27] (see Chapter 8 for further discussion of this issue).

A *density wave* occurs when the electron charge or spin density acquires a periodicity different from that of the underlying lattice. One expects this density modulation to cause an additional periodic potential (energy gap Δ) which is felt by the mobile electrons, which eliminates some of the fermi surface. Just as in superconductors the total weight is expected to be conserved (up to terms of relative order $(\Delta/E_F)^2$) so the formation of the density wave is expected to shift spectral weight up in frequency. This behavior is observed in many density wave materials: an example is shown in the upper panel of Fig (7.3)

[28]. Remarkably, in some materials (most notably the dicalcogenides such as $TaSe_2$) the expected upward shift does not occur [29]: instead, as the temperature is lowered across the density wave transition spectral weight shifts downwards in frequency, apparently because the scattering rate is reduced. This behavior is shown in the lower panel of Fig (7.3). An understanding of the different origins of these two behaviors would be very desirable.

3. Specific Model Calculations

3.1 Direct evaluation of Kubo formula

3.1.1 Formalism: momentum independent self energy.

In this approach one starts from a model of electrons with a given dispersion ε_p (typically interband transitions are neglected), a coupling to the electromagnetic field given by $p \to p - eA/c$ and interactions which are treated by the methods of diagrammatic perturbation theory. Important objects in these calculations are the electron propagator G and self energy Σ:

$$G(p,\omega) = \frac{1}{\omega - \varepsilon_p - \Sigma(p,\omega)} \tag{7.31}$$

A general expression for the conductivity is [30]

$$\sigma_{ab}(i\Omega) = \sigma_{ab}^{dia} + \sigma_{ab}^{para} \tag{7.32}$$

with

$$\sigma_{ab}^{dia}(i\Omega) = \frac{2e^2}{i\Omega} T \sum_{\omega} \int \frac{d^3p}{(2\pi)^3} \frac{\partial^2 \varepsilon_p}{\partial p_a \partial p_b} G(p,\omega) \tag{7.33}$$

and

$$\sigma_{ab}^{para}(i\Omega) = -\frac{2e^2}{i\Omega} T \sum_{\omega} \int \frac{d^3p}{(2\pi)^3} \frac{\partial \varepsilon_p}{\partial p_a} G(p,\Omega+\omega) G(p,\omega) T_b^{\Omega}(p,\omega) \tag{7.34}$$

The vertex function T_b satisfies the integral equation

$$T_b^{\Omega}(p,\omega) = \frac{\partial \varepsilon_p}{\partial p_b} - 2T \sum_{\omega'} \int \frac{d^3p'}{(2\pi)^3} I_{pp'}^{\Omega}(\omega,\omega') G(p',\Omega+\omega') G(p,\omega') T_b^{\Omega}(p',\omega') \tag{7.35}$$

and I is a particle-hole irreducible vertex, whose limit as $\Omega, \omega \to 0$ and $p \to p_F$ is the Landau interaction function. In general an expression for I is difficult to determine and the equation is difficult to solve: quantum Boltzmann equation methods [31] have been more useful in practice.

One instructive special case which can be analysed in detail is a momentum independent self energy, $\Sigma(p,\omega) \to \Sigma(\omega)$. This situation is (to a good approximation) realized in practice for a high density of electrons coupled to phonons [32], in the 'dynamical mean field' or '$d=\infty$' approximation [23] and in the 'marginal fermi liquid' model for high temperature superconductivity [33]. In these cases, the vertex correction vanishes and for a non-superconducting system on the imaginary frequency axis (the 2 is for spin and $\omega_+ = \omega + \Omega$)

$$\sigma_{ab}(i\Omega) = \frac{2e^2}{i\Omega} T \sum_\omega \int \frac{d^3p}{(2\pi)^3} \left[\frac{\partial^2 \varepsilon_p}{\partial p_a \partial p_b} G(p,\omega) - \frac{\partial \varepsilon_p}{\partial p_a} \frac{\partial \varepsilon_p}{\partial p_b} G(p,\omega_+) G(p,\omega) \right] \tag{7.36}$$

A note on units: for simplicity (and because it is the case of greatest experimental relevance) we consider an orthorhombic lattice with lattice constants $d_{a,b,c}$. Making the momentum integrals dimensionless via $p'_a = d_a p_a$ etc and integrating by parts on the first term (recall $\partial\Sigma/\partial p$ is assumed to vanish) leads to

$$\sigma_{ab}(i\Omega) = \frac{2\sigma_0}{i\Omega} T \sum_\omega \int \frac{d^3p'}{(2\pi)^3} \frac{\partial \varepsilon_p}{\partial p'_a} \frac{\partial \varepsilon_p}{\partial p'_b} \left[G(p',\Omega+\omega) G(p',\omega) - G(p',\omega)^2 \right] \tag{7.37}$$

where

$$\sigma_0 = \frac{e^2}{\hbar d_c} \tag{7.38}$$

The conductivity evidently has the dimension of energy and may be converted to conventional units by recalling that $e^2/\hbar = 4k\Omega$.

Eq (7.37) may be easily evaluated numerically if $\Sigma(\omega)$ is known. A widely studied limit arises if, for all relevant frequencies, Σ is small compared to the regime over which ε_p varies. To be precise, if $-W_1 < \varepsilon_p < W_2$ then up to terms of order $\Sigma/(\min(W_1, W_2))$ one may use a pole approximation to perform the integral over the magnitude of ε_p. Performing the standard analytical continuation leads for $\Omega > 0$ to

$$\frac{\sigma_{ab}(i\Omega)}{\sigma_0} = \int \frac{d\omega}{2\pi} \frac{d\cos(\theta)\, d\phi}{4\pi^2} \frac{v_a v_b N_0(\theta,\phi) \left(f(\omega) - f(\omega_+)\right)/\Omega}{\Omega - \Sigma'(\omega_+) + \Sigma'(\omega) - i\left(\Sigma''(\omega_+) + \Sigma''(\omega)\right)} \tag{7.39}$$

Here $\omega_+ = \omega + \Omega$, $v_{a,b}$ are the a, b components of the fermi velocity, f is the fermi function and $N_0(\theta,\phi) = 1/\left|\partial\varepsilon_p/\partial p\right|$ at the fermi surface point specified by the angles θ, ϕ.

3.1.2 Electron-phonon interaction. For electrons interacting with dispersionless optical phonons of frequency ω_0 the self energy

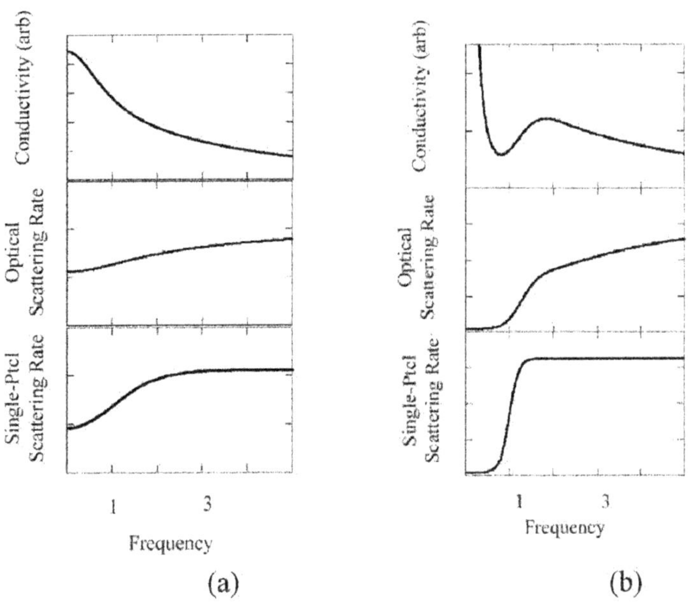

Figure 7.4. Optical conductivity, 'optical scattering rate' and single-particle scattering rate for model of electrons coupled to dispersionless optical phonons, computed from Eqs (7.39, 7.40, 7.41, 7.16). Left panel (a): high temperatures: ($T = \Omega_D/2$) Right panel (b): low temperature ($T = 0.1\Omega_D$). Frequencies measured in units of Ω_D. Coupling $\lambda = 1$. Units of scattering rate and conductivity are arbitrary; frequency is scaled to phonon frequency. The scales for the optical and single particle rates are identical.

on the real frequency axis is [22] (f is the fermi distribution function)

$$\Sigma'_{ep}(\omega) = \lambda \int d\omega' f(\omega' - \omega) \frac{\omega_0^2}{\omega_0^2 - \omega'^2} \tag{7.40}$$

$$\Sigma''_{ep}(\omega) = \frac{\pi\lambda\omega_0}{2} \left(\coth\left[\frac{\omega_0}{2T}\right] - \frac{\sinh\left[\frac{\omega_0}{T}\right]}{\cosh\left[\frac{\omega}{T}\right] + \cosh\left[\frac{\omega_0}{T}\right]} \right) \tag{7.41}$$

The panels of Fig (7.4) show the real part of optical conductivity, the optical and the single-particle scattering rate computed from Eq (7.39) using Eqs (7.40,7.41). The computation was performed for two different temperatures–one much lower than the Debye frequency and one equal to $\omega_D/2$. For $T = \omega_D/2$ (panel a) the conductivity is already not very far from the simple 'Drude' form, as seen from the conductivity and from the frequency dependence of the scattering rate. For the lower temperature (panel b), one sees a large low frequency peak (very weakly scattered electrons, with a mass increased by the electron-phonon interaction) and an extra absorption beginning at the phonon frequency. One sees that the 'optical mass' and scattering rate are roughly speaking 'smoothed' versions of the single-particle mass and scattering rate.

3.1.3 Marginal fermi liquid ansatz. The marginal fermi liquid ansatz [33] is a theoretical prediction for the electron self energy of optimally doped high-T_c superconductors, which seems to have been borne out by recent photoemission experiments [34]. It has not been convincingly derived from any microscopic model. The marginal fermi liquid ansatz for the imaginary part of the electron self energy is that at frequencies ω less than a cutoff frequency ω_c one has

$$\Sigma''(\omega) = \pi\lambda \max(\omega, \pi T) \tag{7.42}$$

If one assumes (as is normally done in the literature) that for frequencies greater than ω_c $\Sigma''(\omega) = \lambda\omega_c$ then

$$\Sigma'_{MFL}(\omega) = \lambda\omega_c \left(\ln\left|\frac{1-\frac{\omega}{\omega_c}}{1+\frac{\omega}{\omega_c}}\right| - \frac{\omega}{\omega_c}\ln\left|\frac{\frac{\omega_c^2}{\omega^2}-1}{\frac{(\pi T)^2}{\omega^2}-1}\right| - \frac{\pi T}{\omega_c}\ln\left|\frac{\pi T-\omega}{\pi T+\omega}\right| \right) \tag{7.43}$$

This self energy has been used with Eq (7.37) to analyse the frequency dependent conductivity of high-T_c superconductors; for results see [35]; but Drew and the author have presented evidence that Landau parameter effects are also important, at least at the factor-of-two level [36].

3.1.4 Weakly coupled lower-dimensional subsystems. An important sub-class of conductivities involves the motion of charge between weakly coupled lower dimensional subsystems; for example the interplane conductivity of high temperature superconductors or the interchain conductivity of quasi-one-dimensional materials. This situation may be described by a Hamiltonian of the form

$$H = -t_\perp \sum_{i,p,\sigma} \left(c^+_{i,p,\sigma} c_{i+1,p,\sigma} + H.c. \right) + H_\| + H_{inter} \tag{7.44}$$

where the label p denotes the momentum in the lower dimensional subsystem (plane or chain), σ is spin, and we have labelled the different planes (or chains) by i. Here $H_\|$ is the Hamiltonian describing the physics in an isolated low dimensional subsystem and H_{inter} denotes any interplane (interchain) interactions. In practice the only important contribution to H_{inter} is likely to be the coulomb interaction. We refer to the direction(s) in which the hybridization is weak as the 'transverse' directions.

The transverse conductivity is typically computed by using the Peierls ansatz to couple the electromagnetic field and then expanding as usual. The crucial new point is that if $t_\perp$ is weak compared to some appropriate in-plane or in-chain energy scale, the conductivity may be computed by perturbation theory in $t_\perp$. At leading nontrivial order one has

$$\sigma_\perp = \frac{t_\perp}{i\Omega}\langle c^+_{i,p,\sigma}c_{i+1,p,\sigma} + H.c.\rangle + \frac{\chi_\perp(i\Omega)}{i\Omega} \tag{7.45}$$

with χ given in the time domain by the usual commutator

$$\chi_\perp = \sum_{p,\sigma}\langle[j_c(t), j_c(0)]\rangle \tag{7.46}$$

with

$$j_c(t) = it_\perp\left(c^+_{i,p,\sigma}c_{i+1,p,\sigma} - H.c.\right) \tag{7.47}$$

Here the momentum sum is over in-plane or in-chain momenta only. To leading order in $t_\perp$ the expectation value which defines $\chi_\perp$ may be calculated assuming $t_\perp = 0$ while to obtain the first term one must calculate to first order in $t_\perp$. If the term H_{inter} may be neglected (or treated in mean field theory) then the results may be simply expressed as products of Green functions pertaining to $H_{||}$. In the absence of superconductivity or density wave order, the only nonvanishing Green function is $G(p,t) = \langle T_t\{c_{p\sigma}(t), c^+_{p\sigma}(0)\}\rangle$ and

$$\sigma_\perp = T\sum_{\omega,p,\sigma}\frac{G(p,\omega+\Omega)G(p,\omega) - G^2(p,\omega)}{i\Omega} \tag{7.48}$$

This formula is useful because in many cases the in-plane (in-chain) Green function is known, so the conductivity may be directly computed. For applications to the interplane conductivity of 'single-layer' high temperature superconductors see [37]; to bilayer cuprates see [38]; to quasi one dimensional materials see [39]. Eq (7.48) is however only the leading term in a perturbative expansion in $t_\perp$. Essler and Tsvelik have recently noted that for a particular form of $t_\perp$ a controlled treatment of the $1d$-$2d$ crossover (including optics) may be constructed [40].

The applicability of this formula has been questioned by Turlakov and Leggett, who argue that 'Coulomb blockade' effects similar to those producing tunnelling anomalies in disordered systems may be important [41]. The issue deserves further analysis.

3.2 The Hubbard Model

The Hubbard model is defined by the Hamiltonian

$$H_{hub} = -\sum_{i,j} t_{i-j} \left(c_{i\sigma}^{+} c_{j_s} + H.c.\right) + U \sum_{i} n_{i\uparrow} n_{i\downarrow} \qquad (7.49)$$

Most studies have assumed a d-dimensional cubic lattice and a hopping which is nonvanishing only between nearest neighbors. The Hubbard model displays a Mott transition: at a density of one electron per site and for a large enough interaction the ground state is insulating, ($D = 0$) and characterized by a gap to charge excitations. The main interest has been in the behavior of the conductivity (and other physical properties) in the vicinity of the Mott insulating phase. It is conventional to describe the carrier density in terms of a doping, $\delta = 1 - n$ away from half filling.

Consider now the kinetic energy and optical properties. For n far from 1, even very strong interactions have a weak effect on K essentially because the electrons can avoid each other; also the model is nearly galilean-invariant. For n near 1 a large U can have profound effects. For a density of one electron per site and sufficiently large U the ground state is insulating ($D = 0$) although band theory would predict it to be metallic. At $n = 1$ and very large U a good approximation to the ground state is one with one electron per site and an insulating gap $\Delta_{opt} = U - \alpha_d t$ and a_d a dimension-dependent constant.

Hopping leads to fluctuations into states with two electrons or no electrons per site and thus if $t \ll U$ to $K \sim t^2/U$. As t/U decreases, K increases, eventually saturating at the band theory value. Depending on the details of the band structure, the insulating behavior may persist down to arbitrarily small U (at $n = 1$) or there may be a critical U at which a Mott transition occurs. Whether or not this happens, one may distinguish large and small U by whether the kinetic energy is substantially (factor of 2) renormalized from the band theory value or not. In essentially all models, the large U regime extends down to $U \approx 2dK$ where d is the spatial dimensionality (recall K was defined as the hopping in one cartesian direction). For large U and n near but not equal to unity there is a small density of holes (if $n < 1$) or doubly occupied sites (if $n > 1$) and these can move more or less freely, leading to $K \sim t^2/U + |1 - n|t$. For $|1 - n|$ not small, the carriers mostly avoid each other; the renormalization of the kintic energy is small and the state is more or less conventional.

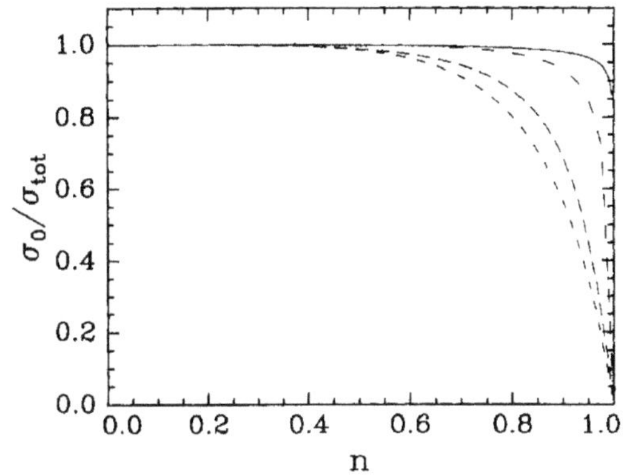

Figure 7.5. Charge stiffness (here denoted σ_0) of one dimensional Hubbard model normalized to total spectral weight, as function of doping for different interaction strengths, from Ref [42] .

In the $d = 1$ Hubbard model an exact solution is available. The zero temperature kinetic energy and Drude weight were computed by Schulz [42] and later studied in more detail by other workers [43]. Representative results are shown in Fig (7.5). Note that as the Mott phase is approached by varying doping, ($\delta \to 0$) D vanishes linearly in δ, with a coefficient which depends on interaction strength, and which may be interpreted as a correlation length ξ (normalized to the lattice constant) for the Mott insulating phase [43].

For not too large U (i.e. $\xi > 1$) the numerical results may be written as $D(\delta, U) = D_0 f(\delta\xi)$ with f a scaling function discussed in detail in [43]. For $\delta\xi \gtrsim 0.4$, $f \to 1$, implying among other things that for $\delta\xi > 1$ the spectral weight rapidly collapses into to the Drude peak. This special feature of one dimensional kinematics occurs even at larger U where the scaling theory does not apply, and is discussed in e.g. [44]. Note also that to observe a significant suppression of total spectral weight one requires a U at least of order the full band-width, and a $\delta \lesssim 0.4$.

For higher dimensions few (see for example [14]) exact results are available. Numerical studies of small clusters are available for $d = 2$ (essentially no useful results exist in $d = 3$); representative results [45] are shown in Fig. (7.6). Little is understood about finite size effects except in $d = 1$ where they have been extensively studied and found to be unpleasantly large [43]. The kinetic energy should be re-examined with the improved computers and algorithms available today.

If the momentum dependence of the self energy is negligible ($\Sigma(p, \omega) \to \Sigma(\omega)$) then a formally exact solution (which must still be implemented numerically) is available. This approximation, which is believed to be reasonably accurate in $d = 3$ is known as the 'dynamical mean field approximation' and has been used to compute the kinetic energy as a function of interaction and doping. Unpublished results obtained by Ferrara and Rozenberg for the kinetic energy are shown in the lower panel of Fig (7.6). Kotliar and collaborators have analysed the conductivity near the Mott transition using this approximation, and compared the results to data on V_2O_3 obtained by G. A. Thomas and co-workers [46]. Fig (7.7) presents theoretical results, along with a qualitative view of the theoretically expected structure, while Fig (7.8) shows the measurements.

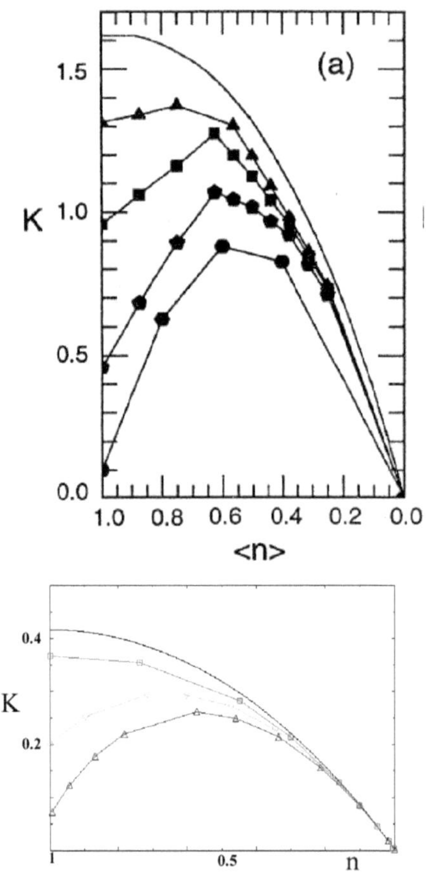

Figure 7.6. Upper panel: Kinetic energy K for Hubbard model in $d = 2$ for $U = 0, 4, 8, 20$; Lower panel: K in dynamical mean field approximation for $U = 0, 1, 2, 4$. The different bandwidth conventions mean that the U-values are not directly comparable; a useful rule of thumb is that when U equals the full bandwidth, $K(n = 1, U) \approx K(n = 1, U = 0)/2$

The electron green function is characterized by three features: a lower Hubbard band, an upper Hubbard band, and a quasiparticle peak in between. The conductivity correspondingly has three peaks corresponding to motion within the quasiparticle band, transitions from the quasiparticle band to the upper Hubbard band (or from the lower Hubbard band to the quasiparticle band) and from the lower to the upper Hubbard band.

One fundamental question which, remarkably, has not yet been fully answered concerns the form of the conductivity in the Mott insulating state. The model has a gap to charge excitations, so one expects a conductivity with vanishing real part at low frequencies, and a conductivity onset for frequencies above the Mott gap energy E_{Mott}. Information should be obtainable in $d = \infty$, but the author is unaware of specific results. In $d = 1$ analytical and numer-

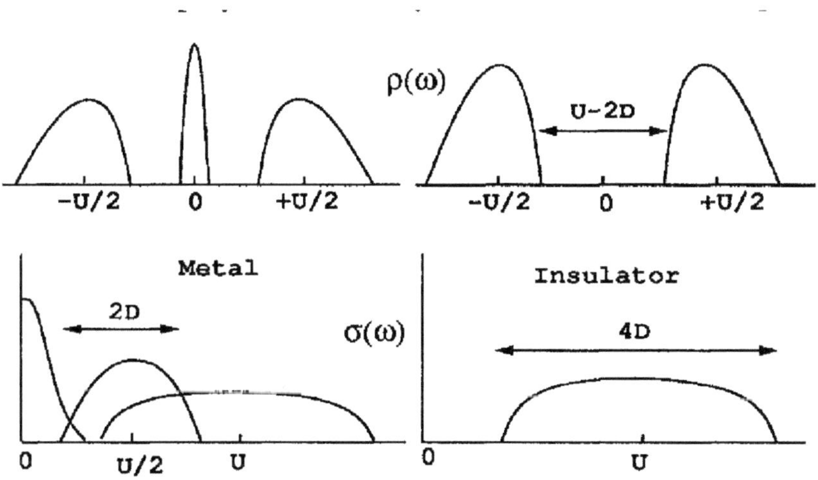

Figure 7.7. Calculated optical conductivity of V_2O_3 on insulating and metallic sides of Mott transition [46] The uppermost curves, labelled $\rho(\omega)$ show the changes in the electron spectral function as the metal is driven (by increasing interaction) from a correlated metal state (left graph) to a Mott insulating state (right graph). The lower graphs give the corresponding changes in conductivity.

ical renormalization group studies of Jeckelmann, Gebhard and Essler [47] indicate that that $lim_{\omega \to E_{Mott}} \sigma(\omega) \propto \sqrt{\omega - E_{Mott}}$ and that the maximum in σ occurs at $\omega \approx 1.25 E_{Mott}$.

3.3 t-J Model

The Hubbard model is an approximation to the physics of transition metal oxides on energy scales of the order of the conduction band width (if this is relatively narrow and relatively well separated from other bands in the solid). A further approximation, valid for large U and low energies, is the '$t - J$' model [48]. This may be derived from the Hubbard model by a formal canonical transformation procedure, and is of the form

$$H_{tj} = P_D \left[-\sum_{i,j} t_{i-j} \left(c^{+}_{i\sigma} c_{j_\sigma} + H.c. \right) + J \sum_{<ij>} \overrightarrow{S}_i \cdot \overrightarrow{S}_j + ... \right] P_D \quad (7.50)$$

Here P_D annihilates states containing doubly-occupied sites, $J \sim t^2/U$ and the ellipsis denotes terms of higher order in t/U.

As defined here the $t - J$ model is an effective model describing the low energy physics of models with strong on-site repulsion. Application

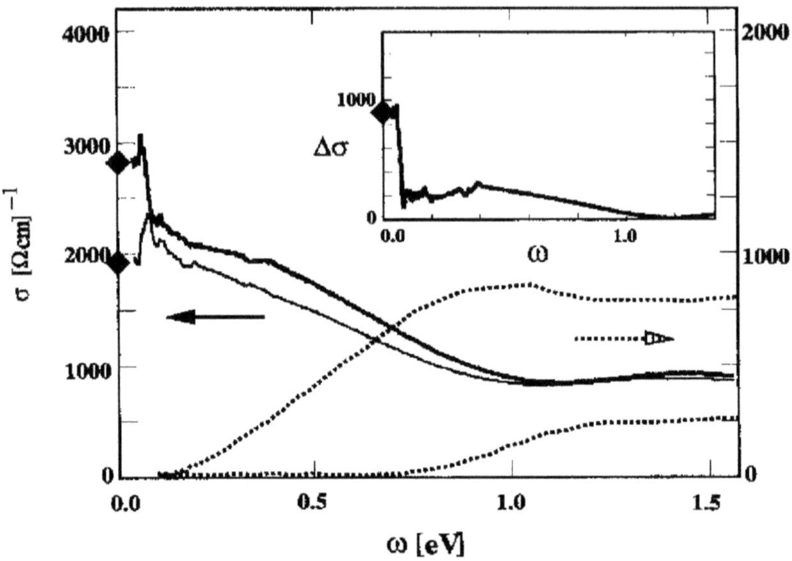

Figure 7.8. Measured optical conductivity of V_2O_3 on insulating and metallic sides of Mott transition. compared to 'dynamical mean field' calculations (from [46]).

of the standard f-sum-rule derivations shows that

$$\frac{2}{\pi e^2}\int_0^\infty d\omega \sigma_{r,s}^{t-J}(\omega) = \frac{1}{V_{cell}}\sum_{i,\delta,\sigma}(t(\delta)\delta_r\delta_s\langle P_D c_{i\sigma}^+ c_{i+\delta,\sigma}P_D\rangle + H.c.) \quad (7.51)$$

The projectors imply that the total spectral weight in the transitions described by the $t-J$ model vanishes as $\delta \to 0$, as therefore, does the charge stiffness D.

Prelovsek has developed an interesting numerical technique to determine the frequency dependent conductivity of the $t-J$ model [49]. Other workers have not followed up on these methods.

3.4 Charge transfer insulators

The Hubbard model involves two energy scales, t and U. It is only a useful representation of low energy physics if *both* t and U are small compared to the band gaps separating the orbitals of interest from other orbitals in the solid. In many presently interesting correlated systems (in particular transition metal oxides) this is not the case: the U is so large that the basic charge transfer process involves shifting a carrier from an orbital on a transition metal site to an orbital on a different ion altogether (most commonly an oxygen ion). Such systems are referred to as 'charge transfer' rather than 'Mott-Hubbard' systems [50].

Charge transfer systems are in many respects similar to Mott-Hubbard systems: in particular, they display a correlation driven metal insulator transition at commensurate densities and for nearby dopings the low energy physics is believed to be described by the $t - J$ model [51]. The charge stiffness and low frequency conductivity are therefore presumably similar to those of the t-J model. However,(and surprisingly, considering the experimental relevance) the form of the conductivity at larger frequencies (for example near the gap energy in the insulator) has not been investigated.

3.5 Kondo Lattice Model

3.5.1 Overview. A wide class of condensed matter phenomena involve carriers interacting with spins, and the basic model describing this situation is the Kondo lattice model:

$$H_{KL} = \sum_{\sigma} \int \frac{d^d p}{(2\pi)^d} \varepsilon_p c^+_{p\sigma} c_{p\sigma} + J \sum_i \vec{S}_i \cdot \vec{\sigma}_{\alpha\beta} c^+_{i\alpha} c_{i\beta} \qquad (7.52)$$

This model has two interesting limits: if the magnitude of the coupling JS is large (relative to the electron band-width) then the carrier spin on site i is 'slaved' to the spin of the local moment. This 'double exchange' (the term is historical) limit is apparently relevant to the 'colossal' magnetoresistance manganites and to a variety of related systems. On the other hand, if the spin magnitude $S_i = 1/2$, the coupling is antiferromagnetic and the magnitude of J is small, then it is possible for the Kondo effect to 'marry' the conduction electrons to the local moments, yielding a 'heavy fermi liquid'. We consider the two limits separately.

3.5.2 Double exchange. The most extensively studied 'double exchange' systems are the 'colossal magnetoresistance' manganese perovskites (and related Ruddlesden-Popper systems). For reviews see, e.g. [52]. The crucial physics here is a very large J which arises from the atomic Hunds coupling. Its magnitude has not been measured directly. Quantum chemical considerations and experience with gas-phase Mn and other Mn compounds suggests [53] that the isolated-ion level splitting (e_g parallel to e_g antiparallel to core spin) $2J_{Hunds}S_c$ is about $2.5eV$. Optical experiments suggest that in the actual CMR materials it is somewhat larger; at least 4eV [54].

The strong coupling of carriers to spins leads to physics called "double exchange" which has very interesting consequences for a number of properties including optics. The essential point is this: the coupling between mobile electrons and core spins is apparently so strong that at physically relevant energies a mobile electron on a given site is constrained to have its spin parallel to the core spin on that site. This implies that the amplitude for an electron to hop from one site to another is modulated by a spin overlap factor which is maximal when the core spins on the two sites are parallel and is minimal when the two core spins are antiparallel.

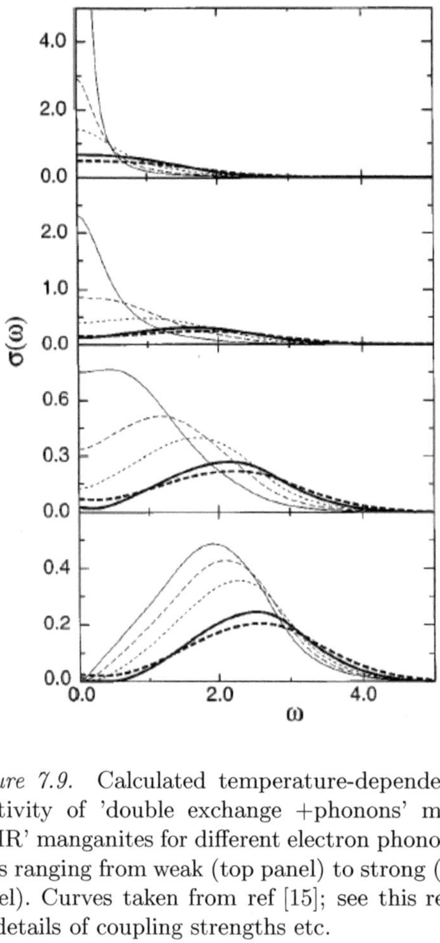

Figure 7.9. Calculated temperature-dependent conductivity of 'double exchange +phonons' model of 'CMR' manganites for different electron phonon couplings ranging from weak (top panel) to strong (bottom panel). Curves taken from ref [15]; see this reference for details of coupling strengths etc.

Ferromagnetic alignment of the core spins increases the electron kinetic energy (and is indeed the driving force for ferromagnetic order in these compounds), antiferromagetic alignment decreases it, and a random spin arrangement reduces it by a factor of approximately $1/\sqrt{2}$. Over a wide range of parameters, 'double exchange' models specified by Eq (7.52). (perhaps supplemented by other interactions) have ferromagnetic ground states with a Curie temperatures relatively small in comparison to the electronic band-width [52]. Thus by varying the temperature over a range small compared to the band-width a large change in the kinetic energy can be seen.

Fig (7.9) shows one example of this phenomenon: results of theoretical calculations (performed using the 'dynamical mean field approximation' and a simplified band structure) of the optical conductivity of a double exchange model with $J_H \to \infty$ and an additional electron-phonon interaction [15]. A strong temperature dependence of the functional form and integrated area is evident.

Figure (7.10) shows the temperature dependence of the kinetic energy (which may be directly computed and for this model is equal to the integrated area under the conductivity). The Curie temperature is evident as a kink in the curve. In the double-exchange-only model ($\lambda = 0$) the approximately $1/\sqrt{2}$ change in K between T_c and $T = 0$ is seen. For increased interactions, a larger change occurs, because the change in the effective t changes the balance between the kinetic energy and the other interaction, and thus the expectation value of the hopping operator. However, for this class of models the largest change which occurs is about 50% of the $T = 0$ kinetic energy. Finally, one can see that for this model at least, the ratio $T_c/K(T_c) \approx 0.2$, relatively independent of interaction strength. Extending this analysis to other interactions would be very important.

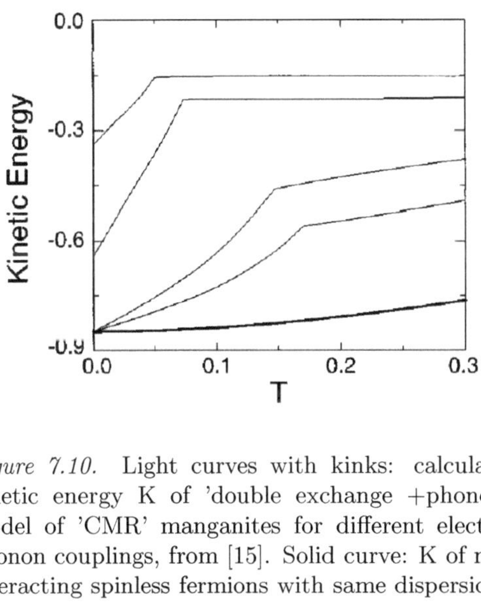

Figure 7.10. Light curves with kinks: calculated kinetic energy K of 'double exchange +phonons' model of 'CMR' manganites for different electron phonon couplings, from [15]. Solid curve: K of non-interacting spinless fermions with same dispersion.

Note that the model studied in ref [15] involves classical phonons, so the 'Migdal parameter' is $T\lambda$. The differences seen between the λ-dependence of $K(T = 0)$ (negligible for $\lambda < \lambda_c$) and that of $K(T > T_c)$ (non-negligible for $\lambda T > 1$) constitute an explicit example of the effect of the electron-phonon interaction on the electron kinetic energy. In the very large J limit, the magnitude and temperature dependence of the observed spectral weight may be related to the magnetic transition temperature; for details see [54], but very recent work [55] indicates that there is no simple relationship for realistic J values.

3.5.3 Heavy fermions. In heavy fermion materials, the carrier-spin interaction leads (via a lattice version of the Kondo effect) at low temperatures to the formation of a 'heavy' fermi liquid characterized by a very long mean free path and a quasiparticle mass which may be as much as 50-100 times the band mass. For a review see e.g. [56].

In a number of cases the enhanced mass has been observed directly via quantum oscillation measurements [57]. The low temperature electronic properties may be described by a self energy which has a very strong frequency dependence and a much weaker (in many cases negligible) momentum dependence [58]. Substitution into Eq (7.39) suggests that the conductivity should exhibit a Drude peak with a very small amplitude (reduced from the band theory value by roughly the same ~ 100 factor as that by which the quasiparticle mass is increased) [59]. This has at least qualitatively been observed [60]. Analyses of the temperature dependence of the superconducting penetration depth [24] suggests that Landau parameter effects are also important on the factor of two level, at least in some heavy fermion materials.

4. Application to Data–CMR and High Tc

4.1 High-T_c superconductors

The high-T_c superconductors are electronically two dimensional materials in which the basic unit is the "CuO_2 plane". This has a basically square symmetry with lattice constant $a \approx 4$Å. It is generally accepted that the important electrons can be described as $Cu\ d_{x^2-y^2}$ electrons (actually, complicated combinations of Cu and O and perhaps other states; the $d_{x^2-y^2}$ should be understood in the $|\psi_L\rangle$ sense discussed above). Superconductivity emerges upon doping an insulating 'parent compound' in which there is one hole per CuO_2 unit. Band theory [61] predicts that the low energy electronic degrees of freedom reside in a single band made up mainly of the antibonding combination of $d_{x^2-y^2}$-symmetry Cu and O orbitals. The best tight binding fit to the two dimensional band structure corresponds to a nearest-neighbor hopping of magnitude $t \approx 0.40eV$ and a second neighbor hopping $t' \approx 0.1eV$ so that the band theory dispersion is

$$\varepsilon_p = -2t(\cos(p_x a) + \cos(p_y a)) + 4t' \cos(p_x a)\cos(p_y a) \tag{7.53}$$

corresponding to a band fermi velocity of about $4eV$–Å and a band kinetic energy (equal to the Drude weight) of about $K_{band} = 0.28eV$.

Consider the antiferromagnetic insulating end materials La_2CuO_4 and Nd_2CuO_4, 'parent compounds' of the hole and electron doped cuprates respectively. Experimental conductivities [62] are shown in Figs (7.11, 7.12). One expects that because the electronic states of the different rare earths lie very far from the chemical potential [61], the optical spectra of these materials should be very similar. In fact, differences are evident in the frequency range $\omega > 3eV$. However, both materials are insulators with gaps of approximately 2eV. The optical absorption beginning

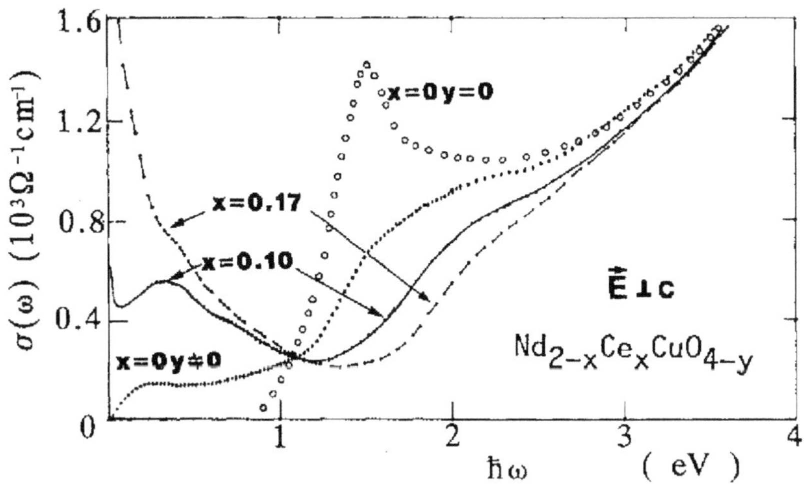

Figure 7.11. Evolution of conductivity of electron doped high temperature superconductors with doping, from [62]

at $\omega = 2eV$ is attributed to optical excitations to the 'upper Hubbard band' or 'charge transfer band'. Evidently a higher energy feature (perhaps at $\omega \sim 5eV$) produces a 'tail' of absorption which extends down to lower energies. It seems reasonable to attribute the $5eV$ absorption to 'non-bonding' oxygen states which are not of fundamental interest for the physics of high T_c, however in the La_2CuO_4 sample it does not seem possible to separate this absorption from the 'upper hubbard band' absorption of physical interest. In the Nd_2CuO_4 material one might argue that the high-frequency absorption in the most highly doped sample is representative of the 'tail' of the $5eV$ absorption. Subtracting this from the measured conductivity, using the in-plane lattice constant of 4Å and an out of plane lattice constant of 6Å and integrating the difference yields a spectral weight of $0.15eV$, approximately half of the band structure value, suggesting a U of about $9t \approx 3eV$. This value of U gives a gap of about $5t$, comparable to the observed gap. Thus one may conclude that (with some ambiguities) the optical data are consistent with that expected from the Hubbard model in the intermediate correlation regime. However, the data are in an awkward frequency regime and are subject to some uncertainties, and the estimates are obviously very rough.

As one dopes away from the insulator, low fequency spectral weight appears as shown in Figs (7.11, 7.12). Some fraction of this weight comes from the $\omega \sim 2eV$ 'upper Hubbard band' region and some fraction ap-

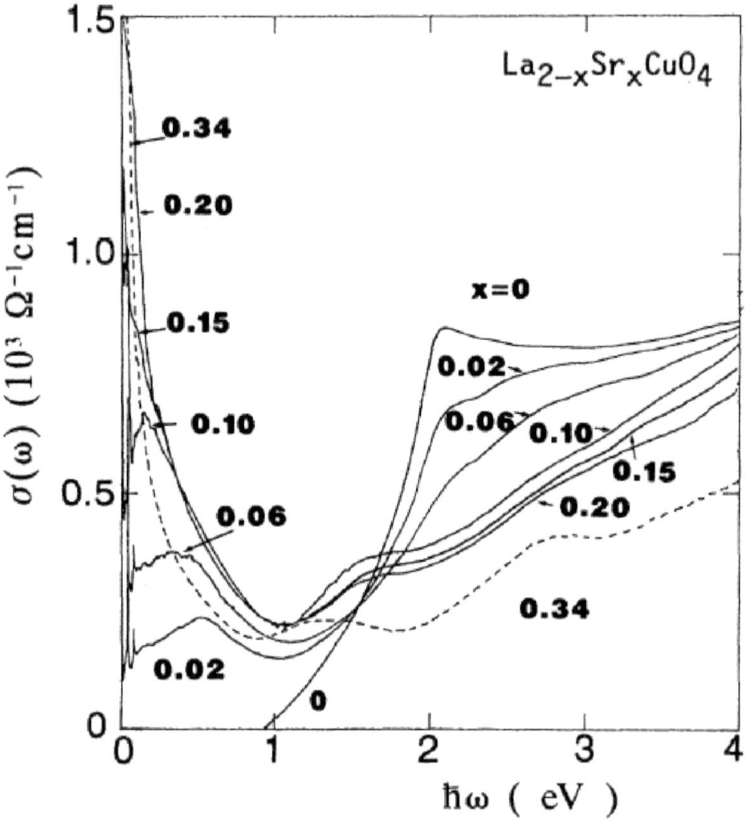

Figure 7.12. Evolution of conductivity of hole doped high temperature superconductors with doping, from [62]

pears to come from much higher energies. A quantitative understanding of the scales over which the spectral weight is redistributed has not yet been achieved.

4.2 CMR materials

The 'colossal' magnetoresistance (CMR) manganites occur in a variety of crystal structures but share the common feature that the mobile electrons arise from $Mn\ e_g$ symmetry d-levels and are very strongly coupled to $S = 3/2$ core spins composed of electrons residing in $Mn\ t_{2g}$ levels. They are important in the present context because (as explained above) the large value of the carrier-core-spin coupling means that the 'kinetic energy' is temperature dependent, allowing nontrivial tests of the Peierls aproximation and the restricted sum rule.

The prototypical compounds are the pseudocubic manganese perovskites $Re_{1-x}Ak_xMnO_3$ (here Re is a rare earth such as La or Pr and Ak is a divalent alkali such as Ca or Sr). The $x = 0$ 'parent compounds' are large gap insulators. The insulating behavior is mainly [16] due to a large-amplitude spatially coherent Jahn-Teller distortion. With doping the distortion is removed (for most choices of Re and Ak–a few compounds remain insulating at all dopings) and for x in the range $0.3 - 0.5$ a ferromagnetic metallic ground state results. Unfortunately, there are important and still ill-understood sample (especially surface) preparation issues which dramatically affect optical data obtained in metallic samples [63]. Roughly speaking, the higher the conductivity (at low temperatures) the better the sample and surface preparation. The data discussed here were obtained by Quijada and Simpson in the group of Drew at the University of Maryland using annealed films [64]. The conductivity of these films is comparable to the best conductivities obtained by other groups, but the data should still be regarded as subject to possible correction.

The band structure has been calculated [65, 66]; the conduction bands are derived from two e_g orbitals on each Mn site and are reasonably well described by a nearest neighbor tight binding model [16, 54] (which, for example, reproduces almost exactly the band theory approximation to the specific heat value quoted for $x = 0.3$ manganite in [66]). (Note also that the value of the Drude plasma frequency $\Omega_p = 1.9eV$ quoted for the $x = 0.3$ manganite in [66] is in error [67]). It is convenient to adopt a Pauli matrix notation in which the up state is the $|\mathrm{x}^2 - y^2\rangle$ orbital and the down state is the $|3z^2 - r^2\rangle$ orbital. Then the basic hopping Hamiltonian is a 2x2 matrix given by

$$\varepsilon = \varepsilon_0(p) + \overrightarrow{\varepsilon} \cdot \overrightarrow{\tau} \tag{7.54}$$

with $\overrightarrow{\tau}$ the usual Pauli matrices and

$$\varepsilon_0(p) = -t(\cos(px) + \cos(py) + \cos(pz)) \tag{7.55}$$

and $\overrightarrow{\varepsilon} = (\varepsilon_x, 0, \varepsilon_z)$ with

$$\varepsilon_x(p) = -\frac{\sqrt{3}t}{2}(\cos(px) - \cos(py)) \tag{7.56}$$

$$\varepsilon_z(p) = t(\cos(pz) - \frac{1}{2}(\cos(px) + \cos(py)) + \Delta_{cf} \tag{7.57}$$

where Δ_{cf} is a 'crystal field' energy splitting arising from a tetragonal distortion away from cubic symmetry (as occurs, e.g. in the layered manganites).

The energy eigenvalues are

$$E_{\pm} = \varepsilon_0 \pm \sqrt{\varepsilon_x^2 + \varepsilon_z^2} \tag{7.58}$$

Note that along the zone diagonals ((1,1,1) and symmetry related) $\varepsilon_{z,x} = 0$ so the two bands are degenerate and along the line to any cube face ((1,0,0) and equivalent) one of the two bands is flat. For this band structure the electron Green function is

$$\mathbf{G}(z,p) = (z - \mathbf{\Sigma}(p,z) - \varepsilon(p))^{-1} \tag{7.59}$$

The current operator following from the Peierls approximation is (for currents in the z direction)

$$\mathbf{j}_z = -t\sin(p_z)(\mathbf{1} + \tau_{\mathbf{z}}) \tag{7.60}$$

The kinetic energy for motion along one of the cartesian directions is

$$K = -\frac{1}{3}Tr[\varepsilon_p \mathbf{G}] \tag{7.61}$$

In the band structure corresponding to the ferromagnetic state of the cubic materials, the chemical potential for $x = 0.3$, $\mu_{0.3}$, ≈ -1 and the corresponding 'band' kinetic energy, obtained by evaluating Eq (7.61) is

$$K_{band}(\mu = -1) = -0.45t = -0.28eV \tag{7.62}$$

In the momentum-independent self energy approximation the conductivity is

$$\sigma(i\Omega) = \frac{1}{i\Omega}T\sum_{i\omega_n}\int\frac{d^3p}{(2\pi)^3}Tr[\overrightarrow{j_p}\mathbf{G}(i\omega + i\Omega, p)\overrightarrow{j_p}\mathbf{G}(\mathbf{i}\omega, p)] \tag{7.63}$$

This conductivity includes both 'Drude' and interband terms. The kinetic energy corresponding to the Drude part of the conductivity is (for the widely studied 1/3 doping level)

$$K_{Drude,x=0.3} = 0.32t = 0.2eV \tag{7.64}$$

Ahn and the author [16] evaluated Eq (7.63) for the insulating 'parent compound' $LaMnO_3$, in which a large-amplitude spatially coherent Jahn-Teller distortion occurs which is sufficient to explain the insulating behavior [65] (although band theory somewhat underpredicts the gap [16]) With doping the distortion is removed and a ferromagnetic metallic ground state results. Their calculation used the nearest neighbor tight binding parametrization of the band theory, along with realistic values for the level splitting caused by the Jahn-Teller distortion, as well as a 'Hubbard U' (treated in the Hartree approximation) and various estimates for the Kondo coupling J. The calculations, while in reasonable agreement with data taken at room temperature, predict a larger than observed increase in spectralweight as the temperature is lowered. The band theory conductivity of $LaMnO_3$ was also calculated by Soloviev et al using $LMTO$ methods to evaluate actual wave function overlaps. Remarkably, the results (while quite close to the tight binding results for the magnitude of the insulating gap) indicate spectral weights about four times smaller than the spectral weights predicted by the Peierls approximation. The source of this discrepancy has not been determined; resolving it is an important issue for future research.

The theoretical conductivity of the metallic materials has not been investigated in such detail. Shiba and collaborators studied the $T = 0$ conductivity of the fully polarized ferromagnetic state [68]. Chattopadhyay and the author [54] considered changes in spectral weight between T_c and $T = 0$ and related these to the magnetic transition temperatures.

Fig (7.13) shows the optical conductivity for several different $x = 0.3$ doped (ferromagnetic metallic ground state) manganites obtained by [64]. A strong dependence of both the form and the integrated area of the low frequency conductivity is evident. Fig. (7.14) shows the integrated area, for different temperatures. A change of spectral weight with temperature is evident; presumably this is related to double exchange.

To analyse the data in a satisfactory manner one must identify the contribution to the observed conductivity coming from the conduction bands. Use of the magnitude and temperature dependence of the spectral weight provides enough information to do this [54, 64]. The temperature dependence must arise from the double-exchange physics and thus from the conduction band. The total conduction band spectral weight must be less than the band kinetic energy, the value of the $T = 0$ spectral weight and the change between low T and $T > T_c$ must be large enough to explain the observed T_c and the change in spectral weight cannot be more than about half of the low-T weight. These considerations led to the conclusion [64] that in the most metallic material, $La_{0.7}Sr_{0.3}MnO_3$ at low T the conduction band contribution essentially exhausts the bound imposed by the band theory, so the '$Hubbard-U$' effects are evidently weak!.

Indications are [15, 64] that an additional interaction (probably the electron-phonon interaction, which causes charge ordering in some materials) leads to additional structure in the conductivity (but not to significant changes in the low frequency spectral weight). Presently avail-

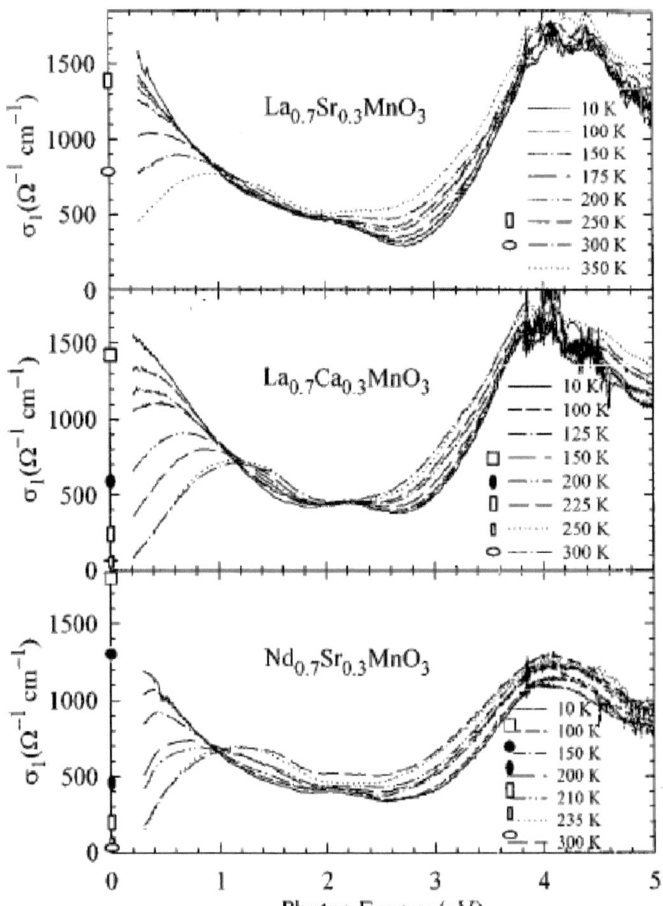

Figure 7.13. Measured conductivity of 'CMR' manganites, from [64]

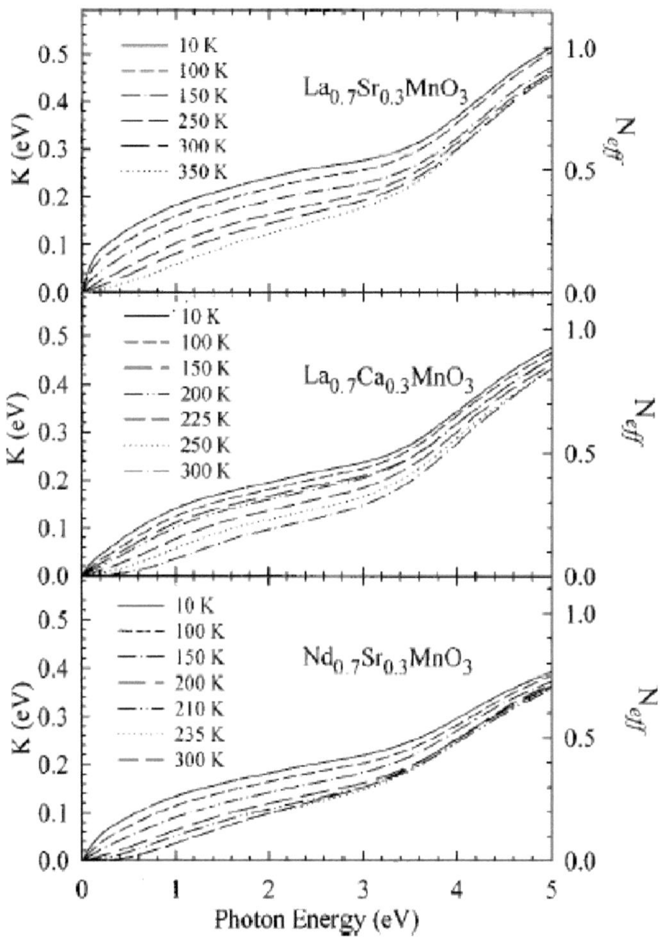

Figure 7.14. Measured spectral weight of 'CMR' manganites, from [64]

able data suggest that these extra interactions are poorly described by a momentum-independent self energy. In particular, if the self energy is momentum independent, then the mass enhancement inferred from the specific heat should be the same as the mass enhancement inferred from the renormalization of the Drude weight. However, recent experimental results suggest that the Drude weight is more strongly reduced (relative to band theory) than is the fermi velocity [69].

5. Conclusions

I hope in this brief survey to have conveyed some of the basic ideas in the theory of optical conductivity of correlated electron systems, along with some of the open theoretical challenges and to have shown how these ideas are used in practice. Most importantly, I hope to have given the reader at least a glimpse of the power of the technique for elucidating correlated electron physics.

Acknowledgments

My understanding of optical conductivity owes much to interactions and collaborations over many years with L. B. Ioffe, B. G. Kotliar, G. A. Thomas, J. Orenstein, H. D. Drew and E. Abrahams and my more recent research has benefitted greatly from outstanding collaborators including K. Ahn, S. Blawid, A. Chattopadhyay and A. J. Schofield. My work in this area has been supported by the US National Science Foundation, most recently through the University of Maryland-Rutgers MRSEC program and NSF DMR0081075.

References

[1] D. van der Marel, A. Tsvetkov, Phys. Rev. **B64** 024530 (2001).

[2] see, e.g. Chapter 7 of J. D. Jackson, *Classical Electrodynamics, 2nd Ed.* (John Wiley and Sons, New York: 1980).

[3] D. Pines and P. Nozieres, *Theory of Quantum Liquids*, Addison Wesley, Reading, MA (1964).

[4] W. Kohn, Phys. Rev. **A133** 171 (1964).

[5] M. Imada, A. Fujimori, and Y. Tokura, Rev. Mod. Phys. **70**, 1039 (1998) .

[6] B. N. Narozhny, A. J. Millis and N. Andrei, Phys. Rev. **B58** 2921-4 (1998).

[7] D. J. Scalapino, S. R. White and S. C. Zhang, Phys. Rev. **B47** 7995 (1993).

[8] N. W. Ashcroft and N. D. Mermin, *Solid State Physics*, (Saunders, Philadelphia: 1980).

[9] A. V. Puchkov, D.N. Basov and T. Timusk, J. Phys. Cond. Mat. **8** 10049 (1996).

[10] O. K. Anderson, C. Arcangeli, R. W. Tank, T. Saha-Dasgupta, G. Kner, O. Jepsen and I. Dasgupta, *Symposium Mater. Res, Soc. 1998* pps. 3-4 Warrendale PA USA.

[11] J. E. Hirsch, Physica **C199** 305 (1992).

[12] P. F. Maldague, Phys. Rev. **B16** 2437 (1977).

[13] D. Baeriswyl, C. Gros and T. M. Rice, Phys. Rev. **B35** 8391 (1987).

[14] A. J. Millis and S. Coppersmith, Phys. Rev. **B42**, 10807-10 (1990).

[15] A. J. Millis, R. Mueller and B. I. Shraiman, Phys. Rev. **B54** 5405 (1996).

[16] K. Ahn and A. J. Millis, Phys. Rev. **B61** 13545 (2000).

[17] *see, e.g.* R. O. Jones and O. Gunnarsson, Rev. Mod. Phys. **61** 689 (1989).

[18] W. G. Aulbur, L. Johnsson and J. W. Wilkins, Solid State Physics **54** pps 1-218 (2000).

[19] L. X. Benedict, E. R. Shirley and R. B. Bohn, Phys Rev. **B57** 9385 (1998) and L. X. Benedict and E. L. Shirley, Phys. Rev. **B59** 5441 (1999).

[20] I. Soloviev, N. Hamada and K. Terakura, Phys. Rev. **B53** 7158 (1996); see also K. Terakura, I Soloviev and H. Sawada in *Colossal Magnetoresistive Oxides,* Y. Tokura, ed., (Gordon and Breach: Tokyo 1999).

[21] X. Zotos and P. Prelovsek, Phys. Rev. **B53** 983 (1997).

[22] A. A. Abrikosov, L. P. Gorkov and I. E. Dzyaloshinskii, *Methods of Quantum Field Theory in Statistical Physics*, (Dover, New York: 1964).

[23] A. Georges, B. G. Kotliar, W. Krauth and M. Rozenberg, Rev. Mod. Phys. **68** 13 (1996).

[24] F. Gross, B. S. Chandrasekhar, D. Einzel, K. Andres, P. J. Hirschfeld, H. R. Ott, J. Bevers, Z. Fisk and J. L. Smith, Z. Phys. **B64** 175 (1986).

[25] M Tinkham, *Superconductivity* (McGraw Hill, New York, 1975).

[26] J. Orenstein, G. A. Thomas, A. J. Millis, S. L. Cooper, D. H. Rapkine, T. Timusk, J. V. Waszczak, Physical Review **B42**, 6342 (1990).

[27] H. J. A. Molegraaf et. al., Science **295** 2239 (2001).

[28] T. Katsufuji, Y. Okimoto, T. Arima, Y. Tojura and J. B. Torrance, Phys. Rev. **B51** 4830 (1995).

[29] V. Vescoli, L. D. Degiorgi, H. Berger and L. Forro, Phys. Rev. Lett. **81** 453 (1998).

[30] A. I. Larkin, J. E. T. P. **46** 1478(1964) and *ibid* 2188

[31] R. Hlubina and T. M. Rice, Phys. Rev **B51** 9253 (1995).

[32] A. A. Migdal, Sov. Phys. JETP **7** 996 (1958).

[33] C. M. Varma, E. Abrahams, S. Schmidt-Rink and A. Ruckenstein, Phys. Rev.Lett. **63** 1936 (1989).

[34] T. Valla, A. V. Feodorov, P. D. Johnson, B. O. Wells, S. L. Hulbert Q. Li, G. D. Gu and N. Koshizuka Science **285** 2110 (1999).

[35] E. Abrahams, J. Phys. I France **6** 2191 (1996).

[36] A. J. Millis and H. D. Drew, Phys. Rev. **B67** 214517 (2003).

[37] L. B. Ioffe and A. J. Millis, Science **285** 1241 (1999) and Phys. Rev. **B61** 9077 (2000).

[38] N. Shah and A. J. Millis, Phys. Rev. **B64** 174506 (2001) and *ibid* **B65** 024506 (2001).

[39] A. V. Lopatin, A. Georges and T. Giamarchi, Phys. Rev. **B63** 75109 (2001).

[40] F. H. L. Essler and A. M. Tsvelik, Phys. Rev. **B65** 115117 (2002).

[41] M. Turlakov and A. J. Leggett, Phys. Rev. **B63** 064518 (2001).

[42] H. J. Schulz, Phys. Rev. Lett. **64** 2831 (1989) and H. J. Schulz, Int. J. Mod. Phys. **B5** 57 (1991).

[43] C. A. Stafford and A. J. Millis, Phys. Rev. **B48** 1409 (1993).

[44] T. Giamarchi and A. J. Millis, Phys. Rev. **B46**, 9325 (1992).

[45] E. Dagotto, A.Moreo, F. Ortolani, D. Poilblanc and J. Riera, Phys. Rev. **B45** 10107 (1992).

[46] M. J. Rozenberg. G. Kotliar, H. Kajueter, G. A. Thomas, D. H. Rapkine, J. M. Honig and P. Metcalfe, Phys. Rev. Lett. **75** 105 (1995).

[47] E. Jeckelmann, F. Gebhard and F. H. L. Essler, Phys. Rev. Lett. **85** 3910(2000); see also E. Jeckelmann, Phys. Rev. **B67** 075106 (2003).

[48] see, e.g. T. M. Rice, chapter 2 in *Les Houches 91, Session LVI* (Elsevier, Amsterdam: 1996); for an explicit and useful derivation of the $t-J$ model from a more fundamental underlying Hamiltonian see A. MacDonald, D. Yoshioka and S. M. Girvin, Phys. Rev. **B41** 2565 (1990).

[49] J. Jaklic and P. Prelovsek, Phys. Rev. **B52** 6903 (1995).

[50] J. Zaanen, G. Sawatzky and J. W. Allen, Phys. Rev. Lett. **55** 418 (1985).

[51] F-C. Zhang and T. M. Rice, Phys. Rev. **B57** 3759 (1988)

[52] See, e.g. the articles in *Colossal Magnetoresistive Oxides,* Y. Tokura, ed (Gordon and Breach: Tokyo, 1999) or those in Phil. Trans. Roy. Soc. **356** no. 1742 (pps 1469-1712) (1998).

[53] G.A. Sawatzky, private communication

[54] A. Chattopadhyay, A. J. Millis and S. Das Sarma, Phys. Rev. **B61** 10738 (2000).

[55] B. Michaelis and A. J. Millis, cond-mat/0212573.

[56] P. A. Lee, T. M. Rice, J. Serene, L. J. Sham and J. W. Wilkins, Comments on Cond. Mat. Phys. **12,** 99 (1986).

[57] For a review see, e.g. S. R. Julian, G. J. MacMullan, C. Pfleiderer, F.Tautz and G. G. Lonzarich, J. Appl. Phys. **76** 6137 (1994).

[58] C. M. Varma. Phys. Rev. Lett. **55** 2723 (1985).

[59] A. J. Millis and P. A. Lee, Physical Review **B35**, 3394 (1987)

[60] For a recent review see L. Degiorgi, Rev. Mod. Phys. **71** 687 (1999).

[61] A. I. Liechtenstein, O. Gunnarsson, O. K. Andersen, and R. M. Martin, Phys. Rev. **B54** 12505 (1996).

[62] S. Uchida, T. Ito, T. Takagi, T. Arima, Y. Tokura and S. Tajima, Phys. Rev. **B43** 7942 (1991).

[63] See, e.g. Fig 4 of K. Takenaka, Y. Sawaki and S. Sugai, Phys. Rev. **B60** 13011 (1999).

[64] M. Quijada, J. Cerne, J. R. Simpson, H. D. Drew, K. H. Ahn, A. J. Millis, R. Shreekala, R. Ramesh, M. Rajeswari and T. Venkatesan, Phys. Rev. **B58** 16093 (1998).

[65] S. Satpathy, Z. S. Popovic and F. R. Vukajilovic, Phys. Rev. Lett. **76** 960 (1996).

[66] W. Pickett and D. Singh, Phys. Rev. **B53** 1146 (1996).

[67] W. E. Pickett, private communication.

[68] H. Shiba, J. Phys. Soc. Jpn. **66** 941 (1997).

[69] J. Simpson, H. D. Drew, V. N. Smolyaninova, R. L. Greene, M. C. Robson, A. Biswas and M. Rajeswari, Phys. Rev. **B60** 16263 (1999).

Chapter 8

OPTICAL SIGNATURES OF ELECTRON CORRELATIONS IN THE CUPRATES

D. van der Marel

Laboratory of Solid State Physics

Materials Science Centre, University of Groningen

Nijenborgh 4, 9747 AG Groningen, The Netherlands

d.van.der.marel@phys.rug.nl

dirk.vandermarel@physics.unige.ch

Abstract The f-sum rule is introduced and its applications to electronic and vibrational modes are discussed. A related integral over the intra-band part of $\sigma(\omega)$ which is also valid for correlated electrons, becomes just the kinetic energy if the only hopping is between nearest neighbor sites. A summary is given of additional sum rule expressions for the optical conductivity and the dielectric function, including expressions for the first and second moment of the optical conductivity, and a relation between the Coulomb energy and the energy loss function. It is shown from various examples, that the optical spectra of high T_c materials along the c-axis and in the ab-plane direction can be used to study the kinetic energy change due to the appearance of superconductivity. The results show, that the pairing mechanism is highly unconventional, and mostly associated with a lowering of kinetic energy parallel to the planes when pairs are formed.

Keywords: Optical conductivity, spectral weight, sum rules, reflectivity, dielectric function, inelastic scattering, energy loss function, inelastic electron scattering, Josephson plasmon, multi-layers, inter-layer tunneling, transverse optical plasmon, specific heat, pair correlation, kinetic energy, correlation energy, internal energy.

1. Macroscopic electromagnetic fields in matter

1.1 Introduction

The response of a system of electrons to an externally applied field is commonly indicated as the dielectric function, or alternatively as the

D. Baeriswyl and L. Degiorgi (eds.), Strong Interactions in Low Dimensions, 237–276.

optical conductivity. The discussion in this chapter is devoted to induced currents and fields which are proportional to the external fields, the so-called linear response. The dielectric and the optical conductivity can be measured either using inelastic scattering of charged particles for which usually electrons are used, or by measuring the absorption of light, or the amplitude and/or phase of light reflected or transmitted by a sample. The two cases of fast particles and incident radiation involve different physics and will be discussed separately.

1.2 Reflection and refraction of electromagnetic waves

Optical spectroscopy measures the reflection and refraction of a beam of photons interacting with the solid. A rarely used alternative is the use of bolometric techniques to measure the absorption of photons directly. A variety of different experimental geometries can be used, depending on the type of sample under investigation, which can be a reflecting surface of a thick crystal, a free standing thin film, or a thin film supported by a substrate. Important factors influencing the type of analysis are also the orientation of the crystal or film surface, the angle of incidence of the ray of photons, and the polarization of the light. In most cases only the amplitude of the reflected or refracted light is measured, but sometimes the phase is measured, or the phase difference between two incident rays with different polarization as in ellipsometry. The task of relating the intensity and/or phase of the reflected or refracted light to the dielectric tensor inside the material boils down to solving the Maxwell equations at the vacuum/sample, sample/substrate, *etc.* interfaces. An example is the ratio of the reflection coefficients (R_p/R_s)and phase differences ($\eta_p - \eta_s$) of light rays with p and s-polarization reflected on a crystal/vacuum interface at an angle of incidence θ. These quantities which are measured directly using ellipsometry

$$e^{i(\eta_p-\eta_s)}\sqrt{\frac{R_p}{R_s}} = \frac{\sin\theta\tan\theta - \sqrt{\epsilon - \sin^2\theta}}{\sin\theta\tan\theta + \sqrt{\epsilon - \sin^2\theta}} \tag{8.1}$$

The real and imaginary part of the dielectric constant can be calculated from such a measurement with the aid of Eq. (8.1). In contrast to a beam of charged particles, the electric field of a plane electromagnetic wave is transverse to the photon momentum. The dielectric tensor elements which can be measured in an optical experiment are therefore transverse to the direction of propagation of the electromagnetic wave. In a typical optical experiment the photon energy is below 6 eV. In vacuum the photon wave number used in optical experiments is therefore $0.0005 Å^{-1}$, or smaller, which is at least three orders of magnitude below the Fermi momentum of electrons in a solid. For this reason it is usually said that optical spectroscopy measures the transverse dielectric constant or the optical conductivity at zero momentum. The optical conductivity tensor expresses the current response to an electrical field

$$\vec{j}(\vec{r},t) = \int d^3\vec{r}' \int dt' \sigma(\vec{r},\vec{r}',t-t')\vec{E}(\vec{r}',t') \tag{8.2}$$

From the Maxwell equations it can be shown that for polarization transverse to the propagation of an electromagnetic wave $d\vec{E}/dt = d\vec{D}/dt + 4\pi\vec{j}$. If the sample has translational invariance, the optical conductivity tensor has a diagonal representation in k-space. Due to the fact that the translational symmetries of a crystalline solid are restricted to a discrete space group, k is limited to the first Brillouin zone. Consequently, as shown by Hanke and Sham[1], the k-space representation of the dielectric tensor becomes a matrix in reciprocal space

$$\sigma(\vec{G},\omega)_{\vec{G},\vec{G}'} = \int d^2\vec{r} \int d^3\vec{r}' \int dt e^{i(\vec{q}+\vec{G})\cdot\vec{r}} e^{-i(\vec{q}+\vec{G}')\cdot\vec{r}'} e^{i\omega t}\sigma(\vec{r},\vec{r}',t) \tag{8.3}$$

The dependence of $\sigma(\vec{q},\omega)_{\vec{G},\vec{G}'}$ on the reciprocal lattice vectors $\vec{G}$, $\vec{G}'$ reflects, that the local fields can have strong variations in direction and magnitude on the length scale of a unit cell. Yet due to the long wavelength of the external light rays the Fresnel equations involve only $\vec{G} = \vec{G}' = 0$. Usually in texts on optical properties the only optical tensor elements considered have $\vec{G} = \vec{G}' = 0$, and in this chapter we will do the same.

Inside a solid the wavelength of the electromagnetic rays can be much shorter than that of a ray with the same frequency travelling in vacuum. Although in this chapter we will not encounter experiments where the finite momentum of the photon plays an important role, we should keep in mind that in principle the photon momentum is non-zero and can have a non-trivial effect on the optical spectra. In particular it may corrupt Kramers-Kronig relations, which is just one out of several reasons why spectroscopic ellipsometry should be favored.

1.3 Inelastic scattering of charged particles

When a fast charged particle, moving at a velocity $\vec{v}_e$, interacts weakly with a solid, it may recoil inelastically by transferring part of its momentum, $\hbar\vec{q}$ and its energy, $\hbar\omega$ to the solid. The fast electron behaves like a test charge of frequency $\omega = \vec{q}\cdot\vec{v}_e$, which corresponds to a dielectric displacement field, $D(\vec{r},t) = eq^{-2}\exp(i\vec{q}\cdot\vec{r} - i\omega t)$. The dielectric displacement of the external charges may be characterized by a density fluctuation, which has no field component transverse to the wave. $D(\vec{r},t)$ is therefore a purely longitudinal field. In a solid mixing of transverse and longitudinal modes occurs whenever fields propagate in a direction which is not a high symmetry direction of the crystal. However, in the long wavelength limit the dielectric properties can be described by only three tensor elements which correspond to the three optical axes of the crystal. Since along these directions no mixing between longitudinal and transverse response occurs, we will consider the situation in this chapter where the fields and their propagation vector point along the optical axis. Inside a material the dielectric displacement is screened by the response of the matter particles, resulting in the screened field $E(\vec{r},t)$ inside the solid[1].

$$\vec{E}(\vec{r},t) = \int d^3\vec{r}' \int dt' \epsilon^{-1}(\vec{r},\vec{r}',t-t')\vec{D}(\vec{r}',t') \tag{8.4}$$

For the same reasons as for the optical conductivity the k-space representation of the dielectric tensor becomes a tensor in reciprocal space

$$\epsilon^{-1}(\vec{q}+\vec{G}',\vec{q}+\vec{G},\omega) = \int d^2\vec{r} \int d^3\vec{r}' \int dt e^{i(\vec{q}+\vec{G}')\cdot\vec{r}} e^{-i(\vec{q}+\vec{G})\cdot\vec{r}'} e^{i\omega t} \epsilon^{-1}(\vec{r},\vec{r}',t) \tag{8.5}$$

where $\vec{G}$ and $\vec{G}'$ denote reciprocal lattice vectors. The relation between the dielectric displacement and the electric field is

$$\vec{E}(\vec{q}+\vec{G}',\omega) = \sum_{\vec{G}} \epsilon^{-1}(\vec{q}+\vec{G}',\vec{q}+\vec{G},\omega)\vec{D}(\vec{q}+\vec{G},\omega) \tag{8.6}$$

The macroscopic dielectric constant, which measures the macroscopic response to a macroscopic perturbation, *i.e.* for vanishingly small $\vec{q}$, is given by[1]

$$\epsilon(\omega) = \lim_{q\to 0} \frac{1}{\epsilon^{-1}(\vec{q},\vec{q},\omega)} \tag{8.7}$$

where it is important, that in this expression first the matrix $\epsilon(\vec{q}+\vec{G}',\vec{q}+\vec{G},\omega)$ has to be inverted in reciprocal space, and in the next step the $(\vec{G}=0,\vec{G}'=0)$ matrix element is taken of the inverted matrix [1]. Energy loss spectroscopy using charged particles can be used to measure the dielectric response as a function of both frequency and momentum. This technique provides the longitudinal dielectric function, *i.e.* the response to a dielectric displacement field which is parallel to the transferred momentum $\vec{q}$. The probability per unit time that a fast electron transfers momentum $\vec{q}$ and energy $\hbar\omega$ to the electrons was derived by Nozières and Pines[2, 3] for a fully translational invariant 'jellium' of interacting electrons

$$P(\vec{q},\omega) = \frac{8\pi e^2}{|q|^2}\text{Im}\left\{\frac{-1}{\epsilon(\vec{q},\omega)}\right\} \tag{8.8}$$

where e is the elementary charge.

1.4 Relation between $\sigma(\omega)$ and $\epsilon(\omega)$

We close this introduction by remarking, that for electromagnetic fields propagating at a long wavelength the two responses, longitudinal and transverse, although different at any nonzero wave vector, are very closely related. We will take advantage of this fact when later in this chapter we extract the energy loss function for $q \approx 0$ from optical data. According to Maxwell's equations for $q \to 0$ the uniform current density is just the time derivative of the uniform dipole field, hence $4\pi j = i\omega(E - D)$. Consequently for $q \to 0$ the conductivity and the dielectric function are related in the following way

$$\epsilon(0,\omega) = 1 + \frac{4\pi i}{\omega}\sigma(0,\omega) \tag{8.9}$$

Throughout this chapter we will use this identity repeatedly.

2. Interaction of light with matter

2.1 The optical conductivity

Let us now turn to the discussion of the microscopic properties of the optical conductivity function. The full Hamiltonian describing the electrons and their interactions is

$$H = \sum_{k\sigma} \frac{\hbar^2 k^2}{2m} c^{\dagger}_{k\sigma} c_{k\sigma} + \sum_G U_G \hat{\rho}_{-G} + \frac{1}{2}\sum_k V_k \hat{\rho}_{\vec{k}} \hat{\rho}_{-\vec{k}} \tag{8.10}$$

$$\hat{\rho}_{\vec{k}} = \sum_{p,\sigma} c^{\dagger}_{p,\sigma} c_{p+k,\sigma} \tag{8.11}$$

In this expression the symbol $c^{\dagger}_{p,\sigma}$ creates a plane wave of momentum $\hbar\vec{p}$ and spin-quantum number σ, U_G represents the potential landscape due to the crystal environment. The third term is a model electron-electron interaction Hamiltonian, representing all electron-phonon mediated and Coulomb interactions, where $\hat{\rho}_{\vec{k}}$ is the k'th Fourier component of the density operator. In addition to the direct Coulomb interaction, various other contributions may be relevant, such as direct exchange terms. As a result the spin- and momentum dependence of the total interaction can have a more complex form than the above model Hamiltonian. Relevant for the subsequent discussion is only, that the interaction term commutes with the current operator. The quantum mechanical expression for the current operator is

$$\vec{j}_{\vec{q}} = \sum_{p,\sigma} \frac{e\hbar\vec{p}}{m} c^{\dagger}_{p-q/2,\sigma} c_{p+q/2,\sigma} \tag{8.12}$$

The current and density operators are symmetric in k-space, satisfying $\hat{\rho}^{\dagger}_{k} = \hat{\rho}_{-k}$ and $j^{\dagger}_{k} = j_{-k}$. In coordinate space the representations of the density and the current are

$$\hat{n}(\vec{r}) = \frac{1}{V}\sum_{q} e^{i\vec{q}\cdot\vec{r}}\hat{\rho}_{\vec{q}} \tag{8.13}$$

$$\vec{j}(\vec{r}) = \frac{e}{V}\sum_{q} e^{i\vec{q}\cdot\vec{r}}\vec{j}_{\vec{q}} \tag{8.14}$$

It is easy to verify, that together $\hat{n}(\vec{r})$ and $\vec{j}(\vec{r})$ satisfy the continuity equation $ie\hbar^{-1}[\hat{n}(\vec{r}), H] + \nabla\cdot\vec{j}(\vec{r}) = 0$.

Let us now consider a many-body system with eigen-states $|m\rangle$ and corresponding energies E_m. For such a system the microscopic expression for the optical conductivity has been explained by A.J. Millis in Chapter 7. The result for finite $\vec{q}$ was derived in 'The theory of quantum liquids' part I, by Nozières and Pines (equation 4.163). For brevity of notation we represent the matrix elements of the current operators as

$$j^{nm}_{\alpha,q} \equiv \langle n|j_{\alpha,q}|m\rangle \tag{8.15}$$

With the help of these matrix elements, and with the definition $\hbar\omega_{mn} = E_m - E_n$ the expression for the optical conductivity is

$$\sigma_{\alpha,\alpha}(\vec{q},\omega) = \frac{ie^2N}{mV\omega} + \sum_{n,m\neq n} \frac{ie^{\beta(\Omega-E_n)}}{V\omega}\left[\frac{j^{nm}_{\alpha,q}j^{mn}_{\alpha,-q}}{\omega-\omega_{mn}+i\eta} - \frac{j^{nm}_{\alpha,-q}j^{mn}_{\alpha,q}}{\omega+\omega_{mn}+i\eta}\right] \tag{8.16}$$

Here N is the number of electrons, V the volume, m the electron mass, q_e the elementary charge, Ω is the thermodynamic potential, $\beta = 1/k_BT$ and η is an infinitesimally small positive number. In principle in the calculation of Eq. (8.16) terms may occur under the summation for which $\omega_{mn} = 0$. As ω_{mn} occurs in the denominator of this expression, these zeros should be cancelled exactly by zeros of the current matrix elements, which poses a special mathematical challenge.

In Eq. (8.16) $\sigma(\omega)$ is represented by two separate terms, a δ-function for $\omega = 0$ and a summation over excited many-body eigen-states. The

δ-function is a diamagnetic contribution of *all* electrons in the system, the presence of which is a consequence of the gauge invariant treatment of the optical conductivity, as explained by Millis in chapter 7. The presence of this term is at first glance rather confusing, since left by itself this δ-function would imply that all materials (including diamond) are ideal conductors! However, the second term has, besides a series of poles corresponding to the optical transitions, also a pole for $\omega = 0$, corresponding to a *negative* δ-function of $\text{Re}\sigma(\omega)$. It turns out, that for all materials except ideal conductors this δ-function compensates exactly the first (diamagnetic) term of Eq. (8.16). This exact compensation is a consequence of the relation [6]

$$\text{For every } n\text{: } \sum_{m\neq n} \frac{j^{nm}_{\alpha,q} j^{mn}_{\alpha,-q}}{\omega_{mn}} = \frac{Ne^2}{2m} \tag{8.17}$$

Experimentally truly 'ideal' conductivity is only seen in superconductors. In ordinary conducting materials the diamagnetic term broadens to a Lorentzian peak due to elastic and/or inelastic scattering. The width of this peak is the inverse life-time of the charge carriers. Often in the theoretical literature the broadening is not important, and the Drude peak is counted to the Dirac-function in the origin. The infrared properties of superconductors are characterized by the presence of both a purely reactive diamagnetic response, and a regular dissipative conductivity [5]. The sum of these contributions counts the partial intra-band spectral weight which we discuss in section 2.4. With the help of Eq. (8.17), the diamagnetic term of Eq. (8.16) can now be absorbed in the summation on the right-hand side

$$\sigma_{\alpha,\beta}(\vec{q},\omega) = \frac{i}{V} \sum_{n,m\neq n} \frac{e^{\beta(\Omega-E_n)}}{\omega_{mn}} \left\{ \frac{j^{nm}_{\alpha,q} j^{mn}_{\alpha,-q}}{\omega - \omega_{mn} + i\eta} + \frac{j^{nm}_{\alpha,-q} j^{mn}_{\alpha,q}}{\omega + \omega_{mn} + i\eta} \right\} \tag{8.18}$$

As explained in section 1.2, usually in optical experiments one assumes $q \to 0$ in the expressions for $\sigma(\omega)$. It is useful at this stage to introduce the generalized plasma frequencies $\Omega^2_{mn} = 8\pi e^{\beta(\Omega-E_n)} |j^{nm}_{\alpha}|^2 \omega^{-1}_{mn} V^{-1}$, with the help of which we obtain the following compact expression for the optical conductivity tensor

$$\sigma_{\alpha\alpha}(\omega) = \frac{i\omega}{4\pi} \sum_{n,m\neq n} \frac{\Omega^2_{mn}}{\omega(\omega + i\gamma_{mn}) - \omega^2_{mn}} \tag{8.19}$$

Although formally the parameter γ_{mn} is understood to be an infinitesimally small positive number, a natural modification of Eq. (8.19) consists of limiting the summation to a set of oscillators representing the main optical transitions and inserting a finite value for γ_{mn}, which in this case represents the inverse lifetime of the corresponding excited state (*e.g.* calculated using Fermi's Golden Rule). With this modification Eq. (8.19) is one of the most commonly used phenomenological representations of the optical conductivity, generally known as the Drude-Lorentz expression.

2.2 The f-sum rule

The expressions Eqs. (8.16), (8.18), and (8.19) satisfy a famous sum rule. This is obtained by first showing with the help of Eq. (8.17), that for each n

$$\sum_{m,m\neq n} \Omega_{mn}^2 \equiv \frac{4\pi e^2 N}{mV} e^{\beta(\Omega - E_n)} \tag{8.20}$$

Second, as a result of Cauchy's theorem in Eq. (8.19) the integral over all (positive and negative) frequencies of $\int \mathrm{Re}\sigma(\omega)$ equals $\sum_{mn} \Omega_{mn}^2/4$. To complete the derivation of the f-sum rule we also use that $\sum_n e^{\beta(\Omega - E_n)} = 1$, which follows from the definition of the thermodynamic potential. Then

$$\int_{-\infty}^{\infty} \mathrm{Re}\sigma(\omega) d\omega = \frac{\pi e^2 N}{mV} \tag{8.21}$$

is the f-sum rule, or Thomas-Reich-Kuhn rule. It is a cornerstone for optical studies of materials, since it relates the integrated optical conductivity directly to the density of charged objects, and the absolute value of their charge and mass. It reflects the fundamental property that also in strongly correlated matter the number of electrons is conserved. Note that the right-hand side of the f-sum rule is independent of the value of $\hbar$. Also the f-sum rule applies to bosons and fermions alike. Because $\mathrm{Re}\sigma(\omega)=\mathrm{Re}\sigma(-\omega)$ the sum rule is often presented as an integral of the conductivity over positive frequencies only. Superconductors present a special case, since $\mathrm{Re}\sigma(\omega)$ now has a δ function at $\omega = 0$: Only half of the spectral weight of this δ-function should be counted to the positive frequency side of the spectrum.

2.3 Spectral weight of electrons and optical phonons

The optical conductivity has contributions from both electrons and nuclei because each of these particles carries electrical charge. The integral over the optical conductivity can then be extended to the summation over all species of particles in the solid with mass m_j, and charge q_j

$$\int_{-\infty}^{\infty} \text{Re}\sigma(\omega)d\omega = \sum_j \frac{\pi q_j^2 N_j}{m_j V} \tag{8.22}$$

Because the mass of an electron is several orders of magnitude lower than the mass of a proton, in many cases the contribution of the nuclei to the f-sum rule is ignored in calculations of the integrated spectral weight of metals. However, important exceptions exist where the phonon contribution cannot be neglected, notably in the c-axis response of cuprate high T_c superconductors. Although Eq. (8.22) is completely general, in practice it cannot be applied to experimental spectra directly. This is due to the fact that the contributions of all electrons and nuclei can only be obtained if the conductivity can be measured sufficiently accurately up to infinite frequencies. In practice one always uses a finite cut-off. Let us consider the example of an ionic insulator: If the integral is carried out for frequencies including all the vibrational modes, but does not include any of the inter-band transitions, then the degrees of freedom describing the motion of electrons relative to the ions is not counted. As a result the large number of electrons and nuclei which typically form the ions are not counted as separate entities. Effectively the ions behave as the only (composite) particles in such a case, and the right-hand side of Eq. (8.22) contains a summation over the ions in the solid. Application of Eq. (8.22) provides the so-called transverse effective charge, which for ionic insulators with a large insulating gap corresponds rather closely to the actual charge of the ions. In the top panel of Fig. 8.1 this is illustrated with the infrared spectrum of MgO. Indeed the transverse effective charge obtained from the sum rule is 1.99, in good agreement with the formal charges of the Mg^{2+} and O^{2-} ions.

Because the mass of the ions is much higher than the free electron mass, the corresponding spectral weight integrated over the vibrational part of the spectrum is rather small. In a metal, even if optical phonons are present, usually the spectral weight at low frequencies is completely dominated by the electronic contributions due to the fact that the free electron mass is much smaller than the nuclear mass. A widely spread misconception is, that the screening of optical phonons in metals leads to a smaller oscillator strength than in ionic insulators. The opposite is true: Due to resonant coupling between vibrational modes and electronic oscillators, the optical phonons in an intermetallic compound often have much *more* spectral weight than optical phonons in insulators. This 'charged phonon' effect was formulated in an elegant way in 1977 by Rice, Lipari and Strässler[9]. They demonstrated, that under resonant conditions, due to electron-phonon coupling, vibrational

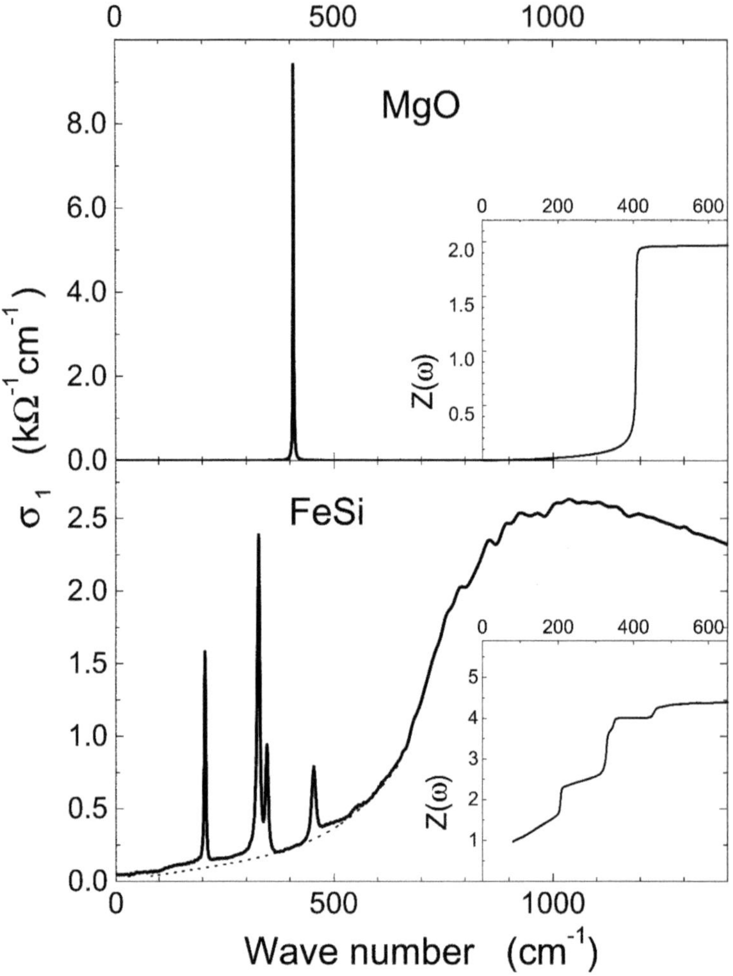

Figure 8.1. Optical conductivity of MgO (top panel) and FeSi at $T = 4$ K (bottom panel). In the insets the function $Z(\omega) = 8\mu(4\pi n_i e^2)^{-0.5}\left\{\int_0^\omega \sigma_{ph}(\omega')d\omega'\right\}^{0.5}$ is displayed. For FeSi the electronic background (dotted curve of the lower panel) was subtracted. For $\omega > 600\text{cm}^{-1}$ $Z(\omega)$ corresponds to transverse effective charge. Data from Ref. [7, 8].

modes borrow oscillator strength from electronic modes, which boosts the intensity of the vibrational modes in the optical conductivity spectra. This effect is now known to be common in many materials, for example in TCNQ-salts,blue bronze, IV-VI narrow-gap semiconductors, FeSi and related compounds, and the beta-phase of sodium vanadate [10, 11, 12, 13, 15, 8, 7, 14, 16]. In Figs. 8.1 and 8.2 the charged phonon effect is illustrated using the examples of FeSi, MnSi, CoSi and RuSi

[8, 7, 14], showing that the transverse charge is between 4 and 5. These compounds are not ionic insulators, because the TM and Si atoms have practically the same electro-negativities and electron affinities. Instead the large transverse charge of these compounds arises from the charged phonon effect predicted by Rice. The strong temperature dependence of the transverse charge of FeSi correlates with the gradual disappearance of the semiconductor gap as the temperature is raised from 4 to 300 K.

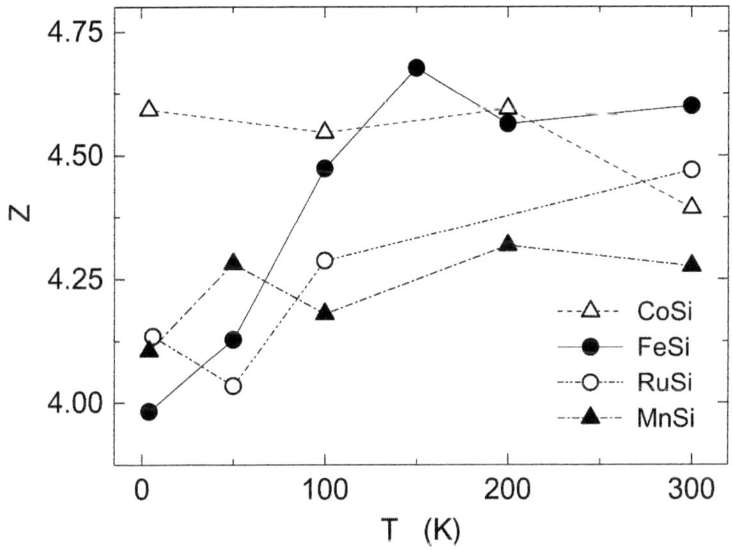

Figure 8.2. Temperature dependence of the transverse effective charge of Co-Si, Fe-Si, Ru-Si, and Mn-Si pairs, calculated from the oscillator strength of the optical phonons. Data from Refs. [7], [8] and [14].

2.4 Partial spectral weight of the intra-band transitions

Often there is a special interest in the spectral properties of the charge carriers. The electrons are subject to the periodic potential of the nuclei, resulting in an energy-momentum dispersion which differs from free electrons. Often one takes this dispersion relation as the starting point for models of interacting electrons. The Coulomb interaction and other (*e.g.* phonon mediated) interactions present the real theoretical challenge. The total Hamiltonian describing the electrons and their interactions is then

$$H = \sum_{k,\sigma} \epsilon_k c^{\dagger}_{k,\sigma} c_{k,\sigma} + \frac{1}{2} \sum_{kqp} V_k \sum_{\sigma\sigma'} c^{\dagger}_{p\sigma} c^{\dagger}_{q\sigma'} c_{q-k\sigma'} c_{p+k\sigma} \tag{8.23}$$

The current operator is in this case

$$\vec{j}_q = \frac{e}{2}\sum_{p,\sigma}\left(v_{p+q/2} - v_{-p+q/2}\right) c^{\dagger}_{p-q/2,\sigma} c_{p+q/2,\sigma} \tag{8.24}$$

where the $v_k \equiv \hbar^{-1}\partial\epsilon_k/\partial\vec{k}$ is the group velocity. The density operator commutes with the interaction part of the Hamiltonian. This has an interesting and very useful consequence, namely that a partial sum rule similar to the f-sum rule exists, which can be used to probe experimentally the kinetic energy term of the Hamiltonian. This partial sum rule for integration of the intra-band conductivity, $\sigma^L_{\alpha\alpha}(\omega)$, yields[17]

$$\int_{-\infty}^{\infty} d\omega \mathrm{Re}\sigma^L_{\alpha\alpha}(\omega) d\omega = \pi\frac{e^2}{V}\sum_{k,\sigma}\langle c^{\dagger}_{k,\sigma}c_{k,\sigma}\rangle\frac{1}{m_k} \tag{8.25}$$

$$\frac{1}{m_k} = \frac{1}{\hbar^2}\frac{\partial^2\epsilon(\vec{k})}{\partial k_\alpha^2} \tag{8.26}$$

Apparently the total spectral weight contained in the inter-band transitions, $\sigma^H_{\alpha\alpha}(\omega)$, is exactly

$$\int_{-\infty}^{\infty} d\omega \mathrm{Re}\sigma^H_{\alpha\alpha}(\omega) = \pi e^2\left(\frac{n}{m} - \frac{1}{V}\sum_k \frac{n_k}{m_k}\right) \tag{8.27}$$

In the limit where the interaction $V_k = 0$, the occupation function n_k in the above summation is a step-function at the Fermi momentum. In this case the summation over k becomes an integration over the Fermi volume with n_k set equal to 1. After applying Gauss's theorem we immediately obtain the well-known Fermi surface integral formula

$$\omega^2_{p,\alpha} = g\int_{S_F}\frac{e^2}{\hbar}v_\alpha(\vec{a})da_\alpha \tag{8.28}$$

where g is the spin degeneracy factor. In the literature two limiting cases are most frequently considered: (i) the free electron approximation, where $m_k = m_e$ is the free electron mass independent of the momentum of the electron, and (ii) the nearest neighbor tight-binding limit. In the latter case the dispersion is $\epsilon_k = -2t_x\cos k_x a_x - 2t_y\cos k_y a_y - 2t_z\cos k_z a_z$, with the effect that $1/m_k = -2t\hbar^{-2}a_\alpha^2\cos(k_\alpha a_\alpha)$, and

$$\sum_\alpha \frac{\hbar^2}{a_\alpha^2\pi e^2}\int_{-\omega_m}^{\omega_m}\mathrm{Re}\sigma_{\alpha\alpha}(\omega)d\omega = -\sum_{k,\sigma}n_k\epsilon_k = \langle -H_{kin}\rangle \tag{8.29}$$

where the integration should be carried out over all transitions within this band, *including* the δ-function at $\omega = 0$ in the superconducting state. The upper limit of the integration is formally represented by the upper limit ω_m. In practice the cutoff cannot always be sharply defined, because usually there is some overlap between the region of transitions within the partially filled band and the transitions between different bands. Hence in the nearest neighbor tight-binding limit the f-sum provides the *kinetic energy* contribution, which depends both on the number of particles and the hopping parameter t[18]. This relation was used by Baeriswyl *et al.* to show, using exact results for one dimension, that the oscillator strength of optical absorption is strongly suppressed if the on-site electron-electron interactions (expressed by the Hubbard parameter U) are increased[19]. The same equation can also be applied to superconductors, examples will be discussed later in this chapter. In the case of a superconductor it is important to realize, that the integration on the left-hand side of Eq. (8.29) should also include the condensate δ-function at $\omega = 0$. As the optical conductivity can only be measured for $\omega > 0$, the spectral weight in the δ-function has to derived from a measurement of the imaginary part of $\sigma(\omega)$, taking advantage of the fact that the real and imaginary part of a δ-function conductivity are of the form

$$\sigma^{singular}(\omega) = \frac{2i\omega_{p,s}^2}{4\pi(\omega + i0^+)}$$

The plasma frequency of the condensate, $\omega_{p,s}$, is inversely proportional to the London penetration depth, $\lambda(T) = c/\omega_{p,s}(T)$ with c the velocity of light. In the literature[20, 21] the δ-function, conductivity integral for $\omega > 0$, and the kinetic energy are sometimes rearranged in the form

$$\frac{\omega_{p,s}^2}{8} = \frac{a^2\pi e^2}{2\hbar^2}\langle -H_{kin}\rangle - \int_{0^+}^{\omega_m} \mathrm{Re}\sigma(\omega)d\omega \qquad (8.30)$$

Whenever the kinetic energy term on the right-hand side changes its value, this expression suggests a 'violation' of the f-sum rule, since the spectral weight in the δ-function now no longer compensates the change of spectral weight in the conductivity integration on the right-hand side. Of course there is no real violation, but part of the optical spectral weight is being swapped between the intra-band transitions and the inter-band transitions. Later in this chapter we will use the relation between kinetic energy as expressed in the original incarnation due to Maldague [18] (Eq. 8.29) to determine in detail the temperature dependence of the *ab*-plane kinetic energy of some of the high T_c superconductors.

It is easy to see, that for a small filling fraction of the band Eq. (8.29) is the same as the Galilean invariant result: The occupied electron states are now all located just above the bottom of the band, with an energy $-t$. Hence in leading orders of the filling fraction $-\langle\psi_g|H_t|\psi_g\rangle = Nt$. Identifying $a^2\hbar^{-2}t^{-1}$ as the effective mass m^* we recognize the familiar f-sum rule, Eq. (8.22), with the free electron mass replaced by the effective mass.

As the total spectral weight (intra-band plus inter-band) should satisfy the f-sum rule, the intra-band spectral weight is bounded from above, *i.e.* $0 \leq \sum_k n_k/m_k \leq n/m$. Near the top of the band the dynamical mass has the peculiar property that it is negative, $m_k < 0$, which in the present context adds a negative contribution to the intra-band spectral weight. On the other hand, the fact that $\text{Re}\sigma(\omega)$ has to be larger than zero, implies that the equilibrium momentum distribution function n_k is subject to certain bounds: If for example n_k would preferentially occupy states near the top of the band, leaving the states at the bottom empty, the intra-band spectral weight would acquire an unphysical negative value. Apparently such momentum distribution functions cannot result from the interactions of Eq. (8.23), regardless of the strength and k-dependence of those interactions.

2.5 Additional sum rules for $\sigma(\omega)$ and $1/\epsilon(\omega)$

Several other sum rule type expressions exist for the optical conductivity and for the dielectric constant. Here we give a summary. In the presence of a magnetic field an optical analogue of the Hall effect exists. The behavior is similar to the DC-limit, resulting in an off-diagonal component of the optical conductivity $\sigma_{xy}(\omega) = -\sigma_{yx}(\omega)$, where the z-axes is parallel to the magnetic field. The optical Hall angle is defined as

$$t_H(\omega) = \frac{\sigma_{xy}(\omega)}{\sigma_{xx}(\omega)} \tag{8.31}$$

The optical (σ_{xx}) and Hall conductivities(σ_{xy}) can be measured directly in optical transmission experiments [22, 23]. Drew and Coleman have shown[24] that this response function obeys the sum rule

$$\frac{2}{\pi}\int_0^\infty t_H(\omega)d\omega = \omega_H \tag{8.32}$$

where the Hall frequency ω_H is unaffected by interactions, and in the Galilean invariant case corresponds to the bare cyclotron frequency, $\omega_H = eB/m$.

A *first moment* sum rule of the optical conductivity is easily obtained for $T = 0$, by direct integration of Eq. (8.18), providing

$$\begin{aligned} &\int_0^\infty \omega\sigma_{\alpha,\alpha}(q,\omega)d\omega = \tfrac{2\pi}{\hbar V}\langle j_{\alpha,q}j_{\alpha,-q}\rangle = \\ &= \tfrac{2\pi e^2\hbar}{m^2 V}\sum_{k,\sigma,\sigma'} k_\alpha^2\langle c^\dagger_{k-q/2,\sigma}c_{k+q/2,\sigma'}c^\dagger_{k+q/2,\sigma'}c_{k-q/2,\sigma}\rangle \end{aligned} \tag{8.33}$$

In free space there is no scattering potential nor a periodic potential causing Umklapp scattering. Hence for electrons moving in free space the right-hand side of Eq. (8.33) is exactly zero. This comes as no surprise: The integral on the left-hand side is also zero, since the optical conductivity of such a system has only a δ-function at $\omega = 0$ due to Galilean invariance. However, in the presence of Umklapp scattering the eigen-states of the electrons with energy-momentum dispersion ϵ_k are no longer the free electron states in the summation of Eq. (8.33). The true eigen-states are superpositions of plane waves. Vice versa the free electron states generated by the $c_k^\dagger$ operators of the above expression can be written as a superposition of the eigen-states of the periodic potential: $c_{k+G,\sigma}^\dagger = \sum_m \alpha_G^m(k) a_{k,m,\sigma}^\dagger$, where the latter operator generates the m'th eigen-state with momentum k in the first Brillouin zone. For brevity we introduce the notation $A_G^m = |\alpha_G^m(k)|^2$, and $\hat{n}_\sigma^j = a_{k,j,\sigma}^\dagger a_{k,j,\sigma}$. Expressed in terms of these band occupation number operators Eq. (8.33) is

$$\lim_{q\to 0}\int_0^\infty \omega\sigma_{\alpha,\alpha}(q,\omega)d\omega = \frac{2\pi e^2\hbar}{m^2V}\sum_{k,G}(k_\alpha + G_\alpha)^2 \sum_{j,m,\sigma} A_G^j A_G^m \langle \hat{n}_\sigma^j (1-\hat{n}_\sigma^m)\rangle \tag{8.34}$$

The summation on the right-hand side strongly suggests an intimate relationship between the optical conductivity and the kinetic energy of the electrons. However, due to the fact that the expression on the right-hand side is rather difficult to calculate, the first moment of $\sigma(\omega)$ is of little practical importance. It's main purpose in the present context is to demonstrate the trend that an increase of the kinetic energy is accompanied by an increase of the first moment of the optical conductivity spectrum. This is consistent with the notion, that an increase of kinetic energy is accompanied by a blue-shift of the spectral weight.

For the energy-loss function a separate series of sum rule type equations can be derived[25, 26, 27]

$$\int_{-\infty}^\infty \mathrm{Im}\frac{-\omega}{\epsilon(\omega)}d\omega = \frac{4\pi^2 e^2 N}{mV} \tag{8.35}$$

which is similar to the f-sum rule for the optical conductivity, Eq. (8.22).

As a result of the fact that the real and imaginary part of the energy loss function are connected via Kramers-Kronig relations, the following relation exists

$$\int_{-\infty}^\infty \mathrm{Im}\frac{-1}{\omega\epsilon(\omega)}d\omega = \pi \tag{8.36}$$

This expression can in principle be used to calibrate the absolute intensity of an energy-loss spectrum, or to check the experimental equipment, since the right-hand side does not depend on any parameter of the material of which the spectrum is taken. We can use the relation between

$\epsilon(\omega)$ and $\sigma(\omega)$, Eq. (8.9), to express Eq. (8 36) as a function of $\sigma(\omega)$. Using Cauchy's theorem, it is quite easy to prove from Eq. (8.36), that

$$\int_0^\infty \mathrm{Re}\frac{1}{\sigma(\omega) - i\lambda\omega} d\omega = \frac{\pi}{2\lambda} \tag{8.37}$$

Often the intra-band optical conductivity is analyzed in terms of a frequency dependent scattering rate $1/\tau(\omega) = (ne^2/m)\mathrm{Re}\{\sigma(\omega)^{-1}\}$, which follows directly from the experimental real and imaginary part of the optical conductivity. Taking the limit $\lambda \to 0$ in Eq. (8.37), we observe that

$$\int_0^\infty \frac{1}{\tau(\omega)} d\omega = \lim_{\lambda \to 0} \frac{\pi}{2\lambda}\frac{ne^2}{m} = \infty \tag{8.38}$$

Hence ultraviolet divergency appears to be a burden of integral formulas of the frequency dependent scattering rate [28, 29, 30] which is hard to avoid.

In section 3.1 we will encounter a relation between the loss-function and the Coulomb energy stored in the electron fluid[26]

$$\int_0^\infty \mathrm{Im}\frac{-1}{\epsilon(\vec{k}, \omega)} d\omega = \frac{4\pi^2 e^2}{\hbar |\vec{k}|^2} \langle \Psi_0 | \hat{\rho}_k \hat{\rho}_{-k} | \Psi_0 \rangle \tag{8.39}$$

This expression is limited to the ground state at $T = 0$, as was also the case for Eq. (8.33). The integrands on the left-hand side of Eq. (8.33) and Eq. (8.39) are *odd* functions of frequency. In contrast the f-sum rule, and the other expressions given in this subsection all involve integrals over an even function of frequency, which is the reason why the latter can be represented as integrals over all (positive and negative) frequencies. The fact that $\hbar$ occurs on the right-hand side of Eqs. (8.33) and (8.39) implies that these expressions are of a fundamental quantum mechanical nature, with no equivalent in classical physics.

Recently Turlakov and Leggett derived an expression for the *third* moment of the energy loss function, which in the limit of $k \to 0$ is a function of the Umklapp potential of Eq. (8.11)

$$\int_{-\infty}^\infty \mathrm{Im}\frac{-\omega^3}{\epsilon_{\alpha\alpha}(\omega)} d\omega = \frac{4\pi^2}{m^2} \left\langle -\sum_G G_\alpha^2 U_G \hat{\rho}_{-G} \right\rangle \tag{8.40}$$

The fact, that the right-hand side of Eq. (8.40) is finite implies, that for $\omega \to \infty$ the loss function of any substance must decay more rapidly than $\mathrm{Im}\{-\epsilon(\omega)^{-1}\} \propto \omega^{-4}$, and that the optical conductivity decays faster than $\mathrm{Re}\{\sigma(\omega)\} \propto \omega^{-3}$. This expression is potentially interesting for the measurement of changes in Umklapp potential, provided that experimental data can be collected up to sufficiently high photon energy, so that the left-hand side of the expression reaches its high frequency limit.

3. The internal energy of superconductors

A necessary condition for the existence of superconductivity is, that the free energy of the superconducting state is lower than that of the non-superconducting state. At sufficiently high temperature important contributions to the free energy are due to the entropy. These contributions depend strongly on the nature of the low energy excitations, first and foremost of all their nature be it fermionic, bosonic or of a more complex character due to electron correlation effects. At $T = 0$ the free energy and internal energy are equal, and are given by the quantum expectation value of the Hamiltonian, which can be separated into an interaction energy and a kinetic energy.

3.1 Interaction energy in BCS theory

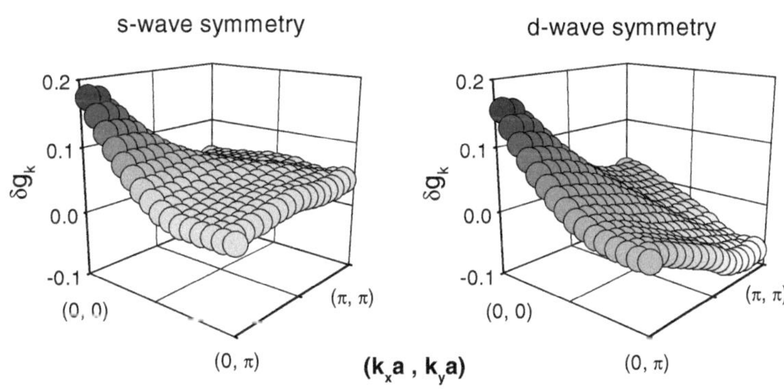

Figure 8.3. The k-space representation of the superconductivity induced change of pair correlation function for the s-wave (left panel) and d-wave symmetry (right panel). Parameters: $\Delta/W = 0.2$, $\omega_D/W = 0.2$. Doping level x = 0.25

We consider a system of electrons interacting via the interaction Hamiltonian given in Eq. (8.23). In the ground state of the system, the interaction energy, including the correlation energy beyond the Hartree-Fock approximation, is just the quantum expectation value of the second (interaction) term of (8.23). Here we are only interested in the difference in interaction energy between the normal and superconducting state.

$$E^s_{corr} - E^n_{corr} = \sum_k V_k \left(\langle \hat{\rho}_k \hat{\rho}_{-k} \rangle_s - \langle \hat{\rho}_k \hat{\rho}_{-k} \rangle_n \right) = \sum_k V_k \delta g_k \qquad (8.41)$$

In BCS theory the only terms of the interaction Hamiltonian which contribute to the pairing are the so-called reduced terms, *i.e.* those terms in the summation of Eq. (8.23) for which the center of mass momentum

$p+q=0$. The quantum mechanical expectation value of the correlation function is

$$\delta g_k = \sum_p (|u_{p+k}|^2 - \theta_{p+k})(\theta_p - |u_p|^2) + \sum_p u_{p+k}v_{p+k}u_p^* v_p^* \tag{8.42}$$

The first term on the right-hand represents the change in exchange correlations, whereas the second term represents the particle-hole mixing which is characteristic for the BCS state. A quantity of special interest is the real space correlation function $\delta g(r,r') = \langle n(r)n(r')\rangle_s - \langle n(r)n(r')\rangle_n$. The Fourier transform of this correlation function is directly related to δg_k appearing in the expression of the interaction energy, Eq. (8.41)

$$\delta g_k = \frac{1}{V^2}\int d^3r \int d^3r' e^{ik(r-r')}\delta g(r,r') \tag{8.43}$$

We see, that if the correlation function $\delta g(r,r')$ could be measured somehow, and the interaction V_k is known, than the interaction energy would follow directly from our knowledge of $\delta g(r,r')$:

$$E^s_{corr} - E^n_{corr} = \int d^3r \int d^3r' V(r-r')\delta g(r,r') \tag{8.44}$$

In a conventional superconductor the quasi-particles of the normal state are also the fermions which become paired in the superconducting state. (Note, that now we are using the concept of Landau Fermi-liquid quasi-particles for the normal state. Later in this manuscript we will explore some consequences of *not* having a Fermi liquid in the normal state, where the quasi-particle concept will be abandoned.) Although the quasi-particle eigen-states of a conventional Fermi liquid have an amount of electron character different from zero, their effective masses, velocities and scattering rates are renormalized. The conventional point of view is, that pairing (enhancement of pair correlations) reduces the interaction energy of the electrons, by virtue of the fact that in the superconducting state the pair correlation function $g(r,r') = \langle\Psi|\hat{n}(r)\hat{n}(r')|\Psi\rangle$ increases at distances shorter than the superconducting coherence length ξ_0. If the interaction energy $V(r- r')$ is *attractive* for those distances, the interaction energy, Eq. (8.44), decreases in the superconducting state, and $V(r-r')$ represents a (or the) pairing mechanism. In Fig. 8.3 we show calculations of δg_k assuming a bandstructure of the form

$$\epsilon_k = \frac{W}{4}[\cos k_x a + \cos k_y a] - \mu \tag{8.45}$$

while adopting an order parameter of the form

$$\Delta_k = \Delta_0 \Theta(|\epsilon_k - \mu| - \omega_D) \tag{8.46}$$

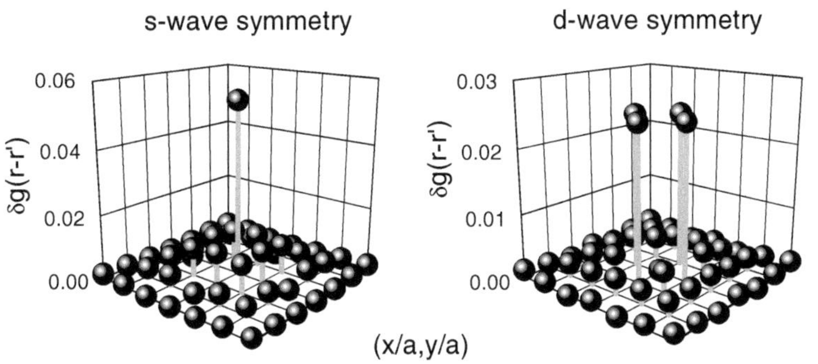

Figure 8.4. The coordinate space representation of the superconductivity induced change of pair correlation function for the s-wave (left panel) and d-wave symmetry (right panel). Parameters: $\Delta/W = 0.2$, $\omega_D/W = 0.2$. Doping level: x = 0.25

for s-wave symmetry, and

$$\Delta_k = \Delta_0 \left[\cos k_x a - \cos k_y a\right] \Theta(|\epsilon_k - \mu| - \omega_D) \tag{8.47}$$

for d-wave symmetry. The parameters used were $\Delta/W = 0.2$, $\omega_D/W = 0.2$, and $E_F/W = 0.43$ corresponding to x=0.25 hole doping counted from half filling of the band. The chemical potential in the superconducting state was calculated selfconsistently in order to keep the hole doping at the fixed value of x=0.25 [31, 32, 33, 34]. From Fig.8.3 we conclude that s-wave pairing symmetry requires a negative V_k regardless of the value of k, whereas the d-wave symmetry can be stabilized either assuming $V_k > 0$ for k in the (π, π) region, or $V_k < 0$ for k near the origin. Both types of symmetry are suppressed by having $V_k > 0$ at small momentum, such as the Coulomb interaction.

In Fig. 8.4 we display the correlation function in coordinate space representation. This graph demonstrates, that d-wave pairing is stabilized by a nearest-neighbor attractive interaction potential. An on-site *repulsion* has no influence on the pairing energy, since the pair correlation function has zero amplitude for $r - r' = 0$. On the other hand, for s-wave pairing the 'best' interaction is an on-site attractive potential, since the s-wave $\delta g(r, r')$ reaches it's maximum value at $r - r' = 0$.

3.2 Experimental measurements of the Coulomb interaction energy

In a series of papers Leggett has discussed the change of Coulomb correlation energy for a system which becomes superconducting[35], and has argued, that this energy would actually decrease in the superconducting state. Experimentally the changes of Coulomb energy can be measured directly in the sector of k-space of vanishing k. The best, and most stable, experimental technique is to measure the dielectric function using spectroscopic ellipsometry, and to follow the changes as a function of temperature carefully as a function of temperature. Because the cuprates are strongly anisotropic materials, it is crucial to measure both the in-plane and out-of-plane pseudo-dielectric functions, from which the full dielectric tensor elements along the optical axes of the crystal then have to be calculated. We followed this procedure for a number of different high T_c cuprates, indicating that the Coulomb energy in the superconducting state *in*creases for k=0. However, for $k \neq 0$ this need no longer be the case. Summarizing the situation[36]: the Coulomb interaction energy increases in the superconducting state for small k. This implies, that the lowering of internal energy in the superconducting state must be caused either by other sectors of k-space (in particular at around the (π, π) point, see Fig. 8.3!), or by a lowering of the kinetic energy in the superconducting state. The latter is only possible in a non Fermi liquid scenario of the normal state.

3.3 Kinetic energy in BCS theory

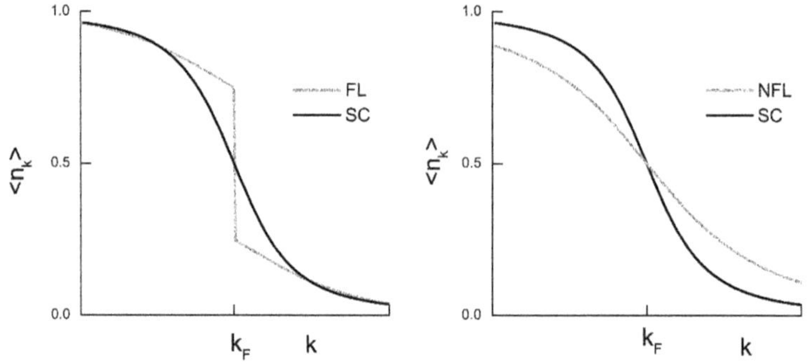

Figure 8.5. Occupation function as a function of momentum in the normal (dash) and the superconducting (solid) state for Fermi liquid (left panel) and an example of a broad distribution function, not corresponding to a Fermi liquid (right panel).

In BCS theory the lowering of the pair-interaction energy is partly compensated by a change of kinetic energy of opposite sign. This can be understood qualitatively in the following way: The correlated motion in pairs causes a localization of the relative coordinates of electrons, thereby increasing the relative momentum and the kinetic energy of the electrons. Another way to see this, is that in the superconducting state the step of n_k at the Fermi momentum is smoothed, as indicated in the left panel of Fig. 8.5, causing E_{kin} to become larger[37].

A pedagogical example where the kinetic energy of a pair is higher in the superconducting state, is provided by the negative U Hubbard model[38]: Without interactions, the kinetic energy is provided by the expression

$$E_{kin} = -t \sum_{<i,j>,\sigma} \langle \Psi | c^{\dagger}_{i\sigma} c_{j\sigma} + H.c. | \Psi \rangle \tag{8.48}$$

Let us consider a 2D square lattice. If the band contains two electrons, the kinetic energy of each electron is $-2t$, the bottom of the band, hence $E_{kin} = -4t$. (In a tight-binding picture the reference energy is the center of the band irrespective of E_F, causing E_{kin} to be always negative). Let us now consider the kinetic energy of a pair in the extreme pairing limit, *i.e.* $U \gg t$, causing both electrons to occupy the same site, with an interaction energy $-U$. The occupation function n_k in this case becomes

$$n_k \approx \frac{1}{N_k} \frac{t}{U} \frac{1}{(1 + 4\epsilon_k/U)^2} \tag{8.49}$$

This implies that the kinetic energy approaches $E_{kin} \to -8t^2/U$. Hence the kinetic energy increases from $E^n_{kin} = -4t$ to $E^s_{kin} = -\frac{8t^2}{U}$ when the local pairs are formed. The paired electrons behave like bosons of charge $2e$. A second order perturbation calculation yields an effective boson hopping parameter[39] $t' = t^2/U$. In experiments probing the charge dynamics, this hopping parameter determines the inertia of the charges in an accelerating field. As a result the plasma frequency of such a model would be

$$\omega^2_{p,s} = 4\pi \frac{n}{2} (2e)^2 \frac{a^2 t^2}{\hbar^2 U} \tag{8.50}$$

whereas if these pair correlations are muted

$$\omega^2_{p,n} = 4\pi n e^2 \frac{a^2 t}{\hbar^2} \tag{8.51}$$

Because the plasma frequency is just the low frequency spectral weight associated with the charge carriers, this demonstrates, that for conventional pairs (*i.e.* those which are formed due to interaction energy lowering) the expected trend is, that in the superconducting state the spectral weight *de*creases. Note, that this argument can only demonstrate the direction in which the plasma frequency changes when the pair correlations become reduced, but it does not correctly provide the quantitative size of the change, since the strong coupling regime of Eq. (8.50) implies the presence of a finite fraction of uncondensed 'preformed' pairs in the normal state. The same effect exists in the limit of weak pairing correlations. In Ref. [40] ((Eq. 29), ignoring particle-hole asymmetric terms) the following expression was derived for the plasma resonance

$$\omega_{p,s}^2 = \frac{4\pi e^2}{V} \sum_k \frac{\Delta_k^2}{\hbar^2 E_k^3} \left[\frac{\partial \epsilon_k}{\partial k}\right]^2 \tag{8.52}$$

where V is the volume of the system, and $E_k^2 = \epsilon_k^2 + |\Delta_k|^2$. Integrating in parts, using that $\Delta_k^2 E_k^{-3} \partial_k \epsilon_k = \partial_k (\epsilon_k / E_k)$, and that $\partial_k \epsilon_k = 0$ at the zone-boundary, we obtain

$$\omega_{p,s}^2 = \frac{4\pi e^2}{V} \sum_k \frac{n_k}{m_k} \tag{8.53}$$

where $m_k^{-1} = \hbar^{-2} \partial^2 \epsilon_k / \partial k^2$, and $n_k = 1 - \epsilon_k / E_k$. For a monotonous band dispersion the plasma frequency of the superconductor is always *smaller* than that of the unpaired system: Because the sign of the band-mass changes from positive near the bottom of the band to negative near the top, the effect of the broadened occupation factors n_k is to give a slightly smaller average over m_k^{-1}, hence ω_p^2 is smaller. Note that the mass of free electrons does not depend on momentum, hence in free space ω_p^2 is unaffected by the pairing.

To obtain an estimate of the order of magnitude of the change of spectral weight, we consider a square band of width W with a Fermi energy $E_F = N_e/(2W)$, where N_e is the number of electrons per unit cell. To simplify matters we assume that $1/m_k$ varies linearly as a function of band energy:$1/m(\epsilon) = (W - 2E_F - 2\epsilon)/(Wm_0)$. We consider the limit where $\Delta << W, E_F$. Let us assume that the bandwidth $\sim$ 1 eV, and $\Delta \sim$ 14 meV corresponding to T_c =90 K. The reduction of the spectral weight is then 0.28 %. If we assume that the bandwidth is 0.1 eV, the spectral weight reduction would typically be 11.4 %.

3.4 Kinetic energy driven superconductivity

If the state above T$_c$ is *not* a Fermi liquid, the situation could be reversed. The right-hand panel of Fig.8.5 represents a state very different from a Fermi liquid, and in fact looks similar to a gapped state. Indeed even for the 1D Luttinger liquid $n(k)$ has an infinite slope at k_F. If indeed the normal state would have a broad momentum distribution like the one indicated, the total kinetic energy becomes lower once pairs are formed, provided that the slope of $n(k)$ at k_F is steeper in the superconducting state. This is not necessarily in contradiction with the virial theorem, even though ultimately all relevant interactions (including electron-phonon interactions) are derived from the Coulomb interaction: The superconducting correlations involve the low energy scale quasi-particle excitations and their interactions. These *effective* interactions usually have characteristics quite different from the original Coulomb interaction, resulting in $E_c/E_{kin} \neq -2$ for the low energy quasi-particles. Various models have been recently proposed involving pairing due to a *reduction* of kinetic energy. In strongly anisotropic materials such as the cuprates, two possible types of kinetic energy should be distinguished: Perpendicular to the planes[41, 42] (along the c-direction) and along the planar directions[43, 44, 45, 46, 47, 48, 49, 50, 51].

4. Experimental studies of superconductivity induced spectral weight transfer

4.1 Josephson plasmons and c-axis kinetic energy

C-axis kinetic energy driven superconductivity has been proposed within the context of inter-layer tunneling, and has been extensively discussed in a large number of papers[43, 44, 41, 52, 42, 21, 53, 54, 55, 56, 57, 20, 58, 59, 60, 61, 62, 63, 64]. One of the main reasons to suspect that superconductivity was c-axis kinetic driven, was the observation of "incoherent" c-axis transport of quasi-particles in the normal state[65] and, rather surprisingly, *also* in the superconducting state[66, 67, 68], thus providing a channel for kinetic energy lowering for charge carriers as soon as pairing sets in. As discussed in section 2.4 a very useful tool in the discussion of kinetic energy is the low frequency spectral weight associated with the charge carriers. In infrared spectra this spectral weight is contained within a the 'Drude' conductivity peak centered at $\omega = 0$. Within the context of the tight-binding model a simple relation exists between the kinetic energy per site, with volume per site V_u, and the low frequency spectral weight[18, 19]

$$E_{kin} = \frac{\hbar^2 V_u}{4\pi e^2 a^2}\omega_p^2 \tag{8.54}$$

Here the plasma frequency, ω_p, is used to quantify the low frequency spectral weight:

$$\frac{\omega_{p,s}^2}{8} + \int_{0+}^{\omega_m} \mathrm{Re}\sigma(\omega)d\omega = \frac{1}{8}\omega_p^2 \tag{8.55}$$

where the integration should be carried out over all transitions within the band, *including* the δ-function at $\omega = 0$ in the superconducting state. The $\delta(\omega)$ peak in $\mathrm{Re}\sigma(\omega)$ is of course not visible in the spectra directly. However the presence of the superfluid is manifested prominently in the London term of $\mathrm{Re}\epsilon(\omega)$ (proportional to $\mathrm{Im}\sigma(\omega)$): $\epsilon_L(\omega) = -\omega_{p,s}^2\omega^{-2}$. In $La_{2-x}Sr_xCuO_4$ the London term is manifested in a spectacular way as a prominent plasma resonance perpendicular to the superconducting planes[69]. This is commonly used to determine the superfluid spectral weight, $\omega_{p,s}^2$, from the experimental spectra. Apart from universal prefactors, the amount of spectral weight of the $\delta(\omega)$ conductivity peak corresponds to the Josephson coupling energy, which in turn is the inter-layer *pair*hopping amplitude. It therefore provides an upper limit to the change of kinetic energy between the normal and superconducting state [41, 52], because the spectral weight transferred from higher frequencies to the $\delta(\omega)$-peak cannot exceed this amount. This allowed a simple experimental way to test the idea of c-axis kinetic energy driven superconductivity by comparing the experimentally measured values of the condensation energy (E_{cond}) and E_J. The inter-layer tunneling hypothesis required, that $E_J \approx E_{cond}$. In the spring of 1996 the first experimental results were presented[53] for Tl2201 (T_c=80 K), showing that E_J was at least two orders of magnitude too small to account for the condensation energy (see Fig. 8.6). Later measurements of λ_c[57] (approximately 17 μm) and the Josephson plasma resonance (JPR)[56] at 28 cm^{-1}, allowed a definite determination of the Josephson coupling energy of this compound, indicating that $E_J \approx 0.3\mu$eV in Tl2201 with $T_c = 80$ K (see Fig. 8.7). This is a factor 400 lower than $E_{cond} \approx 100\mu$eV per copper, based either on c_V experimental data[70], or on the formula $E_{cond} = 0.5N(0)\Delta^2$ with $N(0) = 1eV^{-1}$ per copper, and $\Delta \simeq 15meV$. In Fig. 8.8 the change in c-axis kinetic energy and the Josephson coupling energies are compared to the condensation energy for a large number of high T_c cuprates. For most materials we see that $E_J < E_{cond}$, sometimes differing by several orders of magnitude.

These arguments falsifying the inter-layer tunneling mechanism have been questioned[21], arguing that a large part of the specific heat of Tl2201 is due to 3D fluctuations, and that these fluctuations should be subtracted when the condensation energy is calculated. However, it was recently shown[71] that due to thermodynamical constraints the fluctuation correction can not exceed a factor 2.5 in the case of Tl2201 (as compared to a factor 40 in Ref. [21]). Hence the discrepancy between the Josephson coupling energy and the condensation energy of Tl2201 is still two orders of magnitude.

However, as stressed above, E_J provides only an *upper limit* for ΔE_{kin}. A c-axis kinetic energy change *smaller* than E_J is obtained if we take into account the fact that a substantial part of $\delta(\omega)$-function is just the spectral weight removed from the sub-gap region of the optical conductivity. Usually it is believed that in fact the latter is the *only* source of intensity of spectral weight for the δ-function, known as the (phenomenological) Glover-Tinkham-Ferell[72] sum rule. According to the arguments given

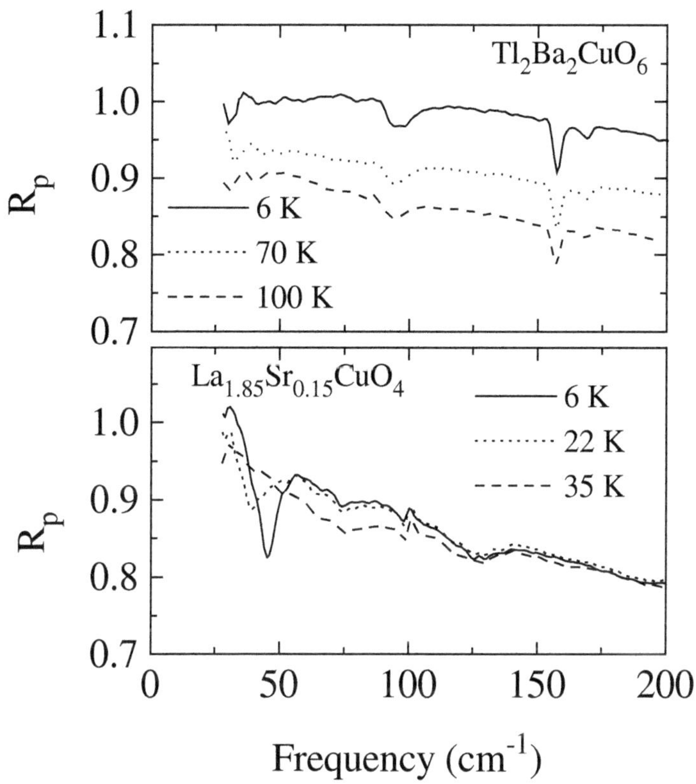

Figure 8.6. Grazing reflectivity of a $Tl_2Ba_2CuO_6$ thin film (upper panel) and $La_{1.85}Sr_{0.15}CuO_4$ single crystal (lower panel) measured with the polarization of the incident light tilted at an angle of 80° relative to the copper-oxygen planes. For LSCO the Josephson plasma resonance can be clearly seen at 40 cm^{-1}. For Tl2212 no the Josephson plasma resonance is observed, indicating that it is located below the lower limit of 30 cm^{-1} of the spectrometer. This implies that the Josephson coupling energy in this compound is at least two orders of magnitude lower than required by the inter-layer tunneling hypothesis. Data from Ref. [54]

in section 3.4 we may conclude that $E_{kin,s} = E_{kin,n}$ when we observe, that *all* spectral weight origins from the far-infrared gap region in agreement with the Glover-Tinkham-Ferrell sum rule. If, on the other hand, superconductivity is accompanied by a lowering of c-axis kinetic energy, part of $\omega_{p,s}^2$ originates from the higher frequency region of inter-band transitions, which begins at typically 2 eV. In other words, we may say that $\omega_{p,s}^2$ is an upper limit to the kinetic energy change

$$0 < E_{kin,n} - E_{kin,s} < \frac{\hbar^2 V_u}{4\pi e^2 a^2}\omega_{p,s}^2 \tag{8.56}$$

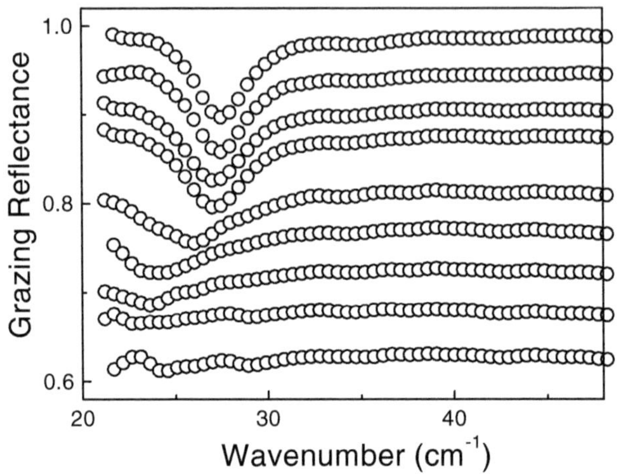

Figure 8.7. P-polarized reflectivity at 80° angle of incidence of $Tl_2Ba_2CuO_6$. From top to bottom: 4K, 10 K, 20 K, 30 K, 40 K, 50 K, 60 K, 75 K, and 90 K. The curves have been given incremental 3 percent vertical offsets for clarity. Data from Ref. [56]

A direct determination of $E_{kin,s} - E_{kin,n}$ is obtained by measuring experimentally the amount of spectral weight transferred to the $\delta(\omega)$ peak due to the passage from the normal to the superconducting state, as was done by Basov *et al.*[20, 60]. These data indicated that for underdoped materials about 60% comes from the sub-gap region in the far infrared, while about 40% originates from frequencies much higher than the gap, whereas for optimally doped cuprates at least 90% originates from the gap-region, while less than 10% comes from higher energy. Experimental artifacts caused by a very small amount of mixing of ab-plane reflectivity into the c-axis reflectivity curves may have resulted in an overestimation of the spectral weight originating from high energies[60], in particular those samples where the electronic $\sigma_c(\omega)$ is very low due to the 2-dimensionality. Optimally doped YBCO is probably less prone to systematic errors due to leakage of R_{ab} into the c-axis reflectivity, since $\sigma_c(\omega)$ of this material is among the largest in the cuprate family. The larger $\sigma_c(\omega)$ causes the c-axis reflectivity to be much larger at all frequencies, thereby reducing the effect of spurious mixing of *ab*-plane reflectivity in the optical spectra on the Kramers-Kronig analyzes.

In summary $\Delta E_{kin,c} < 0.1 E_J$ in most cases. For several of the single-layer cuprates it has become clear now, that ΔE_{kin} significantly undershoots the condensation energy, sometimes by two orders of magnitude or worse, as indicated in Fig. 8.8.

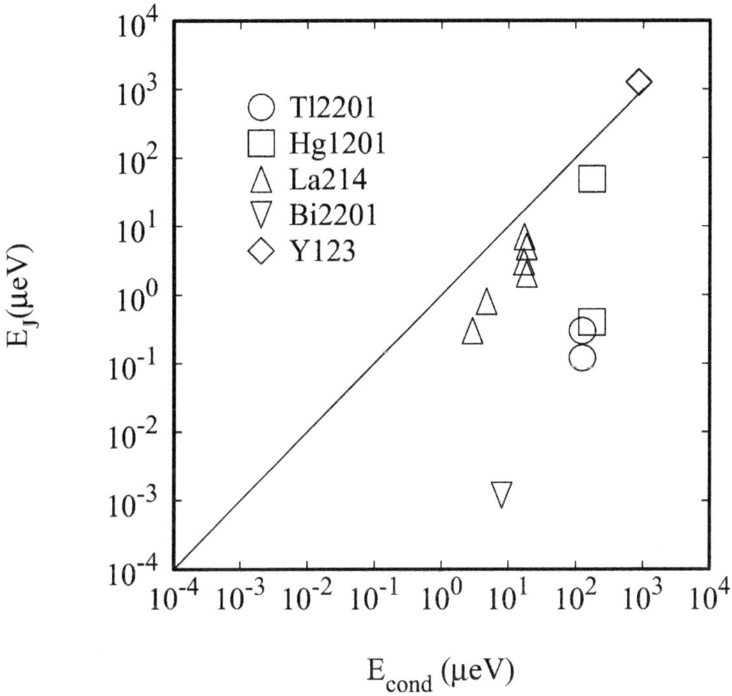

Figure 8.8. Intrinsic Josephson coupling energy [20, 55, 56, 57, 58, 59, 66, 69, 75] versus condensation energy[70, 76]

4.2 Josephson plasmons in multi-layered cuprates

This situation may be different for the bi-layer compounds. In these materials in principle the coupling within the bi-layer may provide an additional source of frustrated inter-layer kinetic energy, which can in principle be released when the material enters the superconductng state. This can in principle be monitored with infrared spectroscopy, because quite generally a stack of Josephson coupled layers with two different types of weak links alternating (in the present context corresponding to inter-bilayer and intra-bilayer) should exhibit three Josephson collective modes instead of one: Two of those modes are longitudinal Josephson plasma resonances, which show up as peaks in the energy loss function $\mathrm{Im}(-1/\epsilon(\omega))$. In between these two longitudinal resonances one expects a transverse optical plasma resonance, which is revealed by a peak in $\mathrm{Re}\sigma(\omega)$. In essence the extra two modes are out-of-phase oscillations of the two types of junctions. This has been predicted in Ref. [77] for the case of a multi-layer of Josephson coupled 2D superconducting layers. Further detailed calculations for the bi-layer case were presented in Refs. [78, 79]. The existence of *two* longitudinal modes and *one* associated transverse plasmon mode at finite frequencies has been confirmed experimentally for the $SmLa_{0.8}Sr_{0.2}CuO_{4-\delta}$ in a series of papers [80, 81, 82, 83, 84] (see Fig. 8.9).

The c-axis optical conductivity of YBCO is one order of magnitude larger than for LSCO near optimal doping. As a result the relative importance of the optical phonons in the spectra is diminished. In the case of optimally doped YBCO, the experiments indicate no significant transfer of spectral weight from high frequencies associated with the onset of superconductivity. C-axis reflectivity data[75] of optimally doped YBCO are shown in Figs.8.10. Above T_c the optical conductivity is weakly frequency dependent, and does not resemble a Drude peak. Below T_c the conductivity is depleted for frequencies below 500 cm^{-1}, reminiscent of the opening of a large gap, but not an s-wave gap, since a relatively large conductivity remains in this range.

There is a slight overshoot of the optical conductivity in the region between 500 and 700 cm^{-1}, due to the fact that the normal state and superconducting state curves cross at 600 cm^{-1}. In the case of the bi-layer cuprates this could be explained as a result of the presence of two superconducting layers per unit cell, resulting in the 'transverse optical' plasma mode mentioned above[62, 63, 64, 73, 74, 75].

For the f-sum rule the presence of this extra mode makes no difference. The extra spectral weight in the superconducting state associated with this mode has in principle the same origin as the spectral weight in the zero-frequency δ-function. In a conventional picture the source would be the spectral weight, removed due to a depletion of $\sigma_c(\omega)$ in the gap-region. The implementation of the sum rule relevant for this case then states that the relative spectral weight function

$$\frac{\Delta A(\omega)}{\omega_{p,s}^2} = \frac{8}{\omega_{p,s}^2} \int_{0+}^{\omega} \left(\sigma_n(\omega') - \sigma_s(\omega')\right) d\omega' \tag{8.57}$$

overshoots the 100 % line close to the 'second plasma' mode, and saturates at 100 % for frequencies far above this mode. This is indeed observed in Fig. 8.10.

Additional studies of the bi-layer (and tri-layer) materials have provided confirmation of the transverse optical plasmon in these materials. In spite of its high frequency, making the assignment to the Josephson effect rather dubious, nevertheless the transverse optical mode either makes its first appearance below T_c, or gains in sharpness and intensity at the temperature where pairs are being formed (which for under-doped cuprates begins already above T_c). Also in at least a number of cases

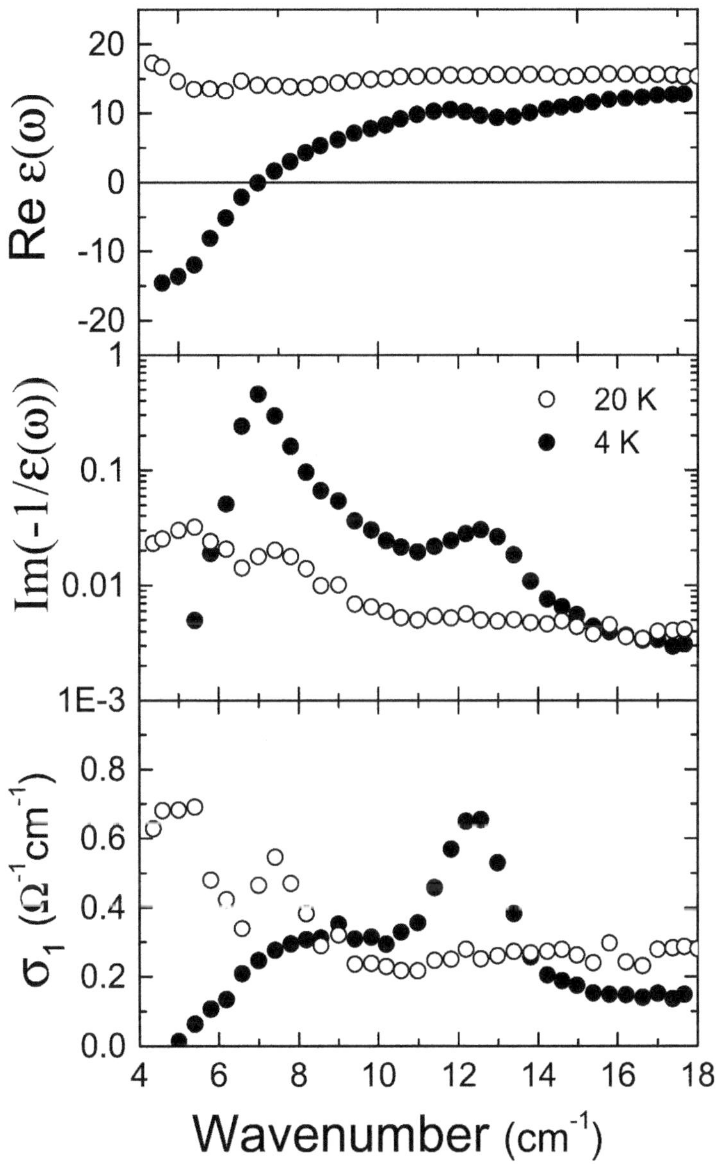

Figure 8.9. (a) Real part of the *c*-axis dielectric function of $SmLa_{0.8}Sr_{0.2}CuO_{4-\delta}$ for 4 K (closed symbols), and 20 K (open symbols) (b) The c-axis loss function, $\text{Im}\epsilon(\omega)^{-1}$. (c) Real part of the *c*-axis optical conductivity. Data from Ref. [83, 85].

the spectral weight of the 'transverse optical' plasmon observed below T_c appears to originate not from the spectral weight removed from the gap region, but from much higher energies[60, 61, 62, 63, 64]. The implic-

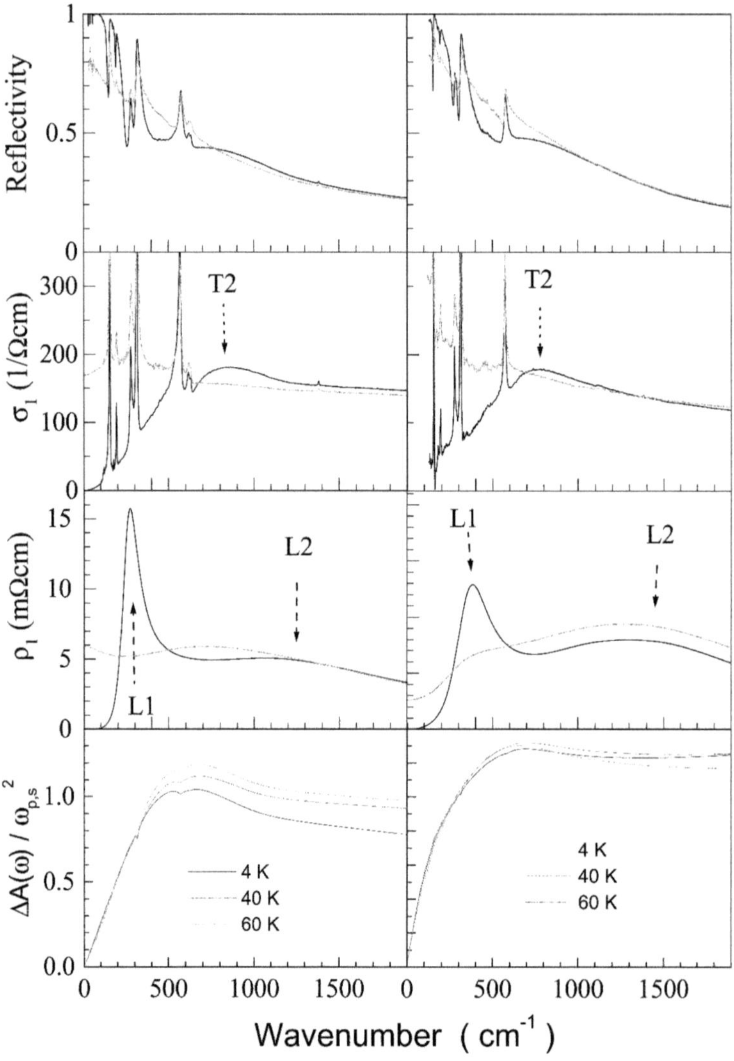

Figure 8.10. C-axis optical spectra of optimally doped (x=6.93) and over-doped (x=7.0) $YBa_2Cu_3O_{7-x}$. From top to bottom: reflectivity, optical conductivity, dynamical impedance and relative spectral weight (Eq. 8.57). The dynamical impedance, $\rho_1(\omega)=\mathrm{Re}4\pi/\omega\epsilon(\omega)$ is proportional to the energy loss function weighted by a factor $1/\omega$. The optical phonons have been subtracted from the loss-functions for clarity. The data are from Ref. [75, 86]

ation of this may be, that a non-negligible fraction of frustrated c-axis kinetic energy is released when these materials become superconducting. This seems to be particularly relevant for the strong intra-bilayer (or tri-layer) coupling of Bi2212, Bi2223 and Y123.

4.3 Kinetic energy parallel to the planes

In-plane kinetic energy driven superconductivity has been proposed by a number of researchers: Hirsch[43, 44, 87] discussed this possibility as a consequence of particle-hole asymmetry. It has also been discussed within the context of holes moving in an anti-ferromagnetic background [88, 89, 90, 91, 92, 93]. More recently the possibility of a *reduction* of kinetic energy associated with pair hopping between stripes has been suggested[47, 51], and an in-plane pair delocalization mechanism have been proposed in the context of the resonating valence band model[49, 50].

A major issue is the question how to measure this. The logical approach would be to measure again $\sigma(\omega, T)$ using the combination of reflectivity and Kramers-Kronig analysis, and then compare the spectral weight function in the superconducting state to the same above T_c. There are several weak points to this type of analysis. In the first place there is the problem of sensitivity and progression of experimental errors: Let us assume, that the change of kinetic energy is of order 0.1 meV per Cu atom (this is approximately the condensation energy of the optimally doped single layer cuprate Tl2201, with T_c = 85 K.). For an inter-layer spacing of 1.2 nm, this corresponds to a spectral weight change $\Delta(\nu_p^2) = 10^5$ cm^{-2}. As the total spectral weight in the far infrared range is of order $\nu_p^2 = 14000^2 \text{cm}^{-2}$, the relative change in spectral weight is of order 0.05 %. Typical accuracy reached for spectral weight estimates using conventional reflection techniques is of order 5%. This illustrates the technical difficulties one has to face when attempting to extract superconductivity induced changes of the kinetic energy.

Experimental limitations on the accuracy are imposed by (i) the impossibility to measure *all* frequencies including the sub-mm range, (ii) systematic errors induced by Kramers-Kronig analysis: The usual procedure is to use data into the visible/ultra-violet range and beyond for completing the Kramers-Kronig analysis in the far infrared, assuming that no important temperature dependence is present outside the far infrared range. Obviously this assumption becomes highly suspicious if the search is concentrated on spectral weight transfer originating from precisely this frequency range.

The remedy is, to let nature perform the spectral weight integral. Due to causality $\text{Re}\epsilon(\omega)$ and $\text{Re}\sigma(\omega)$ satisfy the Kramers-Kronig relation

$$\text{Re}\epsilon(\omega) = 1 - \int_0^\infty \frac{8\text{Re}\sigma(z)}{\omega^2 - z^2} dz \qquad (8.58)$$

The main idea of spectral weight transfer is, that spectral weight is essentially transferred from the inter-band transitions at an energy of several eV, down to the δ-function in $\sigma(\omega)$ at $\omega = 0$. Indeed various groups have reported a change of optical properties in the visible part of the spectrum when the sample becomes superconducting[96, 97, 98, 99]. If this is the case, we have $x = 0$ for the extra spectral weight in relation 8.58. Together with Eq. (8.56) it follows that changes in kinetic energy can be read directly from $\mathrm{Re}\epsilon(\omega)$ using the relation

$$\delta E_{kin}^{eff}(\omega) = \frac{4\hbar^2\omega^2 V_u}{\pi e^2 a^2}\mathrm{Re}\delta\epsilon(\omega) \qquad (8.59)$$

If the spectral weight is transferred to a frequency range ω_0, then the above expression can still be applied for $\omega \gg \omega_0$. If we measure $\mathrm{Re}\epsilon(\omega)$ *directly* using spectroscopic ellipsometry, then indeed nature does the integration of $\sigma(\omega)$ for us at each temperature. This eliminates to a large extent various systematic errors affecting the overall accuracy of the spectral weight sum. It is important to measure the complex dielectric constant for a large range of different frequencies.

The second problem is that already above the superconducting phase transition the optical spectra of these materials have appreciable temperature dependence. What we really like to measure is the spectra of the same material in the superconducting state, and in the 'normal' state, both at the same temperature. Typical magnetic fields required to bring the material in the normal state are impractical, let alone the complications of magneto-optics which then have to be faced. A more practical approach is to measure carefully the temperature dependence over a large temperature range, with small temperature intercepts, and to search for changes which occur at the phase transition.

In Fig. 8.11 the spectral weight from 0 to 10000 cm^{-1} is shown as a function of temperature for the case of Bi2212[99]. Note that this integral corresponds to *minus* the ab-plane kinetic energy. We observe, that in the superconducting state the kinetic energy drops by an amount of about 1 meV per Cu. This is in fact a relatively large effect. This surprising result seems to tell us that in the cuprates the kinetic energy in the superconducting state is *lowered* relative to the normal state. This corresponds to the unconventional scenario depicted in the right-hand panel of Fig.8.5, where the normal state is a non Fermi liquid, whereas the superconducting state follows the behavior of a (more) conventional BCS type wave-function with the usual type of Bogoliubov quasi-particles. The amazing conclusion from this would be, that there is no need for a *lowering* of the interaction energy any more. The condensation energy of optimally doped Bi2212 is about 0.1 meV per Cu atom[76].

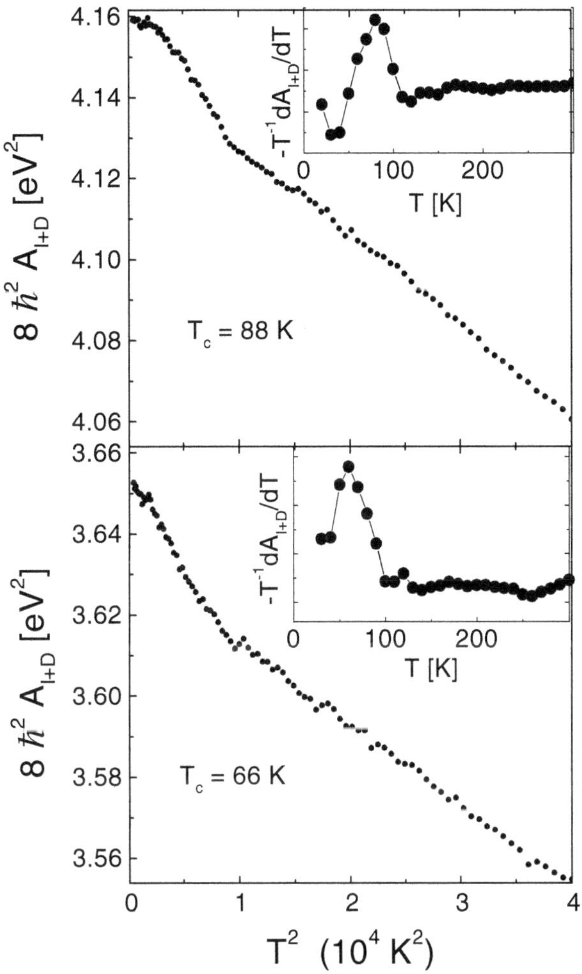

Figure 8.11. Measured values of the quantity $c^2\omega_{p,s}^2 + 8\int_{0+}^{\infty} \text{Re}\sigma_{ab}(\omega)d\omega$ of Bi2212 (T_c=88 K). The data are taken from Ref. [99, 100]. To make the conversion to kinetic energy summed over the two *ab*-plane directions, the numbers along the vertical axis have to be multiplied with a factor $-10^3 V_u/(4\pi e^2 a^2) = -83$ meV / eV^2.

5. Conclusions

The optical conductivity is a fundamental property of solids, contains contributions of vibrational and electronic character. Among the electronic type of excitations the intra-band and inter-band transitions, excitons, and plasmons of different types correspond to the most prominent features in the spectra. In addition multi-magnon excitations or more exotic collective modes can often be detected. The careful study of the optical properties of solids can provide valuable microscopic information about the electronic structure of solids. In contrast to many other spectroscopic techniques, it is relative easy to obtain reliable absolute values of the optical conductivity. As a result sum rules and sum rule related integral expressions can often be applied to the optical spectra. Here we have treated a few examples of sum rule analysis: Application of the f-sum rule to the phonon spectra of transition metal silicides provides information on the resonant electron-phonon coupling in these materials. Integration of the energy-loss function gives the value of the Coulomb energy stored in the material, which is seen to increase when a high T_c cuprate enters the superconducting state. The spectral weight within the partially band of the high T_c cuprates is seen to become larger in superconducting state. This effect exists both perpendicular to the planes and parallel to the planes. This spectral weight change can be associated with a decrease of kinetic energy when the material becomes superconducting. Although the relative spectral weight change along the ab-plane is quite small, it indicates a fairly large change of the ab-plane kinetic energy, large enough to account for the energy by which the superconducting state of these materials is stabilized. In addition the real and imaginary part of the optical conductivity can be used to study the intrinsic Josephson coupling between the superconducting planes. In superconductors with two or more different types of weak links alternating, such as $SmLa_{1-x}Sr_xCu_4$, $YBa_2Cu_3O_{7-x}$, $Bi_2Sr_2CaCu_2O_8$, a rich spectrum of plasma-oscillations is observed in the superconducting state, and sometimes above T_c, due to the multi-layered structure of these materials. This has provided important insights in the nature of the coupling, and it has been used to extract quantitative values of this coupling.

Acknowledgments

Part of the experiments described in this chapter have been carried out by J. H. Kim, J. Schuetzmann, A. Tsvetkov, A. Kuzmenko, G. Rietveld, B. J. Feenstra, H. S. Somal, J. E. van der Eb, A. Damascelli, M. U. Grueninger, D. Dulic, C. Presura, P. Mena, and H. J. A. Molegraaf. The author has benefitted from discussions with, among others, D. I. Khomskii, G. A. Sawatzky, M. J. Rice, P. W. Anderson, A. J. Leggett, Z. X. Shen, S. C. Zhang, A. J. Millis, M. Turlakov, J. E. Hirsch, D. N. Basov, M. R. Norman, and W. Hanke.

References

[1] W. Hanke, and L. J. Sham, Phys. Rev. B **12**, 4501 (1977).

[2] P. Nozières, and D. Pines, Phys. Rev. **111** 442 (1958).

[3] P. Nozières, and D. Pines, Phys. Rev.**113** 1254 (1959).

[4] Note the extra factor of π. In Ref. [2] this factor is absent, but it is recuperated in the final formulas.

[5] The intra-band optical spectrum of a superconductor typically has three contributions: (i) A Dirac-function in the origin corresponding to the supercondensate the presence of which is revealed by a reactive response in the imaginary part of the optical conductivity below the gap diverging as $\sigma_2(\omega) = c^2/(4\pi\lambda(T)^2\omega)$. (ii) A gapped Drude-like peak due absorption of light by pairs of quasi-particles excited above the gap. For $\omega >> 2\Delta$ the spectrum merges with the Drude peak observed in the normal state. (iii) A narrow peak of finite width centered at $\omega = 0$ due to the dissipative response of thermal quasi-particles above the gap.

[6] The proof of Eq. 8.17 follows the same lines as the derivation of the f-sum rule. It employs the fact, that the current operators can be represented as commutators of the hamiltonian with the dipole operator. The expectation value of the hamiltonian is used to cancel out the factor ω_{mn} in the denominator of the expression. In the final step the commutator of the dipole operator and the current operator is calculated in second quantization, which completes the derivation of Eq. 8.17. For a Galilean invariant system at $T = 0$ this summation cannot be carried out directly for $q = 0$. The singular behaviour at this point can be circumvented by considering the limit $q \rightarrow 0$.

[7] A. Damascelli, K. Schulte, D. van der Marel, and A.A. Menovsky, Phys. Rev. B **55**, R4863 (1997).

[8] A. Damascelli, Ph D. Thesis, University of Groningen (1999).

[9] M. J. Rice, N. O. Lipari, and S. Strässler Phys. Rev. Lett. **39**, 1359 (1977).

[10] W. A. Challener and P. L. Richards, Solid State Commun. **52**, 117 (1984).

[11] G.Travaglini and P.Wachter, Phys. Rev. B **30**, 1971 (1984).

[12] L. Degiorgi, P. Wachter and C.Schlenker, Phys. Rev. B **164**, 305 (1990).

[13] L. Degiorgi and G. Grüner, Phys. Rev. B **44**, 7820 (1991).

[14] V. Vescoli, L. Degiorgi, B. Buschinger, W. Guth, C. Geibel, and F. Steglich, Solid State Commun. **105**, 367 (1998).

[15] G. Lucovsky, and R. M. White, Phys. Rev. B **8**, 660 (1973).

[16] C. Presura, M.Popinciuc, P. H. M. van Loosdrecht, D. van der Marel, M. Mostovoy, G. Maris, T. T. M. Palstra, H. Yamada, Y. Ueda, Phys. Rev. Lett., in press (2003).

[17] This expression was originally obtained by R. Kubo, J. Phys. Soc. Japan **12**, 570 (1957). Recently it has been used by M. R. Norman, and C. Pepin, Phys. Rev. B **66**, 100506(R) (2002), to analyze the superconductivity induced transfer of spectral weight of high T_c superconductors. See also Eq. 13 of Ref. [24]. The derivation of this expression proceeds along similar lines as the proof of the f-sum rule, see also the comments in Ref. [6]. Due to the non-parabolic energy-momentum dispersion in this case, the commutator of the dipole operator with the current operator produces an inverse *dynamical* mass of the band-electrons which depends on the canonical momentum, instead of a momentum-independent mass.

[18] Pierre F. Maldague, Phys. Rev. **16**, 2437 (1977).

[19] D. Baeriswyl, J. Carmelo, and A. Luther, Phys. Rev. B **33**, 7247 (1986).

[20] D. N. Basov, S. I. Woods, A. S. Katz, E. J. Singley, R. C. Dynes, M. Xu, D. G. Hinks, C. C. Homes, M. Strongin, Science **283**, 49 (1999).

[21] S. Chakravarty, H. Kee, E. Abrahams, Phys. Rev. Lett. **82**, 2366 (1999).

[22] S. Spielman, B. Parks, J. Orenstein, D. T. Nemeth, F. Ludwig, J. Clarke, P. Merchant, and D. J. Lew, Phys. Rev. Lett. **73**, 1537 (1994).

[23] S. G. Kaplan, S. Wu, H. T. S. Lihn, D. Drew, Q. Li, D. B. Fenner, Julia Phillips, and S. Y. Hou, Phys. Rev. Lett. **76**, 696 (1996).

[24] H. D. Drew, and P. Coleman, Phys. Rev. Lett., Vol. **78**, 1572 (1997).

[25] G. Mahan 'Many-Particle Physics', Plenum press (New York, 1981).

[26] P. Nozières and D. Pines, 'The theory of quantum liquids' part I, W. A. Benjamin 1966.

[27] Misha Turlakov, Anthony J. Leggett, cond-mat/0207422.

[28] D. N. Basov, E. J. Singley, S. V. Dordevic, Phys. Rev. B **65**, 054516 (2002).

[29] N. Shah, A. J. Millis, Phys. Rev. B **64**, 174506 (2001).

[30] A. Abanov, A. V. Chubukov, Phys. Rev. Lett. **88**, 217001 (2002).

[31] D. van der Marel, Physica C **165**, 35 (1990).

[32] D. van der Marel, and G. Rietveld, Phys. Rev. Lett. **69**, 2575 (1992).

[33] G. Rietveld, N. Y. Chen, D. van der Marel, Phys. Rev. Lett. **69**, 2578 (1992).

[34] D. van der Marel, D. I. Khomskii and G.M. Eliashberg, Phys. Rev. B **50**, 16594 (1994).

[35] A. J. Leggett, Proc. Natl. Acad. Sci. USA Vol. **96**, 8365 (1999).

[36] C. N. Presura, Ph D. Thesis, University of Groningen (2003).

[37] M. R. Norman, M. Randeria, B. Janko, J. C. Campuzano, Phys. Rev. B **61**, 14724 (2000).

[38] R. Micnas, J. Ranninger, and S. Robaszkiewicz, Rev. Mod. Phys. **62**, 113 (1990).

[39] P. Nozières, and S. Schmitt-Rink, J. Low Temp. Phys **59**, 195 (1985).

[40] D. van der Marel, Phys. Rev. B. **51**, 1147 (1995).

[41] P. W. Anderson, Science **268**, 1154 (1995).

[42] S. Chakravarty, Eur. Phys. J. B **5**, 337 (1998).

[43] J. E. Hirsch, Physica C **199**, 305 (1992).

[44] J. E. Hirsch, Physica C **201**, 347 (1992).

[45] S. Chakravarty, A. Sudbo, P. W. Anderson, and S. Strong, Science **261**, 337 (1993).

[46] S. Alexandrov and N. F. Mott, High Temperature Superconductors and Other Superfluids (Taylor and Francis, 1994).

[47] V. Emery and S. A. Kivelson, Nature **374**, 4347 (1995).

[48] F. F. Assaad, M. Imada, and D. J. Scalapino, Phys. Rev. Lett. **77**, 4592 (1996).

[49] P. A. Lee, Physica C **317**, 194 (1999).

[50] P. W. Anderson, Physica C **341-348**, 9 (2000).

[51] V. J. Emery, S. A. Kivelson, J. Phys. Chem. Solids **61**, 467 (2000).

[52] A. J. Leggett, Science **274**, 587 (1996).

[53] D. van der Marel, H.S. Somal, B.J. Feenstra, J.E. van der Eb, J. Schuetzmann, Jae Hoon Kim. Proc. 10th Ann. HTS Workshop on Physics, Houston, World Sc. Publ., page 357 (1996).

[54] J. Schuetzmann, H.S. Somal, A.A. Tsvetkov, D. van der Marel, G. Koops, N. Koleshnikov, Z.F. Ren, J.H. Wang, E. Brueck and A.A. Menovski, Phys. Rev. B **55**, 11118 (1997).

[55] C. Panagopoulos, J. R. Cooper, T. Xiang, G. B. Peacock, I. Gameson, and P. P. Edwards, Phys. Rev. Lett. **79**, 2320 (1997).

[56] A. Tsvetkov, D. van der Marel, K. A. Morel, J. R. Kirtley, J. L. de Boer, A. Meetsma, Z. F. Ren, N. Koleshnikov, D. Dulic, A. Damascelli, M. Grüninger, J. Schützmann, J. W. van der Eb, H. S. Somal, and J. H. Wang, Nature **395**, 360 (1998).

[57] K. A. Moler, J. R. Kirtley, D. G. Hinks, T. W. Li, and Ming Xu, Science **279**, 1193 (1998).

[58] J. R. Kirtley, K.A. Moler, G. Villard, A. Maignan, Phys. Rev. Lett. **81**, 2140 (1998).

[59] M.B. Gaifullin, Yuji Matsuda, N. Chikumoto, J. Shimoyama, K. Kishio, and R. Yoshizaki, Phys. Rev. Lett. **83**, 3928 (1999).

[60] D. N. Basov, C. C. Homes, E. J. Singley, M. Strongin, T. Timusk, G. Blumberg, and D. van der Marel, Phys. Rev. B **63**, 134514 (2001).

[61] V. Zelezny, S. Tajima, D. Munzar, T. Motohashi, J. Shimoyama, and K. Kishio, Phys. Rev. B **63**, 060502 (2001).

[62] Dominik Munzar, Christian Bernhard, Todd Holden, Andrzej Golnik, Josef Humlicek, and Manuel Cardona, Phys. Rev. B **64**, 024523 (2001).

[63] Dominik Munzar, Todd Holden, Christian Bernhard, Phys. Rev. B **67**, R020501 (2003).

[64] A.V Boris, D. Munzar, N.N. Kovaleva, B. Liang, C.T. Lin, A. Dubroka, A.V. Pimenov, T. Holden, B. Keimer, Y. -L. Mathis, C. Bernhard, Phys. Rev. Lett. **89**, 277001 (2002).

[65] S. L. Cooper, and K.E. Gray, *Physical Properties of High Temperature Superconductors IV*, ed. D. M. Ginsberg (World Scientific, Singapore 1994).

[66] Jae H. Kim, H.S. Somal, D. van der Marel, A.M. Gerrits, A. Wittlin, V.H.M. Duijn, N.T. Hien and A.A. Menovsky, Physica C **247**, 297 (1995).

[67] D. Dulic, D. van der Marel, A. A. Tsvetkov, W. N. Hardy, Z. F. Ren, J. H. Wang, B. A. Willemsen, Phys. Rev. B **60**, R15051 (1999).

[68] A. Hosseini, S. Kamal, D.A. Bonn, Ruixing Liang, and W.N. Hardy, Phys. Rev. Lett. **81**, 1298 (1998).

[69] K. Tamasaku, Y. Nakamura, and S. Uchida, Phys. Rev. Lett. **69**, 1455 (1992).

[70] J. W. Loram,K. A. Mirza, J. M. Wade, J. R. Cooper, and W. Y. Liang, Physica C **235-240**, 134 (1994).

[71] D. van der Marel, A.J. Leggett, J.W. Loram, J.R. Kirtley Phys. Rev. B **66**, 140501R (2002).

[72] M. Tinkham, and R. A. Ferrell, Phys. Rev. Lett. **2**, 331 (1959).

[73] D. Munzar, C. Bernhard, A. Golnik, J. Humlicek, M. Cardona, Solid State Commun. **112**, 365 (1999).

[74] V. Zelezny, S. Tajima, T. Motohashi, J. Shimoyama, K. Kishio, and D. van der Marel, J. Low Temp. Phys. **117**, 1019 (1999).

[75] M. Grueninger, D. van der Marel, A.A. Tsvetkov, A. Erb, Phys. Rev. Lett. **84**, 1575 (2000).

[76] J. W. Loram, J. Luo, J. R. Cooper, W. Y. Liang and J. L. Tallon, J. Phys. Chem. Solids **62**, 59 (2001).

[77] D. van der Marel, A. Tsvetkov, Czech. J. Phys. **46**, 3165 (1996).

[78] D. van der Marel, and A. A. Tsvetkov, Phys. Rev. B **64**, 024530 (2001).

[79] N. Shah, A. J. Millis, Phys. Rev. B **65**, 024506 (2002).

[80] H. Shibata and T. Yamada, Phys. Rev. Lett. **81**, 3519 (1998).

[81] H. Shibata, Phys. Rev. Lett. **86**, 2122 (2001).

[82] T. Kakeshita, S. Uchida, K. M. Kojima, S. Adachi, S. Tajima, B. Gorshunov und M. Dressel, Phys, Rev. Lett. **86**, 4140 (2001).

[83] D. Dulic, A. Pimenov, D. van der Marel, D.M. Broun, S. Kamal, W.N. Hardy, A.A. Tsvetkov, I.M. Sutjaha, Ruixing Liang, A.A. Menovsky, A. Loidl and S.S. Saxena, Phys. Rev. Lett. **86**, 4144 (2001).

[84] A. Pimenov, A. Loidl, D. Dulic, D. van der Marel, and A. A. Menovsky, Phys. Rev. Lett. **87**, 177003 (2001).

[85] D. Dulic, Ph D. Thesis, University of Groningen (2000).

[86] M. Grueninger, Ph D. Thesis, University of Groningen (1999).

[87] J. E. Hirsch, and F. Marsiglio, Physica C **331**, 150 (1999).

[88] J. Bonca, P. Prelovsek, and I. Sega, Phys. Rev. B **39**, 7074 (1989).

[89] J. A. Riera, Phys. Rev. B **40**, 833 (1989).

[90] Y. Hasegawa, and D. Poilblanc, Phys. Rev. B **40**, 9035 (1989).

[91] E. Dagotto, J. Riera, and A.P. Young, Phys. Rev. B **42**, 2347 (1990).

[92] T. Barnes, A. E. Jacobs, M. D. Kovarik, and W. G. Macready, Phys. Rev.B **45**, 256 (1992).

[93] E. Demler, and S.-C. Zhang, Nature **396**, 733 (1998).

[94] I. Fugol, V. Samovarov, A. Ratner, V. Zhuravlev, G. Saemann-Ischenko, B. Holzapfel, and O. Meyer, Solid State Commun. **86**, 385 (1993).

[95] Y. Yagil, E. K. H. Salje, Physica C **256**, 205 (1996).

[96] M. J. Holcomb, J. P. Collman, and W. A. Little, Phys. Rev. Lett. **73**, 2360 (1994).

[97] W. A. Little, K. Collins, and M. J. Holcomb, J. Superconductivity **12**, 89 (1999).

[98] M. Rübhausen, A. Gozar, M. V. Klein, P. Guptasarma, D. G. Hinks, Phys. Rev. B **63**, 224514 (2001).

[99] H. J. A. Molegraaf, C. Presura, D. van der Marel, P. H. Kes , M. Li, Science **295**, 2239 (2002). Similar conclusions have been obtained in Ref. [100].

[100] A.F. Santander-Syro, R.P.S.M. Lobo, N. Bontemps, Z. Konstantinovic, Z.Z. Li, H. Raffy, cond-mat/0111539.

Chapter 9

CHARGE INHOMOGENEITIES IN STRONGLY CORRELATED SYSTEMS

A. H. Castro Neto
Department of Physics, Boston University
Boston, MA, 02215, USA
neto@bu.edu

C. Morais Smith
Département de Physique, Université de Fribourg, Pérolles
CH-1700 Fribourg, Switzerland
cristiane.demorais@unifr.ch

Abstract We review the problem of stripe states in strongly correlated systems, and some of the theoretical, numerical and experimental methods used in the last few years to understand these states. We compare these states to more traditional charge-ordered states such as charge density waves (CDW) and phase-separated systems. We focus on the origin of stripe states as an interplay between magnetic and kinetic energy, and argue that the stripe state is generated via a mechanism of kinetic energy release that can be described via strongly correlated models such as the t-J model. We also discuss phenomenological models of stripes, and their relevance for magnetism and for the pinning of stripes by the underlying lattice and by disordered impurities. Recent experimental evidence for the existence of stripe states in different cuprate systems is also reviewed.

Keywords: charge inhomogeneities, strongly correlated electronic systems, charge-density-wave states, striped phase, doped Mott insulators, high-T_c superconductors, pattern formation in low-dimensional systems.

D. Baeriswyl and L. Degiorgi (eds.), Strong Interactions in Low Dimensions, 277–320.

1. Introduction

The problem of strong electron correlations in transition-metal oxides and U and Ce intermetallics has been a subject of intense research in the last 20 years. This interest has been driven mostly by puzzling experimental findings in materials such as organic conductors and superconductors, heavy-fermion alloys, and high-temperature superconductors. These systems are characterized by large Coulomb interactions, low dimensionality, strong lattice coupling, and competition between different phases: antiferromagnetism, ferromagnetism, spin density waves (SDW), superconductivity, and charge density waves (CDW). The strong interplay between different order parameters is believed to lead to charge and spin inhomogeneities, and to a myriad of energy and length scales that makes the problem very difficult to treat with the methods and techniques used for the study of Fermi-liquid metals. Although the problem of the quantum critical behavior of metals in the proximity of an isolated zero-temperature phase transition has been subject of much study and heated debate [1], our understanding of the problem of electrons close to multiple phase transitions is still in its infancy. Here we will review both the experimental evidence for the existence of certain inhomogeneous states called "stripe" states and some of the current theoretical approaches used to understand their origin, nature, and importance in the context of magnetism and superconductivity [2].

The formation of static or dynamic spin-charge stripes in strongly correlated electronic systems has been corroborated recently by several experiments, especially in manganites [3, 4], nickelates [5-8], and cuprates [9-34]. The experiments span a large variety of techniques, from scanning tunneling microscopy (STM) [4,9-14], neutron (and x-ray) scattering [5,6,15-25], nuclear magnetic (and quadrupole) resonance (NMR and NQR) [7, 26, 27], muon spin rotation (μSR) [28, 29], optical and Raman spectroscopy [8, 30, 31], transport [32], angle-resolved photoemission (ARPES) [33], and ion channeling [34].

Charge and spin modulated states, such as CDWs, Wigner crystals, SDWs, antiferromagnetism, and ferrimagnetism are common occurrences in many transition-metal compounds. These systems are characterized by an order parameter (such as the charge and/or spin density) that is modulated with a well-defined wave vector **Q**. Because of the modulations and the coupling to the lattice, these states usually present lattice distortions which are easily observed in diffraction experiments such as neutron scattering. In this regard the stripe states discussed here are very much like CDW/SDW instabilities, except that in CDW systems the ordered state is driven by a Fermi-surface instability (usually generated by nesting and/or Van Hove singularities), and Coulomb effects are secondary because of good screening. The clearest example of such Fermi-surface effects occurs in Cr alloys, where the system undergoes a phase transition into a CDW/SDW state [35]. The CDW and SDW transitions occur at the *same temperature*, and the charge order has a period that is 1/2 that of the SDW. The main difference between the phase transition in Cr and the stripe states to be discussed here is that the charge order in stripe systems occurs at higher temperature than the spin order [16]. Thus, on reducing the temperature the onset of charge order occurs first and the spins simply follow. In a weak-coupling analysis of Fermi-surface instabilities, this type of transition is not possible because of reconstruction of the Fermi surface due to the appearance of long-range order. Thus there are different energy scales for the charge and spin order that characterizes the materials discussed here.

We should stress that the CDW and/or SDW instabilities in metallic systems are not trivial, and although we understand the basic mechanisms which drive these instabilities [36] our knowledge of their origin and effects on the electronic degrees of freedom is far from complete. Systems such as transition-metal dichalcogenides [37] have a high temperature CDW transition with a very anomalous metallic phase and show in addition the phenomenon of "stripe formation" [38]. The stripes in these CDW systems are understood, however, because the CDW order is incommensurate with the lattice and therefore phase fluctuations of the CDW order parameter are allowed energetically. Local CDW phase-slips give rise to a filamentary stripe phase which, in fact, has a Fermi-surface origin.

To understand the origin of CDW stripes one may consider the complex order parameter Δ for a CDW with incommensurate ordering wave vector $\mathbf{Q}$. The free energy of the problem may be expressed as [39]

$$F = F_0[\Re(\Delta)] + \int d\mathbf{r} \left\{ \frac{1}{2m^* Q^2} \left[|\mathbf{Q} \cdot (\nabla - i\mathbf{Q})\Delta|^2 + \kappa |\mathbf{Q} \times \nabla\Delta|^2 \right] \right\} \tag{9.1}$$

where $F_0[x]$ is a minimal polynomial of x that respects the symmetry of the lattice and renders the free energy bounded from below. For a triangular lattice, for instance, it can be written as [39]

$$F_0[\Re(\Delta)] = a(\mathbf{r},T)\Re(\Delta)^2 + b(\mathbf{r},T)\Re(\Delta)^3 + c(\mathbf{r},T)\Re(\Delta)^4, \tag{9.2}$$

where the coefficients of the expansion are smooth functions of temperature. In particular for the quadratic term $a(\mathbf{r},T) = a_0(\mathbf{r})(T - T_{ICDW})$ where T_{ICDW} is the transition temperature of the incommensurate CDW (ICDW) state. In (9.1) m^* and κ are parameters specific to the material under consideration and the derivative terms are written such that the free energy of the CDW is minimal when the ordering wave vector lies in the correct direction and has the correct wave length. In the case of an ICDW the order parameter is obtained by minimizing (9.1) to give

$$\Delta_{ICDW}(\mathbf{r}) = \Delta_{0,I} e^{i\mathbf{Q}\cdot\mathbf{r}}, \tag{9.3}$$

where $\Delta_{0,I} = \sqrt{2a_0(T_{ICDW} - T)/(3c_0)}$ for $T < T_{ICDW}$ and zero otherwise. The parameter $c(\mathbf{r}, T) = c_0 =$ constant. Here we have assumed that the parameters in (9.2) may be expanded in a form such as $b(\mathbf{r}) = b_0 + b_1 \exp\{i\mathbf{K}_i \cdot \mathbf{r}\}$ where $\mathbf{K}_i$ are the shortest reciprocal-lattice vectors characteristic of the lattice symmetry.

In the commensurate CDW (CCDW) case the wave vector of the order parameter "locks" with the lattice so that its modulation becomes a fraction of the lattice wave vector $\mathbf{K}_1$. Then one would replace $\mathbf{Q}$ in (9.3) by $\mathbf{K}_1/q$, where q is an integer, and Δ_0 by a value $\Delta_{0,C}$ which must be calculated from the free energy and depends in general on various coefficients of F_0 in (9.2). The transition temperature T_{CCDW} is usually smaller than T_{ICDW} so that the generical behavior of the system consists of two transitions, first into an incommensurate phase and then into a commensurate phase [39]. In many systems the ICDW-CCDW transition does not occur and the system remains incommensurate down to very low temperatures [37].

In order to study the problem of the commensurate-incommensurate transition, and the topological defects which appear due to incommensurability, one must generalize (9.3) to include phase fluctuations. These may be incorporated by writing the order parameter in the form

$$\Delta(\mathbf{r}) = \Delta_0 e^{i\frac{1}{q}\mathbf{K}_1\cdot\mathbf{r} + i\theta(\mathbf{r})}, \tag{9.4}$$

where $\theta(\mathbf{r})$ is the angle variable which determines the commensurability of the system: for the ICDW $\theta(\mathbf{r}) = (\mathbf{Q} - \mathbf{K}_1/q) \cdot \mathbf{r}$, while for a CCDW $\theta = 0$. Because we are considering the simplest problem of a single CDW wave vector the problem becomes effectively one-dimensional if the variables are redefined in a new, rotated, rescaled, reference frame defined by $\mathbf{s} = (x, y) = |\mathbf{Q} - \mathbf{K}_1/q|\mathbf{r}$. In this case it is obvious that $\theta(\mathbf{r}) = \theta(x)$, and by direct substitution of (9.1) we find that the dimensionless free energy per unit of length relative to the commensurate case becomes

$$\delta f = \int dx \frac{1}{2} \left\{ [\partial_x \theta(x) - 1]^2 + g\,[1 - \cos(q\theta)] \right\} \tag{9.5}$$

where g is the coupling constant of the system and depends on the parameters of (9.1). The free energy in (9.5) describes a sine-Gordon model where the cosine term favors the commensurate state ($\theta(x) = 0$) while the gradient term favors the incommensurate state ($\theta(x) = x$). Thus θ is the order parameter and the discrete symmetry $\theta \to \theta + 2\pi/q$ is broken in the ordered phase. There is therefore a critical coupling value g_c that separates these two phases. However, it is easy to see that there are other solutions which minimize the free energy. In fact, variation of (9.5) with respect to θ yields

$$\frac{d^2\theta}{dx^2} = g\,q\,\sin(q\theta) \tag{9.6}$$

which has a particular solution

$$\theta_K(x) = \frac{2}{q}\arctan\left(e^{\sqrt{g}x}/2\right), \tag{9.7}$$

where the boundary conditions are $\theta(x = -\infty) = d\theta(x = -\infty)/dx = 0$. Notice that (9.7) changes smoothly from $\theta = 0$ at $x = -\infty$ to $\theta = 2\pi/q$ when $x \gg 1/\sqrt{g}$. In the context of the sine-Gordon model this is called a topological soliton or kink, while in the CDW literature [39] it is called a discommensuration. In general, the solution of (9.6) is given by [40]:

$$\theta(x) = \frac{2}{q}\arcsin[\eta\mathrm{sn}(\sqrt{g}x/k, k)] \tag{9.8}$$

where $\eta = \pm 1$ and $\mathrm{sn}(u, k)$ denotes the sine-amplitude, which is a Jacobian elliptic function of modulus k. The sine-amplitude is an odd function of its argument u and has period $4K(k)$, where $K(k)$ is the complete elliptic integral of the first kind. By substituting (9.8) in (9.5) and minimizing with respect to k one finds that $k \approx 2\sqrt{g}$. In the limit $g \to 0$ one has also $k \to 0$ and $\mathrm{sn}(u, k) = \sin(u)$, whence $\theta(x) \approx x$ as expected. On the other hand, when $g \gg 1$ one finds $k \approx 1$ and $\mathrm{sn}(u, k) \approx \tanh(u)$, in which case (9.7) is obtained. For a generic value of g the solution has the form of a staircase. The plateaus in the staircase are multiples of $2\pi/q$ and correspond to regions where the CDW is in phase with the lattice, while in the transitions between the plateaus the CDW is not locked, leading to discommensurations. For large values of g the discommensurations are rather narrow and we find stripe states. These states have been observed experimentally in CDW systems [38].

In contrast to these CDW stripes, the stripe systems which we will discuss here have their origin in Mott insulators with very large Coulomb energies, whereas typical CDW/SDW systems are very good metals in their normal phase (Cr is a shiny metal while La_2CuO_4 is opaque and grayish). It is exactly the "mottness" of these systems which complicates the theoretical understanding of their nature. If we take seriously the analogy between dichalcogenides and cuprates we could think of the stripes as phase-slips of an incommensurate order parameter associated with the Mott phase. The primary question concerns the order parameter of a Mott phase. Antiferromagnetic order usually occurs in a Mott insulator but is essentially a parasitic phase (systems of spinless electrons at 1/2-filling with strong next-nearest neighbor repulsion, and frustrated magnetic systems, can be Mott insulators without showing any type of magnetic order [41]). Unfortunately, the order parameter which characterizes the Mott phase in a finite number d of spatial dimensions is not known. Within the dynamical mean-field theory ($d \to \infty$) the order parameter of the Mott transition has been identified with a zero mode of an effective Anderson model [42], but generalizations for the case of finite dimensions have not been established. The search for the order parameter which characterizes "mottness" is one of the important problems of modern condensed matter physics.

The aim of this work is to summarize the current literature on the stripe phase in high-T_c superconductors. Although there is a consensus for the existence of stripes in manganites and nickelates, no agreement has yet been achieved concerning the superconducting cuprates. Despite this controversy, the presence of stripes is now firmly established in $La_{2-x}Sr_xCuO_4$ (LSCO) [16, 17, 30, 31, 32, 33]. In addition, recent experiments suggest that they may also be present in $YBa_2Cu_3O_{7-\delta}$ (YBCO) [19-26,34], as well as in $Bi_2Sr_2CaCu_2O_{8+\delta}$ (BSCCO) [12, 13]. In this work we discuss some of the different experimental techniques which prove or suggest the presence of stripes in cuprates, and we present some theoretical ideas on the existence and relevance of the stripe state.

This review is structured as follows: in section II, a survey of the theoretical derivations of the stripe phase as a ground state of models which are appropriate for describing doped Mott insulators is presented. In section III we discuss the experimental observation of stripes, and in section IV the role of kinetic energy. In section V we consider some phenomenological models, which provide a means to go beyond the question of existence of stripes and allows one to predict measurable quantities. Finally, in section VI, we present the most recent experimental results for YBCO and BSCO, as well as the open questions and topics of debates. In section VII we draw our conclusions.

2. Origin of stripes

One of the main problems in condensed matter theory since the discovery of high temperature superconductors in 1986 [43] is related with the possible dilute phases of Mott insulators [44]. These materials have a large charge transfer gap, so at half-filling are insulating two-dimensional (2D) antiferromagnets well described by the isotropic Heisenberg model [45]. These are trademarks of "mottness" [46] and led Anderson [47] to propose that cuprates may be well described by a Hubbard model with large intra-site repulsion U. Later studies showed that close to half-filling and infinite U the model maps into the t-J model [48]

$$H = -t \sum_{<i,j>,\alpha} P c^{\dagger}_{i,\alpha} c_{j,\alpha} P + J \sum_{<i,j>} \mathbf{S}_i \cdot \mathbf{S}_j, \tag{9.9}$$

where t is the hopping energy and $J \approx 4t^2/U \ll t, U$ is the exchange interaction between neighboring electron spins, $\mathbf{S}_i = c^{\dagger}_{i,\alpha} \vec{\sigma}_{\alpha,\beta} c_{i,\beta}$ ($c_{i,\alpha}$ is the electron annihilation operator at the site i with spin projection $\alpha = \uparrow, \downarrow$, and σ^a with $a = x, y, z$ is a Pauli matrix). In Eq. (9.9) P is the projection operator onto states with only single site occupancy (double occupancy is forbidden). Eq. (9.9) reduces trivially to the Heisenberg model at half-filling and describes the direct interplay between the two main driving forces in the system, magnetism (characterized by J) and kinetic energy (characterized by t).

The existence of a stripe phase in cuprates was first suggested in the context of Hartree-Fock studies of the Hubbard model close to half-filling and at zero temperature ($T = 0$) [49]. This calculation is essentially analogous to the one used to study CDW/SDW transitions in metallic materials. For $U < t$, vertical stripes (parallel to the x- or y-axis of the crystal) were shown to be lower in energy [49, 50, 51], whereas for large $U > t$ diagonal stripes were found to be energetically more favorable [51, 52]. The crossover from vertical to diagonal stripes was calculated numerically to occur at $U/t \sim 3.6$ [51]. Such calculations have recently been generalized to finite temperatures, and the phase diagram was derived as a function of T and doping n_h, (Fig. 9.1 [53]). One important feature of these mean-field calculations is that they predict the formation of charge-ordered domain walls at which the staggered magnetization changes phase by π. The magnetic order parameter is therefore maximal not at $(\pi/a, \pi/a)$ as in an ordinary antiferromagnet, but at an incommensurate vector $(\pi/a \pm \delta, \pi/a)$ where $\delta = \pi/\ell \ll 1$ and ℓ denotes the charge stripe spacing. This incommensurability is an important feature of the stripe problem because, as we will demonstrate below, it leads to a reduction in the *kinetic* energy of the holes. These calculations, however, always predict that the stripe states possess a gap. The simple reason for this effect is that the only way in Hartree-Fock to reduce the energy of the system is by opening a gap at the Fermi surface. Furthermore, Hartree-Fock calculations in strongly interacting systems are not quantitatively reliable because they are unable to take fluctuation effects into account, and therefore should be considered only as providing some qualitative insight into the ground-state properties. They, however, do provide a "high energy" guide (snapshot picture) of the possible phases of the problem, and in fact they have been fundamental for the interpretation of certain experiments such as neutron scattering.

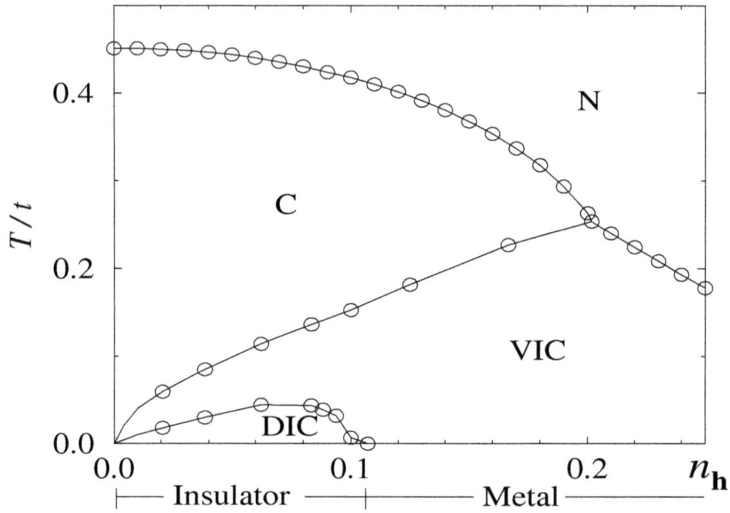

Figure 9.1. Phase diagram in the plane of temperature T and hole concentration x ($n_h = 2x$ for LSCO) obtained by Machida and Ichioka [53] from mean-field studies of the Hubbard model. In the figure, N indicates the normal phase and C denotes the commensurate antiferromagnetic phase. In addition, two other incommensurate phases exist, with vertical (VIC) or diagonal (DIC) stripe order. The VIC phase is metallic, but the DIC is insulating at zero temperature.

Recent theoretical efforts have focused not only on determining the ground-state properties of the Hubbard, but also of the t-J model. Numerical techniques such as density matrix renormalization group (DMRG) [54], quantum Monte Carlo [55], and exact diagonalization [56], have been applied to the 2D t-J model with differing degrees of success. The main problem is that a strongly interacting problem like the t-J model is subject to strong finite-size and boundary-condition effects which are difficult to control.

2.1 Numerical Studies

Early numerical calculations on the t-J model have shown that for physical values of J/t and close to half-filling there is a tendency for phase separation [57]. This phase separation can be pictured in the limit of $t = 0$ (classical limit) as a lowering of the system energy by placing all the holes together in order to minimize the number of broken antiferromagnetic bonds. This simple picture leads to separation into two distinct phases: a commensurate, insulating region and an insulating, hole-rich region. It naturally overestimates the importance of the magnetic energy relative to the kinetic energy, and therefore can be correct only when $J \gg t$. For finite values of t, the hole wave function delocalizes and this picture breaks down. The main question is for which values of J/t a phase separation may arise. Emery, Kivelson and Lin [57] found that phase separation can occur for infinitesimal values of J/t sufficiently close to half-filling. These results, however, have been questioned in the light of more recent numerical data. There is no doubt that the t-J model undergoes phase separation for $J \gg t$ as all numerical calculations indicate. Close to the physical region of $J < t$, the current evidence for phase separation is weak, and so the issue remains controversial.

DMRG calculations in large clusters [54] indicate the presence of stripe correlations in the t-J model. These studies, however, have been criticized on the basis of the special role of boundary conditions. Recent work on the Ising t-J model indicate that stripe formation does occur in this system, independent of the boundary conditions [58]. It was shown via non-perturbative analytical calculations that minimization of the hole kinetic energy is the driving force behind stripe formation. This result has been confirmed by a number of numerical calculations in the t-J model [59], as well as in the t-J_z model [60]. Another important conclusion from these studies is that in the stripe phase the superconducting correlations are extremely weak. In fact DMRG calculations show that the stripe state is a CDW/Luttinger-liquid state with vanishing density of states at the chemical potential, and thus is naturally insulating [54]. The DMRG calculations then support the idea that stripe correlations compete with superconductivity instead of enhancing it. This is consistent with experimental findings in Nd-doped LSCO, where the superconducting transition is reduced when static stripe order sets in [16].

Although the DMRG calculations were originally performed in a t-J model, essentially the same physics is found in the t-J_z model. The reason for the similarity between the t-J (where the spins are dynamical) and the t-J_z (where the spins are static) may be understood on the basis of the fluctuation timescales for each component in the problem. In the t-J model the spins fluctuate with a rate $\tau_s \approx \hbar/J$, while the timescale for hole motion is $\tau_h \approx \hbar/t$. When $J/t < 1$ (the physical regime of the model) one has $\tau_s > \tau_h$, that is, the holes move "faster" than spins. In this case a Born-Oppenheimer approximation is reasonable, since the spins have slow dynamics, and the two problems become essentially identical [56]. The advantage of working with the t-J_z model is that many of its properties are significantly easier to study both numerically and analytically.

The introduction of next-nearest-neighbor hopping t' favors mobile d-wave pairs of holes for $t' > 0$, and single-hole excitations (spin-polarons) for $t' < 0$ [61]. At $t' = 0$ the stripe state is very close in energy to the d-wave pair state, and thus a small change in the boundary conditions or inclusion of small perturbations in the Hamiltonian can easily favor one many-body state relative the other. The quasi-degeneracy of different many-body states is an important characteristic of strongly correlated systems. Moreover, in real materials other effects may also be responsible for the selection of the ground state, that is, for lifting of the quasi-degeneracy. Indeed, by including lattice anisotropies, which arise in the low-temperature tetragonal (LTT) phase of LSCO co-doped with Nd, the stripe state can be easily selected [62, 63, 64]: Hartree-Fock calculations of the Hubbard model have shown that a very small anisotropy (on the order of ten percent) in the hopping parameter t is already sufficient to stabilize the striped phase, independent of the boundary conditions (open or periodic) [63]. Monte Carlo studies of the t-J model have also confirmed these results [64].

2.2 Stripes and phase separation

The problem of the formation of inhomogeneous states in a system with phase separation can be easily understood from a classical point of view by studying the Ginzburg-Landau free energy functional. Let ψ be the order parameter of a system described by a free energy, F, of the form

$$F[\psi] = \alpha(T,x)|\psi|^2 + \frac{\beta(x)}{2}|\psi|^4 + \frac{\gamma}{3}|\psi|^6 \tag{9.10}$$

where ψ may be complex for a superconductor, $\alpha(T,x)$, $\beta(x)$, and $\gamma > 0$ are functions of the temperature T and some parameter x (such as doping or pressure). In the theory of second-order phase transitions, the $|\psi|^6$ term is neglected close to the critical line because $\beta > 0$ in this region of the parameter space. Here, however, we assume that β may be negative, and therefore this term is required so that the free energy is bounded from below. The critical line in the (T,x) plane is given by $T_c(x)$ (as shown in Fig. 9.2) and we assume the existence of a quantum critical point (QCP) at $x = x_a$ (that is, $T_c(x_a) = 0$). Close to the critical line we introduce the parameterization:

$$\begin{aligned} \alpha(T,x) &= \alpha_0\left[T/T_c(x)-1\right], \\ \beta(x) &= \beta_0\left[T_c(x)/T_c(x_s)-1\right], \end{aligned} \tag{9.11}$$

while γ is approximately independent of x and T. Notice that with this choice the parameter $\beta(x)$ is positive for $x > x_s$, signaling that in this regime the transition is of second order. However, $\beta(x)$ vanishes at $x = x_s$ and becomes negative for $x < x_s$, indicating that the nature of the phase transition changes at small x. In fact the point $(x_s, T_c(x_s))$ is a *tricritical* point.

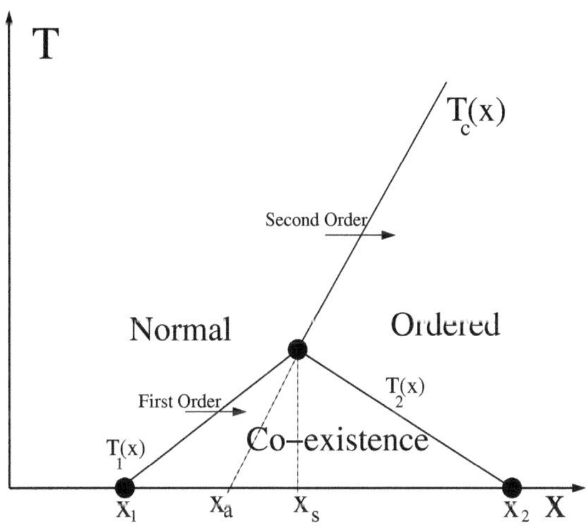

Figure 9.2. Temperature-doping phase diagram for a system close to phase separation. The symbols are explained in the text.

For $x > x_s$ the $|\psi|^6$ term is irrelevant close to the phase transition, and the transition is of second order depending on wether T is greater or smaller than $T_c(x)$. Minimizing the free energy with respect to the order parameter yields

$$|\psi_0(x,T)|^2 \approx \frac{\alpha_0\left(1-T/T_c(x)\right)}{\beta_0\left(T_c(x)/T_c(x_s)-1\right)} \tag{9.12}$$

for $T < T_c(x)$ and $x > x_s$. Notice, however, that at $x \lesssim x_s$ the parameter $\beta(x)$ vanishes, and one must include the $|\psi|^6$ term. In this case, the free energy has two minima (instead of one) at the critical line indicating that the system has two phases, one with $\psi = 0$ (normal) and another with $\psi = \psi_0$ (ordered). Minimization of F with respect to the order parameter provides the condition for the phase transition

$$\beta^2(x) = \frac{16}{3}\alpha(T^*,x)\gamma,$$

$$T^*(x) \;=\; T_c(x) + \frac{3\beta_0^2 T_c(x)}{16\alpha_0\gamma}\left[\frac{T_c(x)}{T_c(x_s)} - 1\right]^2. \qquad (9.13)$$

The solution of these equations gives two critical lines, $T_1(x)$ and $T_2(x)$ on Fig.9.2. These lines terminate at x_1 and x_2, and for $x_1 < x < x_2$ there is a coexistence region with two phases (normal and ordered).

Long-range interactions are readily introduced by modifying the $|\psi|^4$ term in the free energy to

$$F_C \;=\; \int d\mathbf{r} \int d\mathbf{r}' |\psi(\mathbf{r})|^2 \frac{e^2}{|\mathbf{r}-\mathbf{r}'|} |\psi(\mathbf{r}')|^2. \qquad (9.14)$$

In this case of fully phase-separated states the cost in electrostatic energy is too high and phase separation is frustrated to a finite length scale, $\ell_{\rm PS}$, that depends on the coefficient of the $|\psi|^6$ term. The formation of finite droplets with $\psi = \psi_0$ and size $\ell_{\rm PS}$ is therefore more favorable than the separation of the system into two homogeneous phases with $\psi = 0$ and $\psi = \psi_0$. Stripes can also be generated in this model if one adds terms which break the rotational symmetry

$$F_{\pm} = \sum_{\mathbf{q}} \left[\cos(q_x) \pm \cos(q_y)\right] |\psi(\mathbf{q})|^2, \qquad (9.15)$$

depending on whether the interaction with the lattice may be represented in terms of $L = 0$ (plus sign) or $L = 2$ (minus sign) angular momentum states (s- and d-wave, respectively). In the $L = 0$ case a checkerboard state is favored, but even a small d-wave term generates stripes along the x- or y-directions.

Disorder can also frustrate the phase separation as one may show by adding a "random mass" term

$$F_D = \int d\mathbf{r}\; m(\mathbf{r}) |\psi(\mathbf{r})|^2 \qquad (9.16)$$

to the free energy, where $m(\mathbf{r})$ is a gaussian variable with average zero and variance u. Using the replica-technique with n replicas ($n \to 0$ at the end of the calculation) and averaging over disorder gives

$$F = \sum_{a=1}^{n} F_a^{(0)} - u \sum_{a,b=1}^{n} \int d\mathbf{r} |\psi_a(\mathbf{r})|^2 |\psi_b(\mathbf{r})|^2, \qquad (9.17)$$

where $F_a^{(0)}$ is the free energy without disorder and with n fields ψ_n. Notice that in the replica-symmetric case ($\psi_n = \psi$ for all n) the disorder generates a term of the order $-u|\psi|^4$ which decreases the effective value of β, therefore reducing the value of $T_c(x_s)$. In a renormalization-group (RG) sense this term is relevant, and if the disorder is sufficiently strong it will bring the tricritical point to zero temperature (that is, the tricritical point becomes a quantum critical point (QCP) and will completely destroy the first-order phase transition). However, in the ordered phase the system may still possess a coexistence phase with different values of the order parameter. Once again we would have a situation where the system forms droplets of the paramagnetic phase ($\psi = 0$) inside the ordered phase. The size of these droplets depends on the strength of the disorder and it is easy to make them adopt a stripe conformation by adding a term of the form (9.15) that breaks the rotational symmetry.

Although the phenomenology of the problem is quite clear, what is not so evident is how to apply this theory to the cuprates. Emery *et al.* [65] proposed that a model similar to the one discussed here (the Blume-Emery-Griffiths model) may be applied to the cuprates if one defines a pseudo-spin S_i which takes the values $S_i = +1$ and $S_i = -1$ on regions corresponding to hole-rich and hole-poor, respectively, whereas $S_i = 0$ indicates a local density equal to the average value. In this case $\psi(\mathbf{r})$ is the coarse-grained version of S_i and the above discussion is applicable. Notice, once again, that this model completely disregards the kinetic energy of the problem and can be applied only in the situation where $t = 0$. It is therefore not at all surprising that stripes appear. The inclusion of itinerant degrees of freedom is not straightforward. One of the main effects of the presence of itinerant degrees of freedom is the generation of dissipation which can change the dynamical properties (and exponents) of the system. This problem has been the object of recent intensive study in the context of quantum phase transitions when the coupling between the magnetic order parameter and the electrons is weak, and the electrons may be treated as a Fermi liquid [66]. In this case the electronic system serves as a heat bath for the relaxation of the magnetic order parameter. In Mott insulators, the mere existence of a Fermi surface and a Fermi-liquid state can be questioned and thus it is clear that the weak-coupling formalism cannot be applied to these systems. The charge degrees of freedom cannot be modeled purely as a heat bath because their feedback effect in the magnetic system is very strong.

It is interesting to compare the two mechanisms for stripe formation quoted previously. In one mechanism (represented by the Hartree-Fock calculations), stripes are long period CDWs arising from Fermi-surface nesting in a weakly incommensurate system [49, 50, 51, 52]. There are four features which arise from a Fermi-surface instability: 1) the transition is spin driven, *i.e.*, there is a single transition temperature T_c below which the broken-symmetry solution of the Hartree-Fock equations is stable; 2) in the low-T phase there are gaps or pseudo-gaps on the Fermi surface; 3) the spacing between domain walls is equal to π/x, where x is the hole concentration; 4) the high-T phase should be a Fermi liquid.

In the case of stripes arising from Coulomb-frustrated phase separation, the situation is quite different [67]: 1) the transition is charge driven, *i.e.,* local spin order between the antiphase domain walls can develop only after the holes are expelled from the magnetic regions. Ginzburg-Landau considerations indicate either a first-order transition, in which spin and charge order arise simultaneously, or a sequence of transitions in which first the charge order and then the spin order appears as T is lowered [68]; 2) the stripe spacing is not necessarily a simple function of x, and there is no reason to expect the Fermi energy to lie in a gap or pseudo-gap; 3) a high-T Fermi-liquid phase is not a prerequisite.

3. Experimental detection of stripes in high-T_c cuprates: LSCO

Although the first predictions for stripe formation in doped Mott insulators were made 13 years ago, not much attention was paid to these results in connection with high-T_c superconductors until 1995, when experimental data from neutron-scattering measurements in cuprates were interpreted consistently within a stripe picture [16]. Co-doping of cuprates has been extremely important for revealing the modulated charge states. However, the inclusion of co-dopant usually reduces T_c, raising doubts about the coexistence of superconductivity and the striped phase [16, 69]. The first experimental detection of stripes in the cuprates was achieved in a Nd co-doped compound $La_{2-x-y}Nd_ySr_xCuO_4$. For $y = 0.4$ and $x = 0.12$, Tranquada *et al.* [16] found that the commensurate magnetic peak at $\mathbf{Q} = (\pi/a, \pi/a)$ shifts by a quantity $\delta = \pi/\ell$, giving rise to four incommensurate peaks. In addition, new Bragg peaks appear at the points $(\pm 2\delta, 0)$ and $(0, \pm 2\delta)$, indicating that the charges form domain walls separated by a distance ℓ, and that the staggered magnetization undergoes a phase-shift of π when crossing them. The position of the peaks indicates that the stripes are oriented along the vertical and horizontal directions, with a density of one hole per two Cu sites (quarter-filled).

The reason why static stripes could be detected in this compound is based on a structural transition induced by the Nd atoms. Indeed, co-doping with many rare-earth species, including Nd and Eu, produces a buckling of the oxygen octahedra around the Cu sites and a corresponding transition from the low-T orthorhombic (LTO) to a low-T tetragonal (LTT) phase. The critical concentration of Nd needed to destroy superconductivity is a function of the charge-carrier density, *i.e.*, the concentration x of Sr atoms. However, the buckling angle of the octahedra is a universal parameter: for tilts above a critical angle $\theta \sim 3.6°$, superconductivity is completely suppressed in these materials [69]. For the values of Nd co-doping y used in the first neutron-scattering experiments, superconductivity was still present, and the authors claimed that in their samples $T_c \sim 5$ K. However, coexistence of static stripe order and superconductivity in LSCO is an issue that remains controversial, although recent experiments in La_2CuO_{4+y} have shown the coexistence of these two phases [70] in the same volume of the sample.

Another important factor assisting the detection of charge stripes in LSCO systems by elastic neutron scattering was the selected doping concentration $x \approx 1/8$. The 1/8 anomaly was known since 1988, when electrical resistivity measurements in $La_{2-x}Ba_xCuO_4$ (LBCO) were first performed [71]. A mysterious reduction of T_c was detected around $x = 1/8$, but the understanding of this phenomenon was possible only recently, in the light of the stripe picture. Indeed, Ba substitution also induces a structural transition, similar to rare earth co-doping, which probably acts to pin the stripe structure, stabilizing the charge ordering, and hence reducing T_c. Recently, Koike *et al.* have shown that the 1/8 phenomenon is common to all the cuprates [72] and that a similar effect must occur for $x \sim 1/4$ [73].

Though the first neutron scattering experiment was performed in a Nd doped sample with the "magic" hole concentration $x = 1/8$, further measurements on samples with $x = 0.10$ and $x = 0.15$ confirmed the existence of incommensurate peaks in the spin and charge sectors, giving support to the stripe picture (Fig. 9.3 [16]). Moreover, systematic studies of superconducting LSCO samples with a range of doping values x has been performed by inelastic neutron scattering [17]. The detected incommensurability is exactly the same as that obtained in co-doped samples (Fig. 9.4 [16]). Both elastic and inelastic neutron scattering, in addition to NMR [27], NQR, μSR [28], Hall transport [32], and ARPES [33] measurements indicate that stripes are present in LSCO. A linear dependence of the incommensurability δ as a function of the doping concentration x has been detected for $x < 1/8$, indicating that the stripes behave as "incompressible" quantum fluids in this regime, that is, for $0.05 < x < 0.12$ the hole density in each stripe is fixed (one hole per two Cu sites), and by increasing the amount of charge in the system one consequently increases the number of stripes and reduces their average separation $\ell(x)$. Moreover, Yamada *et al.* [74] showed that in this regime T_c is also proportional to δ, *i.e.*, $T_c \propto x \propto \delta \propto 1/\ell(x)$. Above $x = 1/8$,

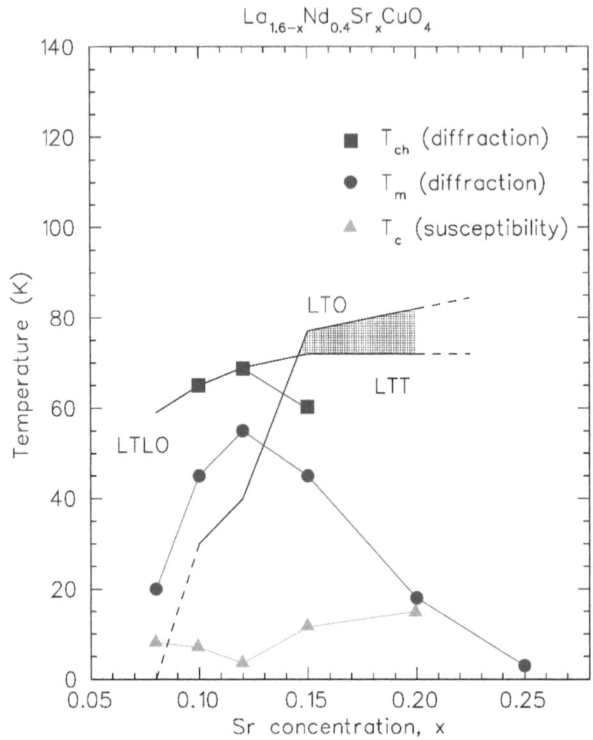

Figure 9.3. Experimental phase diagram obtained by Ichikawa et al. [16] for Nd-doped LSCO. T_{ch} and T_m denote, respectively, the temperatures below which charge and spin ordering could be detected in this system by elastic neutron scattering measurements. The superconducting transition temperature T_c obtained by susceptibility measurements is also shown. In addition, the structural transition lines from the low-temperature orthorhombic (LTO) to the low-temperature tetragonal (LTT) and to the low-temperature-less-orthorhombic (LTLO) phases are displayed.

however, the behavior of the system changes and δ nearly saturates, indicating a transition to a more homogeneous phase. Recently, neutron scattering experiments were performed within the spin-glass regime, for $0.02 < x < 0.05$ [18]. The result was surprising: the incommensurate peaks are rotated by 45° in reciprocal space, suggesting that the stripes are diagonal and half-filled, with one hole per Cu site, analogous to nickelate stripes [18]. However, this conclusion may be premature. Because the incommensurate peaks are observed only in the spin, but not in the charge sector, other explanations of the phenomenon are plausible, such as the formation of a spiral phase [75]. We will discuss this topic below in Secs. 4.2 and 5. A summary of available experimental data concerning the incommensurability is presented in Fig. 9.4.

It is interesting to compare the two mechanisms proposed theoretically [49-53,67] for stripe formation in the light of the experimental results [16]. Charge order indeed appears before spin order, favoring the Emery-Kivelson proposal of frustrated phase separation [67], but the stripe separation clearly displays a linear dependence on the inverse of the hole density, as predicted by the Hartree-Fock analysis [49-53]. Concerning the stripe filling, the Hartree-Fock predictions are observed in the spin-glass regime, whereas the Emery-Kivelson proposition holds within the superconducting underdoped regime. However, recent slave-boson studies of the 3-band Hubbard model have shown that if the oxygen-oxygen hopping integral t_{pp} is finite, quarter-filled stripes are more stable than half-filled ones [76]. DMRG studies of the t-J model found also that quarter-filled stripes are the lowest-energy configuration [54].

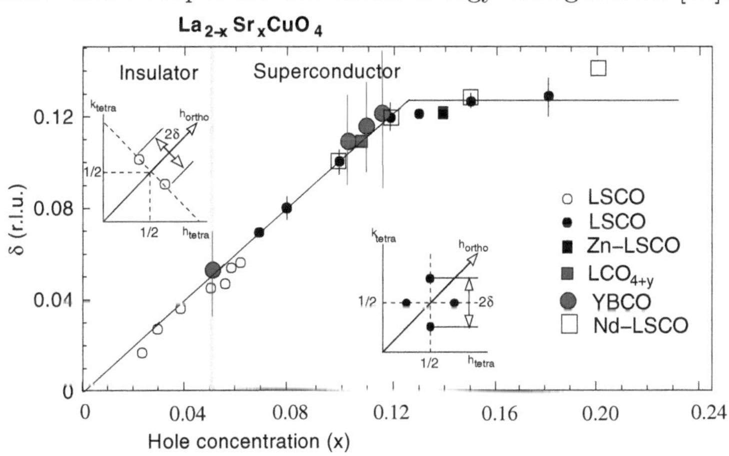

Figure 9.4. Summary of data concerning incommensurability $\delta = \pi/\ell$ as a function of doping concentration x. Data were obtained from neutron-scattering measurements by several groups: open and full small circles are from [18] and [17], respectively; dark squares are from [111], the grey one from [17], and the white one from [16]; large circles are from Refs. [17-22].

4. The single-hole problem: the role of kinetic energy

The problem of a single hole in a 2D antiferromagnet has a long history and it is probably one of the best-studied cases of strongly correlated systems. This problem is by nature single particle because it deals with a single particle interacting with a complex magnetic environment.

Physically this is realized only in physical systems with extremely small carrier densities, and many of their properties can be related to the polaron problem [80]. Here we will not review the problem in any way (there are very good reviews on the subject) but we would like to stress the important role that the hole kinetic energy plays in determining the possible phases. Moreover, as we are going to show, the same physics may be responsible for the stripe phases in transition-metal oxides.

The simplest limit of the t-J model with a single hole is the limit of $J = 0$, that is, $U \to \infty$. This limit was studied by Nagaoka [81], who showed that a single hole makes the system unstable toward a ferromagnetic phase. The origin of ferromagnetism in this case lies in the minimization of the hole kinetic energy: because double occupancy is precluded, the kinetic energy conserves the spin projection, and there is no energy penalty for the formation of ferromagnetic bonds, the kinetic energy is minimal when all the spins have the same direction. This process is essentially the same as that occuring in double-exchange systems such as manganites, where a ferromagnetic coupling between the electron spin and a magnetic host produces a ferromagnetic state by minimizing the kinetic energy. For finite but small J ($J/t \ll 1$) the same effect occurs, but instead of polarizing the entire plane of spins a single hole produces a ferromagnetic polarization cloud of size R: the hole gains a kinetic energy of order $4t - t/(R/a)^2$ by being free to move in the ferromagnetic region but has to pay a magnetic energy cost of order $J(R/a)^{d-1}$ for the generation of a frustrated magnetic surface with ferromagnetic bonds. Minimization of the total energy of the hole indicates that the radius of the ferromagnetic region decreases according to $R/a \sim (t/J)^{1/(d+1)}/(d-1)$ as J/t increases (notice that for the case $d = 1$ this estimate always produces $R = \infty$ for all values of t/J) [82].

As J/t increases the magnetic energy generated by the frustrated magnetic surface becomes too large, and R shrinks to zero. Thus, larger values of J/t lead to a change in the physics. A simple way to reduce the magnetic frustration is to reduce the frustrated surface of the spin configuration. Instead of a frustrated surface of misaligned spins it becomes energetically favorable to create strings of ferromagnetic pairs of spins due to the retraceable motion of the holes. It is clear that in this case the energy of the string grows linearly with its size L, and therefore that the energy required to generate a string is approximately JL. On the other hand, the hole kinetic energy changes from $-t$ to a quantity of order $-t + t/(L/a)^2$, and the problem is essentially equivalent to that of a single particle in a linearly confining potential. The solution of this quantum mechanical problem is straightforward and minimization of the total energy shows that the size of the strings varies according to $L \sim (t/J)^{1/3}$ independent of the dimensionality. Thus, on increasing J/t the single hole case exhibits a crossover from a ferromagnetic polaron to the so-called spin-polaron. The confinement described here is not completely correct because the hole has been considered as a semi-classical entity whereas in fact it is a fully quantum-mechanical object, which could undergo quantum tunneling over classically forbidden regions. This tunneling gives rise to "Trugman loops" where the hole can move diagonally but with a very small tunnel splitting (that is, very large effective mass) [83].

In any case, the true situation lies between these extremes and ferromagnetic polarization is concomitant with string processes. It is clear that the problem of the doped antiferromagnet centers on the compensation of magnetic frustration by reduction of the kinetic energy. Furthermore, the string mechanism provides a clear way to release kinetic energy, namely the retraceable motion of the hole. By incorporating both the kinetic energy (creation of ferromagnetic bonds) and the magnetic energy (generation of strings) one may understand how holes can move in a system with strong antiferromagnetic correlations. We have, however, discussed only the case of a single hole but for superconductivity it is important to understand the situation when the density of holes increases. The first crossover in these systems occurs when a finite *linear* density of holes (say N/L is finite but N/L^2 zero) is reached. The second crossover occurs when N/L^2 becomes finite. Thus, in such strongly correlated systems one has always at least 2 crossovers: from single particle to 1D and from 1D to 2D. In the next section we will discuss the first crossover and show that it is related to the formation of stripes.

4.1 Crossover from single particle to 1D: stripes in the t-J model

Consider an infinite antiphase domain wall oriented along one of the crystal axes directions of the system (Fig.9.5). The cost in energy per hole to create such a state is $J/2(n^{-1}-1)$, where n is the linear density of holes along the stripe. However, the hole wave function is translationally invariant along the stripe and the kinetic energy gain due to longitudinal hopping is $-2t\sin(\pi n)/(\pi n)$. For $J/t = 0.4$ one may show that the energy is minimized for $n = 0.32$ with an energy $E_b \approx -1.255t$ which is larger than the energy of a hole in the bulk (spin polaron), given by $E_{sp} = -2\sqrt{3}t \approx -2.37t$ [56]. Here we have not included the transverse motion of the hole perpendicular to the stripe, which further reduces the kinetic energy of the system but gives also a finite width to the hole wave function. Using a retraceable-path approximation (but ignoring hole-hole interactions) one may calculate analytically the Green function for the holes [58].

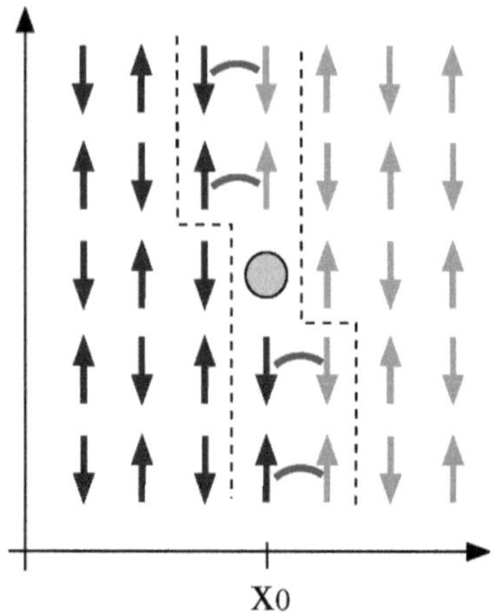

Figure 9.5. Antiphase domain wall with one hole. Thick lines represent broken bonds, while dashed lines mark the position of the topological defect.

Holes are confined to an antiphase domain wall by the potential generated by strings of overturned spins (that is, there is a linearly growing potential transverse to the stripe direction). One may show also that in this configuration the hole is actually a holon, *i.e.,* it carries charge but no spin and any motion of the hole away from the stripe produces a spinon, a particle with a spin of 1/2 but no charge, of energy J. In the bulk the hole carries both spin and charge and therefore is a spin-polaron. Spin-charge separation is thus local, but not macroscopic [58]. Because of this effect, Trugman loops [83], which are responsible for hole deconfinement in the absence of antiphase domain walls, are not effective because the motion of holes away from the wall always produces an excitation of finite energy.

One finds that for $J/t = 0.4$ and $n \approx 0.3$ the energy of the stripe state is $E_0 \approx -2.5t$, and therefore lower than the energy of the spin polaron. Furthermore, the width of the hole wave function has a value on the order of 3 - 4 lattice units [58], and therefore extends a considerable distance from the antiphase domain wall, in contrast to the "cartoon" picture where the stripe has a width of only one lattice spacing [16]. Thus, considering a stripe as a completely 1D object is somewhat misleading because each hole may make long incursions into the antiferromagnetic regions. Moreover, it is clear from these analytic calculations that it is the single-hole kinetic energy which is responsible for the stabilization of the stripe state. These results have been confirmed numerically by DMRG and exact-diagonalization studies [59]. In previous works semi-phenomelogical field theoretical models were proposed to explain the formation of anti- and in-phase domain walls and stripes [84, 85, 86]. However, from the studies on the t-J_z model it becomes clear that stripe formation is a short distance problem (that is, it involves high energy states) and cannot not be properly addressed with the use of field theories that can only describe the low-energy, long-wavelength physics.

We should stress that we are discussing the ground state, that is, the lowest-energy stationary state and therefore the concept of "stripe fluctuations" refers, in this context, to excitations which are separated from the ground state by an energy of order $t(J/t)^{2/3}$ because of confinement in the transverse direction [58]. These results are consistent not only with the DMRG results for the t-J model for small J/t [59] but also with those for the t-J_z model [60]. As a consequence the stripe is metallic, in contrast to Hartree-Fock results which always produce a gap [87]. As in a Luttinger liquid [88] one expects that hole-hole interactions drive the system toward a CDW phase, which would become insulating in the presence of any amount of disorder [89]. Moreover, interactions cause the density of states vanish at the chemical potential following a non universal power of the interaction parameter [88].

4.2 Stripes, magnetism and kinetic energy release

At half-filling, cuprates are antiferromagnetic Mott insulators with the Cu atom carrying a spin of 1/2. La_2CuO_4 is one of the most striking examples of a layered antiferromagnetic Mott insulator. The Néel temperature, instead of taking a value on the order of the planar magnetic exchange J ($\approx$ 1500K), is approximately 300K. This occurs because of the low dimensionality of the system and indeed, an $O(3)$ invariant 2D Heisenberg model may order only at zero temperature due to the breaking of continuous symmetry (Mermin-Wagner theorem) [90]. However, the small inter-planar coupling, $J_\perp$ ($\approx 10^{-4}J$), stabilizes antiferromagnetic order at finite temperature. Magnetism in these systems is evident essentially across the entire phase diagram, although long-range order is lost with only 2% doping by Sr. Since the Sr atoms are located out of the CuO_2 planes and their effect is to introduce holes, the doped holes are essential for the destruction of long-range order. More importantly, it is the minimization of *kinetic energy* of the holes which is the dominant mechanism for the suppression of magnetism. In the following, we discuss several different ways to understand the importance of hole motion in these systems.

A first indication for the importance of hole kinetic energy is given by magnetic measurements for $x < 0.02$ which observe the *recovery* of the magnetization when the system is cooled below the so-called freezing temperature, $T_F(x)$ [28]. The staggered magnetization $M_S(x,T)$ vanishes at the Néel temperature $T = T_N(x)$ and is a smooth function for $T_F(x) < T < T_N(x)$. However for $T < T_F(x)$ the magnetization seems to recover to the full value expected at $x = 0$ and $T = 0$. This effect can be ascribed to localization of the holes after which they affect the magnetization only locally. Exactly where the localization of holes occurs remains unresolved. However, soft X-ray (oxygen K-edge) absorption experiments indicate that the holes are probably in the oxygen sites [91]. NMR experiments appear to indicate that holes would localize preferentially close to the charged Sr atoms for electrostatic reasons [92]. However, because the system is annealed as temperature is reduced, it is quite possible that the unscreened Coulomb interaction between holes plays an important role in the localization process. If this is the case, localized stripe patterns may form even at low doping, although disorder effects from Sr doping are very strong in this region of the phase diagram and one would expect any stripe pattern to be random in the CuO_2 planes [62]. Recent neutron-scattering experiments at low doping find diagonal incommensurate peaks in the magnetic sector [18]. However, as no charge peak has yet been observed in the spin-glass regime, these measurements may be interpreted within the stripe model but the question remains open: the antiferromagnetic peaks could also be interpreted as the formation of a spiral phase [75]. In order to resolve this question unequivocally neutron-scattering experiments in samples heavily doped with spin-zero impurities such as Zn are required. If, as expected for a spiral state [75], the slope of the incommensurability as a function of Sr concentration x changes by a factor $(1 - 2z)$, where z denotes the Zn concentration, the stripe hypothesis would be excluded in the spin-glass regime.

Independent of the pattern of localization, holes may be localized either on the O or the Cu sites. If hole localization occurs at the O sites, one would expect a spin-glass phase to be observed at temperatures below $T_F(x)$ because a localized hole at an O site liberates one spin (configuration p^5) which frustrates the antiferromagnetic coupling between neighboring Cu atoms [93]. If, on the other hand, holes are localized on the Cu sites, the magnetization of the system would be reduced by one quantum of spin. Presumably, because of the delocalization of the hole wave function between O and Cu atoms the two effects can occur simultaneously [48]. The key feature of these experiments is that they indicate the importance of the hole motion for the destruction of long-range order.

The rapid suppression of magnetism with hole doping can be contrasted with the slow suppression of antiferromagnetic order when Cu is replaced by a non-magnetic atoms such as Zn or Mg [94]. In this case, long-range order seems to be lost only at 41% doping, that is, at exactly the classical percolation threshold for a 2D Heisenberg system. As has been shown in recent theoretical and numerical studies of the diluted *quantum* Heisenberg model, magnetic order seems to disappear only close to the classical percolation threshold even in the quantum system [95]. These results have been investigated experimentally and in fact the quantum fluctuations introduced by the dopants apparently are not sufficient to produce a quantum critical point (QCP) below the value of classical percolation [94]. Although the comparison between the problem of hole doping and Zn doping is not at first obvious, from the chemical point of view Zn introduces a static hole in the Cu plane because it has the same valence, but also has one extra proton. The situation on doping by Zn is therefore similar to the problem of holes localized at Cu sites. Indeed the recovered magnetization in the hole-doped case, $M(x, 0)$, is very close to the value which one would obtain on replacing Cu by a density x of Zn atoms. Thus, further confirmation is obtained that the hole kinetic energy is the driving force behind the suppression of antiferromagnetic order in these systems.

5. Phenomenological models: stripes and antiferromagnetism

In the previous sections we have argued that the kinetic energy of holes is fundamental for understanding the stripe phenomenon in cuprates. However, dimensional crossovers are very difficult to measure experimentally. Phase transitions, on the other hand are easy to observe, because they produce strong effects in the thermodynamic properties. Because Mott insulators are usually antiferromagnetic one may ask if such dimensional crossovers in the hole motion affect the antiferromagnetic phase? In the previous section we provided evidence that the kinetic energy of the holes is responsible for the destruction of antiferromagnetism in these systems when the hole concentration is of order of 0.01-0.02. How is it possible that such small doping levels can destroy a robust antiferromagnetic phase, with a Néel temperature which is of order 300K?

A possible way to understand the effect of the hole motion is to consider the formation of infinitely long stripes. The first obvious effect is a breaking of the spatial rotational symmetry. In the antiferromagnetic phase the spin rotational symmetry is also broken, indicating that both symmetries must be broken in the ground state of a striped antiferromagnetic phase. As a consequence the Goldstone modes associated with the broken symmetries must carry information about them. For an ordinary (non-striped) antiferromagnetic phase these are spin-wave modes characterized by an energy dispersion $E(\mathbf{k}) = c_s k$, where c_s is the spin-wave velocity, and a spin-stiffness ρ_s associated with the twist of the order parameter [45]. In a striped antiferromagnetic phase the Goldstone modes remain spin waves, but because of the broken rotational symmetry their dispersion is different if the mode propagates along the direction of broken symmetry or perpendicular to it, *i.e.*, the energy dispersion is not circularly symmetric and $E(k_{||}, k_{\perp}) = \sqrt{c_{||}^2 k_{||}^2 + c_{\perp}^2 k_{\perp}^2}$, where $||$ and $\perp$ refer respectively to the directions parallel and perpendicular to the stripes. At wave lengths longer than the stripe separation ℓ and energies lower than the first spin-wave gap due to the folding of the Brillouin zone, this kind of dispersion is guaranteed by the nature of the broken symmetries.

In an ordinary antiferromagnet the spin-wave velocity is simply related to the lattice spacing a and the exchange constant J by $c_s = SJa^d$. Thus, a striped antiferromagnet may be modeled simply by assuming that the only effect of the stripes is to introduce anisotropy in the exchange constants. Let us consider the case of a spatially anisotropic Heisenberg model with exchange constants J_x and J_y in the x- and y-directions, respectively, in which case the spin-wave velocities in each direction are given by [96]

$$\begin{aligned} c_y^2 &= 2S^2 a^2 J_y (J_x + J_y) \\ c_x^2 &= 2S^2 a^2 J_x (J_x + J_y). \end{aligned} \tag{9.18}$$

Microscopically, one may regard the stripes as causing local modification of the exchange across an antiphase domain wall from J to a value $J'(< J)$. This alteration of J leads to a macroscopic change in the values of the exchange constants in the same way that the introduction of impurities in a solid leads to an average change in the unit-cell volume.

To relate J and J' to J_x and J_y is not a trivial task. One possibility would be to solve the linear spin-wave theory for the striped antiferromagnet and calculate the derivative of the spin-wave energy at the ordering vector $\mathbf{Q}$. For stripes with a separation of N_s lattice sites, this procedure requires the solution of N_s coupled differential equations. Besides being computationally intensive, the solution would not provide significant insight into the origin of J' and would just exchange one phenomenological parameter by another.

The simplest theory describing a striped antiferromagnet is the spatially anisotropic non-linear σ model

$$\begin{aligned} S_{eff} &= \frac{1}{2}\int_0^{\beta\hbar} d\tau \int dx \int dy \left\{ S^2 \left[J_y \left(\partial_y \hat{n}\right)^2 + J_x \left(\partial_x \hat{n}\right)^2 \right] \right. \\ &+ \left. \frac{\hbar^2}{2a^2(J_x+J_y)} \left(\partial_\tau \hat{n}\right)^2 \right\}, \end{aligned} \tag{9.19}$$

where $\hat{n}$ is a unit vector field. The symbols have been chosen to suggest the continuum limit of an underlying effective *integer-spin* Heisenberg Hamiltonian on a square lattice [97]. The underlying anisotropy parameter is the ratio of the two exchange constants, or of the two velocities,

$$\alpha = J_x/J_y. \tag{9.20}$$

The value of α characterizes the theory, but its exact dependence on microscopic parameters is not easy to derive.

We proceed by making use of well-established techniques to analyze the behavior of the field theory described by the action (9.19) to predict the physical properties of the system of interest. It is useful to re-express (9.19) symmetrically by a dimensionless rescaling of variables $x' = (\alpha)^{-1/4} x\Lambda$, $y' = (\alpha)^{1/4} y\Lambda$ ($\Lambda \sim 1/a$ is a momentum cut-off), $\tau' = \sqrt{2(J_x+J_y)\sqrt{J_x J_y}} Sa\tau/\hbar$. The effective action (9.19) becomes

$$S_{eff} = \frac{\hbar}{(2g_0)} \int_0^{\hbar\Lambda\beta c_0} d\tau' \int dx' \int dy' \left(\partial_\mu \hat{n}\right)^2, \tag{9.21}$$

where μ denotes x', y', and τ',

$$g_0(\alpha) = \hbar c_0 \Lambda/\rho_s^0 = \left[2(1+\alpha)/\sqrt{\alpha}\right]^{1/2} (a\Lambda)/S \tag{9.22}$$

is the bare coupling constant, $c_0 = [2(J_x+J_y)\sqrt{J_x J_y}]^{1/2}(aS)/\hbar$ the spin wave velocity and

$$\rho_s^0 = \sqrt{J_x J_y} S^2 \tag{9.23}$$

the classical spin stiffness of the rescaled model. The original anisotropy is now contained in the limits of integration.

Notice that while α depends on the ratio J_x/J_y, the spin stiffness depends on the product $J_x J_y$. Thus, given a microscopic model where J_x

and J_y are expressed in terms of microscopic quantities the field theory is well defined. Unfortunately no calculations yet exist for the microscopic form of these quantities, and certain assumptions are required concerning their behavior. If the spatial rotational symmetry is broken at the macroscopic level, that is, one has infinitely long stripes in the y direction, a simple choice would be $J_y = J$ and $J_x = \alpha J$, whence

$$\rho_s^0 = \sqrt{\alpha}\rho_I \tag{9.24}$$

where

$$\rho_I = JS^2 \tag{9.25}$$

is the spin stiffness of the isotropic system. This choice is valid only when the system is composed of a mono-domain of stripes [98]. If the system is broken into domains, in which case the rotational symmetry is broken micro- but not macroscopically the choice (9.24) may not be the most appropriate. At sufficiently long wave lengths the system is essentially isotropic and therefore the spin stiffness is the same in all directions, [96]

$$\rho_s^0 = \rho_I, \tag{9.26}$$

which is obtained by choosing $J_x = \sqrt{\alpha}J$ and $J_y = J/\sqrt{\alpha}$. We note that these choices are essentially arbitrary and are based on qualitative expectations concerning the nature of the correlations at very long wave lengths. One may show that the choice (9.26) is appropriate very close to the antiferromagnetic phase ($x < 0.02$) where the breaking of the system into domains is quite probably because of disorder effects [96]. However, at larger doping ($x = 1/8$) the choice (9.24) is more appropriate because long-range stripe order is observed [98]. With these two parameterizations one may analyse the problem and calculate physical quantities for comparison with experiments. One general consequence of the anisotropy introduced by the presence of stripes is a growth of quantum fluctuations because of the reduction of effective dimensionality. In fact, by using large N methods and RG calculations, one may demonstrate that the effective spin stiffness is given by

$$\rho_s(\alpha) = \rho_s^0(\alpha)\left[1 - \frac{g_0(1)}{g_c(\alpha)}\right], \tag{9.27}$$

where

$$\begin{aligned} g_c(\alpha) &= \sqrt{8}\pi^2\sqrt{\alpha/(1+\alpha)}\left\{\ln\left(\sqrt{\alpha}+\sqrt{1+\alpha}\right)\right. \\ &+ \left.\sqrt{\alpha}\ln[(1+\sqrt{1+\alpha})/\sqrt{\alpha}]\right\}^{-1} \end{aligned} \tag{9.28}$$

is a critical coupling constant. Notice that (9.27) is reduced from its classical value $\rho_s^0(\alpha)$ for fixed anisotropy α, and that it vanishes at some critical value α_c, whence $g_c(\alpha_c) = g_0$. Hence, as a function of the anisotropy, the model exhibits quantum critical point where the system undergoes a quantum phase transition from an ordered Néel state to a paramagnetic phase. The loss of antiferromagnetic order at $x = 0.02$ can thus be considered as a consequence of the enhancement of quantum fluctuations due to the presence of stripes.

As explained previously, these considerations are valid when the holes move along the stripes and modify the exchange constant across the stripe. However, when localization occurs at low temperatures (as observed in the recovery of magnetization in NQR experiments) the stripes essentially cease to exist and the system undergoes a phase transition into a spin-glass phase. The simplest way to understand this phase is to consider hole localization at the oxygen sites with consequent liberation of one spin 1/2 [93]. This spin frustrates the antiferromagnetic order, because the superexchange interaction of the O spin with the neighboring Cu spins is antiferromagnetic. This problem may be treated by considering the O spin as a classical localized dipole moment [75].

Finally, the theoretical results concerning the existence of stripes and the appropriate model for describing them can be summarized as follows: the proposal that the doped t-J model undergoes a phase separation is supported by variational arguments [77], diagonalization on small clusters [77], and Green-function quantum Monte Carlo calculations [78]. On the other hand, several quantum Monte Carlo calculations [55], series expansions [79], exact diagonalization [56], and DMRG calculations [54] yield results contradicting these claims and supporting the stripe picture. In order to gain more insight into the problem, we begin by considering the simplest possible case, namely, the single-hole problem.

6. Phenomenological models: Transverse fluctuations and pinning of stripes

While much theoretical effort has been concentrated on determining whether stripes are the ground state of models such as the Hubbard and the t-J models, which are supposed to describe high-T_c superconductors, a parallel research direction has also developed which consists of phenomenological studies of the striped phase. In this case, one assumes the existence of stripes and discusses further aspects such as their effect on the antiferromagnetic state (see previous section). Motivated by issues such as the static or fluctuating nature of stripes and the mechanism of stripe pinning, a phenomenological theory for the pinning of stripes has been developed. Zaanen *et al.* have related transverse stripe fluctuations to the restricted solid on solid model (RSOS) which describes the growth of surfaces [99]. In a simplified form of the model, the transverse kink excitations of stripes are mapped to a quantum spin-1 chain model [99], whose Hamiltonian is

$$H = \sum_n \left[-t \left(S_n^x S_{n+1}^x + S_n^y S_{n+1}^y \right) - D S_n^z S_{n+1}^z + J (S_n^z)^2 \right]. \tag{9.29}$$

Here, t has the role of a hopping parameter for transverse kinks, J controls the density of kinks and the D term represents a nearest-neighbor interaction of kinks. The spins take the values $S_n^z = 0, \pm 1$, where $+1$ and -1 are associated respectively with stripe kinks and anti-kinks and $S_n^z = 0$ describes unperturbed (flat) segments. The full phase diagram for this problem was determined numerically by den Nijs and Rommelse [100] after earlier calculations by Schulz [101], who treated the spin-1 problem as two coupled spin-1/2 chains. We have recently reanalysed the calculations of Schulz [102] and derived the correct phase diagram from this formalism, which agrees with the one obtained in Ref. [100] (see Fig. 9.6). Six different phases can be identified, depending on the values of the D and J parameters. If J is positive, the last term of Eq. (9.29) determines that $S_n^z = 0$ and the stripe is straight (flat phase). If J is negative, both values $S_n^z = \pm 1$ are equally favorable with respect to the J term and the D term determines the value of S_n^z. If D is positive, nearest-neighbor segments prefer to be similar and therefore S_n^z and S_n^{z+1} will have the same sign. This gives rise to the ferromagnetic phase (diagonal stripes), with a sequence of kinks or anti-kinks. On the other hand, if D is negative, S_n^z and S_n^{z+1} prefer to have opposite signs, and the stripe will be bond-centered and flat, with a "zig-zag" shape (a kink follows an anti-kink and *vice versa*). In addition, both the flat and the bond-centered flat phases, which are gapped, can undergo a Kosterlitz-Thouless transition to gapless rough or bond-centered rough phases, respectively. The sixth phase, which was not identified by Schulz, corresponds to a gapped, disordered, flat phase (DOF), (Fig. 9.6). In contrast to the flat phase, this phase has a finite density of kinks and anti-kinks, which are positionally disordered, but have an antiferromagnetic order in the sense that a kink $S_z = 1$ is on average followed by an anti-kink $S_z = -1$ (rather than another kink), with any number of $S_z = 0$ states in between them. The DOF phase is the valence-bond phase which is responsible for the Haldane gap.

In the limit of negligible nearest-neighbor interaction ($D \sim 0$), the above model can be related to the t-J model [103] by considering the

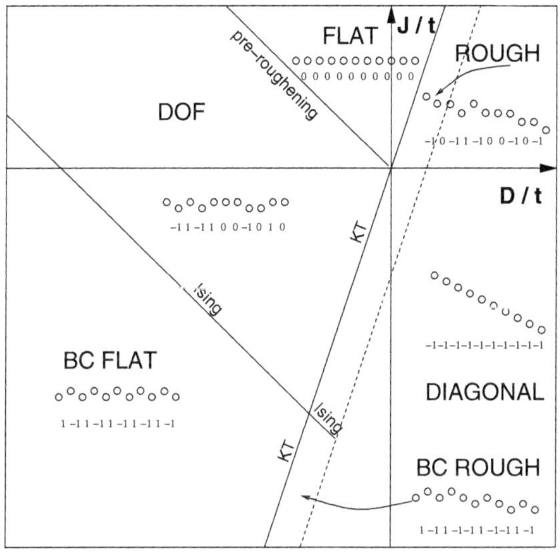

Figure 9.6. Sketch of the phase diagram for the spin-1 chain. The stripe configurations represented by circles, and the corresponding values of S^z are shown below. There are six different phases: 1) a gapped flat phase, corresponding to straight stripes ($S_z = 0$); 2) a gapless rough phase (spins equal to zero and ±1 distributed randomly); 3) a gapped bond-centered (BC) flat phase, which has a long-ranged zig-zag pattern (periodic alternation of $S_z = 1$ and $S_z = -1$); 4) a gapless BC rough phase with a zig-zag pattern (antiferromagnetic correlations with disordered $S_z = \pm1$ but no $S_z = 0$ states); 5) a diagonal stripe phase, corresponding to a ferromagnetic state in the spin language; 6) a gapped disordered flat phase (DOF), where a kink ($S_z = +1$) is followed on average by an anti-kink ($S_z = -1$), but with some $S_z = 0$ states in between.

transverse dynamics of a single vertical stripe in a frozen Néel background. The transverse dynamical properties of holes are described by the t-J_z model, and they move under the condition that the horizontal separation between two neighboring holes cannot be larger than one lattice constant. Notice that this condition does not restrict the motion of the stripe: it may still perform excursions very far from the initial straight-line configuration, but the line cannot be broken. In this case the t-J Hamiltonian describing hole dynamics may be mapped onto a spin-1 chain Hamiltonian analogous to Eq. (9.29), but with $D = 0$ [103]. A duality transformation of the initial quantum string Hamiltonian maps the problem onto one describing a 1D array of Josephson junctions [104], which is known to exhibit an insulator-superconductor transition at $(t/J)_c = 2/\pi^2$ [105]. This Kosterlitz-Thouless transition represents the unbinding of vortex-antivortex pairs in the equivalent XY model, which translates to a roughening transition for the stripe problem. In this way, the transition between the flat and rough phases along the vertical line ($D = 0$) of the phase diagram could be determined precisely by analytical means [104].

These models are related to the sine-Gordon model, and by studying the spectrum of the quantum string in a Hilbert-space sector of zero topological charge, the meaning of the transition in the "string" language may be clarified. At $(t/J)_c$ the (insulating) pinned phase, which has an energy spectrum with a finite gap, turns into a (metallic) depinned phase where the spectrum becomes gapless [104]. This procedure allows the connection of two important and different classes of problems, namely the transverse dynamics of stripes in doped antiferromagnets and a system with the well-known properties of the sine-Gordon model.

In all the models discussed hitherto, the pinning potential arises from the discrete nature of the lattice. However, the introduction of holes into the MO_2 planes (M = Cu or Ni) is not the only consequence of doping a Mott insulator such as La_2MO_4. Doping an antiferromagnetic insulator also introduces disorder into this material due to the presence of counterions, which act as attractive centers for holes. Doping with divalent atoms, such as Sr^{2+} produces quenched disorder, because the ionized dopants are located randomly between the CuO_2 planes. In contrast, doping with excess oxygen generates annealed disorder. Indeed, oxygen atoms have a low activation energy and remain mobile down to temperatures of 200-300 K.

In order to account for the random pinning potential provided by the Sr atoms in nickelates and cuprates, one may add a disorder potential to the previous phenomenological Hamiltonian (9.29) with $D = 0$. This allows a determination of the influence of both disorder and lattice effects on the striped phase of cuprates and nickelates. We consider the problem of a single stripe along the vertical direction confined in a box of size 2ℓ, where ℓ denotes the stripe spacing. The system is described by the phenomenological Hamiltonian

$$\hat{H} = \sum_n \left[-2t\cos\left(\frac{\hat{p}_n}{\hbar}\right) + J\left(\hat{u}_{n+1} - \hat{u}_n\right)^2 + V_n(\hat{u}_n) \right], \tag{9.30}$$

with t the hopping parameter, $\hat{u}_n$ the displacement of the n-th hole from the equilibrium (vertical) configuration, $\hat{p}_n$ its conjugate transversal momentum, J the stripe stiffness, and $V_n(\hat{u}_n)$ an uncorrelated disorder potential satisfying $\langle V_n(u)V_{n'}(u')\rangle_d = d\delta(u-u')\delta_{n,n'}$, where $\langle ... \rangle_d$ denotes the Gaussian average over the disorder ensemble and d is the inverse of the impurity scattering time. Eq. (9.30) may be straightforwardly related to Eq. (9.29) by noting that $S_n^z = u_{n+1} - u_n$ and that the hopping terms S_n^x and S_n^y are connected to translation operators, which can be written in the momentum representation p_n as $\tau_\pm = e^{\pm i p_n/\hbar}$ [104]. We are considering the lattice parameter $a = 1$.

A dimensional estimate provides the dominant features of the phase diagram. At large values of the hopping constant $t \gg J$, the first term may be expanded as $-2t\cos(\hat{p}_n/\hbar) \sim const. + t(\hat{p}_n/\hbar)^2$. In the case of no impurity potential, $V_n(\hat{u}_n) = 0$, hole dynamics is governed by the competition between the kinetic term $t(k_n)^2$, which favors freely mobile holes, and the elastic one, $J(\hat{u}_{n+1} - \hat{u}_n)^2$, which acts to keep them together. When the confinement is determined by the lattice pinning potential, the average hole displacement $\hat{u}_{n+1} - \hat{u}_n$ is of order 1 (the lattice constant is unity) so the wave vector $k_n \sim 1$. A transition from the flat phase, with the stripe pinned by the underlying lattice, to a free phase is then expected at $t/J \sim 1$.

We now consider the opposite limit of strong pinning by impurities. In this case, the potential provided by the lattice is irrelevant and the typical hole displacement is on the order of the separation between stripes, $1/k_n \sim \hat{u}_{n+1} - \hat{u}_n \sim \ell$. By comparing the kinetic $t(1/\ell)^2$ and the elastic $J(\ell)^2$ terms, we observe that a transition should occur at $t/J \sim (\ell)^4$.

Indeed, by deriving the differential renormalization group (RG) equations to lowest nonvanishing order in the lattice and disorder parameters, one obtains a set of flow equations [89], which indicate that the transition from the flat (lattice-pinned) to the free phase occurs at $(t/J)_c = 4/\pi^2$, and the transition from the disorder-pinned to the free phase occurs at $(t/J)_c = (36/\pi^2)\ell^4$. The pinning phase diagram of the striped phase is shown in Fig. 9.7, in which $\delta = 1/2\ell$.

By comparing these results with recent measurements on nickelates and cuprates one concludes that nickelates occupy the lower left corner of the phase diagram, i.e., they have static stripes which are pinned by the lattice and by the impurities. By contrast, cuprates are characterized by freely fluctuating stripes and so appear in the upper right corner. An appropriate treatment of the striped phase in cuprates must therefore include stripe-stripe interactions and the model becomes similar to that for a 2D fluctuating membrane [89, 106]. In this phenomenological framework, both nickelate and cuprate materials can be understood in a unified way, the difference between them being simply the parameter t/J which measures the strength of quantum fluctuations.

Although the number of holes is intrinsically connected with the number of pinning centers, recent experimental developments show that it is

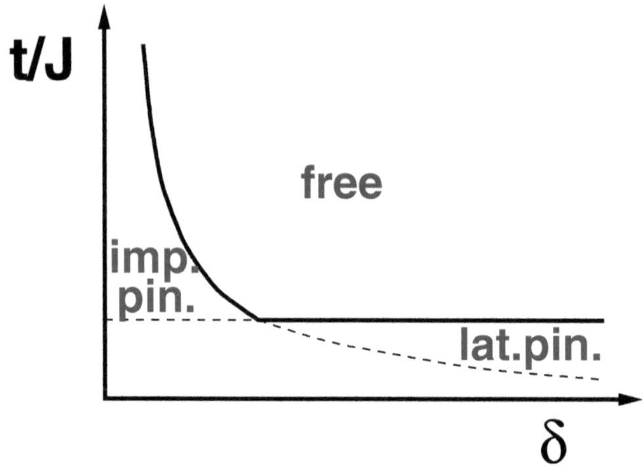

Figure 9.7. Zero-temperature pinning phase diagram of the stripe phase in the presence of lattice and impurity pinning. Three phases can be identified: a quantum membrane phase with freely fluctuating Gaussian stripes, a flat phase with the stripes pinned by the lattice, and a disorder-pinned phase [89].

possible to control these two parameters independently. Co-doping the superconducting cuprate material LSCO with Nd or Zn increases the disorder without modifying the number of charge carriers [16, 72, 107]. On the other hand, growing the superconducting film on a ferroelectric substrate and using an electrostatic field as the control parameter allows the number of charge carriers in the plane to be altered for a fixed Sr concentration x [108]. This class of experiments constitutes an important step towards the control of a normal metal-superconducting transition.

One may then investigate the stripe pinning produced by Zn and Nd co-doping [109]. The two dopants play fundamentally different roles in the pinning process. Nd, as with other rare-earth co-doping, induces a structural transition which produces a correlated pinning potential trapping the stripes in a flat phase. The situation is analogous to the pinning of vortices by columnar defects or screw dislocations [110]. In this case transverse fluctuations are strongly suppressed, long-range order is achieved and thus the incommensurate peaks observed by neutron diffraction become sharper after the introduction of the co-dopant, as observed experimentally [16]. On the other hand, in-plane Zn- and Ni-doping provide randomly distributed point-like pinning centers, similar to oxygen vacancies in the vortex-creep problem. Within the model in which a stripe is regarded as a quantum elastic string, the effect of randomness is to "disorder" the string, promoting line-meandering, destroying the 1D behavior and broadening the incommensurate peaks. A perturbative treatment of the RG equations discussed previously show at the next higher order that this kind of pinning is relevant only in under-doped systems, in agreement with experiments [111].

Finally, it is essential to go beyond the studies of transverse stripe excitations to consider the coupling between longitudinal and transverse modes. Longitudinal modes may be described by a Luttinger-liquid Hamiltonian, and the coupling between longitudinal and transverse fluctuations was investigated using bosonization [102]. One finds that a longitudinal CDW instability can arise if the stripe is quarter-filled and the underlying lattice potential has a zig-zag symmetry. This result has shed additional light on the connection between the formation of a LTT phase and the subsequent appearance of charge order in high-T_c cuprates (Fig. 9.3).

Experimentally, the suppression of superconductivity in LSCO co-doped with Nd (Fig. 9.3), and also the upturn of the resistivity in the normal phase, are correlated with a structural transition from the LTO to the LTT phase. Indeed, at $x = 1/8$ the charge ordering temperature T_{co} reaches its highest value and the superconducting temperature T_c shows a local minimum. Neutron diffraction experiments indicate that in the underdoped phase of LSCO the stripes are quarter-filled [16]. The formation of the LTT phase favors a zig-zag symmetry of the transverse stripe degrees of freedom. Thus, below T_{LTT} the CDW instability discussed above becomes relevant and stabilizes a bond centered string with zig-zag symmetry. If the stripe spacing is exactly commensurate, as at $x = 1/8$, a long-range-ordered CDW (Wigner crystal) can form, leading to the suppression of T_c. On the other hand, for incommensurate doping values, solitonic modes are present in the stripe which prevent long-range charge order. Longitudinal charge order has hitherto not been observed. Nonetheless, the upturn of the in-plane resistivity below T_{LTT} suggests proximity to an insulating phase inside the LTT phase of underdoped cuprates, which is likely to be the bond-centered zig-zag stripe [102].

7. Experimental detection of stripes in YBCO and BSCCO

An important question at this point is whether charge stripes are peculiar to the lanthanates or a generic feature of all the cuprates. The answer is not yet completely settled, and further experiments are required to achieve an unambiguous conclusion. Measurements on YBCO and BSCCO compounds begin to provide a comprehensive picture.

Inelastic neutron scattering experiments recently performed by Mook *et al.* on $YBCO_{6.6}$ [20], which corresponds to a hole doping of $x = 0.10$, detected a dynamical incommensurability in the magnetic sector of $\delta = 0.105 \pm 0.01$, which is exactly the value measured by Yamada *et al.* for the corresponding charge concentration in LSCO [17]. Later, magnetic incommensuration was also observed at $\delta = 0.0625$ in $YBCO_{6.35}$, which corresponds to a doping $x = 1/16$ [21]. In addition, measurements of phonon broadening at a wave vector consistent with the stripe picture (twice the magnetic wave vector) confirmed the previous results for $YBCO_{6.6}$ and $YBCO_{6.35}$ [21]. Eventually, a static charge order peak has been observed in a 21g crystal of $YBCO_{6.35}$, at a wave vector which is exactly double the dynamical magnetic incommensuration, $2\delta = 0.127$, as expected within the stripe picture [22]. Although the charge peaks are small, 6 orders of magnitude below the strongest crystal Bragg peak, their existence is undeniable. However, it should be emphasized that no charge order has been observed in $YBCO_{6.5}$ and $YBCO_{6.6}$, so the situation is not yet resolved.

Another important feature measured recently in YBCO is the 1D nature of the stripes [23]. In a 4g crystal of detwinned $YBCO_{6.6}$, one could observe not four, but only two incommensurate magnetic peaks (the second set of perpendicular peaks have nearly vanished, because the sample was almost completely detwinned). The results suggest stripes aligned along the b−axis, in agreement with far-infrared spectroscopic measurements by Basov *et al.* [31], which indicate that the superfluid density is larger along the b−direction. This behavior cannot be attributed to the chains, because in underdoped materials the missing atoms in the chains would inhibit the chain contribution to superconductivity.

As a conclusion, one could state that neutron scattering experiments in YBCO at several different doping concentration [24] confirm the stripe picture sketched for LSCO and reveal the universality of the previous results.

Concerning BSCCO, the majority of experimental results are obtained from scanning tunneling microscopy (STM) and spectroscopy (STS) [7-12]. The advantage of STM lies in its ability to measure simultaneously, with atomic resolution, both the surface topography and the local density of states (LDOS) of a material. The topographic image can be realized due to the exponential dependence of the tunneling current I on the separation of tip and sample. In addition, the differential conductance $G = dI/dV$, where V is the sample bias voltage, is proportional to the LDOS of the sample at the tip location.

Low-T STS in BSCCO samples ($T_c = 87$K) revealed the existence of a large number of randomly distributed regions, with characteristic lengthscales of order 30 Å, which have anomalous LDOS features. These features were initially referred to as quasiparticle scattering resonances (QPSRs) and were thought to be due to quasiparticle scattering from atomic-scale defects or impurities [9], but later measurements performed in Zn-doped samples indicated that these inhomogeneities do not originate from impurities. The locations of Zn impurities were identified from the zero-bias conductance map and the LDOS map taken simultaneously at the same location showed no correlation between the intensity of the integrated LDOS and the location of the Zn impurities [10].

Spatial variations of the tunneling spectrum and of the superconducting gap can be observed in pure as well as in impurity-doped BSCCO samples, and seem to be intrinsic to the electronic structure. The gap ranges from 25meV to 65meV and has a gaussian distribution [10]. The average gap is very similar to that reported previously from tunneling measurements. The spectra obtained at points with larger integrated LDOS exhibit higher differential conductance, smaller gap values and sharper coherence peaks, which are the characteristic features of spectra taken in samples with high oxygen doping concentration. This observation suggests the interpretation that the inhomogeneities may arise from differences in local oxygen concentrations [10].

Recent experiments in underdoped BSCCO indicate that these high-T_c materials are granular superconductors, with microscopic superconducting grains separated by non-superconducting regions. By doping the material with Ni impurities, it was observed that the position of Ni atoms coincide with regions of small gap ($\Delta < 50$ meV). In underdoped BSCO these small-gap regions, which have large $G(\Delta)$ are separated by percolative regions with large gap Δ and low $G(\Delta)$ [11].

Qualitatively new information is provided by STS measurements in magnetic field up to 7.5T. The quasiparticle states generated by vortices in overdoped BSCCO show a "checkerboard" pattern with a periodicity of four unity cells [12], in agreement with the charge periodicity expected within the stripe picture [16]. Indeed, the magnetic spatial periodicity previously detected by neutron scattering in overdoped LSCO in the presence [112] or absence [17] of a magnetic field is eight unit cells, exactly twice the charge spacing. Shortly after these measurements, the $4a$ periodicity was observed by Kapitulnik *et al.* in nearly optimally doped BSCCO *without* magnetic field [13]. Transformation of the real-space data to reciprocal space showed a periodicity of $4a$ in the randomly distributed regions with anomalous LDOS, which was manifest in four distinct peaks in reciprocal space [13]. Two peaks corresponding to a periodicity of $8a$ along the diagonal due to the periodically missing line of Bi atoms in the BiO plane were also seen clearly, confirming the sample quality. However, later studies by Davis *et al.* have cast doubt on the stripe interpretation: by analyzing the energy dependence of the wave vectors associated with the modulation, they have argued that the checkerboard LDOS modulation in BSCCO is an effect of quasiparticle interference, and not a signature of stripes [14]. Thus there is currently no agreement concerning the interpretation of the STM data.

In summary, one may state that the presence of stripes is now firmly established in LSCO (static in Nd-doped LSCO, dynamical in pure LSCO). Recent neutron scattering experiments indicate that they are also present in YBCO. However, the magnetic YBCO stripes seem to be dynamical, whereas the charge ones are static. Concerning BSCCO, only STM measurements, which are just sensitive to charge and have the disadvantage of being susceptible to surface defects and pinning, were performed hintertho. They suggest that a static charge stripe is present in this compound, although many further measurements are required for a definitive understanding.

8. Conclusions

Doped antiferromagnetic insulators have recently attracted a great deal of attention because many of the materials in this class exhibit novel and interesting behavior. The cuprates, for example, become metallic at low doping concentration, and even superconducting at relatively high temperatures, whereas other systems, such as nickelates, show metallic behavior only at very high doping and are never superconducting. Manganites, on the other hand, can display the phenomenon of giant magneto resistance. Despite the different electric and magnetic properties exhibited by these compounds, a common perovskite structure connects them. Moreover, spontaneous symmetry breaking and stripe formation seems also to be a shared feature.

Stripe-like ground states were first predicted from Hartree-Fock calculations of the Hubbard model [49-53]. Later studies of the $t-J$ model also confirmed the initial results [54]. However, this theoretical work attracted significant attention only when Tranquada *et al.* detected the existence of static spin and charge order in Nd-doped LSCO [16]. The peaks measured by elastic neutron scattering were incommensurate with pure antiferromagnetic order, and suggested that the system had undergone a phase-separation into 1D regions rich in holes (stripes), which were acting as domain walls in the staggered magnetization. The fact that the magnetization changes phase by π when crossing a domain wall has the consequence that the magnetic periodicity is twice the charge-stripe spacing. The associated incommensurability is therefore one half of the charge incommensurability, and the detection of both magnetic and charge incommensuration gave undeniable support to the stripe theory [16]. Later inelastic neutron scattering experiments in pure LSCO showed that there is a spin gap in these materials, but that incommensurate peaks can still be measured at rather low energies. The actual value of the spin gap depends on the doping. The incommensurability measured in these compounds is *exactly* the same as that obtained in Nd-doped LSCO, which is understandable considering replacement of La^{3+} with Nd^{3+} does not add charge carriers to the system, but induces a structural transition that helps to pin the stripes. The presence of stripes in LSCO has been confirmed by several different experimental techniques. Inelastic neutron-scattering experiments [17-23] indicate that incommensurate peaks are also detectable in YBCO, with a linear dependence of the incommensurability as a function of doping concentration similar to that observed in lanthanates. These results seem to indicate that the striped phase could be a generic feature of cuprates, instead of a peculiarity of LSCO. The requirement of very large samples imposed by the neutron scattering renders experimental progress in the field very slow. Concerning BSCCO, at present only STM data is available and while these may support the existence of stripes [12, 13], the question remains open. It is important to note here that STM is a surface probe, whereas neutron scattering measures bulk properties. Thus, surface impurities could play a dominant role in STM experiments, but a secondary one in neutron scattering. The years to come will hopefully show the truth behind all the controversies.

From the theoretical side there has been steady progress in understanding stripe phases. Different approaches have been applied to the problem with differing degrees of success. Mean-field theories of different kinds [113, 114, 115], gauge theories [116], and quantum liquid crystal phenomenology [117] have been employed to describe the stripe state. In this review we focused primarily on studies based on the t-J model, where DMRG calculations have shown the presence of stripe phases [58, 60]. These studies have the advantage that rather few parameters determine the physics of stripe formation. Although this kind of approach describes very well the nature of the stripes, it seems to indicate that the stripe state is essentially insulating and extra degrees of freedom, such as phonons [118], may be required to explain the experimental data in these systems [119, 120].

Phenomenological models have also contributed to a significant evolution in understanding. Appropriate models for describing stripe fluctuations have been developed, and analogies with other known systems established [99, 103, 104]. The effect of stripe pinning by impurities and by the underlying lattice, as well as the differing roles of rare-earth and planar impurities, has been clarified [89, 102, 104, 106, 109].

An important task remaining for the coming years is to show, both experimentally and theoretically whether and how stripes are connected to superconductivity. Systematic investigation of the different ways to suppress superconductivity may yield answers to this complex question.

Acknowledgments

This work is the result of many hours of conversation and discussion with our colleagues and friends. We are particularly indebted to A. Balatsky, D. N. Basov, D. Baeriswyl, A. Bianconi, A. Bishop, A. Caldeira, E. Carlson, A. Chernyshev, E. Dagotto, T. Egami, E. Fradkin, F. Guinea, M. Greven, N. Hasselmann, D. Hone, S. Kivelson, A. Lanzara, G. B. Martins, R. McQueeney, B. Normand, S. H. Pan, L. Pryadko, J. Tranquada, S. White, and J. Zaanen. This review would not have been possible without their support and criticism. C. M. S. acknowledges finantial support from the Swiss National Foundation under grant 620-62868.00.

References

[1] S. Sachdev, *Quantum Phase Transitions* (Cambridge University Press, Cambridge, 1999).

[2] For a recent review with a different perspective see, E. W. Carlson, V. J. Emery, S. A. Kivelson, and D. Orgad, in *The Physics of Conventional and Unconventional Superconductors* ed. by K. H. Bennemann and J. B. Ketterson (Springer-Verlag, 2002); cond-mat/0206217.

[3] S. Mori *et al.*, Nature **392**, 473 (1998); Phys. Rev. Lett. **81**, 3972 (1998).

[4] C. Renner, G. Aeppli, B.-G. Kim, Y.-A. Soh, and S.-W. Cheong, Nature **416**, 518 (2002).

[5] J. M. Tranquada *et al.*, Phys. Rev. Lett. **73**, 1003 (1994); S-H. Lee and S-W. Cheong, Phys. Rev. Lett. **79**, 2514 (1997); H. Yoshizawa *et al.*, Phys. Rev. B **61**, R854 (2000).

[6] C-H. Du *et al.*, Phys. Rev. Lett. **84**, 3911 (2000).

[7] Y. Yoshinari *et al.*, Phys. Rev. Lett. **82**, 3536 (1999).

[8] T. Katsufuji *et al.*, Phys. Rev. B **54**, 14230 (1996); G. Blumberg *et al.*, Phys. Rev. Lett. **80**, 564 (1998); K. Yamamoto *et al.*, Phys. Rev. Lett. **80**, 1493 (1998).

[9] E. W. Hudson, S. H. Pan, A. K. Gupta, K. -W. Ng, and J. C. Davis, Science **285**, 88 (1999).

[10] S. H. Pan *et al.*, Nature **413**, 282 (2001).

[11] K. M. Lang, V. Madhavan, J. E. Hoffman, E. W. Hudson, H. Eisaki, S. Uchida, and J. C. Davis, Nature **415**, 412 (2002).

[12] J. E. Hoffman, E. W. Hudson, K. M. Lang, V. Madhavan, H. Eisaki, S. Uchida, and J. C. Davis, Science **295**, 466 (2002).

[13] C. Howald, H. Eisaki, N. Kaneko, and A. Kapitulnik, cond-mat/0201546; C. Howald, H. Eisaki, N. Kaneko, M. Greven, and A. Kapitulnik, Phys. Rev. B **67**, 014533 (2003).

[14] J. E. Hoffman, K. McElroy, D. -H. Lee, K. M. Lang, H. Eisaki, S. Uchida, and J. C. Davis, Science **297**, 1148 (2002).

[15] S-W. Cheong *et al.*, Phys. Rev. Lett. **67**, 1791 (1991); B. O. Wells *et al.*, Science **277**, 1070 (1997); G. Aeppli *et al.*, Science **278**, 1432 (1998).

[16] J. M. Tranquada *et al.*, Nature **375**, 561 (1995); *ibid.* Phys. Rev. Lett. **78**, 338 (1997); *ibid.* Phys. Rev. B **54**, 7489 (1996); N. Ichikawa *et al.*, Phys. Rev. Lett. **85**, 1738 (2000).

[17] K. Yamada *et al.*, Phys. Rev. B **57**, 6165 (1998); Y. S. Lee *et al.*, Phys. Rev. B **60**, 3643 (1999).

[18] M. Matsuda *et al.*, Phys. Rev. B **61**, 4326 (2000); *ibid.* **62**, 9148 (2000); S. Wakimoto *et al.*, Phys. Rev. B **61**, 3699 (2000).

[19] P. Dai *et al.*, Phys. Rev. Lett. **80**, 1738 (1998); *ibid.* Science **284**, 1344 (1999).

[20] H. A. Mook *et al.*, Nature **395**, 580 (1998).

[21] H. A. Mook *et al.*, Nature **401**, 145 (1999).

[22] H. A. Mook *et al.*, Phys. Rev. Lett. **88**, 097004 (2002).

[23] H. A. Mook *et al.*, Nature **404**, 729 (2000).

[24] P. Dai *et al.*, Phys. Rev. B **63**, 054525 (2001).

[25] M. Arai *et al.*, cond-mat/9912233; *ibid.* Phys. Rev. Lett. **83**, 608 (1999).

[26] B. Grévin, Y. Berthier, and C. Collin, Phys. Rev. Lett. **85**, 1310 (2000).

[27] B. J. Suh *et al.*, Phys. Rev. B **61**, R9265 (2000); M.-H. Julien *et al.*, Phys. Rev. Lett. **84**, 3422 (2000).

[28] F. Borsa *et al.*, Phys. Rev. B **52**, 7334 (1995); B. Nachumi *et al.*, Phys. Rev. B **58**, 8760 (1998); C. Niedermayer *et al.*, Phys. Rev. Lett. **80**, 3843 (1998).

[29] M. Akoshima *et al.*, Phys. Rev. B **62**, 6761 (2000); J. Low Temp. Phys. **117**, 1163 (1999).

[30] A. Lucarelli *et al.*, Phys. Rev. Lett. **90**, 037002 (2003).

[31] D. N. Basov *et al.*, Phys. Rev. Lett. **74**, 598 (1995).

[32] T. Noda *et al.*, Science **286**, 265 (1999). A. Gupta *et al.*, Phys. Rev. B **61**, 11752 (2000).

[33] X. J. Zhou *et al.*, Science **286**, 268 (1999).

[34] R. P. Sharma *et al.*, Phys. Rev. Lett. **77**, 4624 (1996); Nature **404**, 736 (2000).

[35] E. Fawcett, H. L. Alberts, V. Yu. Galkin, D. R. Noakes, and J. V. Yakhmi, Rev. Mod. Phys. **66**, 25 (1994).

[36] G. Grüner, *Density Waves in Solids* (Perseus Publishing, Cambridge, MA, 1994).

[37] R. L. Withers and J. A. Wilson, J. Phys. C **19**, 4809 (1986).

[38] R. M. Fleming, D. E. Moncton, D. B. McWhan, and F. J. DiSalvo, Phys. Rev. Lett. **45**, 576 (1980).

[39] W. L. McMillan, Phys. Rev. B **12**, 1187 (1975).

[40] A. S. Davidov, *Solitons in Molecular Systems* (D. Reidel, Dordrecht, 185).

[41] G. Kotliar, Eur. J. Phys. B **11**, 27 (1999).

[42] A. Georges, G. Kotliar, W. Krauth and M. Rozenberg, Rev. Mod. Phys. **68**, 13 (1996).

[43] J. G. Bednorz and K. A. Muller, Z. Phys. B **64**, 189 (1986).

[44] N. F. Mott, *Metal-Insulator Transitions* (Taylor & Francis, London, 1974).

[45] S. Chakravarty, B. I. Halperin, and D. Nelson, Phys. Rev. Lett. **60**, 1057 (1988); Phys. Rev. B **39**, 2344 (1989).

[46] T. D. Stanescu and P. Phillips, cond-mat/0104478.

[47] P. W. Anderson, Science **235**, 1196 (1987).

[48] F. C. Zhang and T. M. Rice, Phys. Rev. B **37**, 3759 (1988).

[49] J. Zaanen and O. Gunnarsson, Phys. Rev. B **40**, 7391 (1989).

[50] H. J. Schulz, Europhys. Lett. **4**, 609 (1987); Phys. Rev. Lett. **64**, 1445 (1990).

[51] M. Inui and P. B. Littlewood, Phys. Rev. B **44**, 4415 (1991).

[52] D. Poilblanc and T. M. Rice, Phys. Rev. B **39**, 9749 (1989).

[53] K. Machida, Physica C **158**, 192 (1989); K. Machida and M. Ichioka, J. Phys. Soc. Jpn. **68**, 2168 (1999).

[54] S. R. White and D. J. Scalapino, Phys. Rev. Lett. **80**, 1272 (1998); *ibid.* **81**, 3227 (1998); Phys. Rev. B **60**, R753 (1999); Phys. Rev. B **61**, 6320 (2000).

[55] A. Moreo *et al.*, Phys. Rev. B **43**, 11442 (1991); H. Fehske *et al.*, Phys. Rev. B **44**, 8473 (1991); R. Valenti and C. Gros, Phys. Rev. Lett. **68**, 2402 (1992); D. Poilblanc, Phys. Rev. B **52**, 9201 (1995); M. Kohno, Phys. Rev. B **55**, 1435 (1997).

[56] E. Dagotto, Rev. Mod. Phys. **66**, 763 (1994).

[57] V. J. Emery, S. A. Kivelson, and H. Q. Lin, Phys. Rev. Lett. **64**, 475 (1990).

[58] A. L. Chernyshev, A. H. Castro Neto, and A. R. Bishop, Phys. Rev. Lett. **84**, 4922 (2000).

[59] G. B. Martins, C. Gazza, J. C. Xavier, A. Feiguin, and E. Dagotto, Phys. Rev. Lett. **84**, 5844 (2000); G. B. Martins, C. Gazza, and E. Dagotto, Phys. Rev. B **62**, 13926 (2000); G. B. Martins, J. C. Xavier, C. Gazza, M. Vojta, and E. Dagotto, ibid. **63**, 014414 (2001).

[60] A. L. Chernyshev, S. R. White, and A. H. Castro Neto, Phys. Rev. B **65**, 214527 (2002).

[61] S. R. White and D. J. Scalapino, Phys. Rev. B **60**, R753 (1999); B. Normand and A. P. Kampf, Phys. Rev. B **65**, 020509 (2002).

[62] B. P. Stojkoviċ, Z. G. Yu, A. L. Chernyshev, A. R. Bishop, A. H. Castro Neto, and N. Grnbech-Jensen, Phys. Rev. B **62**, 4353 (2000); B. P. Stojkoviċ, Z. G. Yu, A. R. Bishop, A. H. Castro Neto, and N. Grnbech-Jensen, Phys. Rev. Lett. **82**, 4679 (1999).

[63] B. Normand and A. Kampf, Phys. Rev. B **64**, 024521 (2001).

[64] F. Becca, L. Capriotti, and S. Sorella, Phys. Rev. Lett. **87**, 167005 (2001).

[65] M. Blume, V. J. Emery, and R. B. Griffiths, Phys. Rev. A **4**, 1071 (1971); U. Löw, V. J. Emery, K. Fabricius, and S. A. Kivelson, Phys. Rev. Lett. **72**, 1918 (1994).

[66] J. A. Hertz, Phys. Rev. B **14**, 1165 (1976).

[67] V. J. Emery *et al.*, Nature **374**, 434 (1995); *ibid.* cond-mat/0001077. V. J. Emery and S. A. Kivelson, Physica C **209**, 594 (1993); *ibid.* **235-240**, 189 (1994); *ibid.* Phys. Rev. Lett. **74**, 3253 (1995); Löw *et al.*, Phys. Rev. Lett. **72**, 1918 (1994).

[68] L. P. Pryadko *et al.*,Phys. Rev. B **60**, 7541 (1999).

[69] B. Büchner *et al.*, Phys. Rev. Let. **73**, 1841 (1994).

[70] Y. S. Lee, R. J. Birgeneau, M. A. Kastner, Y. Endoh, S. Wakimoto, K. Yamada, R. W. Erwin, S.-H. Lee, G. Shirane, Phys. Rev. B 60, 3643 (1999); B. Khaykovich *et al.*, Phys. Rev. B 67, 054501 (2003); A. T. Savici *et al.*, cond-mat/0202037.

[71] Moodenbaugh *et al.*, Phys. Rev. B **38**, 4596 (1988).

[72] Y. Koike *et al.*, Sol. St. Commun. **82**. 889 (1992); J. Phys. Soc. Jpn. **67**, 3653 (1998); J. Low. Temp. Phys. **105**, 317 (1996); *ibid.*, **117**, 1163 (1999); T. Adachi *et al.*, *ibid.* **117**, 1151 (1999); M. Akoshima *et al.*, Phys. Rev. B **57**, 7491 (1998).

[73] N. Kakinuma, Y. Ono, and Y. Koike, Phys. Rev. B **59**, 1491 (1999).

[74] K. Yamada *et al.*, J. Superconductivity **10**, 343 (1997).

[75] N. Hasselmann, A. H. Castro Neto, and C. Morais Smith, Europhys. Lett. **56**, 870 (2001).

[76] J. Lorenzana and G. Seibold, Phys. Rev. Lett. **89**, 136401 (2002).

[77] V. J. Emery *et al.*, Phys. Rev. Lett. **64**, 475 (1990).

[78] C. S. Hellberg and E. Manousakis, Phys. Rev. Lett. **78**, 4609 (1997).

[79] W. O. Puttika *et al.*, Phys. Rev. Lett. **68**, 538 (1992).

[80] A. L. Chernyshev and R. F. Wood, cond-mat/0208541 (unpublished).

[81] Y. Nagaoka, Phys. Rev. **147**, 392 (1966).

[82] S. R. White, I. Affleck, Phys. Rev. B **64**, 024411 (2001).

[83] S. Trugman, Phys. Rev. B **37**, 1597 (1988).

[84] A. H. Castro Neto, Z. Phys. B-Cond. Matter, **103**, 185 (1997).

[85] O. Zachar, Phys. Rev. B **65**, 174411 (2002)

[86] L. P. Pryadko, S. A. Kivelson, V. J. Emery, Y. B. Bazaliy, E. A. Demler, Phys. Rev. B **60**, 7541 (1999).

[87] I. Martin, G. Ortiz, A. V. Balatsky, and A. R. Bishop, Europhys. Lett. **56**, 849 (2001); Int. J. Mod. Phys. **14**, 3567 (2000).

[88] J. Voit, Rep. Prog. Phys. **58**, 977 (1995).

[89] N. Hasselmann, A. H. Castro Neto, C. Morais Smith, and Yu. A. Dimashko, Phys. Rev. Lett. **82**, 2135 (1999).

[90] P. C. Hohenberg, Phys. Rev. **158**, 383 (1967); N. D. Mermin and H. Wagner, Phys. Rev. Lett. **17**, 1133 (1966).

[91] N. Nucker *et al.*, Phys. Rev. B 37, 5158 (1988); J.-H. Guo *et al.*, Phys. Rev. B 49, 1376 (1994).

[92] P. C. Hammel, B. W. Statt, R. L. Martin, F. C. Chou, D. C. Johnston, and S-W. Cheong, Phys. Rev. B **57**, R712 (1998).

[93] A. Aharony, R. J. Birgeneau, A. Coniglio, M. A. Kastner, and H. E. Stanley, Phys. Rev. Lett. **60**, 1330 (1988).

[94] O. P. Vajk, P. K. Mang, M. Greven, P. M. Gehring, and J. W. Lynn, Science **295**, 1691 (2002).

[95] A. L. Chernyshev, Y. C. Chen, and A. H. Castro Neto, Phys. Rev. Lett. **87**, 067209 (2001); Phys. Rev. B **65**, 214527 (2002).

[96] D. Hone and A. H. Castro Neto, J. of Superconductivity, **10**, 349 (1997); A. H. Castro Neto and D. Hone, Phys. Rev. Lett. **76**, 2165 (1996).

[97] F. D. M. Haldane, Phys. Lett. **93 A**, 464 (1983).

[98] C. N. A. van Duin and J. Zaanen, Phys. Rev. Lett. **78**, 3019 (1997).

[99] H. Eskes, Y. Osman, R. Grimberg, W. van Saarloos, and J. Zaanen, Phys. Rev. B **58**, 6963 (1998); H. Eskes, R. Grimberg, W. van Saarloos, and J. Zaanen, Phys. Rev. B **54**, R724 (1996).

[100] M. den Nijs and K. Rommelse, Phys. Rev. B **40**, 4709 (1989).

[101] H. J. Schulz, Phys. Rev. B **34**, 6372 (1986).

[102] N. Hasselmann, A. H. Castro Neto, and C. Morais Smith, Phys. Rev. B **65**, 220511(R) (2002).

[103] C. Morais Smith, Yu. Dimashko, N. Hasselmann, and A. O. Caldeira, Phys. Rev. B **58**, 453 (1998).

[104] Yu. A. Dimashko, C. Morais Smith, N. Hasselmann, and A. O. Caldeira, Phys. Rev. B **60**, 88 (1999).

[105] R. M. Bradley and S. Doniach, Phys. Rev. B **30**, 1138 (1984).

[106] S. Bogner and S. Scheidl, Phys. Rev. B **64**, 054517 (2001).

[107] G. Xiao, M. Z. Cieplak, J. Q. Xiao, and C. L. Chien, Phys. Rev. B **42**, 8752 (1990).

[108] C. H. Ahn *et al.*, Science **284**, 1152 (1999).

[109] C. Morais Smith, N. Hasselmann, and A. H. Castro Neto, AIP Conf. Proceedings **554**, 209 (2001).

[110] G. Blatter *et al.*, Rev. Mod. Phys. **66**, 1125 (1994).

[111] H. Kimura *et al.*, Phys. Rev. B **59**, 6517 (1999); K. Hirota *et al.*, Physica B **241-243**, 817 (1998).

[112] B. Lake *et al.*, Science **291**, 1759 (2001); Nature **415**, 299 (2002).

[113] C. Di Castro, L. Benfatto, S. Caprara, C. Castellani, and M. Grilli, Physica C **341**, 1715 (2000).

[114] R. S. Markiewicz, J. Phys. Chem. Solids **58**, 1179 (1997).

[115] J. Ashkenazi, Int. J. Mod. Phys. B **13**, 3422 (1999).

[116] H. V. Kruis, Z. Nussinov, and J. Zaanen, cond-mat/0110055.

[117] V. J. Emery, E. Fradkin, S. A. Kivelson, and T. C. Lubensky, Phys. Rev. Lett. **85**, 2160 (2000); D. G. Barci and E. Fradkin, Phys. Rev. B **65**, 45320 (2002).

[118] A. H. Castro Neto, Phys. Rev. B **64**, 104509 (2001).

[119] R. J. McQueeney, Y. Petrov, T. Egami, M. Yethriraj, G. Shirane, and Y. Endoh, Phys. Rev. Lett. **82**, 628 (1999).

[120] Z.-X. Shen, A. Lanzara, S. Ishihara, N. Nagaosa, cond-mat/0108381 and references therein.

Chapter 10

TRANSPORT IN QUANTUM WIRES

Amir Yacoby
Department of Condensed Matter Physics, The Weizmann Institute of Science, Rehovot 76100, Israel.
amir.yacoby@weizmann.ac.il

Abstract In this chapter we review some of the recent experimental studies performed on isolated one-dimensional electron systems. Such systems offer a direct way to test the theoretical predictions of Luttinger liquid theory. We describe here several experimental configurations that enable to probe the conductance of clean and disordered wires, the tunneling density of states and the dispersion of elementary excitations in the wires. The effects of Coulomb interaction on each of these quantities will be discussed in the context of Luttinger liquid theory. We hope this chapter will convey the richness of possibilities that exists in the physics of interacting electrons in one dimension and in its experimental exploration.

Keywords: Cleaved Edge Overgrowth, Tunneling density of states, Spin-Charge separation, GaAs.

1. Introduction

One-dimensional (1D) electronic systems are expected to show unique transport behavior as a consequence of the Coulomb interaction between carriers [1]. Unlike in two and three dimensions [2], where the Coulomb interaction affects the transport properties only perturbatively, in 1D they completely modify the ground state from its well-known Fermi liquid form. The success of Landau Fermi liquid theory in two and three dimensions lays in its ability to lump the complicated effects of the Coulomb interaction into the Fermi surface properties (i.e. mass and velocity) of some newly defined particles known as quasi- particles [3]. Within this new description the quasi- particles are interacting only weakly and thus the underlying transport properties may still be described in terms of single-particle physics. However, in 1D the Fermi

D. Baeriswyl and L. Degiorgi (eds.), Strong Interactions in Low Dimensions, 321–346.

surface is qualitatively altered even for weak interactions [4] and, hence, Landau Fermi liquid theory breaks down. Today, it is well established theoretically that the low temperature transport properties of interacting 1D electron systems are described in terms of a Luttinger liquid (LL) rather than a Fermi-Liquid. This state is characterized by strong correlated electron behavior [5]. Of course, there can be no true long-range order in 1D due to the large quantum mechanical zero-point fluctuations of the electrons. The correlation functions thus decay algebraically in space and in time with exponents that depend continuously on the interaction strength [6, 7, 8, 9].

In recent years there have been numerous experiments that probe the physics of electrons confined to 1D. In this chapter we review some of the central ones that were performed on *isolated* 1D electron systems. We do not aim to review all the experiments on quantum wires but rather describe in detail a select set of experiments through which the manifestation of the intricate nature of Coulomb interactions in 1D will hopefully become clear. The experimental results will be compared to the predictions of Luttinger liquid theory; however, the theory described in detail in chapter 4, will not be derived here.

2. Technological Background

Recent developments in nano fabrication methods and synthesis techniques have opened up the possibility to investigate individual 1D systems. Traditionally, two types of approaches have been taken: 1) Electrostatic confinement of electrons to 1D; and 2) Synthesis of true 1D crystals.

One of the early approaches, based on the first method, was to laterally confine a two-dimensional electron gas (2DEG) into 1D [10, 11]. The high mobility 2DEG is grown by molecular beam epitaxy (MBE) using conventional modulation doping methods. Two metallic gates that are separated by a narrow gap are then evaporated onto the surface of the heterostructure containing the 2DEG. When a negative voltage is applied to the gates the 2DEG underneath them is depleted and a narrow 1D channel of electrons, connected to 2D reservoirs at the ends, is formed. Further increase in the applied bias on the gates will continuously tune the density of the 1D wires all the way to depletion. This approach extends the early work done on quantum point contacts in which two large regions of a 2DEG are connected through a narrow orifice [12]. Using this method, quantum wires, several microns long, were formed. Alternatively, one can etch away the 2DEG and form a 1D mesa structure [13]. This method is particularly useful for optical measurements. It avoids the use of metal gates, which absorb the incident and emitted light. The advantage of the split gate technique is that it allows a rather easy way to contact the 1D wire. However, the minimal width of such wires is determined by the distance of the gates to the 2DEG which in typical samples is larger than 100nm. Furthermore, the typical confining potential that restricts the motion to 1D in this geometry is rather shallow (a few meV in GaAs samples) which makes them rather susceptible to disorder. Another method that takes advantage of the planar geometry for contacting the 1D wire is the growth on v- shaped pre-patterned substrates [14, 15]. However, limitations in the growth process prevent this method from producing high mobility quantum wires.

An alternative approach that provides very strong electrostatic confinement and exceptional quality of wires is the method of cleaved edge overgrowth (CEO) [16, 17]. One begins by growing a modulation-doped quantum well (QW) with a desired width of a few nm (say 20nm). A tungsten metal gate is then evaporated on the top surface. This gate will later enable to separate the 2DEG that has formed in the QW into two parts that would serve as electrical contacts to the wire. The substrate is then reinserted into the MBE and is cleaved along the 110 direction. A second modulation doping sequence is immediately grown on the atomically flat cleavage plane that is perpendicular to the QW. This introduces electrons at the edge, thereby forming one or more edge states along the cleave. Strong overlap between the edge states and the 2DEG couple the two systems intimately along the entire edge. The 1D wire region is obtained by decoupling the edge states from the 2DEG with the help of the tungsten top gate, which after the cleave, extends exactly to the edge of the QW. The wire, thereby obtained, has a length which is determined by the tungsten gate and is, therefore, lithographically defined. The cross-sectional dimension of the wire is determined by the width of the QW and by the electric field binding the electrons to the cleavage plane (typically 10nm wide). Figure 1 shows a blowup of the critical device region under the various bias conditions. The strong confinement in this geometry results in subband separation in excess of 20meV and transport mean free path in excess of 20 μm. Most of the experiments described in this chapter are based on the CEO method.

Another approach to study individual 1D electron systems is the use of 1D crystals. The most mature approach is the use of carbon nanotubes [18, 19]. Recently there has been tremendous improvement in the synthesis of carbon nanotubes and one can reproducibly obtain single wall and multi wall nanotubes. Carbon nanotubes can be viewed as seamless tubes formed out of sheets of graphite. Depending on the way the

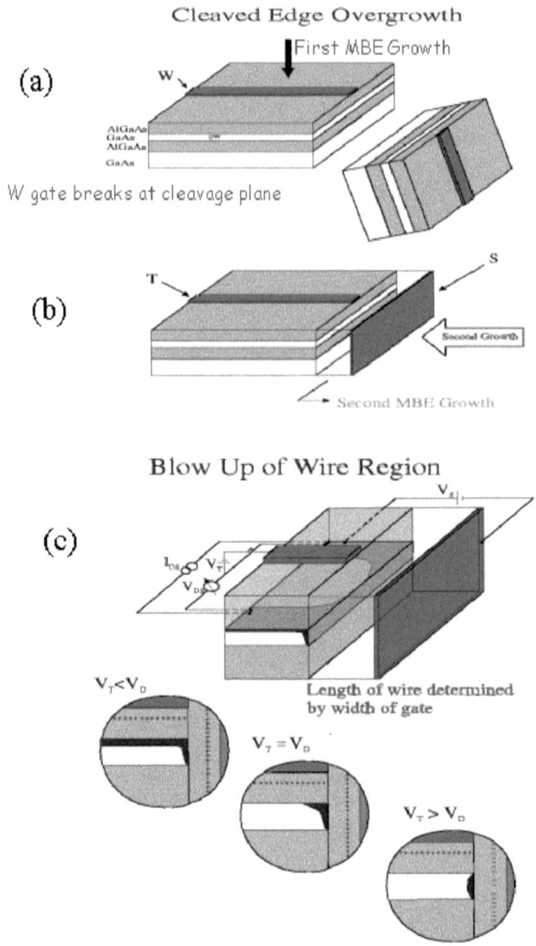

Figure 10.1. The cleaved edge overgrowth scheme - (a) - The first MBE growth of the QW and the in situ cleavage process. (b) - The second MBE growth on the atomically flat cleavage plane. (c) - A blowup of the critical device region showing the two dimensional contacts and the role of the top gate.

sheet is rolled one can obtain insulating, semiconducting and metallic carbon nanotubes. In this chapter we shall focus on measurements done on metallic tubes only.

Recently one dimensional crystals made of a variety of semiconducting material such as GaAs, InAs and ZnSe were formed using metal-catalyzed vapor-liquid-solid growth processes. This type of 1D wires offer many possibilities for basic and applied research in 1D since they can be doped to form n and p type 1D conductors and possibly even a combination of both within one wire. Furthermore, superlattice growth of different semiconducting materials has been demonstrated offering various band-gap engineering possibilities.

It should be emphasized that irrespective of the various realizations of 1D systems the physics described here is universal to all of them. The various wire systems provides access to a broad range of the relevant parameters in 1D, which include density of carriers, wire dimensions (i. e. lengths and cross section), disorder, and the strength of Coulomb interaction.

3. Conductance in clean wires

One of the fingerprints of a clean non-interacting 1D conductor is its quantized conductance in multiples of the universal value $G_Q = \frac{e^2}{h}$. This quantization results from an exact cancellation of the increasing electron velocity and the decreasing density of states as the carrier density increases [20, 21]. Therefore, as subsequent 1D electronic subbands are filled with electrons the conductance increases in a series of plateaus or steps with values equal to G_Q multiplied by the number of partly occupied wire modes (N). Surprisingly, the inclusion of interactions does *not* alter this prediction. Early papers [6, 7, 8, 9], considering infinitely long wires, did in fact predict quantization of conductance with a renormalized, non-universal, value of $G = gG_Q$, where $g \approx (1 + U/2E_F)^{-\frac{1}{2}}$. Here, U is the strength of the Coulomb interaction between neighboring electrons and E_F is the Fermi energy. The ratio U/E_F is proportional to r_s, the electron spacing, divided by the Bohr radius, so that g decreases with decreasing electron density. In practice, however, the wire is of finite length and for transport measurements one must connect the wire to large, Fermi liquid like, reservoirs. In the clean limit, the contact resistance to the reservoirs dominates the dc conductance and the universal value, G_Q , is restored [22, 23, 24]. The reason is that in DC transport, current is being carried by plasmons whose frequency, ω, is given by the external bias, eV. The nature of plasmons with a wavelength smaller than the wire length is indeed significantly modified by the presence of interactions in the wire. However, the wavelength of the emitted plasmons is given by v_F/ω (v_F is the Fermi velocity) and hence in the limit of DC transport such wavelength will exceed the wire length and the plasmons will primarily reflect the behavior in the contacts where Fermi liquid persists. Therefore, in the limit of DC transport one expects to observe the universal value of conductance. Oreg and Finkelstein [25], have further demonstrated that if one correctly calculates the conductance due to the self consistent electric field in the wire rather than the external electric field, the universal value is restored even for the infinite wire case. The universal nature of the conductance of a clean 1D wire makes it extremely difficult to determine the role of Coulomb interaction in experimentally realizable systems.

The first experiment to address the issue of ballistic conductance used the split gate method to study 2μm and 5μm long wires [10]. Their results show that the conductance is nearly quantized (within 5%) to the universal value as expected for clean wires. At low temperatures a weak power law dependence of the conductance could be detected. This power law dependence has been attributed to the presence of impurities in the wire and will be discussed in greater detail in Sec. 5. A different result has been obtained in experiments based on the CEO method [17]. In these experiments, studying the length dependence of the conductance and determining the disorder mean free path to be in excess of 20μm ruled out the presence of disorder. Surprisingly, the measured conductance deviated significantly from the expected universal value and was found to be an integer multiple of $\alpha\frac{e^2}{h}$ where α <1 may vary from 0.6 to 0.9 depending on the cross sectional dimensions of the wire (see Fig. 2).Furthermore, the parameter α decreases as the temperature decreases reaching a finite value when extrapolating to zero temperature. The origin of this non-universal conductance quantization has been traced to the way electrons are being coupled from the 2D reservoirs into the wire [26, 27].

The coupling into a 1D system can be done in one of two ways. The first is adiabatic coupling where the cross section of the wire tapers out continuously to two or three dimensions. Such geometry is possible only for the wires made using split gates or grown on pre-patterned substrates [14, 15]. However, it is not suitable for any of the other wire systems such as the ones made using the CEO method or the single crystal wires. The second alternative is the use of tunnel contacts, which would clearly lower the overall conductance as is being reflected by the parameter α in the CEO experiments. However, the breakdown of single-particle physics in 1D forces the injected electron to decompose into the naturally existing many-body excitations of the wire, which weakens the coupling of the wire to its surroundings. The coupling of electrons into and out of the 1D system, therefore, provides detailed information on the nature of interactions in 1D and will be discussed in detail in Sec. 4. Of course, four terminal measurements can provide a glimpse on the intrinsic conductance of the 1D system provided that the coupling of the voltage

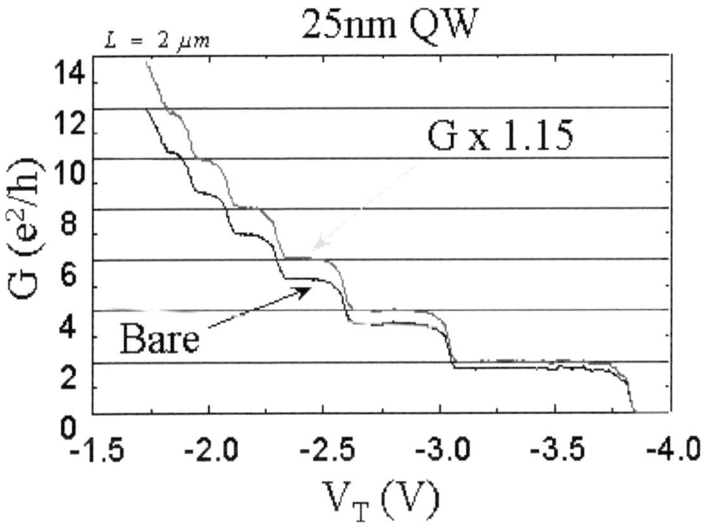

Figure 10.2. Conductance quantization in CEO wires - Conductance of a 2μm CEO wire as a function of the top gate voltage. The bare result is the measured conductance. The curve indicated by G × 1.15 corresponds to the expected universal steps. Therefore the measured conductance is an integer multiple of $\frac{1}{1.15}\frac{e^2}{h}$.

contacts to the wire is very weak. Four terminal measurements in wires fabricated using the CEO method indeed show [28] that despite the fact that the two terminal conductance is reduced below $\frac{e^2}{h}$, the propagation of electrons along the wires is ballistic resulting in a vanishing four terminal resistance.

4. Tunneling Density of States

Coulomb interaction in 1D results in the formation of long-range correlations that decay in space in a power law fashion [7]. In contrast to weakly interacting electrons in 2D and 3D where it is convenient to consider quasi-particles in momentum space, in 1D a real space picture is more appropriate. Heuristically, one can imagine that the electrons, in the strongly interacting limit, order in a Wigner like lattice where each electron tries to maximize its distance from its neighbors [5]. Of course true long-range order cannot exist in 1D due to quantum mechanical zero point motion, however, such real space description encompasses many of the unique features associated with the Luttinger liquid. For example, the tunneling density of states measures the probability for inserting one extra electron into the system. In the case of a true Wigner crystal, the ordered chain of electrons will have to split up and rigidly shift in order to make room for the tunneling electron. Therefore, the tunneling process alters the state of many of the electrons in the wire. Although each electronic state may only be slightly modified due to the tunneling electron, and hence have an overlap matrix element that nearly equals unity, the product of many such overlap matrix elements for all the electrons in the system will be strongly suppressed. It is therefore clear that the Coulomb interaction in 1D will suppress the tunneling density of states [7, 8, 29].

A formal solution to the problem of the tunneling density of states has been derived by Kane and Fisher based on Luttinger liquid theory [7]. For spinless electrons they find that the tunneling density of states into the center of a LL is given by $\nu_c(\varepsilon) \propto \varepsilon^{(\frac{1}{g}+g-2)/2}$ where ε is the energy of the tunneling electron and $g = \left(\sqrt{1+\frac{U}{2E_F}}\right)^{-1}$. Here U is the Coulomb interaction between neighboring electrons. Assuming the electrons tunnel from a Fermi liquid metal into a LL one finds that the corresponding current voltage *(I-V)* characteristics is given by $I \propto \int_o^V \nu_c(\varepsilon)d\varepsilon = V^{(\frac{1}{g}+g)/2}$ and the corresponding differential conductance is given by $\frac{dI}{dV} \propto V^{\alpha_c}$, where $\alpha_c = \frac{1}{2}(\frac{1}{g}+g-2)$. In the case of tunneling into the end of a LL a different exponent is obtained: $I \propto \int_o^V \nu_c(\varepsilon)d\varepsilon = \int_o^V \varepsilon^{\frac{1}{g}-1}d\varepsilon = V^{\frac{1}{g}}$, and the corresponding differential conductance is given by $\frac{dI}{dV} \propto V^{\alpha_e}$, where $\alpha_e = (\frac{1}{g}-1)$. In both cases, the non-interacting limit, given by *g=1*, corresponds to a linear *I-V* characteristics and hence a constant tunneling density of states. Tunneling across a single barrier in a LL can be considered as tunneling from one end of the LL to another. Hence, $I \propto \int_o^V \nu_e(\varepsilon)\nu_e(V-\varepsilon)d\varepsilon = V^{(\frac{2}{g}-1)}$ and $\frac{dI}{dV} \propto V^{\alpha_{e-e}}$, where $\alpha_{e-e} = 2\alpha_e = 2(\frac{1}{g}-1)$.

The tunneling density of states can be studied experimentally by measuring the current voltage characteristics of a tunnel junction to the center or end of a quantum wire or across a barrier within the wire [30, 31, 32, 33]. The parameter g may be deduced by fitting the power law behavior to the predicted one based on LL theory. In carbon nanotubes, there are two propagating modes and each mode has two spin directions [34]. Hence, there are altogether four degenerate modes. The Coulomb interaction produces one charge mode and three neutral modes that are unaffected by the Coulomb interaction. Hence, g in the above expressions should be replaced by $(g+3)/4$ and $1/g$ with $(1/g+3)/4$. The predicted value of g for the (10,10) armchair tubes varies from g=0.2 to 0.3.

A systematic study of the tunneling density of states to the center and to the end of a carbon nanotube was carried out by Yao et al [31]. A carbon nanotube was positioned on top of three metal electrodes. The weak coupling between the electrodes and the tube and the fact that the tube extends to both sides of the metal contact facilitates probing the tunneling density of states into the center of the tube. The measured I-V characteristics between two such metal contacts produced a power-law with an exponent α_c=0.34÷0.35. The deduced value of g=0.22 is in very good agreement with the predicted value [34]. A strong verification of the theory has been obtained by studying the tunneling conductance across an artificially created kink in the nanotube. The kink is produced by manipulating the nanotube using an atomic force microscope (AFM) [30]. Such a kink produces a weak link that dominates the conductance. It can be viewed as connecting two ends of a nanotube with a large tunnel barrier and produces a geometry that probes the tunneling density of states from one end of the nanotube to the other, namely α_{e-e}. The tunneling density of states across such a kink is, therefore, expected to give a power-law that is different from that measured using the metal contact, however, the deduced g is expected to be the same. In the experiments by Yao et al [31] the measured exponent was found to be $\alpha_{e-e} = 0.22$ and the corresponding g=0.18 in reasonable agreement with the one deduced from the tunneling into the center of the tube.

The discussion above assumes that the wire length is infinite and the temperature is lower than the applied voltage. However, in actual experiments these conditions are not necessarily met and one must consider the implications of having a wire with finite length and at finite temperature. The role of temperature is to produce a cutoff in the spatial extent of the Coulomb correlations in the wire [7]. Luttinger liquid theory predicts that the extent of these correlations, L_c, depends on the temperature according to $L_c = \frac{\hbar v_F}{k_B T}$, where v_F is the Fermi velocity. When the applied bias is larger than the temperature, $k_B T$ should be replaced by eV. As the temperature is lowered, a longer wire is needed in order to be able to follow the power-law behavior to lower and lower temperatures. Once $L_c > L$ the differential conductance will saturate at a value corresponding to $\frac{dI}{dV} \propto V^\alpha = \left(\frac{\hbar v_F}{L}\right)^\alpha$ (L being the length of the wire). In the experiments by Tarucha et al [10] such behavior was observed in the temperature dependence of the conductance through 2μm and 5μm long quantum wires. Finally we would like to briefly mention that another model system that behaves according to LL theory is the edge states of a fractional quantum Hall system. Due to the strong magnetic field, charge propagation is chiral and hence these 1D modes are termed chiral LL (CLL) [35]. The tunneling density of states into the edge of a 2DES subject to a strong magnetic field has been studied extensively using the CEO method [36, 37]. Although the measured tunneling exponent agrees with the theory when the filling factor is 1/3 there are significant deviations from the theoretical predictions once the filling factor is tuned away from 1/3. The origin of this discrepancy is a subject of current research.

5. Conductance in disordered wires

Disorder in 1D systems plays a crucial role even in the absence of electron-electron (e-e) interactions [38, 39, 40]. For example, localization of carriers, and the corresponding suppression of conductance is expected in both the interacting and non-interacting cases [39, 40]. Therefore, in order to isolate experimentally the role of interactions we focus here on the properties of a clean (ballistic) 1D system where even a single impurity in an otherwise perfectly clean wire will have a dramatic effect on the transport properties [6, 7, 8, 9].

The difference between a Luttinger liquid and Fermi- liquid becomes dramatic already in the presence of a single impurity even in the practical case of a finite wire [22, 23, 24, 25]. According to Landauer's transport theory [20, 21], the conductance of a single channel wire with a barrier is given by $G = \mid t \mid^2 \frac{e^2}{h}$, where $\mid t \mid^2$ is the transmission probability through the barrier. This result holds even at finite temperatures assuming the transmission probability is independent of energy, as is often the case for barriers that are sufficiently above or below the Fermi energy. In 1D, interactions play a crucial role in that they form charge density correlations. These correlations, similar in nature to charge density waves, are easily pinned by even the smallest barrier, resulting in *zero* conductance at zero temperature. A simple way to understand the role of interaction is to consider the effects of screening of a single impurity by the 1D electrons [8]. The screening is accompanied by Friedel oscillations that form around the impurity due to the sharp cutoff at the Fermi momenta. The oscillations have a $2k_F$ periodicity and in 1D decay only as $1/x$ (x being the distance from the impurity). Electrons propagating in the wire will not only be scattered by the impurity but also by the Friedel oscillations that formed around it. Had the amplitude of the Friedel oscillation been constant as a function of x, a gap in the spectrum would form exactly at the Fermi energy, similar to the Peierls instability [41], resulting in a insulating behavior. The fact that the oscillations decay as $1/x$ results only in a singularity in the density of states given by $\nu(\varepsilon) = \varepsilon^{\frac{2}{g}-2}$. Transport measurements [42], being sensitive to the density of states near the Fermi energy are, therefore, predicted to show a power law behavior of the conductance as a function of temperature, $G(T) = T^{\frac{2}{g}-2}$. In fact the suppression in the tunneling density of states at the end of a wire discussed in Sec. 4 can also be viewed as resulting from the weakly decaying Friedel oscillations formed at the sharp boundary of the wire and, therefore, give exactly the same power law dependence for the differential tunneling conductance.

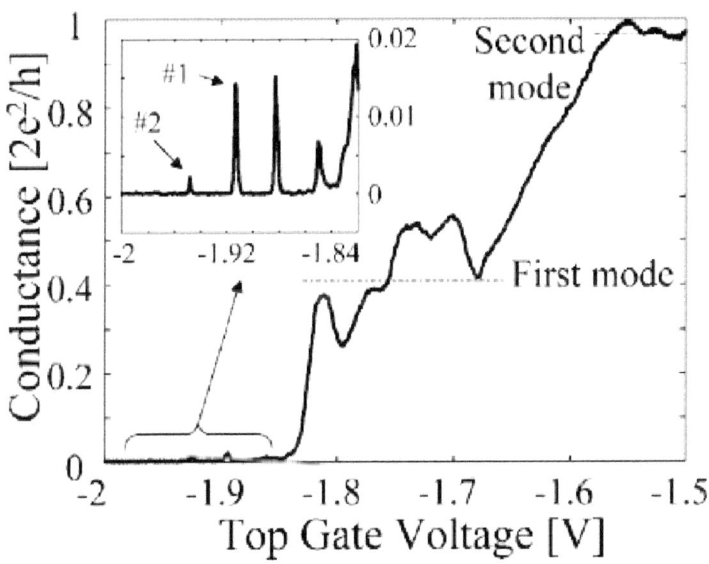

Figure 10.3. Conductance of a disordered CEO wire as a function of the top gate voltage. The disorder in the wire is apparent from the lack of conductance plateaus. Inset: A zoom-in of the conductance of the wire in the subthreshold region.

A particularly interesting case of disorder is the case of a double barrier where resonant tunneling (RT) occurs. The geometry we have in mind is not of a short 1D segment coupled via a tunnel barriers on each side to two long 1D wires. In the case of non interacting electron, the conductance due to RT between two Fermi liquid leads is easily calculated using the Landauer formula [43], $G_{FL} = \frac{e^2}{h} \int \mid t(\varepsilon) \mid^2 \frac{\partial f}{\partial \varepsilon} d\varepsilon$, where $\mid t(\varepsilon) \mid^2$ has the Breit-Wigner line shape centered on the resonant energy, ε_o , $\mid t(\varepsilon) \mid^2 = \frac{\Gamma_i^2}{(\varepsilon-\varepsilon_0)^2+\Gamma_i^2}$, f is the Fermi function and Γ_i is the intrinsic width of the level. When $k_B T >> \Gamma_i$ the case of interest here, one finds that $G_{FL} = \frac{e^2}{h}\Gamma_i \frac{\pi}{4k_BT} \cosh^{-2}(\frac{\varepsilon_0-\mu}{2k_BT})$ where μ is the chemical potential in the leads. The main outcome of this analysis is the line shape of the resonance being the derivative of the Fermi function, its full width at half maximum equals $3.53k_BT$, and the area under the peak (or the peak height multiplied by k_BT) is proportional to Γ_i. In the conventional theory of RT [44], Γ_i depends on the transmission probabilities through the individual barriers, which are independent of temperature and hence should lead to a peak area independent of temperature. In the case of interacting electrons one must consider RT between two Luttinger liquids. Here, the individual transmission probabilities are suppressed as the temperature is lowered (as in the case of tunneling into the end of a LL) [7, 8]. Therefore it is expected, and has been shown theoretically [45, 46], that the extracted Γ_i should drop to zero as $\Gamma_i \propto T^{\frac{1}{g}-1}$. The resonance line shape, however, in the case of $k_BT \gg \Gamma_i$, has been shown [45, 47] to be only slightly modified by the interactions and the change is too small to be detected experimentally. Such RT geometry has been recently realized using the CEO method where a 1D island a few hundreds of nm long is weakly connected to two 1D wire segments [48]. The conductance through several resonant states is shown in Fig. 10.3. Such sequence of resonances is well described by Coulomb blockade theory. The separation between peaks, which measures the charging energy, is estimated to be 2.2meV.

Figure 10.4 shows the extracted Γ_i for the peaks marked in Fig. 10.3. It is clear that Γ_i is not constant but rather drops as a power law of temperature. The extracted values of g for the two peaks are 0.82 (peak #1 in Fig. 10.3) and 0.74 (peak #2 in Fig. 10.3). The change in g results from the change in density induced in the 1D wire when moving from one peak to the next. Due to the spin degeneracy, the Coulomb interaction produces one charge mode characterized by an interaction parameter g_c and one neutral spin mode which is very weakly affected by the interactions with $g_s = 1$. The extracted g is related to g_c of the charge mode and g_s of the spin mode via :$\frac{1}{g} = \frac{1}{2}(\frac{1}{g_s} + \frac{1}{g_c})$. Therefore, we find that g_c=0.69 for peak #1 and g_c=0.59 for peak #2.

At sufficiently high temperatures the assumption of tunneling through a single resonant state breaks down and one would expect an increase in the extracted Γ due to transport through a few excited states of the 1D island [45, 49, 50, 51]. The possibility of an excited state affecting the temperature dependent conductance is of interest since it allows a better test of the Luttinger liquid model. The excited state spectrum of the 1D island is extracted from differential conductance measurements at finite source drain voltage, V_{ds}. Fig. 10.5a shows a gray scale plot of the

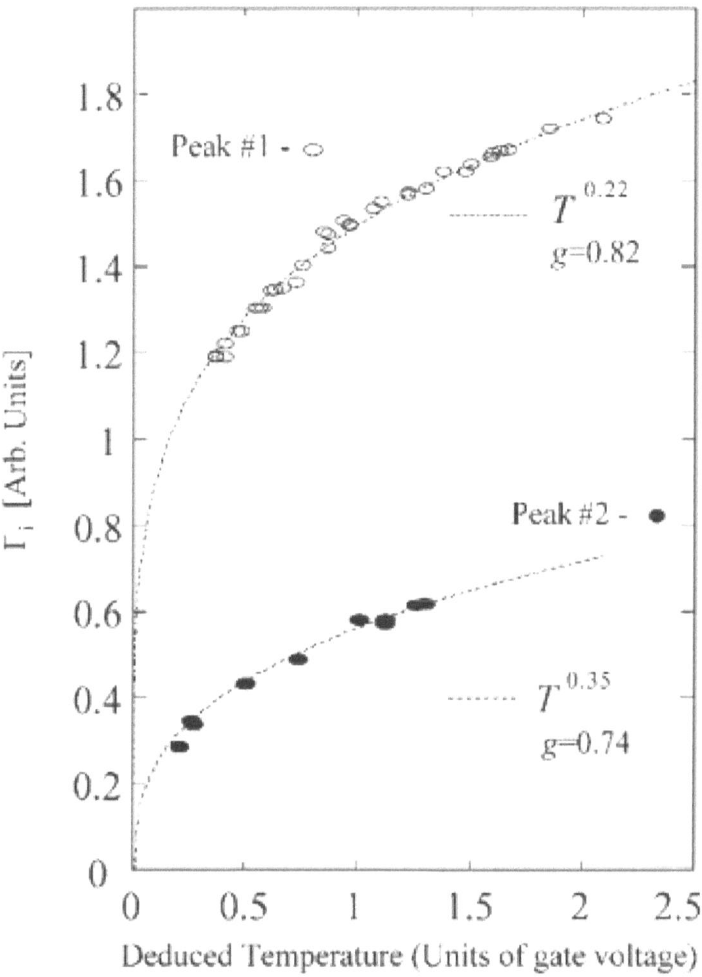

Figure 10.4. The intrinsic line width of the resonance Γi , vs temperature (in units of gate voltage). Both parameters are extracted from a fit to the derivative of the Fermi function. Γi is seen to decrease as a power law of the temperature indicating Luttinger liquid behavior. The dashed lines are a power law fit to the data.

differential conductance as a function of the top gate voltage and V_{ds}. For peak #1 (in Fig. 10.3) several excited states can be observed. The lowest three, at $V_{ds} = -0.4meV, V_{ds} = -0.7meV$, and $V_{ds} = -1.3meV$ are only very weakly coupled (approximately 15% of the intensity of the main peak) and would, therefore, contribute very little to the overall conductance. However, the fourth excited state at $V_{ds} = -1.6meV$ is more strongly coupled. Since an excited state contributes to the conductance only when $4k_BT \geq \Delta E$ (ΔE is the energy of the excited state), within the measured temperature range of 0.25K to 2.5K only the ground state contributes significantly and one expects a single power law behavior as is indeed observed in Fig. 10.4. It should be noted that theoretically, the Luttinger parameter can also be written in terms of the charging energy, U_c, and the level spacing, ΔE, as $g \approx (1+\frac{U_c}{\Delta E})^{-\frac{1}{2}}$ [42]. In the experiment described here the ratio of $\frac{U_c}{\Delta E} \approx 5$ and, hence, one expects $g \cong 0.4$. The large disagreement between the measured Luttinger parameter and the expected one is not understood at this stage.

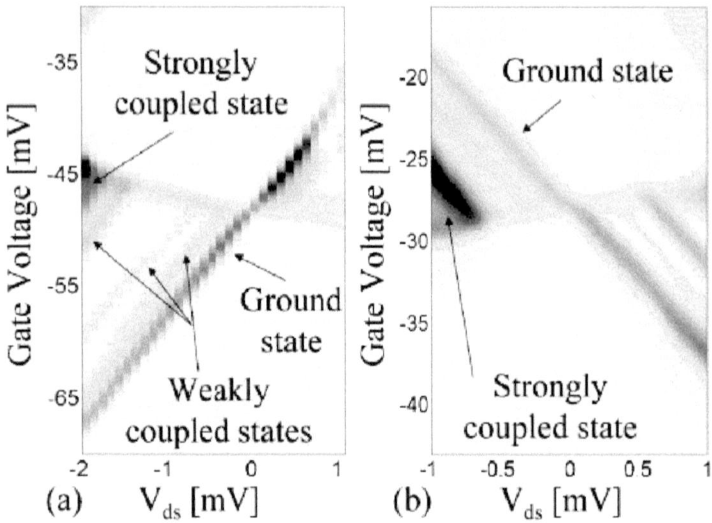

Figure 10.5. Gray scale plots of the nonlinear differential conductance of two peaks. Darker color stands for higher differential conductance. The scale is nonlinear in order to enhance low features. (a) The peak marked as #1 in Fig. 10.3 (V_{ds} is stepped with 100 μV intervals). (b) A different resonance that has a strongly coupled state at V_{ds} = 0.6 meV (V_{ds} is stepped with 20 μV intervals).

A different case is presented in Fig. 10.5b with a strongly coupled excited state at $V_{ds} = -0.6meV$. Here, one expects that at temperatures above 1.2K this excited state would contribute to the conductance. Fig. 10.6 shows the temperature dependence of the extracted Γ_i of this peak. Indeed above 1K, Γ_i deviates from the low temperature power law, indicating a contribution of an additional transport channel to the total conductance. At low temperatures though, only the ground state contributes to the conductance. Therefore, a power-law fit to the low temperature data enables to extract a g value of 0.66 for this wire. Using this g value and the measured energy of the excited state (-0.6meV from

Fig. 10.4) LL theory [45] was used to predict the dependence of Γ over the entire temperature range [48]. The dotted curve in Fig. 10.6 is the result of such a calculation where only the coupling strength to the excited state has been adjusted. We see that the temperature dependence predicted by the model agrees *quantitatively* with the measured dependence, further supporting the fact that Luttinger liquid behavior describes the transport properties of these resonances.

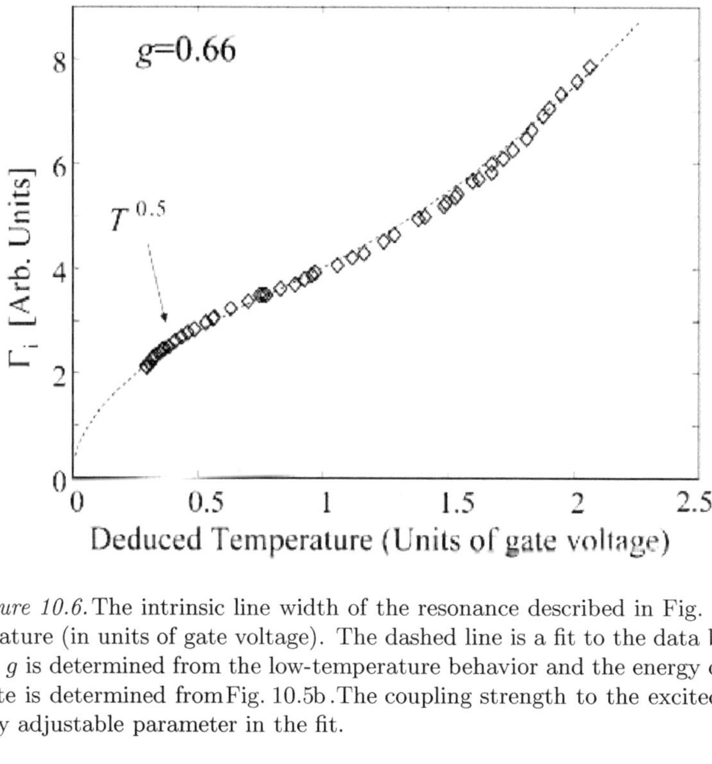

Figure 10.6. The intrinsic line width of the resonance described in Fig. 10.5b vs temperature (in units of gate voltage). The dashed line is a fit to the data based on Ref. 45. g is determined from the low-temperature behavior and the energy of the excited state is determined from Fig. 10.5b. The coupling strength to the excited state is the only adjustable parameter in the fit.

Recently, RT has also been observed in carbon nanotubes [52] and in the tunneling into the 1/3 edge state at the edge of a 2DEG subject to a high magnetic field [53]. Their results are also well described within the framework of LL theory [46, 54].

6. Excitation Spectrum

The validity of Fermi liquid theory in 2D and 3D assures that even in the presence of Coulomb interaction between the electrons, the low-lying excitations are quasi particles with charge e and spin 1/2. The

Coulomb interaction may dress up the single particle excitation such as to modify their propagation velocity, and effective mass compared to the non-interacting Fermi velocity, however, the single particle nature of the excitations remain [2]. Of course, collective modes such as plasmons may also exist in addition to the quasi-particle excitations.

1D electronic systems, on the other hand, have only collective excitations [1]. A unique property of 1D systems is that these collective modes decouple into two kinds: collective spin modes and collective charge modes. Coulomb interactions couple primarily to the latter, and thus strongly influence their dispersion. Conversely, the excitation spectrum of the spin modes is typically unaffected by interactions, and therefore remains similar to the non-interacting case. For example, charge excitations with Fermi momentum $\hbar k_F$ propagate with a velocity $v_p = v_F/g$. Thus, the stronger the Coulomb repulsion, the larger the propagation velocity. However, as was argued in Sec. 3, this unique excitation spectrum is not manifested in the transport properties of clean 1D systems. Furthermore, the decoupling of the spin and charge degrees of freedom will have only subtle effects on the transport properties of disordered wires such as to modify the power laws in the I-V characteristics and modify the excitation spectrum of a quantum dot embedded in a LL [34].

One of the direct ways to probe the excitation spectrum of clean 1D systems is through inelastic light scattering (Raman scattering) where light is scattered off the 1D electron system [55]. The incoming photons excite the elementary excitations of the wire and exit the wire with a different energy and momentum. The difference in energy and momentum of the incident and scattered photons corresponds to the energy and momentum of a specific excitation in the wire. By monitoring the different energies and momenta of the outgoing photons one can map out the dispersion of elementary excitations of the wire. The distinction between charge and spin excitations is achieved by performing the experiment with parallel and perpendicular polarizations of the incident and scattered light respectively. Numerous experiments in GaAs quantum wires have used inelastic light scattering to measure the dispersion of collective excitations in 1D [56, 57, 58, 59, 60]. As expected from LL theory, the collective spin mode was found to propagate with a velocity nearly equal to the Fermi velocity and the collective charge excitation was found to be significantly enhanced [61, 62]. Interestingly, in addition to the spin and charge dispersion branches, a third excitation branch has been observed in all the experiments. This branch was found to be insensitive to the relative polarization of the incident and scattered light and to propagate with the Fermi velocity suggestive of single particle excitations. This branch has been a puzzle for some time since it contradicted the predictions of LL theory, which states that single particle excitations in 1D do not exist. However, it was recently shown [62] that this mysterious branch originates from higher order terms in the inelastic light scattering cross section of the spin density excitation.

Inelastic light scattering is limited to excitations with relatively low momenta. The maximum momentum transfer is given by twice the momentum of the incident photon, which is typically much smaller than the Fermi momentum. Therefore, information on the complete dispersion including $2k_F$ excitations could not be obtained using this technique.

Recently, the complete dispersion curve of the elementary excitations in 1D has been measured in a clean 1D electron systems using a novel transport method [63]. Two parallel 1D wires have been fabricated by CEO from a GaAs/AlGaAs double quantum well (QW) heterostructure. The upper QW is 20nm wide, the lower one is 30nm wide and they are separated by a 6nm $Al_{0.3}Ga_{0.7}As$ barrier. A modulation doping sequence was used that renders only the upper QW occupied by a 2D electron gas with a density $n = 2 \cdot 10^{11} cm^{-2}$ and a mobility $\mu = 3 \cdot 10^6 cm^2/Vs$. The CEO creates two quantum wires in the two QW's along the whole edge of the sample. Both wires are tightly confined on three sides by atomically smooth planes and on the fourth side by a triangular potential formed at the cleaved edge. The 2DEG overlaps the upper wire (UQWR) and is separated from the lower wire (LQWR) by the tunnel barrier, 6nm thick, separating the two QW's (Fig. 10.7).

The measurement of a single, isolated tunnel junction between the wires is facilitated by controlling the density of electrons under the tungsten top gates. The 2μm wide gates are deposited prior to the cleave at a distance of 2μm from each other (see Fig. 10.7). First the gate is biased (g_2 in Fig. 10.7) while monitoring the 2-terminal conductance (G) between contacts to the 2DEG on opposite sides of the gate. When the voltage on g_2 relative to the 2DEG (V_2) is negative enough to deplete the 2DEG, G drops sharply because the electrons have to scatter into the wires in order to pass under g_2. Since tunneling into the LQWR is weak, most of the current is carried at this stage by the UQWR. Decreasing V_2 further depletes the modes in the UQWR under g_2 one by one, causing a stepwise decrease in G. Once again, a deviation of the size of the steps from the universal value is seen (see Sec. 3). When the last mode in the UQWR is depleted, only LQWR modes are left to carry current. This is seen in Fig. 10.8 as a very small step (typically a few percent of the UQWR conductance steps). Decreasing V_2 further, depletes these modes as well and G is suppressed to zero. A value of V_2 on the tunneling step is then chosen, forcing the electrons to tunnel between the wires. In order to

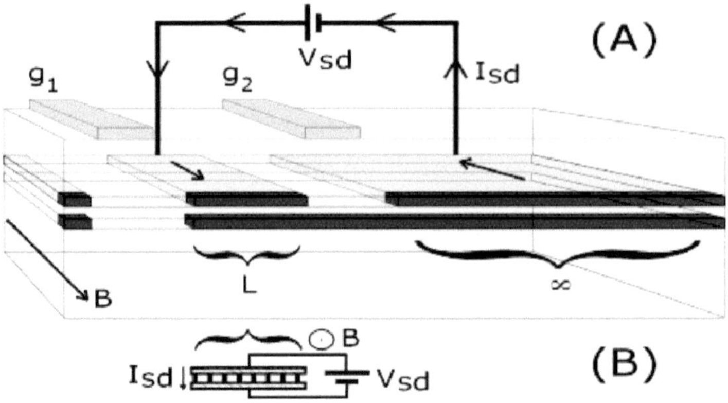

Figure 10.7. (A) Illustration of the double wire setup and the contacting scheme. The 1D wires span along the whole cleaved edge of the sample (front side in the schematic). A barrier separates the lower wire from the upper wire (dark gray) and the 2DEG that overlaps it (light gray). The 2DEG is used to contact both wires. Several tungsten top gates can be biased to deplete the electrons under them (only g_1 and g_2 are shown). A magnetic field is perpendicular to the wires plane. The depicted configuration allows the study of the conductance of a single wire-wire tunnel junction of length L. (B) Equivalent circuit of the measurement configuration: The current flows uniformly between the wires along the whole length of the junction.

focus on one of the two resultant junctions, the length of one of them is reduced by depleting both wires with an additional gate (g_1 in Fig. 10.7). Because the short tunnel junction is much more resistive than the long one, the UQWR between g_1 and g_2 is at electrochemical equilibrium with the source (the 2DEG lying between g_1 and g_2), while the LQWR is at electrochemical equilibrium with the drain (the semi-infinite 2DEG to the right of gate g_2 in Fig. 10.7). As a result, V_{sd} drops across an isolated wire-wire tunnel junction of length L (the distance between g_1 and g_2).

Measuring the tunneling current through a single tunnel junction allows to determine experimentally the dispersion relations of the elementary excitations in the wires [64, 65, 66, 67, 68, 69]. Being separated by a thick barrier, the wires are essentially independent of each other. The junction is long enough for tunneling to be treated as spatially invariant to a good approximation ($L \gg \lambda_F \sim 10$nm, where λ_F is a typical Fermi wavelength in the wires). Therefore, when an electron tunnels between the wires, not only its energy is conserved, but also its momentum. The energy of a tunneling electron is controlled by changing V_{sd}, while its momentum is controlled by changing B. To first order, B shifts the dispersions of the modes in the UQWR- $E_{ui}(B,k)$ - by $k_B = eBd/\hbar$ (ui enumerates the modes of the UQWR, d is the distance between the wires centers) relative to the dispersions of the modes in the LQWR- $E_{lj}(B,k)$ (lj enumerates the modes of the LQWR). Tunneling between the wires is suppressed unless there is a k that satisfies the tunneling condition $E_{ui}(B,k-k_B) = E_{lj}(B,k) - eV_{sd}$, for which one wire is occupied while the other is not. Thus, current is appreciable only in regions of the (V_{sd}, B) plane that are bounded by $B(V_{sd})$ that satisfies the tunneling condition with $|k|$ or $|k-k_B|$ equal to one of the Fermi wave numbers of one of the modes in the wires, $k_{F,ui}$ or $k_{F,lj}$. These boundaries show up as peaks in the nonlinear differential tunneling conductance $G(V_{sd}, B)$, i.e. the measurement directly probes the dispersion of one wire with the help of the Fermi points of the other wire (see Fig. 10.9).

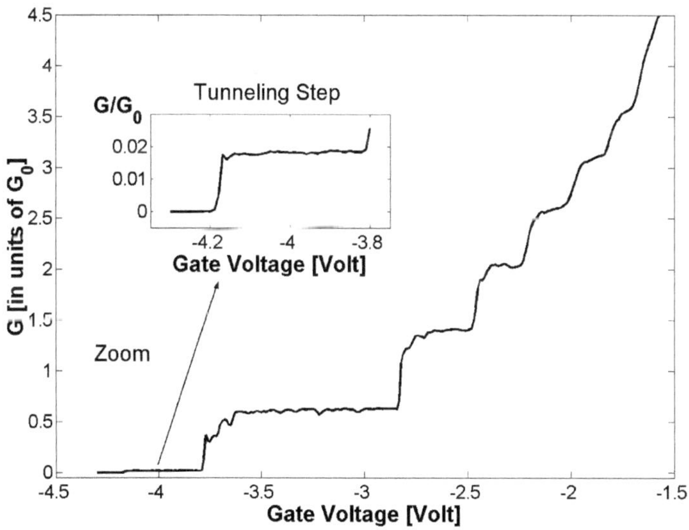

Figure 10.8. Conductance as a function of the top gate voltage. The conductance steps correspond to successive depletion of the upper wire modes. Inset - Once the upper wire is fully depleted (beyond the last conductance step) the remaining conductance is due to tunneling between the wires. The inset shows a blowup of this tunneling conductance.

It has been shown theoretically [67, 68] that in the presence of interactions this measurement technique provides direct information on the single-particle spectral functions within each wire. The spectral function, which is the Fourier transform of the one-particle retarded Green's function, measures the integrity of an electron as an elementary excitation in a many-body system. For non-interacting electrons, the spectral function is given by $A(k,\varepsilon) = 2\pi\delta(\varepsilon - \varepsilon_k)$ where ε_k is the dispersion of the electronic band. Coulomb interaction will typically broaden the spectral features. However, in systems obeying Fermi liquid theory, a distinct single-particle like spectral feature remains. In 1D, such single particle spectral features are completely absent. Instead, two spectral features appear, corresponding to the spin and charge excitations of the system. Thus measuring the spectral function of the wires corresponds to directly measuring the collective excitation spectrum of the interacting system. In contrast to the method of inelastic light scattering, the k-difference that one can induce between the wires has no fundamental limitation - it is easy to attain $k_B > 2k_F$ with reasonable B's. The result of a typical measurement is presented in Figs. 10.9 a,c. The most prominent features in such scans are the dispersion curves of the elementary excitations in the wires.

In order to determine the effect of the Coulomb interactions on the excitation spectrum, the measured excitation spectrum $G(V_{sd}, B)$ was compared with the expected excitation spectrum of non-interacting electrons. The latter is solely determined by the band mass m and by the density of electrons. Consider, for example, tunneling of non-interacting electrons between mode ui and mode lj. At $B = 0$, tunneling is expected to be significant only if the dispersion ui overlaps that of lj, otherwise it is impossible for an electron to tunnel whilst conserving its energy and momentum. To compensate for a density mismatch between the modes, a bias $eV_{sd} = E_{F,lj} - E_{F,ui}$ must be applied for the dispersions to overlap ($E_{F,ui}$ and $E_{F,lj}$ are the Fermi energies in ui and lj). Thus a point of enhanced current is expected on the V_{sd}-axis (see Fig. 10.9b). Several such points can be seen in Fig. 10.9 a,c, each corresponding to tunneling amongst a different pair of modes. This is especially clear in panel (c), where a smooth background has been removed from the raw data in order to improve the visibility of the dispersions. Tunneling between the modes is also possible near $V_{sd} = 0$. As B is ramped up the spectra are shifted by k_B relative to one another. Initially, tunneling remains rare because the available initial states are unoccupied. This persists until $k_{B1}^{i,j} \equiv edB_1^{i,j}/\hbar = |k_{F,ui} - k_{F,lj}|$, when initial states become available at the Fermi level, allowing tunneling between states propagating in the same direction. Increasing B further blocks tunneling because now all available final states are occupied until B satisfies: $k_{B2}^{i,j} \equiv edB_2^{i,j}/\hbar = |k_{F,ui} + k_{F,lj}|$, when empty final states become available at the Fermi level, allowing tunneling between states propagating

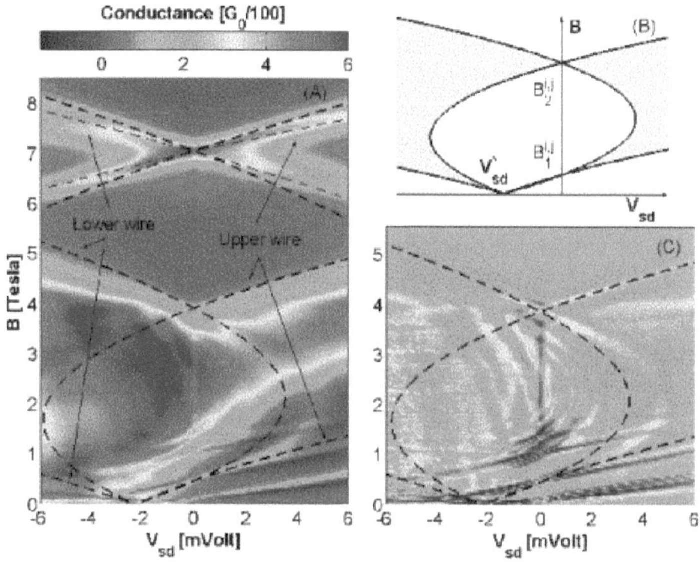

Figure 10.9. (A) G(V_{sd} ,B) as a function of V_{sd} and B for a 6μm long tunnel junction. The bar above the figure gives the color scale. Some of the dispersions can be easily discerned. The calculated noninteracting dispersions (thick dashed lines) do not describe the data well. A reduced mass model yields a superior fit everywhere, exemplified by the thin dashed lines at high B. The evident enhancement of G(V_{sd}, B) at $V_{sd} < 0$ and B $<$ 5T is attributed to tunneling directly from the 2DEG in the upper QW to the LQWR. Also visible is the suppression of G(V_{sd} ,B) near $V_{sd} = 0$. (B) A schematic of the regions of momentum conserving tunneling (gray) for which we expect the tunneling current to be enhanced. At $V^*{}_{sd}$, the density mismatch between mode ui and mode lj is compensated for and the spectra overlap. Tunneling between the Fermi points of the modes occurs at $B_1^{i,j}$ and $B_2^{i,j}$. The boundaries, which show up as peaks in G(V_{sd},B), give the dispersions of the electrons. (C) The lower part of (A) after subtracting a smoothed background. The scale has been optimized to improve the visibility. The dispersions of the various bands in both wires are easily discerned, as well as the suppression of G(V_{sd} ,B) around $V_{sd} = 0$.

in opposite directions. Thus enhanced tunneling is expected between the wires at $B_1^{i,j}$and $B_2^{i,j}$(Fig. 10.9b). These crossing points determine the Fermi momenta in each wire, hence allowing an independent measurement of the density of electrons in each mode, the only parameter needed to completely determine the non- interacting dispersion. Determining the density in this method does not depend on the strength of the Coulomb interactions. Their presence merely smears the crossing points in the (V_{sd}, B) plane, but does not shift them to different values of B. In reality such crossing points can be seen very clearly, as for example in Figs. 10.9a,c.

In practice the analysis is slightly complicated by the small changes of density in each mode induced by B as it is ramped up - in wires with finite cross-section the lower modes are populated at the expense of the higher ones. This effect is very small for most of the modes in Figs. 10.9a,c, but has a noticeable effect on the modes that make up the upper crossing point at $B_2^{1,1} = 7T$. The simplest quantitative interpretation of the $G(V_{sd}, B)$ scans requires the solution of the single particle Schrödinger equation for the levels in each of the two wires in the presence of B. From the solution one obtains the non- interacting dispersion of the electrons in each of the wires. Examples of such non-interacting dispersions are overlaid on the data in Figs. 10.9 a, c. One can clearly see that the density is determined correctly since the calculated dispersions fit the measured $B_{1,2}^{i,j}$'s. In spite of this, they clearly deviate everywhere else from the measured curves. Similar mismatches between measured and calculated dispersions are always observed, suggesting that the non-interacting behavior does not describe the excitation spectrum. In order to quantify the deviations from the non- interacting behavior, the data was fitted again with a non-interacting model with a renormalized mass $m* = 0.75m$. The main conclusion is that a substantial enhancement of the velocity of the collective excitations, namely $v_p/v_F = m/m*$ is observed. This enhancement is in line with the predictions of Luttinger liquid theory stating: $v_p/v_F = 1/g$ and hence g =0.75. This value is in agreement with the values obtained from the resonant tunneling experiments (Sec. 5) also done on similar wires. Near the crossing point at $B_2^{1,1} 7T$ one finds that the calculated non-interacting curves bound the regions of enhanced conductance (see Fig. 10.9a). Such behavior is expected to result from spin-charge separation [70, 71], in which the velocity of the spin degrees of freedom is given by the non-interacting velocity.

Recently, further verification for spin-charge separation was obtained from the intricate oscillatory patterns observed as a result of the finite length of the tunnel junction [72]. In CEO samples, where the elastic mean free path exceeds the length of the junction, the finite length of the junction is the main cause of breaking of translation invariance and hence momentum relaxation. The oscillations in B can be viewed as Aharonov-Bohm oscillations resulting from flux penetrating the area of the junction and are thus given by: $\Delta B \cdot A = \Delta B \cdot d \cdot L = \phi_0$, where L is the length of the junction and $\phi_0 = h/e$. The oscillations in V_{sd} can be viewed as resulting from an increase in momentum: $\Delta k \cdot L = \frac{e\Delta V_{sd} L}{\hbar v_c} = 2\pi$, where v_c is the velocity of the charge excitations. In case of spin-charge separation, two periodicities in V_{sd} are expected corresponding to the two excitation velocities. The oscillatory pattern will therefore form an additional beating pattern, ΔV_{slow} with $\frac{e\Delta V_{slow}L(v_c - v_s)}{\hbar v_c v_s} = 2\pi$. Such a beating has been recently observed [72] and the deduced charge and spin velocities agree once more with a $g = \frac{v_c}{v_s} = 0.7$.

7. Summary

In this chapter we reviewed some of the experimental work performed on isolated 1D systems in order to unravel the intricate nature of e-e interactions in 1D. Recent technological breakthroughs have clearly paved the way for even more controlled and sophisticated experiments but at the same time left a vast number of fundamental phenomena to be explained. One of the main topics that has received rather little attention experimentally and yet is at the heart of Luttinger liquid theory is the role of spin and the phenomena of spin-charge separation on 1D. Many theoretical predictions such as the dependence of the spin velocities on interaction strength remains to be tested and would therefore be a subject of fruitful research in the future.

References

[1] D. C. Mattis, in 'The Many Body Problem', World Scientific Publishing Co., 1992.

[2] B. L. Altshuler, A. G. Aronov in: Electron- Electron Interactions in Disordered Systems, Edited by A. L. Efros and M. Pollak.

[3] P. Nozières, ' Interacting Fermi Systems', Adison Willey Publishing Co., 1997.

[4] F. D. M. Haldane, J. Phys. **C 14**, 2585 (1981).

[5] L. I. Glazman, I. M. Ruzin, and B. I. Shklovskii, Phys. Rev. **B 45**, 8454 (1992).

[6] W. Apel and T. M. Rice, Phys. Rev. **B 26**, 7063 (1992).

[7] C. L. Kane and M. P. A. Fisher, Phys. Rev. Lett. **68**, 1220 (1992); C. L. Kane and M. P. A. Fisher, Phys. Rev. **B 46**, 15233 (1992).

[8] D. Yue, L. I. Glazman, and K. A. Matveev, Phys. Rev. **B 49**, 1966 (1994).

[9] M. Ogata and H. Fukuyama, Phys. Rev. Lett. **73**, 468 (1994).

[10] S. Tarucha, T. Honda, and T. Saku, Solid State Comm. **94**, 413 (1995).

[11] B. E. Kane, G. R. Facer, A. S. Dzurak, N. E. Lumpkin, R. G. Clark, L. N. Pfeiffer, and K. W. West, Appl. Phys. Lett. **72**, 3506 (1998).

[12] B. J. van Wees, H. van Houten, C. W. J. Beenakker, J. G. Williamson, L. P. Kouwenhoven, D. van der Marel, and C. T. Foxon, Phys. Rev. Lett. **60**, 848 (1988).

[13] J. S. Weiner, G. Danan, A. Pinczuk, J. Valladares, L. N. Pfeiffer, and K. West, Phys. Rev. Lett. **63**, 1641 (1989).

[14] D. Scheiner, Y. Hanein, and M. Heiblum, Semi. Sci. Tech. **12**, 1046 (1997).

[15] D. Kaufman, Y. Berk, B. Dwir, A. Rudra, A. Palevski, and E. Kapon, Physica **E 7**, 756 (2000).

[16] L. N. Pfeiffer, H. L. Stormer, K. W. Baldwin, K. W. West, A. R. Goni, A. Pinczuk, R. C. Ashoori, M. M. Dignam and W. Wegscheider, Journal of Crystal Growth, **849**, 127 (1993).

[17] A. Yacoby, H. L. Stormer, Ned S. Wingreen, K. W. Baldwin, L. N. Pfeiffer, and K. W. West, Phys. Rev. Lett. **77**, 4612 (1996).

[18] S. Iijima, Nature **354**, 56 (1991).

[19] Carbon Nanotubes, Edited by: M. Endo, S. Iijima, and M. S. Dresselhaus, Pergamon Press (1996).

[20] C. W. J. Beenakker and H. von Houten, in Solid State Physics, Semiconductor Heterostructures and Nanostructures, H. Ehrenreich and D. Turnbull editors, (Academic Press, New York 1991).

[21] Y. Imry, 'Introduction to Mesoscopic Physics', Oxford university Press, 1997.

[22] I. Safi and H. J. Schulz, Phys. Rev. **B 52**, R17040 (1995).

[23] V. V. Ponomarenko, Phys. Rev. **B 52**, R8666 (1995).

[24] D. L. Maslov, Phys. Rev. **B 52**, R14368(1995).

[25] Y. Oreg and A. M. Finkel'stein, Phys. Rev. **B 54**, 14255 (1996).

[26] R. de Picciotto, H. L. Stormer, A. Yacoby, L. N. Pfeiffer, K. W. Baldwin, K.W. West, Phys. Rev. Lett. **85**, 1730 (2000).

[27] M. Governale, M. Grifoni, and G. Schön, Phys. Rev **B 62**, 15996 (2000).

[28] R. de Picciotto, H. L. Stormer, L. N. Pfeiffer, K. W. Baldwin, K. W. West, Nature **411**, 51 (2001).

[29] S. Eggert, Phys. Rev. Lett. **84**, 4413 (2000).

[30] H. W. Ch. Postma, M. de Jonge, Z. Yao, and C. Dekker, Phys. Rev. **B 62**, 10653 (2000).

[31] Z. Yao, H. W. Ch. Postma, L. Balents, and C. Dekker, Nature **402**, 273 (1999).

[32] R. Tarkiainen, M. Ahlskog, J. Penttila, L. Roschier, P. Hakonen, M. Paalanen, and E. Sonin, Phys. Rev. **B 64**, 195412 (2001).

[33] E. Graugnard, P. J. de Pablo, B. Walsh, A. B. Ghosh, S. Datta, and R. Reifenberger, Phys. Rev. **B 64**, 125407 (2001).

[34] C. L. Kane, L. Balents, and M. P. A. Fisher, Phys. Rev. Lett. **79**, 5086 (1997).

[35] X. G. Wen, Phys. Rev. **B 43**, 11 025 (1991); Phys. Rev. Lett. **64**, 2206 (1990); Phys. Rev **B 44**, 5708 (1991); Int. J. Mod. Phys. **B 6**, 1711 (1992).

[36] M. Grayson et al., Phys. Rev. Lett. **80**, 1062 (1998).

[37] A. M. Chang, M. K. Wu, C. C. Chi, L. N. Pfeiffer, and K. W. West, Phys. Rev. Lett. **86**, 143 (2001).

[38] D. J. Thouless, Phys. Rep. **13 C**, 93 (1974).

[39] E. Abrahams, P. W. Anderson, D. C. Licciardello, and T. V. Ramakrishnan, Phys. Rev. Lett. **42**, 673 (1979).

[40] T. Giamarchi and H. J. Schulz, Phys. Rev. **B 37**, 325 (1988).

[41] R. E. Peierls, in 'Quantum theory of Solids', Oxford University Press, 2001.

[42] M. Bockrath, D. Cobden, J. Lu, A. G. Rinzler, R. E. Smalley, L. Balents, and P. L. McEuen, Nature **397**, 598 (1999).

[43] H. Grabert, M.H. Devoret, 'Single Charge Tunneling: Coulomb Blockade Phenomena in Nanostructures', H. Grabert and M.H. Devoret Editors, NATO Asi Series. Series B, Physics, Vol 294 (1992).

[44] L. P. Kouwenhoven et al., in Mesoscopic Electron Transport, edited by L. L. Sohn, L. P. Kouwenhoven, and G. Schön, NATO ASI (Kluwer Academic Publishers, Dordrecht,1996).

[45] A. Furusaki, Phys. Rev. **B 57**, 7141 (1998).

[46] Yu. V Nazarov and L. I. Glazman, condmat/0209090.

[47] I. J. Maasilta and V. J. Goldman, Phys. Rev. **B 55**, 4081 (1997).

[48] O. Auslaender, A. Yacoby, R. de Picciotto, K. W. Baldwin, L. N. Pfeiffer, and K. W. West, Phys. Rev. Lett. **84**, 1764 (2000).

[49] T. Kleimann, M. Sassetti, B. Kramer, and A. Yacoby, Phys. Rev. **B 62**, 8144 (2000).

[50] A. Braggio , M. Grifoi, M. Sassetti and F. Napoli, Euro Phys. Lett. **50**, 236 (2000).

[51] Yi-feng Yang and Tsung-han Lin, Phys. Rev. **B 64**, 233314 (2001).

[52] H. W. Ch. Postma, T. Teepen, Z. Yao, M. Grifoni and C. Dekker, Science **293**, 76 (2001).

[53] M. Grayson, D. C. Tsui, L. N. Pfeiffer, K. W. West, and A. M. Chang, Phys. Rev. Lett. **86**, 2645 (2001).

[54] C. Chamon and E. Fradkin, Phys. Rev. **B 56**, 2012 (1997).

[55] G. D. Mahan, 'Many-Particle Physics', Kluewer Academic/ Plenum Publishers, 2000.

[56] R. Goni et al., Phys. Rev. Lett. **67**, 3298 (1991).

[57] A. Schmeller et al., Phys. Rev. **B 49**, 14778 (1994).

[58] R. Schtrenz et al., Phys. Rev. Lett. **73**, 3022 (1994).

[59] C. Schuller et al., Phys. Rev. **B 54**, 17304 (1996).

[60] F. Perez et al., Phys Rev. **B 60**, 13310 (1999).

[61] M. Sassetti and B. Kramer, Phys. Rev. Lett **80**, 1485 (1998).

[62] B. Kramer and M. Sassetti, Phys. Rev. **B 62**, 4238 (2000).

[63] O. Auslaender, A. Yacoby, R. de Picciotto, K.W. Baldwin, L.N. Pfeiffer, and K.W. West, Science **295**, 825 (2002).

[64] J. P. Eisenstein, L. N. Pfeiffer, K. W. West, Appl. Phys. Lett. **58**, 1497 (1991).

[65] W. Kang, H. L. Stormer, L. N. Pfeiffer, K. W. Baldwin, K. W. West, Nature **403**, 59 (2000).

[66] A. Altland, C. H. W. Barnes, F. W. J. Hekking, and A. J. Schofield, Phys. Rev. Lett. **83**, 1203 (1999).

[67] U. Zulicke and M. Governale, Phys. Rev. **B 65**, 205304 (2002).

[68] D. Carpentier, C. Peça, and L. Balents, Phys. re. bf B 66, 153304 (2002).

[69] J. Voit, Journal of Electron Spectroscopy and Related Phenomena **117-118**, 469 (2001).

[70] L. Balents and R. Egger, Phys. Rev. **B 64**, 35310 (2001).

[71] M. Governale, D. Boese, U. Zülicke, and C. Schroll, Phys. Rev. **B 65**, 140403 (2002).

[72] Y. Tserkovnyak, B. I. Halperin, O. M. Auslaender, and A. Yacoby, Phys. Rev. Lett. **89**, 136805 (2002).

Chapter 11

TRANSPORT IN ONE DIMENSIONAL QUANTUM SYSTEMS

X. Zotos
University of Fribourg
and IRRMA (EPFL - PPH), 1015 Ecublens, Switzerland
Xenophon.Zotos@epfl.ch

P. Prelovšek
Faculty of Mathematics and Physics, University of Ljubljana,
and J. Stefan Institute, 1000 Ljubljana, Slovenia
peter.prelovsek@ijs.si

Abstract In this Chapter, we present recent theoretical developments on the finite temperature transport of one dimensional electronic and magnetic quantum systems as described by a variety of prototype models. In particular, we discuss the unconventional transport and dynamic - spin, electrical, thermal - properties implied by the integrability of models as the spin-1/2 Heisenberg chain or Hubbard. Furthermore, we address the implication of these developments to experimental studies and theoretical descriptions by low energy effective theories.

Keywords: one dimensional quantum many body systems, transport, integrability

1. Introduction

The electronic and magnetic properties of reduced dimensionality materials are significantly modified by strong correlation effects. In particular, over the last few years, the physics of quasi-one dimensional electronic systems, has been the focus of an ever increasing number of theoretical and experimental studies. They are realized as three dimen-

D. Baeriswyl and L. Degiorgi (eds.), Strong Interactions in Low Dimensions, 347–382.

sional (3D) compounds composed of weakly interacting chains or, in a latest and very promising development, as monoatomic width chains fabricated by self-assembly on surfaces.

Experimentally, recent studies made possible by the synthesis of new families of compounds characterized by very weak interchain coupling and low disorder, indicate unconventional transport and dynamic behavior; for example, unusually high thermal conductivity in quasi-one dimensional magnetic compounds [1, 2, 3], ballistic spin transport in magnetic chains [4, 5] or optical conductivity in quasi-1D organic conductors, showing a low frequency very narrow "Drude peak" even at relatively high temperatures [6, 7].

Theoretically, it is well known that one dimensional (1D) systems of interacting electrons do not follow the phenomenological description of the ordinary Landau Fermi liquids, but rather they are characterized by a novel class of collective quantum states coined Luttinger liquids [8]. Furthermore, it has quite recently been realized that several prototype models commonly used to describe 1D materials imply ideal transport properties (dissipationless) even at high temperatures. This phenomenon is the quantum analogue of transport by nondecaying pulses (solitons) in 1D classical nonlinear integrable systems [9].

Of course, 1D electronic and magnetic systems have, since the sixties, been a favorite playground where theoretical ideas were confronted with experimental results on an ever improving quality and variety of quasi-1D compounds. We are now in a position to claim reliable theoretical analysis on the thermodynamics, quantum phase transitions and spectral functions of prototype many body Hamiltonians used to describe 1D materials. The tools at hand range from exact analytical solutions (e.g. using the Bethe ansatz (BA) method) [10], the low energy Luttinger liquid approach and powerful numerical simulation techniques as the Quantum Monte Carlo (QMC) [11] and Density Matrix Renormalization Group (DMRG) [12] method. In particular, the ground-state properties as well as the low-temperature behavior of the correlation functions and of most static quantities in the scaling (universal) regime of Luttinger liquids have been extensively studied in recent decades and are well understood by now [13, 14, 15, 16].

On the other hand, although most experimentally relevant, less studied and understood are the transport and dynamic properties of 1D interacting electronic or magnetic systems. Paradoxically, while the equilibrium properties of prototype integrable models as the spin-1/2 Heisenberg or Hubbard model have been extensively analyzed, their finite temperature transport has attracted little attention; most studies till a few years ago, have basically focused on the low energy description in terms of the Luttinger model. A noticeable exception has been the issue of diffusive versus ballistic behavior and thermal conductivity of spin chains, a long standing and controversial issue. The difficulties encountered with the transport quantities can be attributed to the fact that the scattering and dissipation in clean 1D fermionic systems are not dominated by low-energy processes and thus the transport properties are not universal.

Presently, transport properties are at the focus of intense theoretical activity; in particular, prototype integrable models (as the Heisenberg, t-J, Hubbard, nonlinear-σ) are studied by exact analytical techniques (e.g. Bethe ansatz, form factor method) and numerical simulations. However, the complexity of these methods often renders the resulting behavior still controversial. Furthermore, the transport of quasi-one dimensional systems is (re-) analyzed within the effective Luttinger liquid theory or by semiclassical, Boltzmann type, approaches.

It is fair to say that the study of finite temperature/frequency conductivities in strongly correlated systems presents at the moment fundamental conceptual as well as technical challenges. Development of new analytical and numerical simulation techniques is required, as well as progress on the basic understanding of scattering mechanisms and their effects.

In the following, we will mostly concentrate on the conductivity of bulk, clean systems where the scattering mechanism is due to electronic or magnetic interactions (Umklapp processes). In particular, we will not address issues on the transport of mesoscopic systems (e.g. nanowires, nanotubes) or other dissipation mechanisms as coupling to phonons or disorder.

In section 2 we start by presenting some elements of linear response theory (or Kubo formalism), the theoretical framework commonly used for describing transport properties. Then, in section 3 and 4 we continue with a presentation of the state of the transport properties of prototype systems, in particular the Heisenberg and Hubbard model. In section 5, we present a short overview of alternative approaches based on low energy effective field theories as the Luttinger liquid, sine-Gordon and quantum nonlinear-σ models. Finally, in section 6, we close with a critical assessment of the present status and a discussion of open issues.

This presentation is definitely not an exhaustive account of theoretical studies on the transport properties of one dimensional quantum systems but it rather aims at presenting a coherent and self-contained view of some recent developments.

2. Linear response theory

In this section we introduce the basic definitions and concepts that will be used in the later development. The framework of most transport studies is linear response theory where the conductivities are given in terms of finite temperature (T) dynamic correlations calculated at thermodynamic equilibrium [17]. For instance, the real part of the electrical conductivity at frequency ω is given by the corresponding dynamic current correlation $\chi_{jj}(\omega)$,

$$\sigma'(\omega) = 2\pi D\delta(\omega) + \sigma_{reg}(\omega) \tag{11.1}$$

$$\sigma_{reg}(\omega) = \Re\frac{1}{i\omega}\chi_{jj}(\omega) \tag{11.2}$$

$$\chi_{jj}(\omega) = i\int_0^{+\infty} dt e^{izt}\langle[j(t),j]\rangle, \quad z=\omega+i\eta \tag{11.3}$$

with j the appropriate current operator. In a spectral representation the conductivity is,

$$\sigma_{reg}(\omega) = \frac{1-e^{-\beta\omega}}{\omega}\frac{\pi}{L}\sum_n p_n \sum_{m\neq n} \mid \langle n \mid j \mid m\rangle \mid^2 \delta(\omega-(\epsilon_m-\epsilon_n)), \tag{11.4}$$

$\mid n\rangle, \epsilon_n$ denoting the eigenstates and eigenvalues of the Hamiltonian, p_n the corresponding Boltzmann weights and β the inverse temperature (in the following we take $\hbar = \kappa_B = e = 1$); the *dc* conductivity is given by the limit $\sigma_{dc} = \sigma_{reg}(\omega \to 0)$. We will mostly discuss one dimensional tight-binding models on L sites where the current operator does not commute with the Hamiltonian.

To define the current operators we use the continuity equations of charge, magnetization or energy for the electrical, magnetic and thermal conductivity, respectively. We will explicitly present them below in the discussion of the Heisenberg and Hubbard models.

A quantity that presently attracts particular attention is the prefactor D of the $\delta-$function, named the Drude weight or charge stiffness. This quantity was introduced by W. Kohn in 1964 as a criterion of (ideal) conducting or insulating behavior [18] at $T = 0$ in the context of the Mott-Hubbard transition. This meaning becomes clear by noting that D is also the prefactor of the low frequency, imaginary (reactive - non-dissipative) part of the conductivity,

$$D = \frac{1}{2}[\omega\sigma''(\omega)]_{\omega\to 0} = \frac{1}{L}\left(\frac{1}{2}\langle -T\rangle - \sum_n p_n \sum_{m\neq n} \frac{|\langle n \mid j \mid m\rangle|^2}{\epsilon_m - \epsilon_n}\right); \tag{11.5}$$

here $\langle T\rangle$ denotes the thermal expectation value of the kinetic energy, generalizing the zero temperature expression to $T > 0$ by considering a thermal average. Thus, a finite Drude weight implies an "ideal conductor", a freely accelerating system. The second definition of the Drude weight follows from the familiar optical sum-rule [19, 20, 21] using eq.(11.1),

$$\int_{-\infty}^{+\infty} \sigma'(\omega) d\omega = \frac{\pi}{L}\langle -T\rangle, \tag{11.6}$$

with the average value of the kinetic energy replacing for nearest neighbor hopping tight binding models the usual ratio of density over mass of the carriers for systems in the continuum.

At $T = 0$ the Drude weight $D_0 = D(T = 0)$ is the central quantity determining charge transport. As already formulated by Kohn in a very physical way, D_0 can also be expressed directly as the sensitivity of the ground state energy ϵ_0 to an applied flux $\phi = eA$ $(e = 1)$,

$$D_0 - \frac{1}{2L}\frac{\partial^2 \epsilon_0}{\partial \phi^2}|_{\phi\to 0}. \tag{11.7}$$

For a clean system, since at $T = 0$ there cannot be any dissipation, one expects that $\sigma_{reg}(\omega \to 0) = 0$ and we have to deal with two fundamentally different possibilities with respect to D_0:

$D_0 > 0$ is characteristic of a *conductor* or *metal*,

$D_0 = 0$ characterizes an *insulator*.

The insulating state can originate from a filled electron band (usual band insulator) or for a non-filled band from electron correlations, that is the Mott-Hubbard mechanism; the latter situation is of interest here. Note, that the same criterion of the sensitivity to flux has been applied to disordered systems, in the context of electron localization theory [22].

The theory of the metal-insulator transition solely due to Coulomb repulsion (Mott transition) has been intensively investigated in the last decades by analytical and numerical studies [23] of particular models of correlated electrons and it is one of the better understood parts of the physics of strongly correlated electrons.

At finite temperatures, within the usual Boltzmann theory for weak electron scattering, the relaxation time approximation represents well the low frequency behavior,

$$\sigma(\omega) = \sigma_{dc}/(1 + i\omega\tau), \tag{11.8}$$

where the relaxation time τ depends on the particular scattering mechanism and is in general temperature dependent. In the following, we consider only homogeneous systems without any disorder, so the relevant processes in the solid state are electron-phonon scattering and the electron-electron (Coulomb) repulsion. When the latter becomes strong it is expected to dominate also the transport quantities.

Even in a metal with $D_0 > 0$ it is not evident which is the relevant scattering process determining $\tau(T)$ and $\sigma_{dc}(T)$. In the absence of disorder and neglecting the electron-phonon coupling the standard theory of purely electron-electron scattering would state that one needs Umklapp scattering processes to obtain a finite τ. That is, the relevant electron Hamiltonian includes the kinetic energy H_{kin}, the lattice periodic potential V and the electron-electron interaction H_{int},

$$H = H_{kin} + V + H_{int}. \tag{11.9}$$

Then, in general, the electronic current density j is not conserved in an Umklapp scattering process as the sum of ingoing electron momenta equals the sum of outgoing ones only up to a nonzero reciprocal vector G, $\sum_i k_i = mG$. In other words, the noncommutativity of the current with the Hamiltonian, $[H, j] \neq 0$, leads to current relaxation and thus, by the fluctuation-dissipation theorem, to dissipation. The interplay of V and H_{int}, however, turns out to be fairly involved in the case of strong electron-electron repulsion. This will become clear in examples of integrable tight binding models of interacting systems that we will discuss below, which have anomalous (diverging) transport coefficients.

Experiments on many novel materials, - strange metals - with correlated electrons, question the validity of the concept of a current relaxation rate $1/\tau$. Prominent examples are the superconducting cuprates with very anisotropic, nearly planar, transport [23] where the experimentally observed $\sigma(\omega)$ in the normal state can be phenomenologically described only by strongly frequency (and temperature T) dependent $\tau(\omega, T)$. The experiments on $\sigma(\omega)$ in quasi-1D systems are covered elsewhere [7].

With this background, we will now discuss different possible scenaria for the behavior of the $T > 0$ conductivity. A clean metallic system at $T = 0$ is characterized by a $\delta-$function Drude peak and a finite frequency part that vanishes, typically with a power law dependence, implying zero

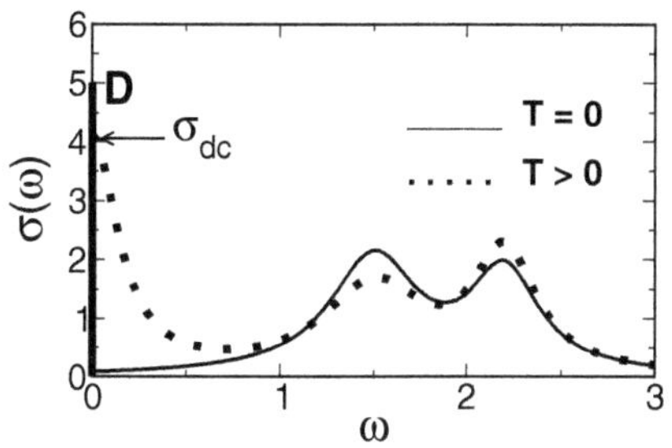

Figure 11.1. Schematic representation of the typical behavior of the conductivity for a clean metal.

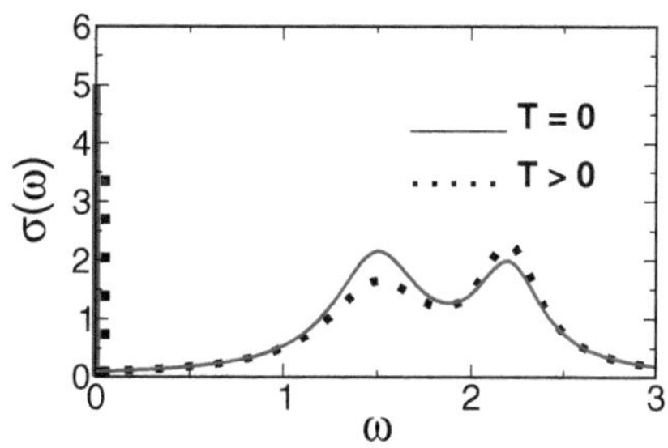

Figure 11.2. Illustration of the conductivity of a clean metal remaining an ideal conductor at finite temperatures.

dc regular conductivity. In the common sense scenario, at finite temperatures the $\delta-$function broadens to a "Drude peak" of width inversely proportional to a characteristic scattering time and thus a finite $\omega \to 0$ limit implying a finite *dc* conductivity. The scattering mechanisms can be intrinsic, due to interactions, or extrinsic due to coupling to other excitations, phonons, magnons etc. This typical behavior is shown in Fig. 11.1.

Actually, for a finite size system (as often studied in numerical simulations) D is nonzero even at finite T; it only goes to zero, typically exponentially fast, as the system size tends to infinity. Physically, this expresses the situation where the thermal scattering length is less than the system size.

But it is also possible that constraints on the scattering mechanisms limit the current decay, so that the system remains an ideal conductor ($D > 0$) even at $T > 0$. A schematic representation of a system remaining an "ideal conductor" at finite T is shown in Fig. 11.2.

In a system with disorder, D vanishes even at zero temperature and the *dc* residual conductivity is finite (provided the disorder is not strong enough to produce localization).

For an insulating system, e.g. due to interactions or the band structure as we discussed above, D vanishes at zero temperature; in the conventional case, D remains zero at $T > 0$, while activated carriers scattered via different processes give rise to a finite *dc* conductivity. But it is also possible that D becomes finite, turning a $T = 0$ insulator to an ideal conductor; for instance, a system of independent particles (e.g. one described within a mean field theory scheme), insulating due to the band structure, turns to an ideal conductor at $T > 0$. Finally, it is also conceivable that both D and σ_{dc} remain zero at $T > 0$, a system that can be called an "ideal insulator".

To the above scenaria we should add the possibility that the low frequency conductivity at finite temperatures is anomalous, e.g. diverging as a power law of the frequency, resulting in an infinite *dc* conductivity. Actually, as we will discuss later (Discussion section), this kind of behavior is fairly common in classical one dimensional nonlinear systems.

Thus, the first step in characterizing a system is the evaluation of the Drude weight at $T = 0$ in order to find out whether the system is conducting or insulating. The peculiarity that has recently been noticed is that most prototype models, assumed faithful representations of the physics of several quasi-one dimensional materials, have finite Drude weight also at finite temperatures (even $T \to \infty$), thus implying intrinsically ideal conductivity. In other words, interactions do not present a sufficient scattering mechanism to turn these systems into normal conductors. This behavior is unlike the one observed in the higher dimensional version of the same models, that become normal conductors at finite temperatures [24]. This unconventional behavior has been attributed to the integrability of these models.

To evaluate the Drude weight is not an easy matter as, although frequency independent, it represents a transport property and thus it cannot be obtained via a thermodynamic derivative (e.g. of the free energy). Direct calculation using the optical sum rule eq.(11.6) is obviously involved requiring the value of all current matrix elements. A very convenient and physical formulation is the one by W. Kohn, eq.(11.7), that generalized at finite temperatures [25] reads,

$$D = \frac{1}{2L} \sum_n p_n \frac{\partial^2 \epsilon_n(\phi)}{\partial \phi^2} |_{\phi \to 0}. \tag{11.10}$$

By considering the change of the free energy as a function of flux (that vanishes in the thermodynamic limit as it is proportional to the susceptibility for persistent currents) we can also arrive at an expression for the Drude weight as the long time asymptotic value of current-current correlations [26],

$$D = \frac{\beta}{2L} \sum_n p_n \langle n|j|n\rangle^2 = \frac{\beta}{2L} \langle j(t)j\rangle_{t\to\infty} \equiv \beta C_{jj}. \tag{11.11}$$

As an example, for a 1D tight binding free spinless fermion system with nearest neighbor hopping t, the application of a flux ϕ modifies the single particle dispersion to $\epsilon_k = -2t\cos(k+\phi)$ giving,

$$D_0 = \frac{t}{\pi}\sin(\pi n) = N(\epsilon_F) j_F^2$$
$$D(T) \sim D_0 - \frac{\pi t}{12}(\frac{T}{t})^2 \quad (n = \frac{1}{2}). \tag{11.12}$$

Here, n is the fermion density, $N(\epsilon_F)$ the density of states and j_F the current at the Fermi energy. Notice the quadratic decrease with temperature of the Drude weight that, as we will see later, is generic even for interacting one dimensional fermionic systems out of half-filling.

In the recent literature, that we will discuss below, the Drude weight of integrable systems is evaluated by the BA technique at zero or finite temperatures using the Kohn expression (11.10). The difficulty in this approach is the need for the estimation of finite size energy corrections of the order of $1/L$, a rather subtle procedure within this method.

Another approach, proved particularly efficient in establishing that systems with a finite Drude weight at finite temperature exist, uses an inequality proposed by Mazur [27]. This inequality states that if a system is characterized by conservation laws Q_n then:

$$\lim_{\mathcal{T}\to\infty} \frac{1}{\mathcal{T}} \int_0^{\mathcal{T}} \langle A(t)A\rangle dt \geq \sum_n \frac{\langle AQ_n\rangle^2}{\langle Q_n^2\rangle}. \tag{11.13}$$

Here $\langle \quad \rangle$ denotes a thermodynamic average, the sum is over a subset of conserved quantities Q_n orthogonal to each other in the sense $\langle Q_n Q_m\rangle = \langle Q_n^2\rangle \delta_{n,m}$, $A^\dagger = A$ and we take $\langle A\rangle = 0$.

Thus, for time correlations $\langle A(t)A\rangle$ with non-singular low frequency behavior we can obtain a bound for $C_{AA} = \lim_{t\to\infty}\langle A(t)A\rangle$,

$$C_{AA} \geq \sum_n \frac{\langle AQ_n\rangle^2}{\langle Q_n^2\rangle}. \tag{11.14}$$

For integrable systems, such as the spin-1/2 Heisenberg or Hubbard model that are known to possess nontrivial conservation laws because of their integrability, useful bounds can be obtained by considering just the first non-trivial conservation law. We should stress however that this approach has not provided yet a complete picture of the Drude weight behavior as we will discuss below in concrete examples.

Finally, another argument relating the behavior of the Drude weight to the (non-) integrability of a model is by the use of Random Matrix Theory [28, 29, 25]. It is known that integrable systems are characterized by energy level crossings upon varying a parameter and so Poisson statistics in the energy level spacing; thus it can be argued that the typical value of diagonal current matrix elements (slope of energy levels with respect to an infinitesimal flux) is of the order of one, plausibly implying a finite Drude weight according to eq.(11.11). On the contrary, nonintegrable systems, due to level repulsion, are described by Wigner or GOE statistics and thus the characteristic value of diagonal current matrix elements is of the order of e^{-L} (inversely proportional to the density of many body states) implying now a vanishing Drude weight as $L \to \infty$.

Besides electrical transport, the thermal conductivity of 1D systems has recently attracted particular interest; within linear response theory it is given by the analogous Green-Kubo formula expressed in terms of the energy current - energy current dynamic correlation function,

$$\kappa(\omega) = \Re \frac{\beta}{i\omega} \chi_{j^E j^E}(\omega). \tag{11.15}$$

Unlike the conductivity, there is no "mechanical force" (as the flux ϕ) that can be applied to the system in order to deduce expressions similar to the Drude weight, but the long time asymptotic value of energy current correlations has an analogous meaning.

Finally, in magnetic systems, the "spin conductivity" (spin diffusion constant) can be probed, for instance, by NMR experiments that measure at high temperatures the Fourier transform of spin-spin autocorrelations at the Larmor frequency ω_N,

$$S(\omega_N) = \int_{-\infty}^{+\infty} dt \int dq e^{i\omega_N t} \langle S_q^z(t) S_{-q}^z \rangle. \tag{11.16}$$

By using the continuity equation,

$$\omega^2 \langle S_q^z S_{-q}^z \rangle_\omega = q^2 \langle j_q^z j_{-q}^z \rangle_\omega \tag{11.17}$$

for a system where the total spin $z-$component is conserved, the spin-spin dynamic correlations can be analyzed via the corresponding spin-current correlations in analogy to electrical transport [30]; the role of local charge is played by the $z-$ component of the local magnetization (see next section for a more detailed discussion on this point).

We will now briefly discuss different methods, analytical and numerical, that are available for the study of finite temperature dynamic correlations in strongly interacting systems. Among the analytical approaches that have been used for the study of transport and dynamic properties of 1D systems, each has its own advantages and drawbacks. The traditional memory function approach [31] provides a complete picture of the temperature/frequency dependence but it is a perturbative method based on the assumption of a regular relaxation behavior that might be dangerous in 1D systems. The high temperature moment expansion provides useful information on the possibility of anomalous transport but the extraction of transport coefficients is also based on the phenomenological assumption of regular, diffusive behavior [32]. Progress in the exact evaluation of dynamic correlations in integrable systems has recently been achieved in the calculation of the Drude weight by the Bethe ansatz technique and of the frequency dependent conductivity by the form factor method. The Drude weight studies however are still controversial as they involve the calculation of finite size corrections, while the form factor approach has so far been limited to the calculation of zero temperature correlations and mostly in gapped systems. It is expected however that progress in BA techniques will provide a full picture of the dynamic properties of integrable systems. It is amusing to remark the paradoxical situation where the only strongly correlated systems for which we can probably have a complete solution of their dynamics are the integrable ones, which however, exactly because of their integrability, show unconventional behavior.

Among numerical simulation techniques, the ED (exact diagonalization) provides exact answers over the full temperature/frequency range but of course only on finite size systems [33]. Due to the exponentially growing size of Hilbert space, this limits the size of systems that can be studied to only about 20 to 30 sites, depending on the complexity of the Hamiltonian. We should also remark that, in principle, the full excitation spectrum is required for the evaluation of finite temperature correlations [1] Furthermore, the obtained frequency spectra are discrete, $\delta-$functions corresponding to transitions between energy levels, so that some ad-hoc smoothing procedure is needed; this is particularly crucial in attempting to extract the low frequency behavior. Nevertheless, finite size scaling in 1D systems can provide very useful hints on the macroscopic behavior, particularly at high temperatures where all energy levels are involved. This regime is the most favorable in attempting to simulate the physical situation where the scattering length is less or comparable to the system size.

The Quantum Monte Carlo techniques allow the study of far larger systems and they provide directly the dynamic correlations at finite temperatures but in imaginary time [35]. By analytical continuation, using

[1] In a recent advance, finite temperature dynamic correlations for a prototype model have been successfully evaluated using only one quantum state (microcanonical ensemble) [34].

for instance the Maximum Entropy procedure, one is able in principle to extract the main features of the frequency dependence; experience shows however, that fine issues as the temperature dependence of the Drude weight or the presence of diffusive behavior which is a low frequency property, are difficult to establish reliably.

Finally, the DMRG method that has been so successful in the study of ground state and thermodynamic properties of 1D systems, has only recently been extended to the reliable study of zero temperature conductivities in gapped systems [36, 37]. At finite temperatures it is also possible to obtain very high accuracy data on autocorrelation functions in imaginary time by the use of the transfer matrix DMRG [38]. However, similarly to QMC methods, it is very difficult to extract subtle information on the finite T dynamics because of the extremely singular nature of analytic continuation that hides the useful information even for practically exact imaginary time data.

3. Heisenberg model

The prototype model for the description of localized magnetism is the Heisenberg model. For a one dimensional system the minimal Hamiltonian describing magnetic insulators is,

$$H = \sum_l h_l = J \sum_{l=1}^{L} (S_l^x S_{l+1}^x + S_l^y S_{l+1}^y + \Delta S_l^z S_{l+1}^z) \tag{11.18}$$

where S_l^α $(\alpha = x, y, z)$ are spin operators on site l ranging from the most quantum case of spin S=1/2 to classical unit vectors. For S=1/2 the system is integrable by the Bethe ansatz method and its ground state, thermodynamic properties and elementary excitations have well been established [10]. As a brief reminder to the discussion that follows, note that for $J > 0$, $\Delta > 0$ corresponds to an antiferromagnetic coupling while $\Delta < 0$ to a ferromagnetic one; a canonical transformation maps $H(\Delta)$ to $-H(-\Delta)$. Further, the anisotropy parameter Δ describes two regimes, the "easy-plane" for $|\Delta| < 1$ or the "easy-axis" for $|\Delta| > 1$. The isotropic case, occuring in most materials for symmetry reasons, corresponds to $\Delta = 1$. For $|\Delta| \leq 1$ the system is gapless and characterized by a linear spectrum at low energies, while for $\Delta > 1$ a gap opens; in particular, at $\Delta = 1$ the elementary excitation spectrum is described by the "des Cloiseaux-Pearson" dispersion $\epsilon_q = \frac{J\pi}{2}|\sin q|$. For $\Delta < -1$ there is a transition to a ferromagnetic ground state.

In general, other types of terms appear in the description of quasi-1D materials such as longer range or on site anisotropy interactions, but in this review we will focus on the prototype model eq.(11.18).

At this point we should mention that the spin-1/2 Heisenberg model is equivalent to a model of interacting spinless fermions (the "t-V" model) obtained by a Jordan-Wigner transformation [39];

$$H = (-t)\sum_{l=1}^{L}(c_l^\dagger c_{l+1} + h.c.) + V\sum_{l=1}^{L}(n_l - \frac{1}{2})(n_{l+1} - \frac{1}{2}), \qquad (11.19)$$

where $c_l(c_l^\dagger)$ denote annihilation (creation) operators of spinless fermions at site l and $n_l = c_l^\dagger c_l$.

The correspondence of parameters is $V/t = 2\Delta$ and the opening of a gap at $\Delta \geq 1$ corresponds to an interaction driven metal-insulator (Mott-Hubbard type) transition.

3.1 Currents and dynamic correlations

Regarding the transport and dynamic properties of the Heisenberg model, three cases have mostly been discussed: the classical one, the spin S=1 and the spin S=1/2; the S=1 case has been extensively analyzed by mapping its low energy physics to a field theory [40], the nonlinear-σ model (see section 5). In connection to experiment, the main issue is the diffusive vs. ballistic character of spin transport as probed for instance by NMR experiments and recently the contribution of magnetic excitations to the thermal conductivity of quasi-one dimensional materials [1].

To discuss magnetic transport, we must first define the relevant spin j^z and energy j^E currents by the continuity equations of the corresponding local spin density S_l^z (provided the total S^z component is conserved) and local energy h_l;

$$S^z = \sum_l S_l^z, \qquad \frac{\partial S_l^z}{\partial t} + \nabla j_l^z = 0, \qquad (11.20)$$

gives for the spin current,

$$j^z = \sum_l j_l^z = J\sum_l (S_l^x S_{l+1}^y - S_l^y S_{l+1}^x). \qquad (11.21)$$

Here and thereafter, $\nabla a_l = a_l - a_{l-1}$ denotes the discrete gradient of a local operator a_l. In general ($\Delta \neq 0$) the spin current does not commute with the Hamiltonian, $[j^z, H] \neq 0$, so that nontrivial relaxation is expected and thus finite spin conductivity at $T > 0$.

Similarly, the energy current j^E is obtained by,

$$j^E = \sum_l j_l^E, \quad \frac{\partial h_l}{\partial t} + \nabla j_l^E = 0, \tag{11.22}$$

$$\begin{aligned} j^E &= J\sum_l (S^x_{l-1}S^z_l S^y_{l+1} - S^y_{l-1}S^z_l S^x_{l+1}) + \Delta(S^y_{l-1}S^x_l S^z_{l+1} - S^z_{l-1}S^x_l S^y_{l+1}) \\ &+ \Delta(S^z_{l-1}S^y_l S^x_{l+1} - S^x_{l-1}S^y_l S^z_{l+1}) \end{aligned} \tag{11.23}$$

We will now briefly comment on the framework for discussing spin dynamics and in particular how it is probed by NMR experiments. According to the spin diffusion phenomenology (for a detailed description see ref. [41]) when we consider the (q,ω) correlations of a conserved quantity $A = \sum_l A_l$, such as the magnetization or the energy, it is assumed that it will show a diffusive behavior in the long-time $t \to \infty$, short wavelength $q \to 0$ regime [2]. In the language of dynamic correlation function, diffusive behavior means that the time correlations decay as,

$$\langle\{A_l(t), A_0(0)\}\rangle = 2\chi_A T \int \frac{dq}{2\pi} e^{iql - D_A q^2|t|} \tag{11.24}$$

where D_A, χ_A are the corresponding diffusion constant and static susceptibility, respectively. For a 1D system, this behavior translates to a characteristic $1/\sqrt{t}$ dependence of the autocorrelation function.

Fourier transforming the above expression we obtain,

$$S_{AA}(q,\omega) = \int_{-\infty}^{+\infty} dt\; e^{i\omega t}\frac{1}{2}\langle\{A_q(t), A_{-q}(0)\}\rangle \sim \frac{\chi_A D_A q^2}{(D_A q^2)^2 + \omega^2}. \tag{11.25}$$

By using the continuity equation (11.20), this Lorentzian form can be further modified to obtain the current-current correlation function,

$$S_{j^A j^A}(q,\omega) \sim \frac{\chi_A D_A \omega^2}{(D_A q^2)^2 + \omega^2} \tag{11.26}$$

which gives the diffusion constant D_A by taking the $q \to 0$ limit first and then $\omega \to 0$.

On the other hand, a ballistic behavior is signaled by a $\delta-$function form, $S_{j^A j^A}(q,\omega) \sim \delta(\omega - cq)$, where c is a characteristic velocity of

[2] This phenomenological statement goes under the name of Ohm's law in the context of electrical transport, Fourier's law for heat or Fick's law for diffusion.

the excitations. This $\delta-$function peak moves to zero frequency as $q \to 0$ and its weight is proportional to the long time asymptotic of the current-current correlations

$$C_{j^A j^A} = S_{j^A j^A}(q=0, t \to \infty). \tag{11.27}$$

The above anticommutator correlations are related to the imaginary part of the susceptibility $\chi(q,\omega)$, that describes the dissipation, by,

$$S_{AA}(q,\omega) = \coth(\frac{\beta\omega}{2})\chi''_{AA}(q,\omega). \tag{11.28}$$

In relation to the experimental study of spin dynamics, the NMR has developed to a very powerful tool; for instance, the $1/T_1$ relaxation time is directly related to the spin-spin autocorrelation by,

$$\frac{1}{T_1} \sim |A|^2 \int_{-\infty}^{+\infty} dt \cos(\omega_N t)\langle\{S_l^z(t), S_l^z(0)\}\rangle \tag{11.29}$$

where $|A|^2$ is the hyperfine coupling [4] and ω_N the Larmor frequency. Using the relation (11.28), $1/T_1$ gives information (in the high temperature limit, $\beta\omega_N \to 0$) on $\chi''(q,\omega)$ as,

$$\frac{1}{T_1} \sim T|A|^2 \sum_q \frac{\chi''(q,\omega_N)}{\omega_N}. \tag{11.30}$$

The diffusive behavior, characterized by the $1/\sqrt{t}$ decay of the spin correlations, is extracted in an NMR experiment by analyzing the $q \to 0$ contribution [5]. It gives a $1/\sqrt{\omega_N}$ behavior that is detected as a $1/\sqrt{H}$ magnetic field dependence,

$$\frac{1}{T_1} \sim \frac{1}{\sqrt{\omega_N}} \sim \frac{T\chi(q=0)}{\sqrt{D_s H}}, \tag{11.31}$$

considering that the Larmor frequency $\omega_N \sim H$, D_s being the spin diffusion constant and $\chi(q=0)$ the static susceptibility.

3.2 Spin and energy dynamics

Returning now to the state of spin and energy dynamics, the classical Heisenberg model has been extensively studied by numerical simulations, the first studies dating from the 70's [42]. Nevertheless, the issue of diffusive behavior (even at $T=\infty$ where most simulations are carried out) still seems not totally clear, the energy and spin showing distinctly different dynamics. On the one hand, simulations clearly indicate that energy transport is diffusive [43] but on the other hand, the decay of spin autocorrelations is probably inconsistent with the expected $1/\sqrt{t}$ law [44, 43] exhibiting long-time tails.

On the other extreme, for the fully quantum spin S=1/2 model, the simplest case is the $\Delta = 0$, so called XY limit. Here, the spin current commutes with the Hamiltonian resulting in ballistic transport; this can also be seen in the fermionic, t-V, version of model that corresponds to free spinless fermions ($V/t = 0$ in eq.(11.19)) where now the charge current is conserved. In the infinite temperature limit ($\beta = 0$) the spin and energy autocorrelations can be calculated analytically using the Jordan-Wigner transformation and are of the form [45]:

$$\langle S_l^z(t) S_l^z \rangle = \frac{1}{4} J_0^2(Jt) \tag{11.32}$$

$$\langle h_l(t) h_l \rangle = \frac{J^2}{8} \left(J_0^2(Jt) + J_1^2(Jt) \right) \tag{11.33}$$

which both behave as $1/t$ for $t \to \infty$, unlike the $1/\sqrt{t}$ form in the diffusion phenomenology (J_0, J_1 are Bessel functions). Actually the $\beta = 0$ limit, often theoretically analyzed for simplicity, is not unrealistic as the magnetic exchange energy J can be of the order of a few Kelvin in some materials.

For $|\Delta| < 1$ the Drude weight at $T = 0$ has been calculated using the BA method [46, 21] and is given by,

$$D_0 = \frac{\pi}{8} \frac{\sin(\pi/\nu)}{\frac{\pi}{\nu}(\pi - \frac{\pi}{\nu})}, \tag{11.34}$$

where $\Delta = \cos(\pi/\nu)$ [3]. For $\Delta > 1$, $D(T = 0) = 0$ as the system is gapped.

At finite temperatures, several numerical and analytical studies indicate that for $|\Delta| < 1$ the spin transport is ballistic [47, 48, 49, 50, 51], in accord with the conjecture that this behavior is related to the integrability of the model [52, 53, 25]. Pursuing this conjecture, one can attempt to use the Mazur inequality eq. (11.14) in order to obtain a bound on the Drude weight and thus establish that the transport is ballistic. Inspection of the known conservation laws for the Heisenberg model [54]

[3]The parametrization of Δ in terms of ν is common in the BA literature as the formulation greatly simplifies for ν =integer.

shows that already the first nontrivial one, Q_3, has a physical meaning; it corresponds to the energy current, $Q_3 = j^E$ and it can be used to establish a bound for D [26],

$$D(T) \geq \frac{\beta}{2L} \frac{\langle j^z Q_3 \rangle^2}{\langle Q_3^2 \rangle}. \tag{11.35}$$

This expression can be readily evaluated in the high temperature limit ($\beta \to 0$),

$$D(T) \geq \frac{\beta}{2} \frac{8\Delta^2 m^2 (1/4 - m^2)}{1 + 8\Delta^2 (1/4 + m^2)}, \qquad m = \langle S_l^z \rangle, \tag{11.36}$$

where m is equal to the magnetization density in the Heisenberg model or to $n - 1/2$ in the equivalent fermionic $t - V$ model (n is the density). It establishes that ballistic transport is possible at all finite temperatures in the Heisenberg ($t - V$) model; notice however, that the right hand side vanishes for $m = 0$, that corresponds to the specific case of the antiferromagnetic regime at zero magnetic field or to the $t - V$ model at half-filling. Of course this does not mean that D is indeed zero in these cases as this relation provides only a bound. It should also be remarked that the obtained bound is proportional to Δ^2 and so we do not recover the simple result that $D(T) > 0$ in the XY-limit. Furthermore, it can be shown, using a symmetry argument, that even by taking into account all conservation laws the bound remains zero at $m = 0$ [26].

A BA method based calculation of $D(T)$ for $|\Delta| < 1$ was also performed [55], using a procedure proposed for the Hubbard model [56], that relies upon a certain assumption on the flux dependence (see eq.(11.10)) of bound state excitations ("rigid strings"). The resulting behavior is summarized in Figs. 11.3 and 11.4. From this analysis the following picture emerges:

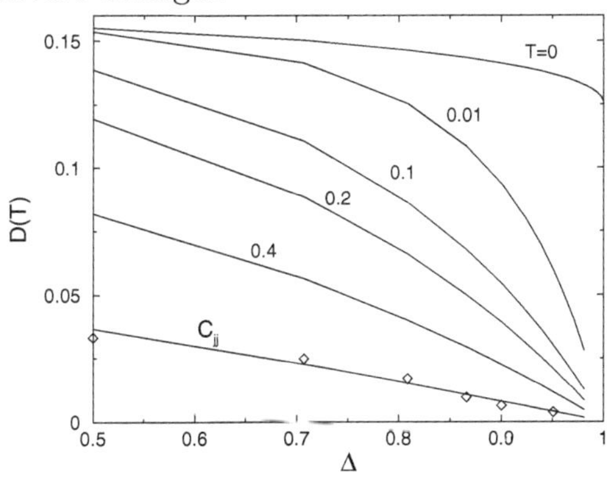

Figure 11.3 $D(\Delta)$ at various temperatures. The lowest line is the high temperature proportionality constant $C_{jj} = D/\beta$. The symbols indicate exact diagonalization results [30].

(i) at zero magnetization, in the easy plane antiferromagnetic regime $(0 < \Delta < 1)$, the Drude weight decreases at low temperatures as a power law $D(T) = D_0 - \text{const}.T^{\alpha}, \quad \alpha = 2/(\nu - 1)$;
(ii) in the ferromagnetic regime, $-1 < \Delta < 0$, $D(T)$ decreases quadratically with temperature (as in a noninteracting, XY-system);
(iii) the same low temperature quadratic behavior is true at any finite magnetization;
(iv) for $\beta \to 0$, $D(T) = \beta C_{jj}$ and it can be shown that $D(-\Delta) = D(\Delta)$ by applying a unitary transformation in the expression eq.(11.11);
a closed expression for C_{jj} can be obtained by analytic calculations [57], $C_{jj} = (\pi/\nu - 0.5\sin(2\pi/\nu))/(16\pi/\nu)$ for $|\Delta| < 1$ while $C_{jj} = 0$ for $\Delta > 1$;
(v) at the isotropic antiferromagnetic point $(\Delta = 1)$, $D(T)$ seems to vanish, implying non ballistic transport at all finite temperatures.

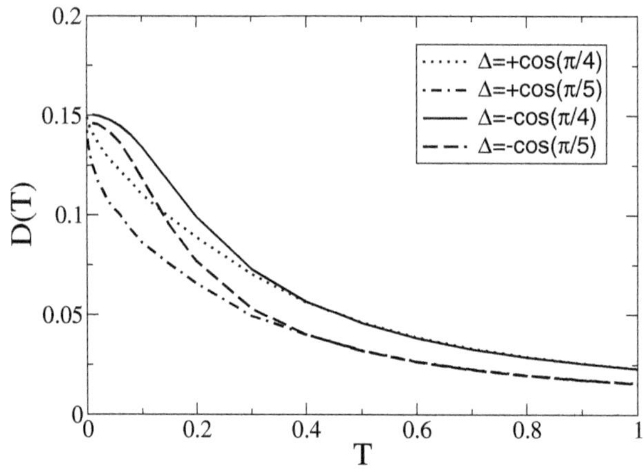

Figure 11.4 Temperature dependence of the Drude weight D vs. T [55]

This last result seems in accord with the most recent NMR data [5]. Of course, the low frequency conductivity must also be examined in order to determine whether there is no anomalous behavior (e.g. power law divergence) that precludes a normal diffusive behavior; such unconventional behavior is presently debated in classical nonlinear 1D systems (see final section of Discussion). It should not be surprising if future rigorous studies reveal that the isotropic Heisenberg exhibits a singular behavior, as it lies at the transition between a gapless and gapped phase.

In this context, we should also mention that the power law decrease of $D(T)$ for $0 < \Delta < 1$ is not corroborated by recent QMC simulations [58]. The disagreement might be due either to the "rigid string" assumption used in the BA analysis or to the very low temperatures, of the order of the energy level spacing, that are studied in the QMC simulations [4].

Considering the limited results obtained so far using the Mazur inequality compared to the exact BA analysis, it remains an open question whether the behavior of the Drude weight can be fully accounted for solely by a proper consideration of conservation laws present in the Heisenberg model.

For $\Delta > 1$ numerical simulations [48] and analytical arguments [51] indicate that the Drude weight vanishes at all temperatures. In this regime, based on ED numerical simulations, it was proposed that a new phase might exist, an "ideal insulator", characterized by vanishing Drude weight and diffusion constant (*dc* conductivity in the fermionic version). This conjecture remains presently still rather tentative, due to the small size of the systems that have been studied so far.

On the other hand, a semiclassical field theory approach [59] concluded that gapped systems are diffusive. This approach is based on a mapping of the massive excitations to impenetrable classical particles of two or more charges (corresponding to different spin directions) that propagate diffusively (see section 5) and it has mostly been used for the analysis of gapped spin-1 systems.

In parallel to these developments, the spin S=1/2 Heisenberg model was studied in the scaling limit using conformal invariance arguments [60, 61]. This field theory approach amounts to considering a linearized spectrum and thus neglecting the effects of curvature, a point that we will further discuss below in section 5. In particular, it was shown that the uniform dynamic susceptibility describes ballistic behavior, the corresponding $1/T_1$ relaxation time was evaluated and the theory was extensively compared to experimental data [62]. Notice, however, that a later experimental NMR work [5] concludes that the $q = 0$ mode of spin transport is ballistic at the $T = 0$ limit, but has a diffusion-like contribution at finite temperatures even for $T << J$. We should remark that, over the years, the most common interpretation of NMR experiments was within the diffusion phenomenology, as for instance for the $S = 5/2$ TMMC compound [63].

Finally, the finite (q, ω) response functions of the S=1/2 model at $T = 0$ were studied by the bosonization technique [64] after mapping it

[4]Reliable results for the Drude weight can be obtained by QMC simulations only at low temperatures because a sufficiently fine spacing of Matsubara frequencies is required for the extrapolation to zero frequency.

to spinless fermions (eq.(11.19)). For $\Delta < 1$, the conductivity shows the typical ballistic form; for $\Delta > 1$ it vanishes below the gap, showing a square-root frequency dependence above.

Turning now to energy transport, it is easy to see that the energy current is a conserved quantity [65, 26] for all values of the anisotropy Δ implying that the currents do not decay and so the thermal conductivity is infinite. This peculiarity has also been noticed by an earlier analysis of moments at infinite temperature [66]. So the quantity characterizing thermal transport is the equal time correlation $\langle j^E j^E \rangle$ that represents the weight under the low frequency peak that will develop from the zero frequency $\delta-$function when a dissipative mechanism is introduced. This picture is analogous to that of the electrical conductivity illustrated in Fig. 1.1. It implies that, given an estimate of the temperature dependence of the characteristic scattering time one is able to extract the value of the *dc* thermal conductivity, further assuming some form (e.g. eq.(11.8)) for the low frequency behavior.

This quantity has also recently been exactly calculated using the BA method [67]; it is shown in Fig. 11.5.

Furthermore, the experimental observation of unusually high thermal conductivity in ladder compounds [1] motivated the theoretical study of the thermal Drude weight in 1D anisotropic, frustrated and ladder spin-1/2 systems [68, 69]; the proposal of unconventional thermal transport in these systems is still debated.

Finally, the S=1 Heisenberg chain shows a qualitatively different behavior characterized by the presence of an energy (Haldane) gap at low energies. The S=1 Heisenberg model is not integrable but the physics at low energies is usually mapped onto the quantum nonlinear-σ model that is again an integrable system. The results known on this model will be briefly discussed in section 5 along with a semiclassical approach to describe this type of gapped systems. The same low energy mapping is used for the analysis of "ladder" compounds.

As a guide to experimental investigations and theoretical studies, we can recapitulate the above discussion of the dynamics of the Heisenberg S=1/2 model as follows. It seems clear that ballistic behavior at all temperatures should be expected in the easy-plane regime and at all finite magnetizations, while the isotropic point is a subtle borderline case. The behavior in the easy-axis antiferromagnetic regime might be particularly interesting and it is not settled at the moment. Exceptionally high thermal conductivity should be expected in all regimes.

To complete the above picture, we should stress that not much is known on the low frequency behavior of the conductivities at finite tem-

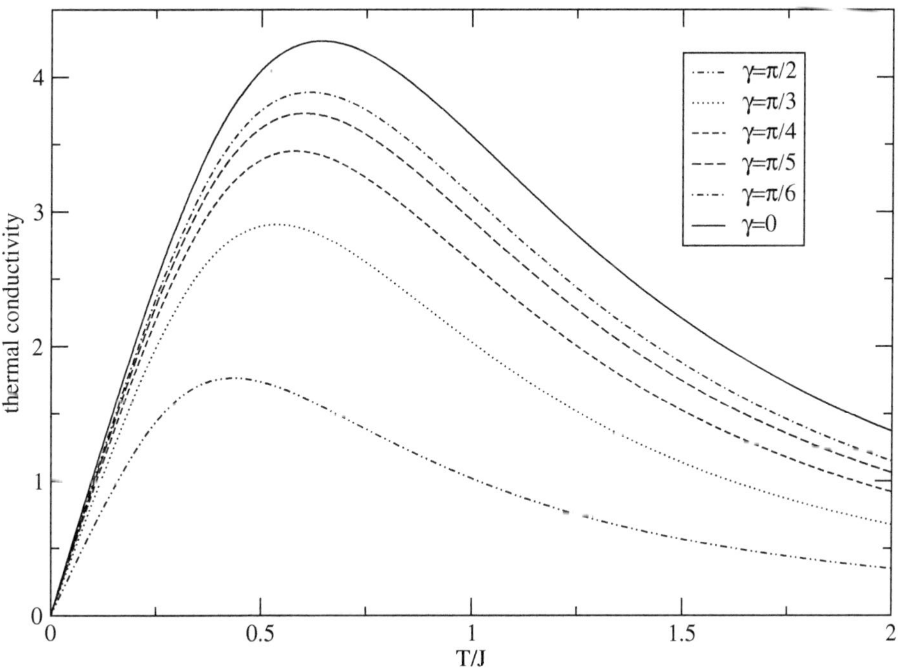

Figure 11.5. Thermal conductivity, $\langle j^E j^E \rangle$ in units of J^2, for various anisotropy parameters $\Delta = \cos(\gamma)$ [67].

perature. This leaves open the possibility of unconventional behavior, neither ballistic nor simple diffusive but one characterized by long time tails, giving rise to power law (or logarithmic) behavior at low frequencies.

4. Hubbard model

The prototype model for the description of electron-electron correlations is the Hubbard model given by the Hamiltonian,

$$H = \sum_l h_l = (-t) \sum_{\sigma,l} (c_{l\sigma}^\dagger c_{l+1\sigma} + h.c.) + U \sum_l n_{l\uparrow} n_{l\downarrow} \tag{11.37}$$

where $c_{l\sigma}(c_{l\sigma}^\dagger)$ are annihilation (creation) operators of fermions with spin $\sigma = \uparrow, \downarrow$ at site l and $n_{l\sigma} = c_{l\sigma}^\dagger c_{l\sigma}$.

At half-filling (n=1, 1 fermion per site) it describes a Mott-Hubbard insulator for any value of the repulsive interaction $U > 0$, while it is a metal at any other filling.

The one dimensional Hubbard model is also integrable by the Bethe ansatz method and its phase diagram, elementary excitations, correlation functions have been extensively studied [10, 70].

4.1 Currents

Similarly to the Heisenberg model, we can discuss the electrical, spin and thermal conductivity by defining the charge j, spin j^s and energy j^E currents from the respective continuity equations of the local particle density n_l,

$$\frac{\partial n_l}{\partial t} + \nabla j_l = 0, \quad j = \sum_l j_l = \sum_{l\sigma} j_{l\sigma} = (-t) \sum_{\sigma,l} (i c_{l\sigma}^\dagger c_{l+1\sigma} + h.c.), \tag{11.38}$$

spin density $n_{l\uparrow} - n_{l\downarrow}$,

$$\frac{\partial (n_{l\uparrow} - n_{l\downarrow})}{\partial t} + \nabla j_l^s = 0, \quad j^s = \sum_l j_l^s = \sum_l j_{l\uparrow} - j_{l\downarrow} \tag{11.39}$$

and energy density h_l,

$$\frac{\partial h_l}{\partial t} + \nabla j_l^E = 0, \quad j^E = \sum_{l,\sigma} j_{l\sigma}^E \tag{11.40}$$

$$j_{l\sigma}^E = (-t)^2 (i c_{l+1\sigma}^\dagger c_{l-1\sigma} + h.c.) - \frac{U}{2} j_{l,\sigma} (n_{l,-\sigma} + n_{l+1,-\sigma} - 1).$$

4.2 Electrical and thermal transport

With respect to the electrical conductivity the interaction U and density dependence of the Drude weight D at zero temperature has been established using the BA method [71, 72, 73] (see Fig. 11.6). There are two simple limits:
(i) The free fermion case $U = 0$ where j is conserved and $D_0 = \frac{2t}{\pi} \sin \frac{\pi n}{2}$ where n is the density of fermions ($n = 2k_F/\pi$). Here D_0 vanishes for an empty band $n = 0$ and a filled band $n = 2$, being maximum at half filling, $n = 1$.
(ii) Another simple limit is $U = \infty$. Since in this case the double occupation of sites is forbidden, fermions behave effectively as spinless fermions and the result is $D_0 = \frac{t}{\pi} |\sin(\pi n)|$; here D_0 vanishes also at half filling.

Analytical results in 1D indicate that the $D_0 = 0$ value at half filling persists in the Hubbard model for all $U > 0$, whereby the density dependence $D(n)$ is between the limits $U = 0$ and $U = \infty$. The insulating state at half filling is a generic feature of a wider class of 1D models characterized by repulsive interactions, such as the t-V model (discussed above), the t-J model etc.

In Fig. 11.6, along with the Drude weight, the zero temperature (ballistic) Hall constant R_H of a quasi-1D system is also shown. According to a recent formulation [74], R_H can be expressed in terms of the derivative of the Drude weight with respect to the density,

$$R_H = -\frac{1}{D}\frac{\partial D}{\partial n}. \tag{11.41}$$

The Hall constant is the classical way for determining the sign of the charge carriers. For a strictly one dimensional system of course it makes no sense to discuss the Hall effect; but if we consider a quasi-one dimensional system with interchain coupling characterized by a hopping $t' \to 0$, then within this formulation we recover a simple picture for the behavior of the sign of carriers as a function of interaction. In agreement with intuitive semiclassical arguments, the Hall constant behaves as $R_H \simeq -1/n$ at low densities changing to $R_H \simeq +1/\delta(\delta = 1 - n)$ near half-filling, with the turning point depending on the strength of the interaction U. Notice that if $D \propto n$ with a small proportionality constant, that would be interpreted within a single particle picture as indicative of a large effective mass, then we would still find $R_H \simeq -1/n$. This observation might be relevant in the context of recent optical and Hall experiments [75, 7] where a small Drude weight is observed although the Hall constant indicates a carrier density of order of one.

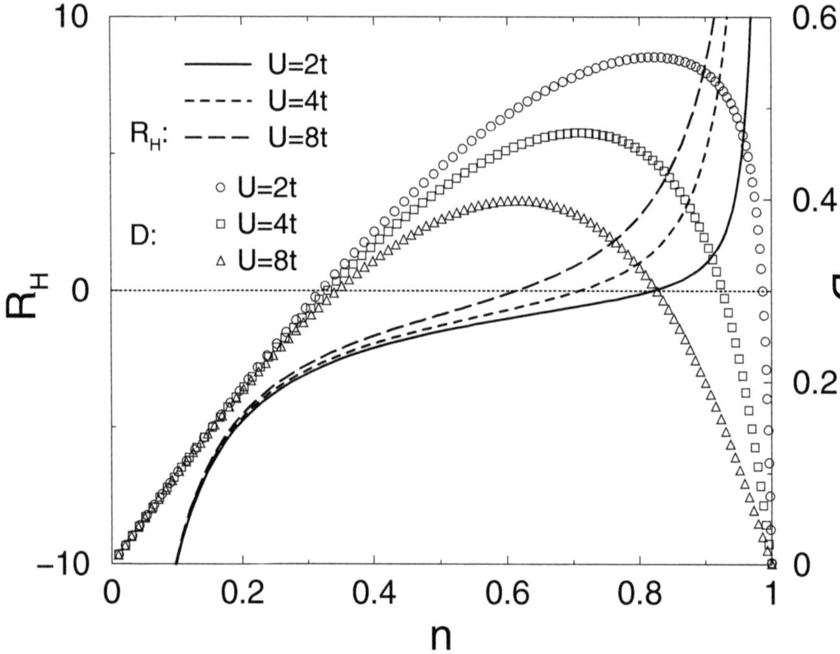

Figure 11.6. Drude weight D and R_H for the quasi-1D Hubbard model from expression (11.41).

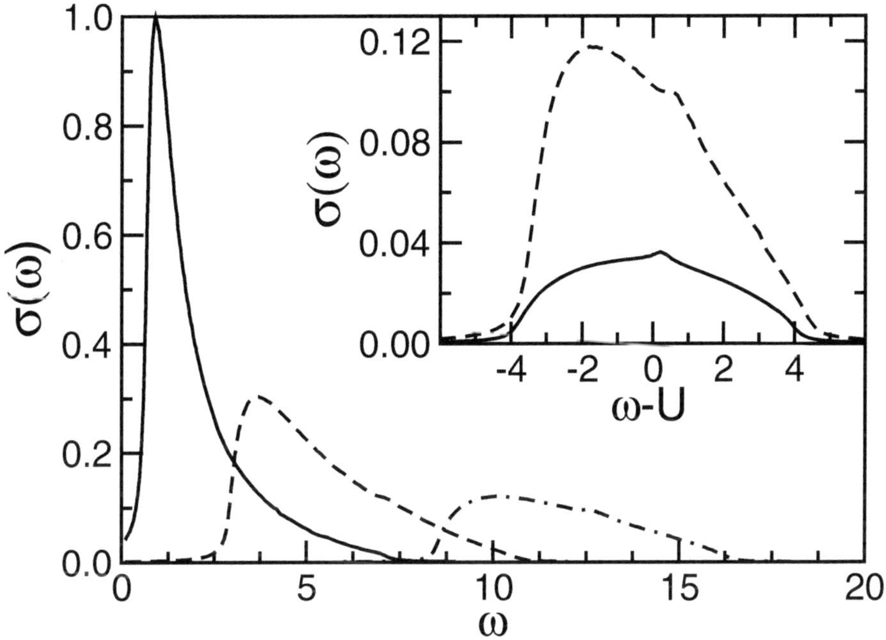

Figure 11.7. Optical conductivity at $T = 0$ for $U/t = 3, 6, 12$ (from left to right) calculated with DMRG on a 128-site lattice [76]. Inset: $\sigma(\omega)$ for $U/t = 12$ (dashed) and 40 (solid) calculated on a 64-site chain.

Recently, using the form factor and DMRG methods the frequency dependence of the conductivity at half filling and at $T = 0$ has also been studied [76] and is shown in Fig. 11.7. The DMRG method provided the entire absorption spectrum for all but very small couplings where the field theoretical approach was used; the two methods are in excellent agreement in their common regime of applicability. As expected, the Drude weight is zero, signaling an insulating state (for a detailed analysis of the scaling of D with system size at and close to half-filling, see [77]) and the finite frequency conductivity vanishes up to the gap. Above the gap, a square root dependence is observed but not a divergence; this behavior is in contrast to that obtained by the Luttinger liquid method [78] and it is typical of a Peierls (band) insulator where a divergence occurs. This absence of a singularity is also in agreement with a rigorous analysis of the sine-Gordon (sG) field theory (see section 5), the generic low energy effective model for the description of a Mott-Hubbard insulator.

To complete the zero temperature picture, the frequency dependent conductivity of the Hubbard model out of half-filling has been studied using results from the BA method and symmetries [79]. A broad absorption band was found separated from the Drude peak at $\omega = 0$ by a pseudogap; this pseudo-gap behavior is in contrast to the ω^3 dependence found within the Luttinger liquid analysis [78].

Again, at all finite temperatures the transport is ballistic characterized by a finite Drude weight. In an identical way to the Heisenberg model, this can easily be established by the Mazur inequality using the first nontrivial conservation law Q_3. For the Hubbard model Q_3 differs from the energy current j^E by the replacement of U by $U/2$ [26]. Evaluating $\langle jQ_3\rangle^2/\langle Q_3^2\rangle$ for $\beta \to 0$ we obtain,

$$D(T) \geq \frac{\beta}{2L}\frac{\langle jQ_3\rangle^2}{\langle Q_3^2\rangle} = \frac{\beta}{2}\frac{[U\sum_\sigma 2n_\sigma(1-n_\sigma)(2n_{-\sigma}-1)]^2}{\sum_\sigma 2n_\sigma(1-n_\sigma)[1+U^2(2n_{-\sigma}^2-2n_{-\sigma}+1)]}, \tag{11.42}$$

where n_σ are the densities of $\sigma =\uparrow,\downarrow$ fermions.

By inspection we can again see that from this inequality we cannot obtain a finite bound for $D(T)$ for $n_\uparrow + n_\downarrow = 1$. Nevertheless, a full BA calculation [56] seems to show that the Drude weight at half-filling is exponentially activated $D(T) \sim \frac{1}{\sqrt{T}}e^{-E_{gap}/T}$ at low temperatures and decreases as T^2 out of half-filling. Thus the zero temperature insulator turns to an ideal conductor at finite temperatures. Notice that this behavior is different from the one in the Heisenberg (or "t-V") model in the gapped phase ($\Delta > 1$) where the Drude weight seems to vanish at all finite temperatures. We can conjecture that this distinct behavior of insulating phases can be understood in the framework of the corresponding low energy sine-Gordon field theory as these two models map to different parameter regimes of the sG model [64, 13]. A very similar calculation, using the Mazur inequality, can also be carried out for the long time asymptotics of the spin current, j^s, correlations. It gives a finite bound, and thus ballistic spin transport for $n_\uparrow - n_\downarrow \neq 0$; no BA calculation has so far been performed for the spin conductivity.

On the thermal conductivity we find similar results, namely a finite value on the long time decay of energy current correlations, which can readily be evaluated for $\beta \to 0$ [26],

$$\lim_{t\to\infty}\langle j^E(t)j^E\rangle = C_{j^Ej^E} \geq \frac{\langle j^E Q_3\rangle^2}{\langle Q_3^2\rangle}. \tag{11.43}$$

Again this inequality gives a finite bound for a system out of half-filling as long as $n_\uparrow + n_\downarrow \neq 0$ and this for any magnetization. For this model the actual temperature dependence of $C_{j^E j^E} = \lim_{t\to\infty}\langle j^E(t) j^E\rangle = C_{j^E j^E}$ has not yet been evaluated. Finally, the low temperature thermoelectric power was studied using the Bethe ansatz picture for the charge (holons) and spin (spinons) excitations [80]. The resulting sign of the thermopower close to the Mott-Hubbard insulating phase is consistent with the one derived from the Hall constant above, $S \sim sign(1-n)T|m^*|/|1-n|^2$.

In summary, we have shown that the prototype model for describing electron correlations in one dimensional systems, the Hubbard model, shows unconventional, ballistic charge, spin and thermal transport at all finite temperatures. Of course real quasi-one dimensional materials are presumably characterized by longer range than the Hubbard U interactions. So, although the above picture should be taken into account in the interpretation of experiments, (quasi-) one dimensional magnetic compounds might presently appear as better candidates for the experimental observation of these effects. Theoretically, the full frequency dependence of the conductivities at finite temperatures remains to be established.

5. Effective field theories

An alternative to analyzing the transport of quasi-one dimensional materials within microscopic models, as described in previous sections, is to approach the problem within effective low energy models for interacting electrons, i.e. starting with the Luttinger liquid Hamiltonian. This path is very attractive since it represents the counterpart of the usual Landau phenomenological approach to Fermi liquid in higher-D electronic systems. It should be pointed out that even in a 3D system the continuum field theory is not enough to describe a current decay and Umklapp processes are finally responsible for a finite intrinsic resistivity $\rho(T) \propto T^2$ [81].

In an effective (low energy) field theoretical model for 1D interacting electrons the band dispersion around the Fermi momenta $k = \pm k_F$ is linearized and left- and right- moving excitations are defined. Apart from Umklapp terms, the model of interacting fermions can then be mapped onto the well known Luttinger liquid Hamiltonian [16, 13] and analyzed via the bosonization representation. In particular one obtains for the charge sector,

$$H_0 = \frac{1}{2\pi}\int dx\Big[u_\rho K_\rho(\pi\Pi_\rho)^2 + \frac{u_\rho}{K_\rho}(\partial_x\phi_\rho)^2\Big], \qquad (11.44)$$

where the charge density is $\rho(x) = \partial_x \phi_\rho$ and Π_ρ is the conjugate momentum to ϕ_ρ. Interactions appear only via the velocity parameter u_ρ and Luttinger exponent K_ρ. The charge current in such a Luttinger model, $j = \sqrt{2} u_\rho K_\rho \Pi_\rho$, is clearly conserved in the absence of additional terms.

Umklapp terms can as well be represented with boson operators,

$$H_{\frac{1}{2m}} = g_{\frac{1}{2m}} \int dx \cos(m\sqrt{8}\phi_\rho(x) + \delta x), \tag{11.45}$$

where m is the commensurability parameter ($m = 1$ at half-filling - one particle per site, $m = 2$ for quarter filling - one particle for two sites etc) and δ the doping deviation from the commensurate filling. In principle, the mapping of a particular (tight binding) microscopic model onto a field theory model, e.g. via perturbation theory, generates terms H_m with arbitrary m. While Umklapp terms are irrelevant in the sense of universal scaling of the static properties, they appear to be crucial for transport. They drive a metal at half-filling to an insulator, while at an arbitrary (incommensurate) filling they should cause a finite resistivity since the current is not conserved any more (for an overview of the transport properties emerging within the Luttinger liquid picture see [82]).

However, the proper treatment of transport within the Luttinger picture in the presence of Umklapp processes is quite involved and even controversial. Giamarchi [78] first calculated the effect of Umklapp scattering within lowest order perturbation theory for the memory function $M(\omega)$; he thus determined the low-ω behavior of the dynamical conductivity $\sigma(\omega) \propto 1/(\omega + M(\omega))$ that yielded a nonzero finite temperature conductivity. At the same time he realized, by using the Luther-Emery method, that the Umklapp term can be absorbed in the Hamiltonian in such a way as to conserve the current and pointed out the possibility of infinite *dc* conductivity even in the presence of Umklapp. A similar lowest-order analysis [83] for general commensurate filling predicts at $T = 0$ that $\sigma(\omega) \propto \omega^{\nu-2}$ and the resistivity $\rho(T) \propto T^\nu$ with $\nu = 4n^2 K_\rho - 3$. On the other hand, Rosch and Andrei [84] pointed out that even in the presence of general Umklapp terms there exist particular operators, linear combinations of the translation operator and number difference between left- and right- moving electrons, which are conserved. Since in general such operators have a nonvanishing overlap with the current operator j, this leads to finite $D(T > 0) > 0$ if only one Umklapp term is considered. At least the interplay of two noncommuting Umklapp processes is needed to yield a finite resistivity $\rho(T > 0) > 0$.

From a different perspective Ogata and Anderson [85] argued that because of spin-charge separation in 1D systems an effect analogous to phonon drag (in this case spinon-holon drag) appears that leads to a finite dissipation. Using a Landauer like semi-phenomenological approach they concluded the existence of a linear-T resistivity and linear frequency dependence of the optical conductivity.

The bosonization of the Luttinger liquid model leads [13] to the quantum sine-Gordon model (eq.(11.45)) which is an integrable system and has extensively been studied as a prototype nonlinear quantum (or classical) field theory. It is the generic field theory for describing the low energy properties of one dimensional Mott insulators. The thermodynamic properties and excitation spectrum consisting of solitons/antisolitons and breather states have been established by semi-classical and BA techniques [10]. Presently, there is an effort to determine the transport properties of this model rigorously. In particular, the frequency dependence of the zero temperature conductivity in the commensurate (insulating) phase, zero soliton sector, has been evaluated using the form factor approach [86]. The main result is that the square root singularity at the optical gap, characteristic of band insulators, is generally absent and appears only at the Luther-Emery point; furthermore, the perturbative result [78] is recovered only at relatively high frequencies. Besides these studies, the Drude weight and optical response near the metal-insulator transition, in the incommensurate phase at zero temperature, have also been studied by Bethe ansatz [87] and semiclassical methods [88]. Still, a rigorous evaluation of the Drude weight and frequency dependence of the conductivity at finite temperatures is missing; nevertheless, we can plausibly argue that because of the integrability of the sine-Gordon model, it will turn out that also this model describes an ideal conductor at least over some interaction range. Thus, it might remain an open question which scattering processes and/or band curvature must be taken into account in order to recover a normal, diffusive behavior at finite temperatures.

Finally, it is well known [89] that the spectrum of integer spin and even-leg ladder systems is gapped and that the low energy physics is described by the one-dimensional quantum $O(3)$ nonlinear sigma model [40].

In imaginary time τ the action at inverse temperature β is given by

$$S=\frac{c}{2g}\int_0^\beta d\tau\Big[(\partial_x n_\alpha)^2+\frac{1}{c^2}(\partial_\tau n_\alpha)^2\Big], \tag{11.46}$$

where x is the spatial coordinate, c a characteristic velocity, $\alpha = 1, 2, 3$ is an $O(3)$ vector index and $n_\alpha(x, \tau)$ a unit vector field $n_\alpha^2(x, \tau) = 1$.

In a series of works, Sachdev and collaborators [90, 91, 92] developed a picture of the low and intermediate temperature spin dynamics based on the idea that the spin excitations can be mapped to an integrable model describing a classical gas of impenetrable particles (of a certain number of species depending on the spin), a problem that can be treated analytically. Within this framework they have extensively analyzed NMR experiments on S=1 compounds [93] and they concluded that these systems behave diffusively. In contrast to this semiclassical approach, using the Bethe ansatz solution of the quantum nonlinear$-\sigma$ model [94], Fujimoto [95] found a finite Drude weight, exponentially activated with temperature, and he thus concluded that the spin transport at finite temperatures is ballistic. The origin of this discrepancy is not clear at the moment and can be due either to a subtle role of quantum effects on the dynamics that is neglected in the semiclassical approach or to a particular limiting procedure (the magnetic field going to zero) in the BA solution.

6. Discussion

We hope that the above presentation demonstrated that the transport theory of one dimensional quantum systems is a rapidly progressing field, fueled by both theoretical and experimental developments. Still, on the question, what is the finite temperature conductivity of bulk electronic or magnetic systems described by strongly interacting one dimensional Hamiltonians, it is fair to say that no definite answer has so far emerged nor there is a clear picture of the relevant scattering mechanisms.

In this context, it is interesting and instructive to draw an analogy with the development of the respective field in classical physics, namely the finite temperature transport in one dimensional nonlinear systems. Interestingly, in this domain we are also witnessing a flurry of activity after several decades of studies. Again, the issue of ballistic versus diffusive (usually energy) transport in a variety of models and the necessary ingredients for observing normal behavior is sharply debated [96, 97]. Similarly to the quantum systems, numerical simulations are intensely employed along with analytical approaches and discussions on the conceptual foundations of transport theory.

For quantum systems it is reasonable to expect that the finite temperature transport properties of integrable models will, in the near future, be amenable to rigorous analysis by mathematical techniques, for instance in the framework of the Bethe ansatz method. At the same time, as we mentioned earlier, it is amusing to notice that the integrable systems that we can exactly analyze, present singular transport properties presumably exactly because of their integrability.

To obtain normal behavior, it is reasonable to invoke perturbations destroying the integrability of the model, as for instance longer range interactions, interchain coupling, coupling to phonons, disorder etc. In this scenario, it is then necessary to find ways to study the effect of perturbations around an integrable system and in particular to determine the vicinity in parameter space around the singular-integrable point where unconventional transport can be detected. This issue is also extensively studied in classical systems as it is the most relevant in the interpretation of experiments and in estimating the prospects for technological realizations. It is worth keeping in mind the possibility that integrable interactions actually render the system more immune to perturbations, an effect well known and exploited in classical nonlinear systems [9].

Related to this line of argument is the following question. If integrable models show ballistic transport and low energy effective theories like the sine-Gordon model are also integrable, then which mechanisms are necessary to obtain dissipative behavior ?

Of course it is also possible that the conventional picture according to which only integrable systems show ballistic transport might well be challenged. One dimensional nonintegrable quantum systems could also show singular transport in the form either of a finite Drude weight or low frequency anomalies. This behavior has been observed in classical nonintegrable nonlinear systems where the current correlations decay to zero in the long time limit but too slowly, so that the integral over time (giving the *dc* conductivity) diverges. The opposite behavior might also be realized, namely that integrable quantum systems show normal diffusive transport in some region of interaction parameter space (this possibility was raised in the case of gapped systems as the easy-axis spin 1/2 Heisenberg model or the quantum nonlinear$-\sigma$ model, see section 5). Furthermore, the issue of the crossover of the dynamics between quantum and classical systems has, at the moment, very little been explored and in particular the question whether quantum fluctuations might stabilize ballistic transport behavior.

To address all the above open issues there is a clear need for the development of reliable analytical and numerical simulation techniques (as the DMRG or QMC) to tackle the evaluation of dynamic correlations at low temperatures. In particular, progress is needed to include the coupling between the different, magnetic, electronic and phononic, excitations.

In summary, one of the most fascinating aspects in this field is to understand the extent to which the so successful physics, experimental and technological realizations of classical (integrable) nonlinear systems can be carried over to the quantum world of many body (quasi-) one dimensional electronic or magnetic strongly interacting systems. This effort is accompanied by the experimental challenge to synthesize novel materials/systems that realize this physics.

Acknowledgments

It is a pleasure to acknowledge discussions over the last few years on this problem with many colleagues, in particular H. Castella, F. Naef, A. Klümper, C. Gros, A. Rosch, D.Baeriswyl, H.R. Ott, H. Beck, T. Giamarchi, M. Long, N. Papanicolaou and a careful reading of the manuscript by D. Baeriswyl. This work was supported by the Swiss National Foundation, the University of Fribourg, the University of Neuchâtel and the EPFL through its Academic Guests Program.

References

[1] "Energy transport in one-dimensional spin systems", A.V. Sologubenko and H.R. Ott, Chapter 12 of present volume.

[2] A.V. Sologubenko et al., Phys. Rev. Lett. **84**, 2714 (2000); A. V. Sologubenko et al., Phys. Rev. B**64**, 054412 (2001).

[3] C. Hess et al., Phys. Rev. B**64**, 184305 (2001).

[4] M. Takigawa et al., Phys. Rev. Lett. **76**, 4612 (1996).

[5] K.R. Thurber et al., Phys. Rev. Lett. **87**, 247202 (2001).

[6] M. Dressel, A. Schwartz, G. Grüner and L. Degiorgi, Phys. Rev. Lett. **77**, 398 (1996).

[7] "Electrodynamic response in "one-dimensional" chains", L. Degiorgi, Chapter 6 of present volume.

[8] F.D.M. Haldane, J. Phys. C**14**, 2585 (1981).

[9] "Optical solitons: theoretical challenges and industrial perspectives", editors: V.E. Zakharov and S. Wabnitz, Les Houches Workshop, Springer (1998).

[10] "Quantum Inverse Scattering Method and Correlation Functions", V.E. Korepin, N.M. Bogoliubov and A.G. Izergin, Cambridge Univ. Press (1993).

[11] H.G. Evertz, Adv. Phys. **52**, 1 (2003).

[12] "Density-matrix renormalization : a new numerical method in physics", Workshop Proceedings, Dresden, I. Peschel et al. eds., Springer (1999).

[13] V.J. Emery in *Highly Conducting One-dimensional Solids*, ed. by J. Devreese *et al.* (Plenum, New York, 1979), p.247.

[14] J. Solyom, Adv. Phys. **28**, 209 (1979).

[15] J. Voit, Rep. Prog. Phys. **58**, 977 (1995)

[16] "Luttinger liquids: the basic concepts", K. Schönhammer, Chapter 4 of present volume.

[17] see also "Optics of correlated systems", A.J. Millis, Chapter 7 of present volume.

[18] W. Kohn, Phys. Rev. **133**, A171 (1964).

[19] P.F. Maldague, Phys. Rev. B**16**, 2437 (1977).

[20] D. Baeriswyl, C. Gros and T.M. Rice, Phys. Rev. B**35**, 8391 (1987).

[21] B.S. Shastry and B. Sutherland, Phys. Rev. Lett. **65**, 243 (1990).

[22] J.T. Edwards and D.J. Thouless, J. Phys. C**5**, 807 (1972).

[23] M. Imada, A. Fujimori and Y. Tokura, Rev. Mod. Phys. **70**, 1039 (1998).

[24] J. Jaklič and P. Prelovšek, Adv. Phys. **49**, 1 (2000).

[25] H. Castella, X. Zotos, P. Prelovšek, Phys. Rev. Lett. **74**, 972 (1995).

[26] X. Zotos, F. Naef and P. Prelovšek, Phys. Rev. B**55** 11029 (1997).

[27] P. Mazur, Physica **43**, 533 (1969).

[28] M. Wilkinson, Phys. Rev. B**41**, 4645 (1990).

[29] B.D. Simons and B.L. Altshuler, Phys. Rev. Lett. **70** 4063 (1993).

[30] F. Naef and X. Zotos, J. Phys. C. **10**, L183 (1998); F. Naef, Ph. D. thesis no.2127, EPF-Lausanne (2000).

[31] W. Götze and P. Wölfle, Phys. Rev. B**6**, 1226 (1972).

[32] A.G. Redfield and W.N. Yu, Phys. Rev. **169**, 443 (1968); erratum, Phys. Rev. **177**, 1018 (1969).

[33] J. Jaklič, P. Prelovšek, Phys. Rev. B**49**, 5065 (1994).

[34] M.W. Long, P. Prelovšek, S. El Shawish, J. Karadamoglou and X. Zotos, cond-mat/0302211.

[35] M. Jarrel and J.E. Gubernatis, Phys. Rep. **269**, 135 (1996).

[36] T.D. Kühner and S.R. White, Phys. Rev. B**60**, 335 (1999).

[37] E. Jeckelmann, Phys. Rev. B**66**, 045114 (2002).

[38] F. Naef, X. Wang, X. Zotos, W. von der Linden, Phys. Rev. **B60**, 359 (1999).

[39] E. Lieb, T. Schultz and D. Mattis, Ann. Phys. (N.Y.) **16**, 407 (1961).

[40] F.D.M. Haldane, Phys. Lett. **93A**, 464 (1983).

[41] L.P. Kadanoff and P.C. Martin, Ann. Phys. **24**, 419 (1963).

[42] N.A. Lurie, D.L. Huber and M. Blume, Phys. Rev. **B9**, 2171 (1974).

[43] O. F. de Alcantara Bonfim and G. Reiter, Phys. Rev. Lett. **69**, 367 (1992).

[44] G. Müller, Phys. Rev. Lett. **60**, 2785 (1988); N. Srivastava, J-M. Liu, V.S. Viswanath and G. Müller, J. Appl. Phys. **75**, 6751 (1994) and references therein.

[45] T. Niemeijer, Physica **36**, 377 (1967).

[46] F.D.M. Haldane, Phys. Rev. Lett. **45**, 1358 (1980).

[47] K. Fabricius and B.M. McCoy, Phys. Rev. B**57**, 8340 (1998) and references therein.

[48] X. Zotos and P. Prelovšek, Phys. Rev. B**53**, 983 (1996).

[49] B.N. Narozhny, A.J. Millis and N. Andrei, Phys. Rev. B**58**, 2921 (1998).

[50] B.N. Narozhny, Phys. Rev. B**54**, 3311 (1996).

[51] N.M.R. Peres, P.D. Sacramento, D.K. Campbell and J.M.P. Carmelo, Phys. Rev. B**59**, 7382 (1999).

[52] B.M. McCoy, in *Statistical Mechanics and Field Theory*, ed. V.V. Bazhanov and C.J. Burden (World Scientific), 26 (1995).

[53] K. Saito, S. Takesue and S. Miyashita, Phys. Rev. E**54**, 2404 (1996).

[54] M. P. Grabowski and P. Mathieu, Ann. Phys. **243**, 299 (1996).

[55] X. Zotos, Phys. Rev. Lett. **82**, 1764 (1999).

[56] S. Fujimoto and N. Kawakami, J. Phys. A**31**, 465 (1998).

[57] A. Klümper, private communication.

[58] J. V. Alvarez and C. Gros, Phys. Rev. Lett. **88**, 077203 (2002).

[59] K. Damle and S. Sachdev, Phys. Rev. B**57**, 8307 (1998).

[60] H.J. Schulz, Phys. Rev. B**34**, 6372 (1986).

[61] S. Sachdev, Phys. Rev. B**50**, 13006 (1994).

[62] M. Takigawa, O.A. Starykh, A.W. Sandvik and R.R.P. Singh, Phys. Rev. B**56**, 13681 (1997) and references therein.

[63] D. Hone, C. Scherer ad F. Borsa, Phys. Rev. B**9**, 965 (1974).

[64] R. Shankar, Int. J. Mod. Phys. B**4**, 2371 (1990).

[65] T. Niemeijer and H.A.W. van Vianen, Phys. Lett. **34A**, 401 (1971).

[66] D.L. Huber and J.S. Semura, Phys. Rev. B**182**, 602 (1969).

[67] A. Klümper and K. Sakai, J. Phys. A. **35**, 2173 (2002).

[68] J. V. Alvarez and C. Gros, Phys. Rev. Lett. **89**, 156603 (2002).

[69] F. Heinrich-Meisner, A. Honecker, D.C. Cabra and W. Brenig, Phys. Rev. B**66**, 140406 (2002).

[70] T. Deguchi, F.H.L. Essler, F. Göhmann, A. Klümper, V.E. Korepin, K. Kusakabe, Phys. Rep. **331** (5), 197 (2000).

[71] F.D.M. Haldane, Phys. Lett. **81A**, 153 (1981).

[72] H.J. Schulz, Phys. Rev. Lett. **64**, 2831 (1990).

[73] N. Kawakami and S.K. Yang, Phys. Rev. B**44**, 7844 (1991).

[74] X. Zotos, F. Naef, M. Long, and P. Prelovšek, Phys. Rev. Lett. **85**, 377 (2000).

[75] G. Mihaly et al., Phys. Rev. Lett. **84**, 2670 (2000).

[76] E. Jeckelmann, F. Gebhard and F.H.L. Essler, Phys. Rev. Lett. **85**, 3910 (2000).

[77] C.A. Stafford and A.J. Millis, Phys. Rev. B**48**, 1409 (1993).

[78] T. Giamarchi, Phys. Rev. B **44**, 2905 (1991); Phys. Rev. B **46**, 342 (1992).

[79] J.M.P. Carmelo, N.M.R. Peres and P.D. Sacramento, Phys. Rev. Lett. **84**, 4673 (2000); J. M. P. Carmelo, J. M. E. Guerra, J. M. B. Lopes dos Santos, and A. H. Castro Neto, Phys. Rev. Lett. 83, 3892 (1999)

[80] C.A. Stafford, Phys. Rev. B**48**, 8430 (1993).

[81] "Fundamentals of the Theory of Metals", A.A. Abrikosov, North-Holland (1988).

[82] T. Giamarchi, Physica B **230-232**, 975 (1997).

[83] T. Giamarchi and A. Millis, Phys. Rev. B**46**, 9325 (1992).

[84] A. Rosch and N. Andrei, Phys. Rev. Lett. **85**, 1092 (2000).

[85] M. Ogata, P.W. Anderson, Phys. Rev. Lett. **70**, 3087 (1993).

[86] D. Controzzi, F.H.L. Essler and A.M. Tsvelik, Phys. Rev. Lett. **86**, 680 (2001).

[87] E. Papa and A.M. Tsvelik, Phys. Rev. B**63**, 085109 (2001).

[88] D.N. Aristov, V.V. Cheianov and A. Luther, Phys. Rev. B**66**, 073105 (2002).

[89] E. Dagotto and T.M. Rice, Science **271**, 618 (1996).

[90] S. Sachdev and K. Damle, Phys. Rev. Lett. **78**, 943 (1997); K. Damle and S. Sachdev, Phys. Rev. B**57**, 8307 (1998).

[91] "Quantum Phase transitions", S. Sachdev, Cambridge University Press (1999).

[92] C. Buragohain and S. Sachdev, Phys. Rev. B**59**, 9285 (1999).

[93] M. Takigawa et al., Phys. Rev. Lett. **76**, 2173 (1996).

[94] P.B. Wiegmann, Phys. Lett. **152B** 209 (1985); JETP Lett. **41** 95 (1985).

[95] S. Fujimoto, J. Phys. Soc. Jpn., **68**, 2810 (1999).

[96] F. Bonetto, J.L. Lebowitz and L. Rey-Bellet, "Mathematical Physics 2000", 128, Imp. Coll. Press, London (2000).

[97] S. Lepri, R. Livi and A. Politi, Phys. Rep. **377**, 1 (2003).

Chapter 12

ENERGY TRANSPORT IN ONE-DIMENSIONAL SPIN SYSTEMS

A.V. Sologubenko and H.R. Ott

Laboratorium für Festkörperphysik, ETH Hönggerberg, CH-8093 Zürich, Switzerland

ott@solid.phys.ethz.ch, sologubenko@solid.phys.ethz.ch

Abstract We review the recent progress in studies of the thermal conductivity in quasi one-dimensional spin systems. After a brief outline of the existing theoretical understanding of the energy transport in one-dimensional spin systems, we present and discuss recently obtained experimental results on materials of this type. These include compounds containing Heisenberg antiferromagnetic $S = 1/2$ spin chains and spin ladders, spin-Peierls compounds, as well as spin chain materials where the itinerant magnetic excitations may be regarded as solitons.

Keywords: Energy transport, thermal conductivity, spin chains

1. Introduction

The appeal of one-dimensional (1D) physical systems for theorists is the perspective to work out exact solutions for a number of problems related with 1D model systems that have no analogues in higher dimensions. From the experimental point of view, a lot of progress has been made in material preparation techniques and good physical realizations of various 1D model systems have been achieved. Low dimensional systems also offer the possibility to study quantum effects that are masked in 3D systems but emerge most significantly in one-dimensional materials. Relevant examples are the absence of long-range order in 1D spin systems, and the breakdown of the Fermi-liquid theory in 1D systems of interacting electrons.

Both theoretical and experimental interests in energy (heat) transport in 1D systems have initiated related work for many years but more recently the field experienced a renaissance because of the availability of new theoretical methods and of experimental observations of anomalous

D. Baeriswyl and L. Degiorgi (eds.), Strong Interactions in Low Dimensions, 383–417.

features of heat transport in a few classes of new materials which provide good physical realizations of idealized 1D model systems.

Quite generally, energy transport can experimentally be probed by measurements of the thermal conductivity κ as a function of temperature T and magnetic field H. The interpretation of respective data often meets complications but most of these difficulties can be mastered in reliable ways. If this applies, a lot of information concerning a variety of interactions in solids can be extracted from $\kappa(T, H)$ data.

In this review, we consider only one, yet very important type of 1D systems, namely one-dimensional spin systems. Rather comprehensive discussions of heat transport in 1D systems via electrons and phonons can be found in recent review articles of Lepri *et al.* [1], Rego [2], and Ciraci *et al.* [3]. Although we briefly outline the current theoretical understanding of the subject, we intend to focus mainly on the experimental progress in this field of research. An extensive presentation of many theoretical aspects of transport in low-dimensional systems is given in the chapter by X. Zotos and P. Prelovšek.

2. Energy transport in solids

2.1 Theoretical background

The thermal conductivity coefficient relates the heat flux $\dot{\mathbf{Q}}$ and the temperature gradient ∇T by

$$\dot{\mathbf{Q}} = -\kappa \nabla T. \tag{12.1}$$

As is the case for the electrical conductivity σ, κ is generally a tensor, but in many situations it is sufficient to restrict the analysis to one of the main crystallographic axes, e.g. α, and thus to consider only one component $\kappa_{\alpha\alpha} \equiv \kappa$. The most common method to measure the thermal conductivity relies on the proportionality between the heat flux and the temperature gradient (Fourier's law). Although this law has many times been verified experimentally, there is, as far as we know, no rigorous theoretical justification for it until now. In some special cases Fourier's law is claimed to be invalid, and special conditions are required for the proportionality between $\dot{\mathbf{Q}}$ and ∇T. The problem concerning the existence of diffusive heat transport implied by Eq. (12.1), recently became the subject of intensive theoretical investigations. Many details of the current status of this topic can be found, e. g., in Ref. [1].

In analogy with the heat transport in an ideal gas, the simplest approximation for the thermal conductivity is given by the kinetic equation

$$\kappa = \frac{1}{3} C v \ell, \tag{12.2}$$

where C is the specific heat, v the velocity, and ℓ the mean free path of the entities that are carrying the energy. It is assumed that these entities all move with the same velocity and experience the same mean free path.

The most widely used method in analyzing thermal conductivity data of solids relies on the kinetic Boltzmann equation. This approach is based on the assumption that the transfer of heat is accomplished by quasiparticles (phonons, electrons, magnons, etc.) on the branch p of possible modes with energy $\hbar\omega$ and quasi-momentum $\hbar\mathbf{k}$, and that the problem can be treated with second order perturbation theory. Another simplifying assumption usually employed in connection with this method is that any deviation $\delta n(\mathbf{k},p)$ from the equilibrium distribution $n(\mathbf{k},p)$ of occupied modes relaxes exponentially, such that

$$\frac{\partial \delta n(\mathbf{k},p)}{\partial t} = -\frac{\delta n(\mathbf{k},p)}{\tau(\mathbf{k},p)}. \tag{12.3}$$

with $\tau(\mathbf{k},p)$ representing the relaxation time of the quasiparticle mode $(\mathbf{k},p)$. In this relaxation time approximation the corresponding equation for the thermal conductivity can be written as

$$\kappa_{\alpha\beta} = \frac{1}{V}\sum_p\sum_{\mathbf{k}} \hbar\omega(\mathbf{k},p)\frac{\partial n(\mathbf{k},p)}{\partial T} v_\alpha(\mathbf{k},p)v_\beta(\mathbf{k},p)\tau(\mathbf{k},p,T), \tag{12.4}$$

where v_i is the group velocity of the $(\mathbf{k},p)$ mode along the direction i and V is the volume. The total relaxation time is determined by the simultaneous action of different relaxation mechanisms. Examples are interactions with defects and other quasiparticles. In most cases, the individual scattering mechanisms act independently from each other and, therefore, the total relaxation rate τ^{-1} can be represented as a sum of individual terms

$$\tau^{-1} = \sum_i \tau_i^{-1}, \tag{12.5}$$

where τ_i^{-1} represents the ith scattering mechanism. Eq. 12.5 is also known as Matthiessen's rule. In some special cases, however, interference effects can be important [4]. Any type of defects or quasiparticles can, in principle, scatter other quasiparticles. The relative strength of a particular scattering mechanism depends on the magnitude of the perturbation in the Hamiltonian of the system and can vary with temperature, concentration of the scattering centers or external parameters, such as a magnetic field. Detailed discussions of heat transport in solids based on a Boltzmann-type kinetic theory and taking into account various scattering processes, can be found, e.g., in relevant books [5, 6] or review articles [4, 7, 8].

The approach described above has proven to be very successful in analyses of heat transport in 3D solids by quasiparticles such as phonons, electrons and magnons. With reasonable choices of possible scattering channels, important information about relaxation rates and types of defects involved in the scattering has been extracted from the experimental $\kappa(T, H)$ data of various types of materials. However, the Boltzmann formalism is only applicable if the excitation of quasiparticles may be regarded as a small perturbation, and this requirement is often invalid in low-dimensional systems. The second difficulty in our context is that the application of the relaxation time approximation is questionable in strongly correlated systems. Strong correlations are, however, an intrinsic property of low-dimensional spin systems.

That is why for many cases of 1D systems it is more suitable to use the Kubo formalism of the linear response theory [9]. In this scheme, the thermal conductivity is given by

$$\kappa_{\alpha\beta} = \frac{V}{k_B T^2} \int_0^\infty < J_\alpha(t) J_\beta(0) > dt, \qquad (12.6)$$

where J_i is the energy flux along the i-direction and $< \ldots >$ is the equilibrium average at the temperature T. This approach does not require such simplifying assumptions as the weakness of interactions or those of quasiparticle models. It relies on the direct calculation of the energy correlation function. The main task is to accurately calculate the energy correlations for a large-size system. The Kubo formalism has recently become the preferred method for calculations of the energy transport in various low-dimensional systems.

2.2 Energy transport by magnetic excitations from experiment

The most often used method for measurements of the energy transport is a straightforward application of Eq. (12.1). This so-called steady-state method is based on monitoring the temperature gradient that is established by applying a constant heat flux through the sample. It is, of course, not possible to realize an experiment where the energy transport of an isolated, purely 1D spin system is probed. In reality the measurements are made with 3D materials where the crystal lattice contains ions with nonzero spin and where the exchange interaction between the spins in one direction is much stronger than along other directions. Therefore, the heat transport via crystal lattice excitations (phonons) is omnipresent in steady-state experiments. Besides phonons, any type of quasiparticles, such as electrons, excitons, polarons etc can, in principle, participate in the transport of energy. To separate the contribution of itinerant spin excitations κ_s from all other contributions, which in the simplest case is at least the phonon conductivity κ_{ph}, is not an easy task, and in some instances even impossible. In general, the separation of the spin contribution can be achieved by either reducing the relative magnitude of all other contributions or by accurate calculations of all other contributions using realistic theoretical models and inserting experimentally accessible parameters. The reduction of κ_{ph} can be realized, e. g., by introducing additional centers of phonon scattering that, however, should have no detrimental effect on the magnetic heat transport.

Heat transport by itinerant spin excitations is expected to be much smaller in directions perpendicular to the spin chains than along them. Therefore a comparison of the thermal conductivities parallel and perpendicular to the spatial extension of the 1D spin system is usually helpful for the evaluation of the background contributions. Nevertheless, some precautions are indicated, because the phonon thermal conductivity as well as other possible channels of heat transport may be anisotropic as well. If this anisotropy is absent or weak, a conventional and very useful way to assess κ_{ph} is to invoke the Debye model, which assumes the same group velocities and relaxation rates for phonons of different polarizations. The phonon thermal conductivity is then given by

$$\kappa_{\mathrm{ph}} = \frac{k_{\mathrm{B}}}{2\pi^2 v}\left(\frac{k_{\mathrm{B}}}{\hbar}\right)^3 T^3 \int_0^{\Theta_D/T} \frac{x^4 e^x}{(e^x-1)^2}\tau(x,T)dx, \tag{12.7}$$

where $x = \hbar\omega/k_{\mathrm{B}}T$, ω is the frequency of a phonon, $\tau(x,T)$ is the total relaxation time, and Θ_D is the Debye temperature.

In a typical thermal conductivity experiment, the heat flux through a sample is created by a heater which generates only lattice excitations and therefore, some amount of spin-phonon interaction is required for providing the transfer of energy from the lattice to the spin system. Naturally this spin-lattice interaction reduces both the phonon- and the spin-related thermal conductivities because of scattering processes involving both types of quasiparticles but, if this interaction is absent or very weak, the spin-related heat conduction is not observable via a thermal conductivity experiment. The effective thermal conductivity κ_{eff} which is experimentally accessible in a coupled spin-lattice system is [10]

$$\kappa_{\mathrm{eff}} = (\kappa_s + \kappa_{\mathrm{ph}})\left(1 + \frac{\kappa_s}{\kappa_{\mathrm{ph}}}\frac{\tanh(AL/2)}{AL/2}\right)^{-1}, \tag{12.8}$$

where L is the sample length, and

$$A = \left(\tau_{s-p}^{-1} \frac{\kappa_s^{-1} + \kappa_{\rm ph}^{-1}}{C_{\rm ph}^{-1} + C_s^{-1}} \right)^{1/2} . \qquad (12.9)$$

In (12.9) τ_{s-p} is the spin-lattice relaxation time, and $C_{\rm ph}$ and C_s are the lattice and the spin specific heats, respectively. For very short spin-lattice relaxation times ($\tau_{s-p}^{-1} \rightarrow \infty$), $\kappa_{\rm eff} = \kappa_{\rm ph} + \kappa_s$. In the opposite case of a very weak spin-lattice interaction ($\tau_{s-p}^{-1} \rightarrow 0$), however, $\kappa_{\rm eff} = \kappa_{\rm ph}$ and the thermal transport by the spin system cannot be observed in such experiments, irrespective of how large the intrinsic κ_s of the sample is.

Two types of spin-phonon interactions can be distinguished. First, single-ion interactions cause resonant scattering of phonons at paramagnetic ions. The second possibility is of magnetostrictive type where the interaction is provided by changes of the exchange interaction due to lattice deformations. It was shown by Kawasaki [11] and Stern [12] that for the case of the magnetostrictive spin-phonon interaction, the same correlation functions which determine the spin thermal conductivity are also involved in the scattering of phonons. Those effects are especially pronounced in the vicinity of magnetic phase transitions, where the magnetic specific heat is anomalously enhanced. This offers an indirect method for the evaluation of the spin thermal conductivity. The feasibility of this approach was demonstrated by Rives, Walton and Dixon [13] who measured $\kappa(T)$ of the 3D uniaxial antiferromagnet $MnCl_2 \cdot 4H_2O$ in the region of the Néel temperature and identified the influence of the spin-phonon interaction. No transition to a long-range ordered phase is expected for ideal 1D spin systems and the transitions observed for real quasi-1D systems are all caused by interchain interactions. Nevertheless, an analysis in the spirit of Ref. [13] can be an alternative for the evaluation of κ_s in cases where a direct separation of spin and lattice contributions is, for some reasons, impossible.

There are other possible methods to probe the energy transport in spin systems. In the experiment of Hunt and Thompson [14], a heat pulse was introduced into the spin system of solid ^{3}He via applying RF power and the energy diffusion was monitored by employing NMR techniques. This method has the advantage that it does not rely on the existence of spin-phonon coupling, but, on the other hand, it is indirect and very difficult to realize.

An interesting effect, specific to the class of 1D spin systems, was predicted by Rácz [15]. He showed that the energy flux in an integrable 1D spin system should lead to a shift δk in the characteristic wavevector of the spin-spin correlations. This shift is estimated to be large enough to be observed in an inelastic neutron scattering experiment, thus providing yet another possibility to measure the magnitude of the energy flux in the spin system.

3. Energy transport in 3D spin systems (main features)

Before going into the details of heat transport in 1D spin systems, it seems reasonable to give a brief account of the most important results concerning the spin-mediated thermal conductivity in 3D magnetic systems, intended to provide a reference frame for the interested reader.

The thermal conductivity of weakly anisotropic 3D magnetic materials in ferro- or antiferromagnetically ordered phases, i. e., at temperatures below the Curie temperature T_C or the Néel temperature T_N, respectively, is well described by the quasiparticle (magnon) model which relies on the general equation (12.4). The original idea of Sato [16], claiming that spin waves can contribute to the heat conduction in solids was supported by pioneering experimental observations of magnon heat transport in ferrimagnetic YIG [17, 18] and was subsequently elaborated in more details in a number of theoretical papers [19, 20, 21, 22]. At very low temperatures, when the magnons are mostly scattered by sample boundaries and, therefore, the magnon mean free path is constant, the typical temperature dependence of the magnon-mediated thermal conductivity is $\kappa_s \propto T^2$ for ferromagnetic substances and $\kappa_s \propto T^3$ for antiferromagnets. This is valid for cases of gapless magnon spectra or of gaps Δ_s that are much smaller than k_BT. However, at temperatures much lower than Δ_s/k_B, an exponential decrease of $\kappa_s(T)$ is expected. At higher temperatures, the temperature dependence of κ_s depends on the type of the dominant scattering mechanisms.

Besides providing a channel for heat transport, magnetic excitations also scatter phonons and thus reduce the phonon thermal conductivity. The net effect in magnetic substances can be either positive or negative, depending on the relative strength of both effects, and the situation can be different in different temperature regions. The application of external magnetic field changes both the magnitude of the magnon thermal conductivity and the strength of the phonon-spin scattering, leading to peculiar field dependences $\kappa(H)$ of the total thermal conductivity. Since the subject of this paper is heat transport in 1D spin systems where intrinsic 3D ordering is absent, we refrain from giving a full account of existing experimental results on the thermal conductivity of 3D systems in magnetically ordered phases. A review of this topic may be found, e.g., in Refs. [10, 23].

By enhancing the temperature in the ordered phase and approaching T_C (or T_N), the magnon mean free path decreases rapidly because of the interaction between individual magnons. In the vicinity and above the transition temperature, the concept of considering magnons as quasiparticles looses its meaning and thus the Boltzmann-type formalism becomes invalid. At these temperatures, the magnetic energy transfer occurs via diffusion at the rate of $\sim J/\hbar$ and the distance between collisions is close to the distance a between neighboring spins. Under these circumstances the thermal conductivity is given by [11]

$$\kappa_s = 8n_s\tau a^2 S^3(S+1)^3(zJ^2)^2(27k_B\hbar^2T^2)^{-1}, \tag{12.10}$$

where n_s is the number of spins per unit volume, and z is the number of nearest neighbors. The relaxation time τ is given via

$$\tau^{-2} = \frac{8}{3}\xi S(S+1)zJ \tag{12.11}$$

and

$$\xi \equiv 1 - \frac{39}{5z^2}\{1 + \frac{3}{26S(S+1)}\}. \tag{12.12}$$

The energy diffusion constant $D_E = \kappa/C_s$, where C_s is the specific heat per unit volume, is, at high temperatures, given by

$$D_E = KJ[S(S+1)]^{1/2}a^2/\hbar, \tag{12.13}$$

where K is a constant of order unity. The diffusive thermal conductivity in the paramagnetic regime appears to be considerably smaller than the phonon conductivity at the same temperatures, and therefore has not directly been observed in 3D spin systems with this type of experiments.

4. One-dimensional spin systems

4.1 Types of quasi-1D spin systems

Spin chains are probably the best suited systems for studying quantum effects that are related with low dimensionality. Upon reducing the dimensionality of the spin system, quantum effects gain in importance and cause both qualitative and quantitative differences in the physical behavior of 1D systems as compared to their 3D counterparts. As mentioned above, unlike in 3D and 2D systems, there is no long-range order at non-zero temperature in purely one-dimensional spin systems. Similarly, quantum effects are much more pronounced for smaller values S of the spins. For example, $S = 5/2$ systems exhibit, in many respects, classical behavior. Reducing S, however, leads to considerable renormalisations of the excitation spectrum and qualitative differences between systems with different values of S. Chains of antiferromagnetically coupled spins with $S = 1/2$, e.g., adopt a gapless ground state with an algebraic decay of spin correlations, whereas the same variety of spin chains with $S = 1$ form a spin-liquid ground state, separated by an energy gap from the spectrum of excited states and an exponential decay of spin correlations.

Apart from the magnitude of the spin value S, magnetic systems may also be distinguished by their spin dimensionality ν, i. e., by the number of components of the spins. Depending on this number, one distinguishes between model systems of Heisenberg type ($\nu = 3$), XY type ($\nu = 2$), and Ising type ($\nu = 1$). If, as is often the case, only nearest-neighbor interactions between individual spins are taken into account, the corresponding general Hamiltonian is

$$H = \sum_{i}^{N} \left(J_x S_x^i S_x^{i+1} + J_y S_y^i S_y^{i+1} + J_z S_z^i S_z^{i+1} \right) - g\mu_B \mathbf{H} \cdot \sum_{i}^{N} \mathbf{S}_i, \quad (12.14)$$

where the sums are over N spins in the chain. The second term considers the influence of an external magnetic field $\mathbf{H}$. Obviously, positive or negative values of J_α correspond to antiferromagnetic (AFM) or ferromagnetic (FM) interactions, respectively. Depending on the relative values of the exchange integrals J_α, the Ising model applies when $J_x = J_y = 0$, $J_z \neq 0$, the XY (or planar) model Hamiltonian corresponds to $J_x, J_y \neq 0$, $J_z = 0$, and for the Heisenberg-type situation, $J_x, J_y, J_z \neq 0$).

If Eq. (12.14) is extended by inclusions of additional terms, such as single-ion anisotropy $D\sum_i (S_z^i)^2$) or higher-order neighbor interactions and frustration effects, many different ground states may be achieved. The same is true if mixed-spin systems or different combinations of spin chains, such as ladders or zig-zag chains, are considered. Introducing a coupling to the lattice can result in the appearance of new phases such as spin-density-wave states, or in the occurrence of spin-Peierls transitions. We do not attempt to give a full account of all possible types of magnetic 1D systems, because the theory of energy transport has only been developed for a few types of 1D spin systems. Relevant experimental data sets are available for materials where the spin chain subsystems are best described by three types of Hamiltonians. These are isotropic $S = 1/2$ Heisenberg AFM chain systems, almost isotropic $S = 5/2$ Heisenberg AFM chains with weak planar anisotropy, and almost isotropic $S = 1/2$ and $S = 1$ Heisenberg FM chains with weak planar anisotropy.

Most of the experiments discussed in this brief review were made on materials containing $S = 1/2$ Heisenberg-type antiferromagnetic chains

(HAFC). Therefore, a brief description of the excitation spectrum of the corresponding model system seems useful.

The Bethe ansatz method allows to find exact solutions for the ground state of the model system [24]. Although no long-range order exists for the $S = 1/2$ HAFC, the spin-spin correlations decay only slowly as a function of the distance r between spins, proportional to $1/r$. The elementary excitations of an isotropic $S = 1/2$ HAFC are unbound pairs of fermionic $S = 1/2$ excitations [25], called spinons. The dispersion relation for spinons is gapless and is given by [25]

$$\varepsilon(k) = \frac{J\pi}{2} \sin ka, \tag{12.15}$$

where ε is the spinon energy. The possible values of the wave vector k are restricted to one half of the Brillouin zone $0 \leq k \leq \pi/a$. Since spinons can only be created in pairs, they form a continuum of two-spinon states with wave vectors $q = k_1 + k_2$ and energy $\epsilon(q) = \varepsilon(k_1) + \varepsilon(k_2)$ [25], where k_1 and k_2 are the wavevectors of the two individual spinons, respectively. Thermodynamic quantities, such as the specific heat or the static magnetic susceptibility have been calculated using the Bethe ansatz solution [26, 27]. The specific heat is linear in T at low temperatures ($T \ll J$) and decreases as $1/T^2$ at high temperatures $T \gg J$.

By introducing a planar anisotropy, such that in Eq. (12.14) $J_x = J_y \equiv J$ and $-J < J_z \leq J$, the spectrum of spin excitations in this XXZ model system remains gapless. On the contrary, AFM chains with an Ising-type anisotropy ($J_z > J$) exhibit a gapped spectrum of spin excitations. There are a number of other reasons for the formation of a spin gap in AFM chains, and several examples, related to energy transport will be given below. These are, e.g., the magnetoelastic interaction in spin-Peierls systems, the geometric arrangement of chains in even-leg spin ladders, the charge ordering in quarter filled ladders, and the dimerization due to the ordering of holes doped into chains. In $S = 1/2$ systems with a spin gap, the corresponding excitations are described in terms of either bound pairs of spinons or $S = 1$ magnons.

4.2 Theoretical aspects of energy transport in 1D spin systems

As a result of the enhanced influence of quantum fluctuations, no long range order phenomena are expected in 1D spin chains even at $T = 0$ K. Early theoretical estimates have shown that for a particular class of 1D spin systems, instead of diffusive energy flow typical for paramagnetic 3D substances, the energy propagation occurs ballistically. The character of this energy transport differs for different anisotropies of the spin-spin interactions.

Huber *et al.* [28, 29] calculated the second and forth moments of the Fourier transforms of the energy density correlation functions and showed that the diffusion of energy is absent for all wave vectors k for both XY and Heisenberg $S = 1/2$ chains. This is not so, however, for $S > 1/2$ where, at least for small k's, the diffusive energy transport dominates. The same authors also calculated the energy diffusion constant D_E for Heisenberg spin chains. For large S (classical limit), Eq. (12.13) is valid at high-temperatures, but for $S = 1/2$, D_E diverges. This result was confirmed by others using different theoretical methods [30, 31]. It was also shown that in an external magnetic field, the energy transport remains non-diffusive in the XY $S = 1/2$ chain [30] but energy diffusion is expected for isotropic $S = 1/2$ chains [31]. If next-nearest-neighbor interactions are not negligible, a diffusive spin-mediated energy transport is reestablished [31].

Computer-simulation studies of energy transport in classical 1D Heisenberg chains, with either FM or AFM nearest-neighbor interactions, were made by Lurie *et al* [32]. The authors obtained high-temperature values of D_E, and the spin diffusion constant D_S in the high temperature limit.

The analysis of dynamical correlation functions of the spin and energy densities for the 1D $S = 1/2$ Ising model [33] also indicates the absence of energy diffusion. This result is closely related to the fact that the energy density (for $q = 0$) is a constant of motion for the considered model. It was pointed out that this relies on the fact that an $S = 1/2$ chain can be mapped onto a system of noninteracting quasiparticles. The lack of interaction between quasiparticles of course implies an infinite mean free path and hence the absence of diffusive processes.

Recently, transport properties of various low-dimensional quantum systems were investigated by considering the integrability of corresponding models [34, 35, 36, 37, 38, 15]. This subject is discussed in detail in the chapter authored by X. Zotos and P. Prelovšek. Integrable models are characterized by a macroscopic number of conservation laws. One of them applies to the energy current, thus implying, in the absence of external perturbations, an infinite thermal conductivity [35]. Many important 1D model spin systems, such as the anisotropic Heisenberg model with nearest-neighbor interaction, were shown to be integrable, and the prediction of ballistic energy transport in those idealized systems seems to be undisputed [35, 38]. However, for real and experimentally accessible spin chain systems the influence of perturbations which can lead to deviations from integrability is hardly negligible. Such perturbations are almost inevitably introduced by impurities, phonons, 3D coupling, or next-nearest neighbor interactions. It then needs to be examined to what extent the predictions for the idealized systems have to be revised. With respect to experiment, the influence of phonons is especially important since, as we have pointed out in 2.2, some level of spin-phonon interaction is necessary for the observability of spin-related heat transport.

Klümper and Sakai [39] and Alvarez and Gros [40] considered the thermal conductivity of an integrable system, given by (in the notation of Ref. [40])

$$\kappa = \kappa^{(th)}\tau \tag{12.16}$$

with a finite thermal Drude weight $\kappa^{(th)}$. The relaxation time τ is set by extrinsic perturbations. Klümper and Sakai [39] calculated the temperature dependence of $\tilde{\kappa} \equiv \pi\kappa^{(th)}$ for the XXZ model. It was predicted that $\tilde{\kappa} \propto T$ for $T \ll J$ and $\tilde{\kappa} \propto 1/T^2$ at $T \gg J$. A maximum of $\tilde{\kappa}(T)$ is expected at roughly half the temperature of the specific-heat maximum. Calculations of Alvarez and Gros [40] claim the same results for the isotropic Heisenberg chain. They also calculated $\kappa^{(th)}(T)$ for two models with a gap in the excitation spectrum, namely for the isotropic Heisenberg chain with dimerized nearest and homogeneous next-nearest neighbor exchange and for the two-leg Heisenberg ladder. The calculations also give a peak of $\kappa^{(th)}(T)$ at temperatures below the temperature of the specific heat maximum, and an asymptotical $1/T^2$ behavior at high temperatures.

The influence of a gap in the spin excitation spectrum on the thermal conductivity in a 1D spin system was investigated by Saito and Miyashita [41]. The gap was introduced by an alternation of bonds in the Heisenberg chain, leading to different values of J between neighboring sites. For the zero-gap situation, these authors found a peak in $\kappa(T)$, in good agreement with the results of Klümper and Sakai [39]. It was found that the opening of the gap upon increasing the difference between different J values leads to an enhancement of the thermal conductivity and to a shift of the $\kappa(T)$ peak to higher temperatures, concomitant with a shift of the peak in the temperature dependence of the specific heat. From this observation it is concluded that bond alternation should have little influence on the mean free path of spin excitations.

The thermal conductivity of spin chains has also been of theoretical interest because of the possibility to study nonlinear excitations (solitons), typical for 1D systems in general and in magnetic spin systems in particular [42]. The motion of solitons is generally unaffected by interactions with other quasiparticles including other solitons and therefore, a robust transport of energy via solitons is expected. Wysin and Kumar [43] developed a theory for the thermal conductivity of a one-dimensional easy-plane classical ferromagnet in an external magnetic field. They considered the heat transport in a two-component ideal gas of magnons and solitons. They predicted a peak in the thermal conductivity vs. H, whose position and height both increase with temperature as T^2. Although the authors treated the soliton system in terms of classical mechanics, they also discussed possible quantum effects.

5. Experimental results

Any real three-dimensional solid containing chain-type structural elements can only approximately be regarded as a realization of a 1D system because albeit weak yet non-negligible interchain interactions cannot be avoided. The obvious signature of these interactions, for chains of half-integer spins, is the existence of 3D ordering phenomena at low temperatures. For some spin-chain materials, however, the ratio between the interchain and the intrachain interaction is very small. As a result, the corresponding phase transitions appear only at very low temperatures and there are extended temperature regimes above the critical temperature where the systems exhibit (quasi) one-dimensional features and may in many respects be considered as good physical realizations of various 1D model systems.

5.1 Heisenberg $S = 1/2$ AFM chains

$KCuF_3$. One of the most completely studied materials which represents very well an $S = 1/2$ isotropic HAFC system is $KCuF_3$. The compound has a tetragonal crystal structure, and the Cu^{2+} ions carrying the spins form chains running along the c-axis. The antiferromagnetic intrachain interaction $J = 34$ meV is relatively strong in comparison with the ferromagnetic interchain interaction $J' \approx 0.3$ meV [44, 45]. Because of the modest ratio $J'/J \sim 10^{-2}$, a 3D AFM ordering sets in at relatively high temperatures of $T_N = 38$ K or 22 K for polytype-a and polytype-d material, respectively. In both cases, the ordered moments of $\mu_0 = 0.48\mu_B$ per Cu ion are oriented along the c-axis [46]. Inelastic neutron scattering measurements [47] revealed that the spin excitation spectrum of $KCuF_3$ at $T > T_N$ is consistent with the spinon concept. Below T_N, the energy excitation spectrum does not change much for energies higher than k_BT_N, but at low energies, two gapless branches of transverse spin excitations with a linear dispersion near the AFM zone center develop [48]. In addition, a longitudinal mode with an energy gap at $q = 0$ was observed in inelastic neutron scattering experiments [48].

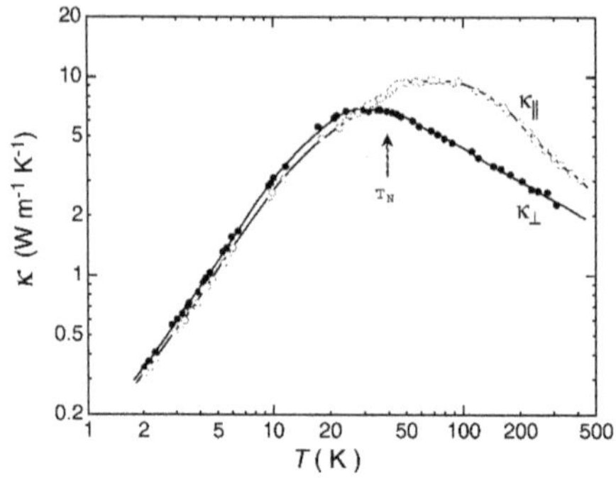

Figure 12.1. The temperature dependences of the thermal conductivities both along and perpendicular to the chains in $KCuF_3$ [50].

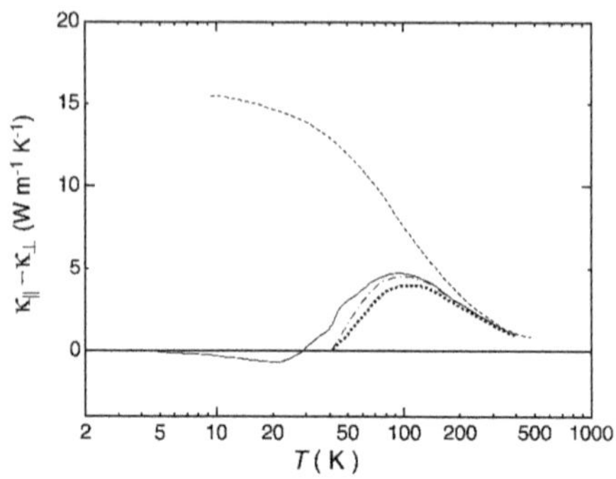

Figure 12.2. Spin contribution to the thermal conductivity along the chain direction (solid line) [50]. The broken line represents calculations for HAFC without corrections for the critical behavior near T_N. The dotted and the dashed-dotted lines represent calculations for HAFC, incorporating the critical behavior of Heisenberg- or Ising-type, respectively.

Results concerning the heat transport in $KCuF_3$ are available in Ref. [49] and, in more detail, in Ref. [50]. The temperature dependence of the thermal conductivity was measured on a nearly cube-shaped sample with the heat flow directed along the [001] direction (parallel to the chains, $\kappa_{\parallel}$) and also along the [110] direction (perpendicular to the chains, $\kappa_{\perp}$). The anisotropy ratio $\kappa_{\parallel}/\kappa_{\perp}$ is close to 1 below $T_N = 39.8$ K, but increases rapidly above T_N (Fig. 12.1). The authors of Ref. [50] assumed that $\kappa_{\perp}$ is exclusively due to phonon mediated heat transport. Since the difference between the lattice parameters c and a is relatively small for $KCuF_3$, the phonon thermal conductivity was assumed to be isotropic. The difference $\kappa_{\parallel} - \kappa_{\perp}$ at temperatures close to and above T_N, shown in Fig. 12.2 by the solid line, was attributed to the heat transport via diffusive spin modes $\kappa_{\rm dif}$. In the same work it was suggested that the diffusive character of the energy transport is restored by weak interchain interactions and an Ising component of the interchain exchange. It is obvious that, in spite of the predictions for integrable HAFC, the spin contribution to κ is relatively small above T_N and is completely absent below T_N. The broken line in Fig. 12.2, representing an estimate for the spin mediated heat transport $\kappa_{1D\text{-}\rm HAF}$ of HAFC [51], deviates considerably from the experimental curve. This was interpreted to be the result of critical scattering in the region of T_N. Fits to the experimental data for $T \geq T_N$ using the equation

$$\kappa_{\mathrm{dif}} = \kappa_{1D\cdot\mathrm{HAF}} \left| \frac{T - T_N}{T} \right|^{\rho} \tag{12.17}$$

with values of ρ compatible with the critical exponents that are valid for Heisenberg (dotted line in Fig. 12.2) and Ising systems (dashed-dotted line in Fig. 12.2) suggested that the Ising exponent is more adequate for approximating the experimental data. As may be seen, however, the overall agreement between experiment and theory is rather poor. Thus, because of the strong influence of fluctuations near the ordering transition, $KCuF_3$ does not appear to be a very suitable material for studying the spin-related thermal conductivity of 1D systems. In addition it is not clear why no thermal transport carried by magnons is observed far below T_N, where the influence of fluctuations should be much reduced.

Yb_4As_3. An interesting realization of an $S = 1/2$ HAFC was found in Yb_4As_3 (see a recent review article of Schmidt *et al.* [52]). In this compound, the low dimensionality is caused by a charge-ordering transition. Above the transition temperature $T_{co} = 295$ K, Yb_4As_3 is an intermediate-valent metal with the cubic anti-Th_3P_4 structure, where Yb ions form interpenetrating chains along the four diagonals of the unit cell. At T_{co}, the unit cell contracts along one of the diagonals, thus forming chains of Yb^{3+} ions along this diagonal while the rest of ytterbium ions adopt the 2+ configuration. The crystal electric field splits the ground state multiplet of an Yb^{3+} ion into four Kramers doublets. The separation between the lowest doublet and the excited states is rather large (≥ 14 meV) and therefore, at low temperatures, only the ground state with an effective spin $S = 1/2$ needs to be considered. In the low-temperature phase, the material is a semimetal with a concentration of about 10^{-3} itinerant holes per formula unit. Inelastic neutron scattering measurements by Kohgi *et al.* [53] demonstrated that the spin excitation spectrum is a spinon continuum, typical for an $S = 1/2$ isotropic 1D HAF Hamiltonian with $J \approx 2.2$ meV. A linear term in the temperature dependence of specific heat was observed at low temperatures and was attributed to excitations in the spin system [54].

The thermal conductivity κ of Yb_4As_3 was measured on a multidomain sample where in some fraction of the domains the spin chains were oriented along the heat flux direction [54, 52]. The $\kappa(T)$ data in zero magnetic field are presented in Fig. 12.3. At temperatures between 0.5 and 6 K, $\kappa(T)$ can be approximated by the sum of two terms $a_1 T + a_2 T^2$ (a_1 and a_2 are constants). Below 0.5 K, however, $\kappa(T)$ decreases much faster than predicted by this relation. This is emphasized in the inset

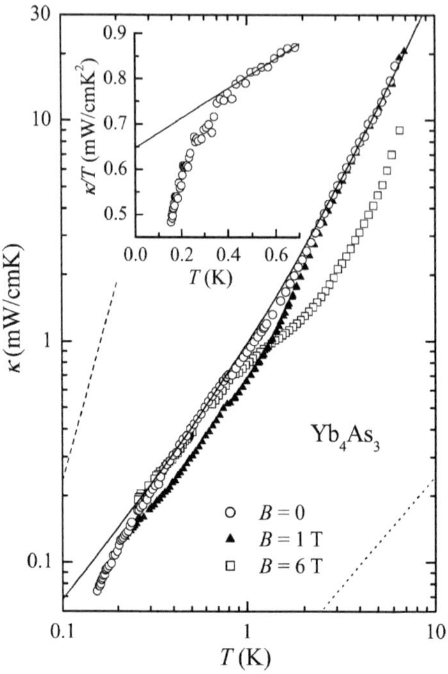

Figure 12.3. The temperature dependence of the thermal conductivity of Yb_4As_3 [54]. The solid line is the fit to the equation $\kappa = a_1 T + a_2 T^2$. The dotted line is the electronic thermal conductivity calculated from the electrical resistivity data. The dashed line is an estimate for $\kappa_{\text{ph}}(T)$ if only boundary scattering is considered.

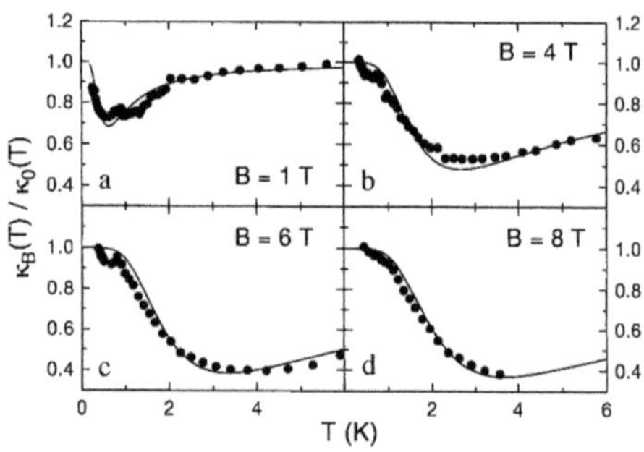

Figure 12.4. Field dependences of the normalized thermal conductivity in Yb_4As_3[54]. The solid lines are fits employing a model that considers phonon-soliton scattering.

of Fig. 12.3. The first term was attributed to the contribution of 1D spin-wave-like excitations and/or holes and the second term to phonons scattered off those excitations/holes [54]. Assuming that the linear in T terms of both the specific heat and the thermal conductivity are exclusively due to spin excitations, the mean free path of those excitations turns out to be constant and equal to 500 Å [52]. The behavior of $\kappa(T)$ below 0.5 K, which correlates with similar features of the specific heat, was attributed to spin-glass freezing.

External magnetic fields cause a sizable reduction of the thermal conductivity of Yb_4As_3 [54], illustrated in Fig. 12.4. This reduction was attributed to an additional scattering of phonons by solitons that are induced by the magnetic field. In Ref.[54] the data were analyzed by

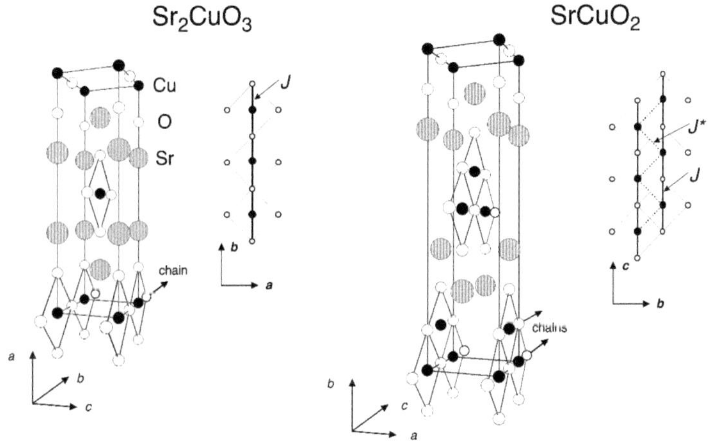

Figure 12.5. Crystal structures of Sr_2CuO_3 and $SrCuO_2$, respectively.

using a phenomenological model ascribing the phonon relaxation rate to resonant phonon-soliton scattering, as suggested in Ref. [55] for $S = 5/2$ spin chains. The model agrees fairly well with experiment, but one of the fit parameters, the soliton rest energy, exhibits a nonlinear field dependence. This anomalous behavior was considered as a manifestation of the limitations of the essentially classical model if applied to quantum dominated $S = 1/2$ chains.

Sr_2CuO_3 and $SrCuO_2$. Recently, some excellent physical realizations of different low-dimensional spin models have been found in cuprate compounds. Among them, Sr_2CuO_3 is considered as a nearly ideal realization of an $S = 1/2$ isotropic HAFC. The crystal structure of the material contains chains of Cu^{2+} ions sitting in the centers of CuO_4 squares (see Fig 12.5). The neighboring CuO_4 squares share corners, thus forming 180° Cu-O-Cu bonds with an extremely strong AFM exchange J of 200-300 meV. The interchain interaction J' is very weak ($\alpha \equiv J'/J \sim 10^{-5}$) [56], such that the temperature of the 3D ordering transition T_N is about 5 K or even lower [57], and the ordered moment is extremely small $\mu_0 = 0.06\mu_B$ [58]. A related material is $SrCuO_2$ which is built by double Cu-O chains with J as large as in Sr_2CuO_3 but arranged in pairs forming Cu-O zig-zag ribbons (see Fig. 12.5). The interaction J^* between the two chains of the same ribbon occurs via 90° Cu-O-Cu bonds providing a weak ($|J^*| = 0.1$-$0.2\ J$) frustrating ferromagnetic interaction [59]. Because of the low $T_N \approx 5$ K, the interchain

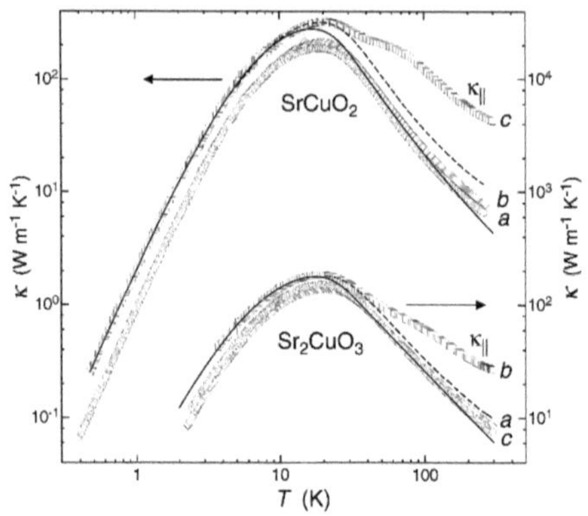

Figure 12.6. The temperature dependences of the thermal conductivities of Sr_2CuO_3 and $SrCuO_2$ along different crystallographic directions [64]. The solid and dashed lines represent different evaluations of the phonon contribution to $\kappa_\parallel$ (see Ref. [64]).

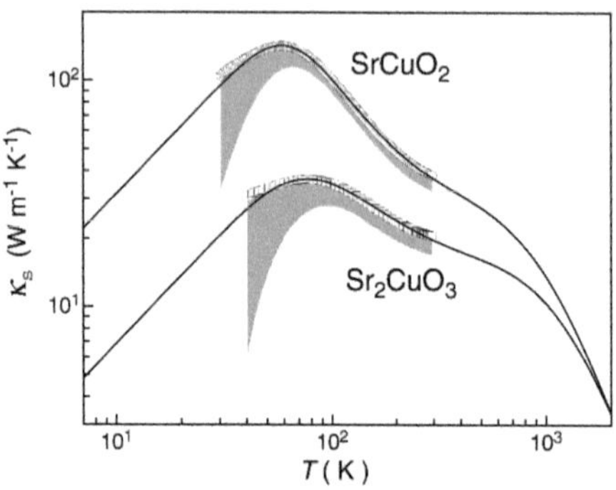

Figure 12.7. Spinon thermal conductivities of $SrCuO_2$ and Sr_2CuO_3. The solid lines are calculated considering a free fermion model. The shaded areas indicate possible errors caused by the uncertainties in the estimate of the phonon thermal conductivity.

interaction J' for $SrCuO_2$ must be as low as for Sr_2CuO_3. Although no attempts have been made to obtain direct evidence of the spinon continuum by inelastic neutron scattering measurements, other experiments are consistent with the spinon scenario [60, 61, 62]. Because the interaction between the chains forming the Cu-O ribbons in $SrCuO_2$ is ferromagnetic, the excitation spectrum should remain gapless [63].

The low-temperature thermal conductivities of Sr_2CuO_3 and $SrCuO_2$ [57, 64] are displayed in Fig. 12.6. The measured total thermal conductivity is not isotropic (see Fig. 12.6), but the anisotropy ratio $\kappa_\parallel/\kappa_\perp$ is only weakly T-dependent at $T \leq 30$ K. This residual anisotropy was attributed to the phonon contribution $\kappa_{\rm ph}$. An anomalous contribution to $\kappa_\parallel$, the conductivity along the chain direction, may readily be identified in Fig. 12.6. Based on a detailed analysis of phonon scattering mechanisms, the phonon contribution $\kappa_{\rm ph}$ and its extrapolation to temperatures above 30 K was established for $\kappa_\parallel$. The excess contribution $\kappa_s = \kappa_\parallel - \kappa_{\rm ph}$ was attributed to heat transport by spinons. This spinon contribution, shown in Fig. 12.7, is rather large; the maximum value of κ for $SrCuO_2$ is more than 20 times larger than the corresponding maximum value for $KCuF_3$. The calculation of the expected diffusive energy transport $\kappa_{s,\rm dif} = D_E C_s$, where the the high-temperature limit of the energy diffusion constant D_E is given by Eq. (12.13), and hence $\kappa_{s,\rm dif} \sim n_s a^2 k_B^2 T/\hbar$, results in much smaller values than κ_s from experiment. This discrep-

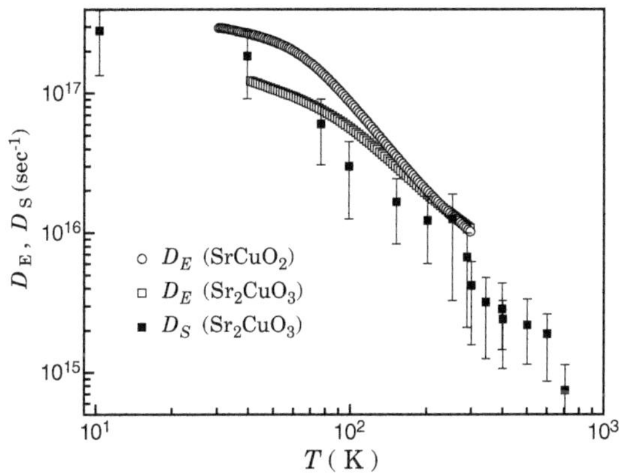

Figure 12.8. The energy diffusion constants $D_E(T)$ of Sr_2CuO_3 and $SrCuO_2$, estimated from thermal conductivity data [57, 64], and the spin diffusion constant $D_S(T)$ of Sr_2CuO_3 [65].

ancy was considered as an indication for a quasiballistic (non-diffusive) nature of the energy transport in Sr_2CuO_3 and $SrCuO_2$. The $\kappa_s(T)$ data were analyzed in terms of a fermionic model for spinons which, for $T \ll J$, gives

$$\kappa_s = \frac{2n_s k_B^2 a}{\pi\hbar} T \int_0^{J\pi/2k_BT} \frac{x^2 c^x}{(e^x+1)^2} \ell_s(\varepsilon, T)dx, \tag{12.18}$$

where $x = \varepsilon/k_BT$. The numerical evaluation of (12.18) reveals that the spinon mean free path ℓ_s increases with decreasing temperature, up to a few thousand Å in the region of the peak of $\kappa_s(T)$.

Sr_2CuO_3 provides the opportunity for comparing the experimental results for the effective energy diffusion constant D_E, which can be calculated from $\kappa_s(T)$ data [57, 64], and the spin diffusion constant D_S, established by NMR measurements [65]. As shown in Fig.12.8, both parameters exhibit similar temperature dependences. At high temperatures, D_E slightly exceeds D_S. Theoretical calculations and numerical modeling typically give values of the ratio D_E/D_S in the interval between 1.4 and 3, in fair agreement with the data on Sr_2CuO_3, if the experimental uncertainty is taken into account. For comparison, $D_E/D_S \approx 2.1$ in the region $0.05 < T < 0.12$ K for ^{3}He [14].

The observation of spin-mediated energy transport in $S = 1/2$ chain cuprates, strongly enhanced in comparison with the expected diffusive behavior, is in principle consistent with theoretical predictions of ballistic energy transport for the relevant integrable models. In the ballistic limit one would expect the mean free paths to be of the order of the sample dimensions. The observed values of ℓ_s are still considerably smaller, however. The obvious question in this context is which processes may cause the reduction of the mean free path. It was suggested that the scattering at defects and phonons reduces ℓ_s [57, 64]; the solid lines in Fig. 12.7 correspond to calculations based on a simple phenomenological equation for $\ell_s(T)$ [57]. It is clear that a more accurate theory treating the interactions of spinons with phonons and various defects is needed to explain the temperature dependence of the spin-related thermal conductivity in real $S = 1/2$ chain materials. Such calculations are especially important because a qualitatively different behavior is expected for quasiparticles with different wavevectors, diffusive for small k and ballistic for large k, not accounted for by the simple approach of Eq. (12.18).

5.2 S=1/2 spin-Peierls compounds

$CuGeO_3$. The above mentioned materials with rather large values of J, may be regarded as close to ideal $S = 1/2$ HAFC's with gapless spectra of spin excitations. The vibrations of the crystal lattice do not strongly influence the magnetic properties of these compounds and vice versa. If the magnetoelastic coupling is strong, however, a spin-Peierls-type transition may lead to a dimerized state characterized by a spin gap and a lattice deformation. This type of transition has previously been observed in organic spin-chain materials, but in recent years the common interest in this field was focused onto the only known inorganic spin-Peierls compound $CuGeO_3$ [66].

Subunits of the crystal structure of $CuGeO_3$ are both CuO_2 and GeO_4 chains, directed along the c-axis of an orthorhombic unit cell. The Cu-O-Cu bond angle is close to 90° (about 98°) which provides a superexchange J of 10.4 meV in the chain direction [67]. The interaction between the chains is considerable: $J_b \sim 0.1J$ and $J_a \sim -0.01J$ [67]. The spin-Peierls transition occurs at $T_{SP} \approx 14$ K. Below this temperature, a gap of about 2 meV is established in the spectrum of spin excitations (AFM magnons). Above T_{SP}, the magnetic properties of $CuGeO_3$ can be rather well described by a Heisenberg model taking into account, besides the nearest-neighbor interaction J, also a next-nearest-neighbor interaction αJ, with $\alpha \approx 0.36$ [68].

Results of measurements of the thermal conductivity of $CuGeO_3$ were reported in Refs. [69, 70, 71, 72, 73, 74, 75, 76, 77, 78]. The temperature dependence of the zero-field thermal conductivity along the chain direction $\kappa_c(T)$ exhibits two distinct maxima, one of them above T_{SP}, at temperatures between 15 and 20 K, and the other distinctly below

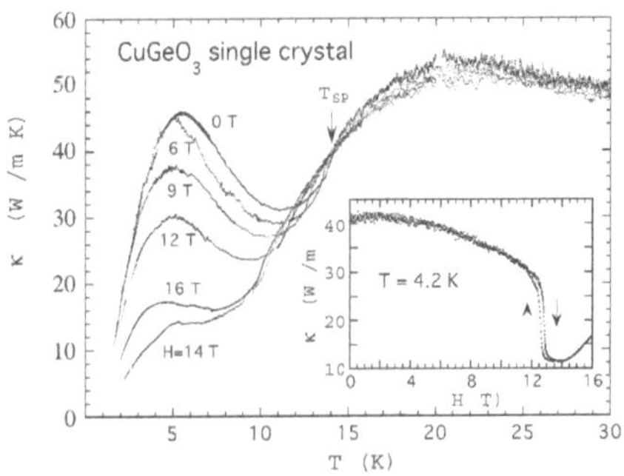

Figure 12.9. The thermal conductivity of $CuGeO_3$ along the c-axis in magnetic field [70]. Inset: κ as a function of H at $T = 4.2$K

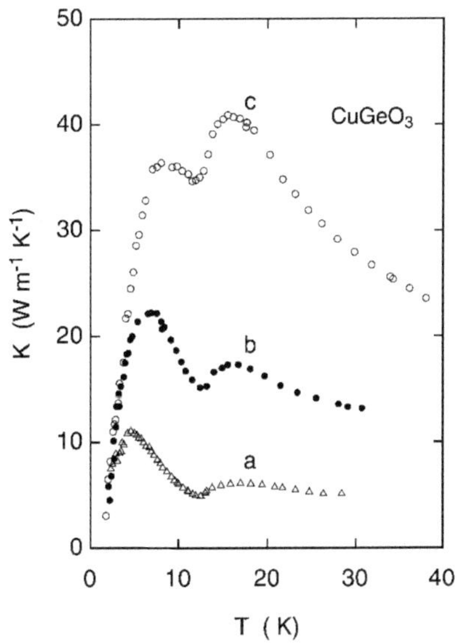

Figure 12.10. The thermal conductivity of $CuGeO_3$ along the a, b, and c-axes [71].

T_{SP}, as shown in Fig. 12.9. The scenario put forward by Ando *et al.* [70] and, independently, by Salce *et al.* [71] suggests that $\kappa(T)$ above T_{SP} comprises both phonon and spinon contributions, the spinons being responsible for the high-temperature peak. With decreasing T, κ_s decreases rapidly below T_{SP} as a result of the opening of the spin gap, while the phonon contribution increases because of the reduction of the phonon-spin interaction. This interpretation of the low-T peak is compatible with the strong suppression of this peak by magnetic fields, as may be seen in Fig. 12.9. Magnetic fields reduce the gap and thus enhance the phonon-magnon scattering. The high-temperature maximum is much less affected by external magnetic fields, as expected, if the above interpretation is valid. Takeya *et al.* [74, 76] investigated $\kappa_c(T)$ of $Cu_{1-x}Mg_xGeO_3$ and found that substituting Cu by Mg progressively suppresses κ_c until the critical concentration $x_c \sim 0.025$ is reached and then, the thermal conductivity is not reduced further with increasing x. Assuming that, for $x > x_c$, $\kappa_s \approx 0$ and that Mg-doping has little influence on $\kappa_{\rm ph}$, Takeya *et al.* [74, 76] estimated the spinon contribution $\kappa_s(T)$ for $x < x_c$. They found that the spinon mean free path of undoped $CuGeO_3$, just above T_{SP}, is $\ell_s \approx 1300$ Å [74] and decreases as $1/T$ at higher temperatures [76]. Estimates of Salce *et al.* [71] give values of l_s between 300 and 1000 Å in the same temperature region.

Another interpretation of the thermal conductivity of $CuGeO_3$ attributes both maxima of $\kappa(T)$ to phonons [69, 72, 73, 77]. The maximum of $\kappa_{\mathrm{ph}}(T)$ above T_{SP} is thought to be caused by an increasing scattering of phonons by spin excitations in the vicinity of the subsequent phase transition at T_{SP}. A strong argument in favor of this interpretation is the fact that similar two-peak features are observed for $\kappa(T)$ measured along all the crystallographic directions (see Fig. 12.10). It is difficult to attribute the high-temperature maximum of $\kappa_a(T)$ to a spinon contribution because the velocity of spin excitations v_s in this direction is vanishingly small (as $v_{s,a}/v_{s,c} = |J_a/J| \sim 0.01$). In contrast, a weakly anisotropic phonon scattering by spin fluctuations near T_{SP} provides a natural explanation for the obvious similarity of the anomalies in $\kappa(T)$ along all directions.

Although the existence and the magnitude of the spinon thermal conductivity in $CuGeO_3$ above T_{SP} remains an open question, convincing evidence for thermal transport via AFM magnons at $T \ll T_{SP}$ in $Cu_{1-x}Mg_xGeO_3$ was provided in Ref. [75]. Within a certain range of x and well below T_{SP}, a number of experimental results indicate the coexistence of a spin-singlet and an AFM-ordered state. Takeya *et al.* [75] observed that at $T < 0.58$ K, $\kappa_c(T)$ of $Cu_{1-x}Mg_xGeO_3$ ($x = 0.016$) significantly exceeds that of the pure ternary compound which does not order magnetically. Likewise a significant enhancement of the magnon specific heat was observed. The mean free path of the magnons was found to be of the order of the sample's dimensions, a clear sign of ballistic magnon propagation.

5.3 $S = 1/2$ spin ladders

$(La,Sr,Ca)_{14}Cu_{24}O_{41}$. The m-leg spin ladders are formed by m chains ("legs") with a coupling J between the neighboring spins of the same leg and $J_\perp$ between the neighboring spins on adjacent chains ("rungs") [79]. It has been demonstrated [80] that 2-leg $S = 1/2$ AFM Heisenberg ladders exhibit a spin gap for any nonzero $J_\perp$, in striking contrast to the gapless spectrum of single $S = 1/2$ chains. A 3-leg ladder, however, has again a gapless excitation spectrum and exhibits a quasi-long-range order, similar to the case of an $S = 1/2$ HAFC. This trend holds for any number of legs in the ladder: gapless spectra for odd-leg ladders and the formation of spin gaps in even-leg ladders. For the same ratio $J/J_\perp$, the size of the gap rapidly decreases with increasing m.

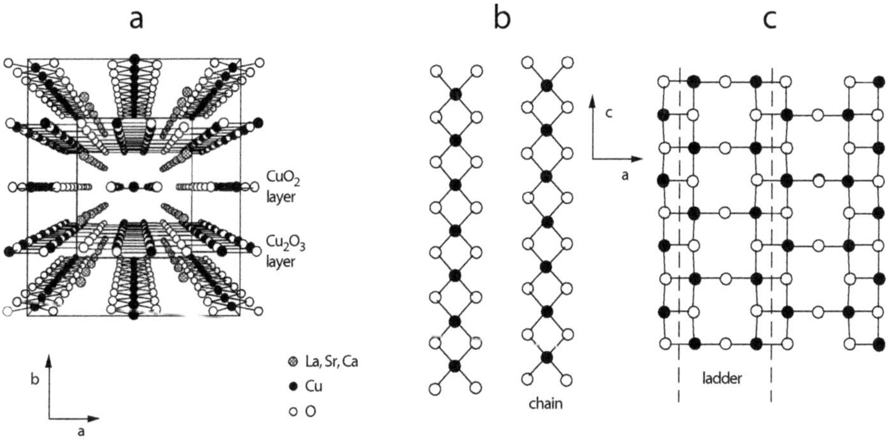

Figure 12.11. Crystal structure of (La,Sr,Ca)$_{14}$Cu$_{24}$O$_{41}$.

A number of materials, adopting a crystal structure containing spin ladders, such as $SrCu_2O_3$, $Sr_2Cu_3O_5$, (La,Sr,Ca)$_{14}$Cu$_{24}$O$_{41}$, CaV_2O_5, (La,Sr)CuO$_{2.5}$, and $Cu_2(C_5H_{12}N_2)_2Cl_4$, have been synthesized in recent years [81], but large enough single crystals, suitable for probing the anisotropy of the thermal conductivity, are available for (La,Sr,Ca)$_{14}$Cu$_{24}$O$_{41}$ only. The crystal structure of stoichiometric $Sr_{14}Cu_{24}O_{41}$, shown in Fig. 12.11, consists of CuO_2 and Cu_2O_3 layers, alternatingly stacked along the b-axis and separated by Sr layers.

The Cu_2O_3 subunit, as shown in Fig. 12.11 (c), contains two-leg ladders in which the Cu ions are linked by 180° Cu-O-Cu bonds which provide a strong antiferromagnetic coupling J of 110-130 meV between the Cu ions along the legs [82, 83]. From inelastic scattering measurements, the ratio $J/J_\perp$ was initially estimated to be 0.55 [82]; however, the analysis of more recent experimental data takes into account an additional ring interaction and suggests that $J \approx J_\perp$ [83]. The copper spins on the same rung of a ladder adopt a singlet ground state separated from the triplet excited state by an energy gap Δ_{ladder} of 30 - 33 meV [82, 84]. The dispersion $\omega(k)$ of the ladder spin excitations is very steep along the ladder direction but flat in the rung direction [85].

The CuO_2 layers contain linear chains of Cu ions linked by two nearly 90° Cu-O-Cu bonds (see Fig. 12.11 (b)), similar to $CuGeO_3$ but, due to the presence of a large number of holes, the magnetic excitation spectrum is different here. The formal valence of Cu in $Sr_{14}Cu_{24}O_{41}$ is +2.25, suggesting six holes per formula unit of the stoichiometric compound. Only a small fraction of them is situated on the ladders. Those holes are

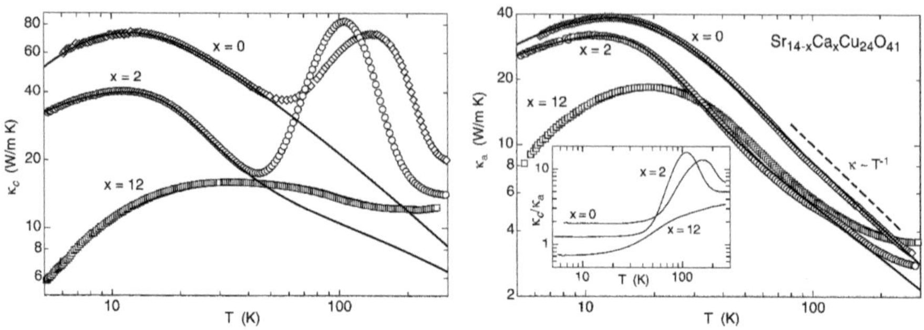

Figure 12.12. Thermal conductivity of $(Sr)_{14-x}Ca_xCu_{24}O_{41}$ along (left panel) and perpendicular (right panel) to the ladder direction [88].

mobile and responsible for the electrical conductivity of the material. Most holes are localized at oxygen sites in the CuO_2 chains, and are coupled to a copper spin to form a Zhang-Rice singlet. The remaining magnetic Cu ions in the chains form dimers which develop a long-range ordered structure. The ground and excited states of the dimers are separated by a gap Δ_{chain} ~11 meV, with a small dispersion amplitude of the order of 1 meV [86, 87].

Replacing Sr by isovalent Ca initiates a transfer of holes from the chains to the ladders, leading to a change of the temperature dependence of the c-axis resistivity from semiconducting to metallic at $x \sim 6$–8. The substitution does not alter Δ_{chain}, but the ordered state of dimers in the chains is less stable [86].

The temperature dependences of the thermal conductivities $\kappa_{\parallel}$ along the c direction of $Sr_{14-x}Ca_xCu_{24}O_{41}$ ($x = 0, 2, 12$) [88] exhibit the most distinct anomalies observed for any material containing 1D spin systems. Typical features of phonon heat transport are observed for $\kappa_a(T) \equiv \kappa_{\perp}(T)$ in the whole covered temperature range (see right panel of Fig. 12.12) Below 30 K, also $\kappa_c(T) \equiv \kappa_{\parallel}(T)$ is dominated by the phonon conduction and hence $\kappa_{\parallel}/\kappa_{\perp}$ is of the order of 1 in this temperature range (see the inset of Fig. 12.12, right panel). At higher temperatures, $\kappa_c(T)$ is qualitatively different from $\kappa_a(T)$, especially for semiconducting materials with $x = 0$ and $x = 2$, where an excess heat conduction leads to distinct maxima in $\kappa_c(T)$ above 100 K. These features are even more pronounced than those of the chain cuprates Sr_2CuO_3 and $SrCuO_2$, shown in Fig. 12.6. The shoulder-type anomalies are replaced by an anomalous second maximum of $\kappa_{\parallel}(T)$ at elevated temperatures. The authors of Ref. [88] attributed the low temperature maxima of $\kappa_{\parallel}(T)$ and $\kappa_{\perp}(T)$ to phonons and the excess high-temperature contribution to $\kappa_{\parallel}(T)$, to itinerant magnetic excitations.

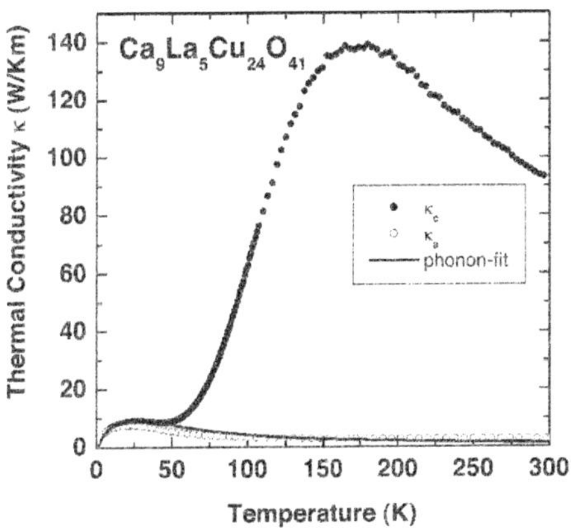

Figure 12.13. Thermal conductivity of $Ca_9La_5Cu_{24}O_{41}$ (after Ref. [89]).

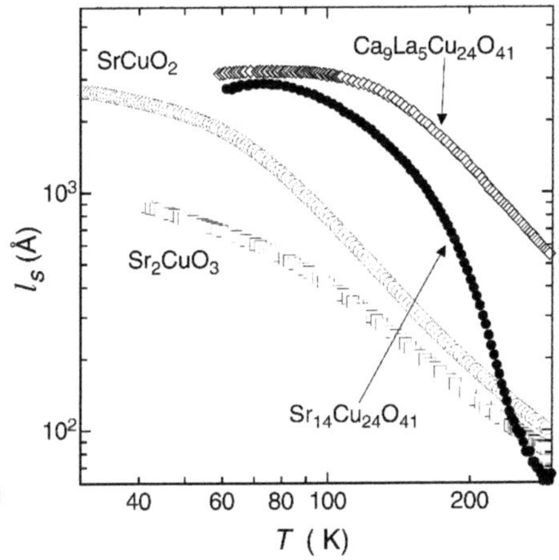

Figure 12.14. Temperature dependences of the mean free paths of the spin excitations in spin ladders $Sr_{14}Cu_{24}O_{41}$ and $Ca_9La_5Cu_{24}O_{41}$ [89] and in the spin chain systems Sr_2CuO_3 and $SrCuO_2$ [64].

The magnetic origin of an analogous high-T anomaly was even more convincingly demonstrated in Ref. [89] by investigating the anisotropy of $\kappa(T)$ in $Ca_9La_5Cu_{24}O_{41}$. A random distribution of Ca^{2+} and La^{3+} ions enhances the lattice disorder in this compound and hence a sizeable reduction of the phonon contribution without strongly altering the magnon contribution is anticipated. As shown in Fig. 12.13 for $Ca_9La_5Cu_{24}O_{41}$, the low-temperature maximum in $\kappa(T)$, presumably due to phonons, is indeed strongly reduced in comparison with $Sr_{14}Cu_{24}O_{41}$ for all crystallographic directions. The high temperature peak in $\kappa_{c(T)}$, however, is even more pronounced than in the parent compound.

Since there are two independent magnetic 1D systems in $Sr_{14-x}Ca_xCu_{24}O_{41}$, namely spin chains and spin ladders, the obvious problem is to identify the subsystem which carries an excessive amount of heat. The spectrum of magnetic excitations is gapped for both systems, and the minimum gap is about three times larger for the ladder than for the chains. However, the magnon excitations of the chains exhibit a rather flat dispersion, in contrast to the very high group velocity of ladder magnons along the c-direction, and, since the square of the group velocity appears in Eq. (12.4) for the quasiparticle thermal conductivity, the ladder spin excitations are expected to dominate the energy transport, in spite of the larger gap. This was confirmed via the analysis of the magnon contribution κ_s at low temperatures, where it exhibits an exponential increase with T. The value of the gap extracted from the $\kappa_s(T)$ data [88, 89] is close to those of Δ_{ladder} established by neutron scattering experiments [82, 84].

The magnon mean free path of hole-doped ladders at temperatures of the order of 300 K is found to be mostly limited by the scattering of magnons by holes. If the number of holes on the ladders is considerably reduced, such as in $Ca_9La_5Cu_{24}O_{41}$, the temperature dependence of the corresponding mean free path is very similar to $\ell_s(T)$ in the independent-chain cuprates Sr_2CuO_3 and $SrCuO_2$ (see Fig. 12.14). This suggests that most likely the same type of scattering limits the magnon heat transport, namely the spin-phonon interaction, as suggested in Ref. [88]. At low temperatures, $\ell_s(T)$ saturates to values of several thousand Å, remarkably similar to the the case of Sr_2CuO_3 and $SrCuO_2$. The origin of this low-temperature saturation is not yet well understood.

NaV_2O_5. This material represents an example of the so-called "quarter filled ladder" [90]. The two-leg ladders are formed by corner sharing VO_5 pyramids. The neighboring ladders weakly interact via common edges of adjacent VO_5 pyramids. The ladders are oriented along the b-axis and the rungs point along the a-axis. The formal valence of V is +4.5, and at high temperatures, all vanadium sites are equivalent and hence NaV_2O_5 may be considered as an intermediate valence compound. Two vanadium atoms situated on each rung share one electron and the effective spin per rung is 1/2. Therefore, at high temperatures, NaV_2O_5 exhibits a behavior typical of an $S = 1/2$ AFM Heisenberg linear chain with $J \approx 48$ meV [91]. A spin gap $\Delta \sim 9$ meV opens below $T_c \approx 35$ K, which was initially attributed to be caused by a spin-Peierls transition [91]. It has later been found that this transition corresponds to a charge ordering in the system of the vanadium ions.

Measurements of the temperature dependence of the thermal conductivity of single-crystalline NaV_2O_5 along the b-axis revealed an enormous enhancement of $\kappa(T)$ below T_c (see Fig. 12.15) [72, 73]. The heat transport in NaV_2O_5 was assumed to be of purely phononic origin, and the anomaly in this material and a similar anomaly below T_{SP} of $CuGeO_3$ were explained in terms of a reduction of phonon scattering below T_c or T_{SP}, respectively. This reduction follows naturally from the conjecture that the scattering of phonons by spin fluctuations is strong above T_c but, because of the formation of a spin gap, is significantly reduced below T_c. The reduction of the anomaly below T_c in Na-deficient samples (Fig. 12.15) supports this scenario. Measurements of the magnetic susceptibility indicate that the spin gap in $Na_{1-x}V_2O_5$ is progressively filled

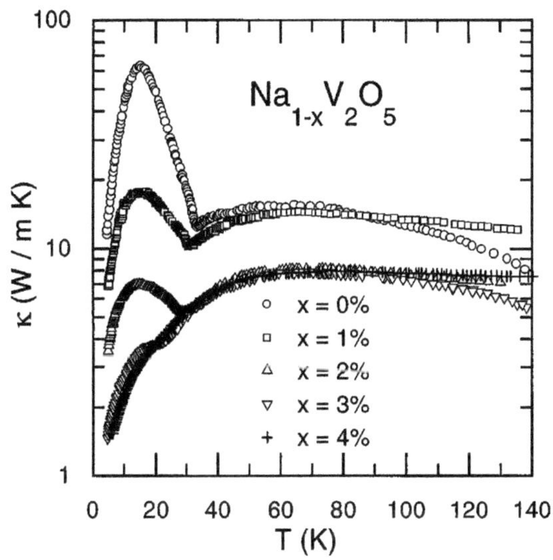

Figure 12.15. Temperature dependence of thermal conductivity in $Na_{1-x}V_2O_5$ [72].

by magnetic excitation states with increasing x. The authors of Ref. [72] point out, however, that also the charge ordering at T_c might be responsible for the anomaly in $\kappa(T)$.

5.4 Soliton-carrying chains

A few well-studied cases of materials, containing Heisenberg-type chains with planar anisotropy may be found in the literature [42]. The main interest in these systems is caused by the possibility of studying nonlinear magnetic excitations (solitons). If the magnetic field is applied in the easy plane, the equations of motion can be transformed into nonlinear equations whose solutions are both plane waves (magnons) and topologically stable nonlinear excitations (solitons) [42].

5.4.1 Antiferromagnetic chains.

TMMC and DMMC. Tetramethylammonium manganese trichloride (TMMC) and dimethylammonium manganese trichloride (DMMC) have similar crystal structures containing $MnCl_3$-chains. The spin of the Mn^{2+} ions is, with $S = 5/2$, rather large. Hence the magnetic properties of these compounds exhibit classical behavior. The AFM exchange between the Mn^{2+} is isotropic. The appropriate Hamiltonian is given by Eq. (12.14) with an added single-ion anisotropy term. The parameters adopt similar values for both materials with $J \approx 1.1$ meV

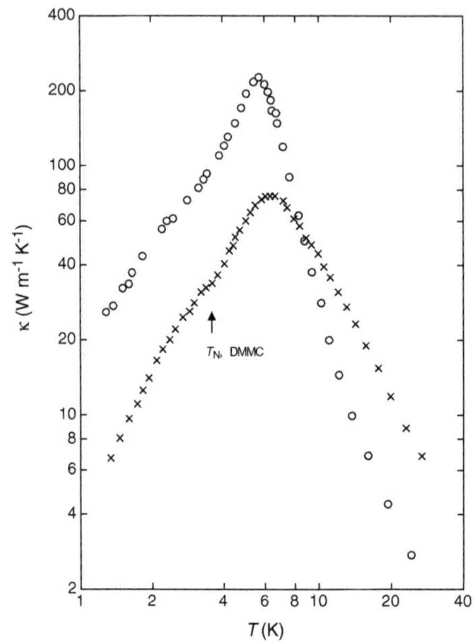

Figure 12.16. Zero-field thermal conductivity of TMMC (circles) and DMMC (crosses) along to the chain direction [55].

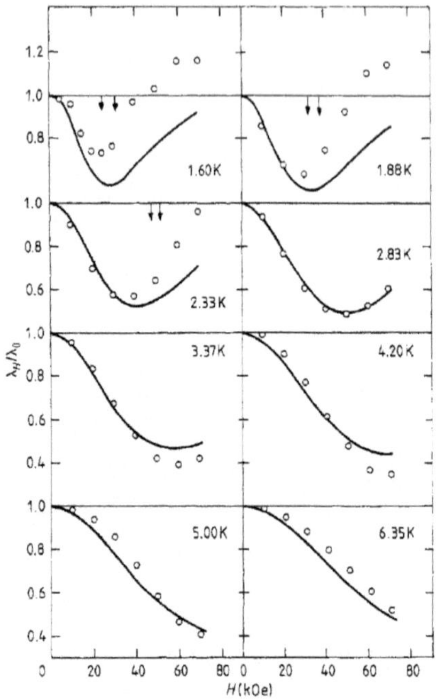

Figure 12.17. Isothermal field dependence of the thermal conductivity of TMMC along the chain direction, normalized with respect to the zero-field value [55]. The solid lines are fits employing the phonon-soliton scattering model. The arrows indicate the limits for the onset of the 3D ordering as described in the text.

and $D \approx 0.01$ meV. The easy plane is perpendicular to the chain direction, i.e., the c-axis. The interchain interaction is weaker in TMMC ($J'/J \sim 10^{-4}$) than in DMMC ($J'/J \sim 10^{-3}$), and correspondingly, the transition temperatures of 3D ordering are different for the two compounds, with $T_N = 0.85$ K for TMMC and $T_N = 3.60$ K for DMMC. Below T_N, an additional anisotropy appears in the ab-plane because of a small interchain interaction, such that the a-axis becomes the easy direction.

The zero-field low-temperature thermal conductivities $\kappa_{\parallel}(T)$ parallel to the c-axis of TMMC and DMMC are shown in Fig. 12.16 [55, 92]. In the cited references, these curves are not discussed in detail. More attention was given to the magnetic field dependences $\kappa(H)$. A selection of $\kappa(H)$ data for TMMC, taken at several temperatures and with the field oriented parallel to the easy plane, are presented in Fig. 12.17. Due to the presence of domains with different orientations within the ab-plane in TMMC, the field-induced ordering transition is not well defined and occurs somewhere between the two limits denoted by arrows in Fig. 12.17. Similar behavior of $\kappa(H)$ was found for the thermal conductivity of DMMC. Minima in $\kappa(H)$ were observed at temperatures below 4 K and in fields up to 70 kOe. The results agree with the interpretation that the heat is mainly transported by phonons, whereby the magnetic solitons act as effective scattering centers. The number of solitons n_{sol} depends on both temperature and magnetic field and has a maximum at a particular value of H/T, in the case of TMMC at $H/T = 14.7$ kOe/K. If the scattering of phonons by solitons dominates, a minimum of $\kappa_{\mathrm{ph}}(H)$ is expected at the field where, at the respective temperature, n_{sol} has a maximum. The $\kappa(H)$ data were fitted to the Debye model in the form of Eq. (12.7) and with the total relaxation rate

$$\tau^{-1}(x) = \tau_B^{-1} + \tau_R^{-1}\frac{n_{\mathrm{sol}}}{n_{\mathrm{ph}}}\frac{x^2}{(x^2 - x_0^2)^2}. \qquad (12.19)$$

The parameter x_0 represents the ratio $E_s(0)/k_BT$, where $E_s(0)$ is the soliton rest energy, and n_{ph} is the number of phonons. The first and the second term on the right-hand side of Eq. (12.19) represent the field-independent boundary scattering and the resonant phonon-soliton scattering, respectively. In this way, $\kappa(H)/\kappa(0)$ can be approximated at different temperatures with a single adjustable parameter, i.e., the ratio τ_R/τ_B. The resulting curves in Fig. 12.17 are in good agreement with the experimental data below the field-induced 3D ordering transitions denoted by arrows in Fig. 12.17. The disagreement between theory and experiment at higher fields was attributed to the failure of the pure 1D soliton model in the 3D ordered phase.

5.4.2 Ferromagnetic chains. Measurements of the thermal conductivity in spin-chain materials with ferromagnetic exchange are documented for two materials, i. e., $[C_6H_{11}NH_3]CuBr_3$ [93] and $CsNiF_3$ [94].

$[C_6H_{11}NH_3]CuBr_3$ and $CsNiF_3$. The structure of $[C_6H_{11}NH_3]CuBr_3$ (CHAB) contains linear $CuBr_3$-chains directed along the c-axis, with a relatively strong intrachain interaction between the $S = 1/2$ spins of Cu^{2+}. The magnetic properties can be described by the Heisenberg FM Hamiltonian of Eq. (12.14), including a weak XY anisotropy, such that $J \equiv J_x = J_y = -9.5$ meV and $J_z \approx 0.98J$ [95].

The y-direction coincides with the crystallographic c-axis and the x-direction lies in the ab-plane at an angle of 25° from the b-axis. A weak AFM interchain interaction $J' \sim 10^{-3}J$ induces a 3D ordering below $T_N = 1.5$ K.

The other soliton-carrying FM chain compound is $CsNiF_3$ [42]. The individual $S = 1$ chains formed by arrays of Ni^{2+} ions can be described by the Heisenberg Hamiltonian (12.14), including a single-ion anisotropy term and the parameters $J = -2.0$ meV and $D = 0.28$ meV. The chains are directed along the c-axis, and the easy plane is perpendicular to the c-axis. The AFM interchain interaction is rather weak $|J'/J| \sim 10^{-3}$. The compound undergoes a 3D ordering transition at $T_N = 2.7$ K and exhibits typical quasi-1D features at higher temperatures. Because of this planar anisotropy, $CsNiF_3$ is a standard material for investigations probing magnetic solitons [42].

The thermal conductivities of CHAB [93] and $CsNiF_3$ [94] were measured in the temperature range between 2 and 10 K with the heat flow along the chain direction. For $CsNiF_3$, $\kappa(T)$ increases monotonously in the whole covered temperature range, and a peak of $\kappa(T)$ at about 4 K is observed for CHAB. The field dependences $\kappa(H)$ at several selected temperatures were investigated in external magnetic fields $H < 80$ kOe which were oriented along the main crystallographic axes. With the field direction in the easy plane, thus allowing for the creation of magnetic solitons, a gradual increase of κ with H was observed. For $CsNiF_3$, the initial increase of $\kappa(H)$ is intercepted by a trend to saturation at $H > 3$kOe. Various scenarios were considered for the interpretation of the data. Finally, the authors of Ref. [94] concluded that the observed $\kappa(H)$ curves are consistent with a dominant scattering of phonons by magnons.

If the external field is pointing along the direction perpendicular to the easy plane, no solitons are present and κ increases with increasing H, but only above approximately 30 kOe. Since the number of magnons decreases with increasing H, a reduction of phonon scattering by magnons may be expected, consistent with the data.

6. Summary and outlook

From this survey of existing results on the thermal transport in 1D or quasi-1D spin systems, it is obvious that the number of open questions exceeds that of definite answers.

In spite of the enormous amount of theoretical work devoted to low-dimensional spin systems during the last several decades, the transport of energy has been treated in only a relatively small number of publications.

In most of these, the considered conditions are restricted to limiting cases, such as very high or very low temperatures, zero external magnetic field, large values of spins etc. The achieved results are often rather general statements concerning the nondiffusive energy propagation and an anomalous behavior of heat transport in spin chains. Only very recently, new theoretical calculations considered realistic systems and seem potentially useful for the analysis of experimental results.

This unsatisfactory situation concerning the theory of energy transport in low-dimensional spin systems has a reason. It is mainly due to the previous lack of reliable experimental data on the transport of heat via magnetic excitations in low-dimensional spin systems. Although a reasonable amount of information was contained in existing results of experiments probing the thermal conductivity, most of the observed anomalies had been attributed to unusual phonon heat transport affected by phonon-spin interactions. It seems that the significant impact of the scattering of phonons by magnetic solitons is well established experimentally but a theory for this process has not yet been developed.

Only recently, thanks to the progress in synthesizing new spin-chain type compounds and observations of both linear temperature dependences of the low-temperature thermal conductivity in insulators and double-peak features of $\kappa(T)$ along the directions of the 1D units at elevated temperatures, provided unambiguous evidence for the heat transport by itinerant magnetic excitations. The available experimental results reveal the importance of the interactions between spin excitations and defects, charge carriers and, most importantly, lattice excitations. Up to now the mean free path of spin excitations has been treated phenomenologically at best. Further progress of understanding clearly depends on more sophisticated analyses of the processes that limit the energy transport in real low-dimensional spin systems. In all spin systems where relevant data are available, the mean free path of itinerant spin excitations typically increases with decreasing temperature. For some unexplained reason, however, the mean free path never exceeds distances of the order of a few thousand Å (see Fig.12.14). It is not clear whether this is accidental or a typical phenomenon.

Neither experimental nor theoretical studies of the thermal conductivity in $S = 1$ chains with a Haldane gap have appeared in the literature. According to Fujimoto [96], the ideal $S = 1$ Heisenberg chain can be mapped onto the quantum 1D nonlinear σ-model which is integrable and, therefore, the spin transport should be ballistic. Corresponding features are thus expected for the spin mediated energy transport.

Another open issue is the detrimental influence of the interaction between spin excitations and charge carriers on the spin-related energy transport. Experiments on spin-ladder compounds have shown that hole doping reduces the thermal conductivity via magnons. Regarding other 1D conductors, the situation is not clear. Very recent results on Bechgaard salts, however, give evidence for a significant heat transport via spinons in compounds with metallic conductivity [97]. In view of the possibility of a spin-charge separation in 1D systems, such investigations seem very promising.

Acknowledgments

We acknowledge useful discussions with J. V. Alvarez, C. Gros, C. Hess, T. Lorenz, and J. Takeya. This work was financially supported in part by the Schweizerische Nationalfonds zur Förderung der Wissenschaftlichen Forschung.

References

[1] S. Lepri, R. Livi, and A. Politi, Phys. Rep. **377**, 1 (2003).

[2] L. G. C. Rego, Phys. Status Solidi A **187**, 239 (2001).

[3] S. Ciraci, A. Buldum, and I. P. Batra, J. Phys.: Condens. Matter **13**, R537 (2001).

[4] P. Carruthers, Rev. Mod. Phys. **33**, 92 (1961).

[5] J. M. Ziman, *Electrons and phonons* (Clarendon Press, Oxford, 1960).

[6] R. Berman, *Thermal conduction in solids* (Clarendon Press, Oxford, 1976).

[7] P. G. Klemens, in *Solid State Physics*, edited by F. Seitz and D. Turnbull (Academic Press, Inc., New York, 1958), Vol. 7, p. 1.

[8] G. A. Slack, in *Solid State Physics*, edited by H. Ehrenreich, F. Seitz, and D. Turnbull (Academic Press, Inc., New York, 1979), Vol. 34, p. 1.

[9] R. Kubo, Rep. Prog. Phys. **29**, 255 (1966).

[10] D. J. Sanders and D. Walton, Phys. Rev. B **15**, 1489 (1977).

[11] K. Kawasaki, Progr. Theor. Phys. **29**, 801 (1963).

[12] H. Stern, J. Phys. Chem. Solids **26**, 153 (1965).

[13] J. E. Rives, D. Walton, and G. S. Dixon, J. Appl. Phys. **41**, 1435 (1970).

[14] E. R. Hunt and J. R. Thompson, Phys. Rev. Lett. **20**, 249 (1968).

[15] Z. Rácz, J. Stat. Phys. **101**, 273 (2000).

[16] H. Sato, Progr. Theor. Phys. **13**, 119 (1955).

[17] B. Lüthi, Phys. Chem. Solids **23**, 35 (1962).

[18] R. L. Douglass, Phys. Rev. **129**, 1132 (1963).

[19] A. I. Akhiezer, V. G. Bar'yakhtar, and S. V. Peletminskii, *Spin waves* (North-Holland Publishing Company, Amsterdam, 1968).

[20] J. Callaway, Phys. Rev. **132**, 2003 (1963).

[21] J. Callaway and R. Boyd, Phys. Rev. **134**, A1655 (1964).

[22] P. Erdös, Phys. Rev. **139**, A1249 (1965).

[23] G. S. Dixon, Phys. Rev. B **21**, 2851 (1980).

[24] H. A. Bethe, Z. Phys. **71**, 205 (1931).

[25] L. D. Faddeev and L. A. Takhtajan, Phys. Lett. **85A**, 375 (1981).

[26] M. Takahashi, Progr. Theor. Phys. **50**, 1519 (1973).

[27] A. Klümper and D. C. Johnston, Phys. Rev. Lett. **84**, 4701 (2000).

[28] D. L. Huber and J. S. Semura, Phys. Rev. **182**, 602 (1969).

[29] D. L. Huber, J. S. Semura, and C. G. Windsor, Phys. Rev. **186**, 534 (1969).

[30] T. Niemeijer and H. A. W. van Vianen, Phys. Lett. **34A**, 401 (1971).

[31] D. A. Krueger, Phys. Rev. B **3**, 2348 (1971).

[32] N. A. Lurie, D. L. Huber, and M. Blume, Phys. Rev. B **9**, 2171 (1974).

[33] T. N. Tommet and D. L. Huber, Phys. Rev. B **11**, 450 (1975).

[34] H. Castella, X. Zotos, and P. Prelovšek, Phys. Rev. Lett. **74**, 972 (1995).

[35] X. Zotos, F. Naef, and P. Prelovšek, Phys. Rev. B **55**, 11029 (1997).

[36] K. Saito, S. Takesue, and S. Miyashita, Phys. Rev. E **54**, 2404 (1996).

[37] B. N. Narozhny, A. J. Millis, and N. Andrei, Phys. Rev. B **58**, R2921 (1998).

[38] F. Naef and X. Zotos, J. Phys.: Condens. Matter **10**, L183 (1998).

[39] A. Klümper and K. Sakai, J. Phys. A **35**, 2173 (2002).

[40] J. V. Alvarez and C. Gros, Phys. Rev. Lett. **89**, 156603 (2002).

[41] K. Saito and S. Miyashita, J. Phys. Soc. Jpn. **71**, 2485 (2002).

[42] H.-J. Mikeska and M. Steiner, Adv. Phys. **40**, 191 (1991).

[43] G. M. Wysin and P. Kumar, Phys. Rev. B **36**, 7063 (1987).

[44] H. Ikeda and K. Hirakawa, J. Phys. Soc. Jpn. **35**, 722 (1973).

[45] S. K. Satija *et al.*, Phys. Rev. B **21**, 2001 (1980).

[46] M. T. Hutchings, E. J. Samuelsen, and G. Shirane, Phys. Rev. **188**, 919 (1969).

[47] D. A. Tennant, T. G. Perring, R. A. Cowley, and S. E. Nagler, Phys. Rev. Lett. **70**, 4003 (1993).

[48] B. Lake, D. A. Tennant, and S. E. Nagler, Phys. Rev. Lett. **85**, 832 (2000).

[49] K. Hirakawa, H. Miike, and H. Hayashi, J. Phys. Soc. Jpn. **33**, 266 (1972).

[50] H. Miike and K. Hirakawa, J. Phys. Soc. Jpn. **38**, 1279 (1975).

[51] D. L. Huber, Progr. Theor. Phys. **39**, 1170 (1968).

[52] B. Schmidt *et al.*, Physica B **300**, 121 (2001).

[53] M. Kohgi *et al.*, Phys. Rev. B **56**, R11388 (1997).

[54] M. Köppen *et al.*, Phys. Rev. Lett. **82**, 4548 (1999).

[55] J. A. H. M. Buijs and W. J. M. de Jonge, J. Phys. C **15**, 6631 (1982).

[56] N. Motoyama, H. Eisaki, and S. Uchida, Phys. Rev. Lett. **76**, 3212 (1996).

[57] A. V. Sologubenko *et al.*, Phys. Rev. B **62**, R6108 (2000).

[58] K. M. Kojima *et al.*, Phys. Rev. Lett. **78**, 1787 (1997).

[59] T. M. Rice, S. Gopalan, and M. Sigrist, Europhys. Lett. **23**, 445 (1993).

[60] O. V. Misochko *et al.*, Phys. Rev. B **53**, R14733 (1996).

[61] H. Suzuura *et al.*, Phys. Rev. Lett. **76**, 2579 (1996).

[62] J. Lorenzana and R. Eder, Phys. Rev. B **55**, R3358 (1997).

[63] S. R. White and I. Affleck, Phys Rev. B **54**, 9862 (1996).

[64] A. V. Sologubenko *et al.*, Phys. Rev. B **64**, 054412 (2001).

[65] K. R. Thurber, A. W. Hunt, T. Imai, and F. C. Chou, Phys. Rev. Lett. **87**, 247202 (2001).

[66] K. Uchinokura, J. Phys.: Condens. Matter **14**, R195 (2002).

[67] M. Nishi, O. Fujita, and J. Akimitsu, Phys. Rev. B **50**, 6508 (1994).

[68] J. Riera and A. Dobry, Phys. Rev. B **51**, 16098 (1995).

[69] A. M. Vasil'ev *et al.*, JETP Lett. **66**, 868 (1997).

[70] Y. Ando *et al.*, Phys. Rev. B **58**, R2913 (1998).

[71] B. Salce *et al.*, Phys. Lett. A **245**, 127 (1998).

[72] A. N. Vasil'ev *et al.*, Phys. Rev. Lett. **81**, 1949 (1998).

[73] V. V. Pryadun *et al.*, Physica B **259-261**, 990 (1999).

[74] J. Takeya *et al.*, Phys. Rev. B **61**, 14700 (2000).

[75] J. Takeya *et al.*, Phys. Rev. B **62**, R9260 (2000).

[76] J. Takeya *et al.*, Physica B **284-288**, 1645 (2000).

[77] M. Hofmann *et al.*, Physica B **312-313**, 597 (2002).

[78] J. Takeya *et al.*, Phys. Rev. B **63**, 214407 (2001).

[79] E. Dagotto and T. M. Rice, Science **271**, 618 (1996).

[80] E. Dagotto, J. Riera, and D. Scalapino, Phys. Rev. B **45**, 5744 (1992).

[81] E. Dagotto, Rep. Prog. Phys. **62**, 1525 (1999).

[82] R. S. Eccleston *et al.*, Phys. Rev. Lett. **81**, 1702 (1998).

[83] M. Matsuda *et al.*, Phys. Rev. B **62**, 8903 (2000).

[84] T. Imai *et al.*, Phys. Rev. Lett. **81**, 220 (1998).

[85] S. Katano *et al.*, Phys. Rev. Lett. **82**, 636 (1999).

[86] M. Matsuda, T. Yosihama, K. Kakurai, and G. Shirane, Phys. Rev. B **59**, 1060 (1999).

[87] L. P. Regnault *et al.*, Phys. Rev. B **59**, 1055 (1999).

[88] A. V. Sologubenko *et al.*, Phys. Rev. Lett. **84**, 2714 (2000).

[89] C. Hess *et al.*, Phys. Rev. B **64**, 184305 (2001).

[90] H. Smolinski *et al.*, Phys. Rev. Lett. **80**, 5164 (1998).

[91] M. Isobe and Y. Ueda, J. Phys. Soc. Jpn. **65**, 1178 (1996).

[92] J. A. H. M. Buijs, Ph.D. thesis, TH Eindhoven, 1983.

[93] H. A. M. de Gronckel, W. J. M. de Jonge, K. Kopinga, and L. F. Lemmens, Phys. Rev. B **37**, 9915 (1988).

[94] H. A. M. de Gronckel, W. J. M. de Jonge, K. Kopinga, and L. F. Lemmens, Phys. Rev. B **44**, 4654 (1991).

[95] K. Kopinga, A. M. C. Tinus, and W. J. M. de Jonge, Phys. Rev. B **25**, 4685 (1982).

[96] S. Fujimoto, J. Phys. Soc. Jpn. **68**, 2810 (1999).

[97] T. Lorenz *et al.*, Nature **418**, 614 (2002).

Chapter 13

DUALITY IN LOW DIMENSIONAL QUANTUM FIELD THEORIES

Matthew P.A. Fisher
Kavli Institute for Theoretical Physics, University of California, Santa Barbara, CA 93106-4030
mpaf@kitp.ucsb.edu

Abstract In some strongly correlated electronic materials Landau's quasiparticle concept appears to break down, suggesting the possibility of new quantum ground states which support particle-like excitations carrying fractional quantum numbers. Theoretical descriptions of such exotic ground states can be greatly aided by the use of duality transformations which exchange the electronic operators for new quantum fields. This chapter gives a brief and self-contained introduction to duality transformations in the simplest possible context - lattice quantum field theories in one and two spatial dimensions with a global Ising or XY symmetry. The duality transformations are expressed as exact operator change of variables performed on simple lattice Hamiltonians. A Hamiltonian version of the Z_2 gauge theory approach to electron fractionalization is also reviewed. Several experimental systems of current interest for which the ideas of duality might be beneficial are briefly discussed.

Keywords: Duality, quantum Ising model, quantum XY model, rotors, Z2 gauge theory, fractionalization

1. Introduction

At the heart of quantum mechanics is the wave-particle dualism. Quantum particles such as electrons when detected "are" particles, but exhibit many wavelike characteristics such as diffraction and interference. In condensed matter physics one is often interested in the collective behavior of 10^{23} electrons, which must be treated quantum mechanically

D. Baeriswyl and L. Degiorgi (eds.), Strong Interactions in Low Dimensions, 419–438.

even at room temperature [1]. Fluids of light atoms such as He-3 and He-4 also exhibit collective quantum phenomena in the Kelvin temperature range [2, 3], and these days heavier atoms can be much further cooled to exhibit Bose condensation. The two low temperature phases of He-4 are a beautiful manifestation of this wave-particle dichotomy in the many-body context - the superfluid at ambient pressures behaving as a single collective "wavefunction" [4] and the crystalline solid at high pressures best thought of in terms of the "particles".

The collective behavior of such many-particle quantum systems is usually discussed in terms of "particles" rather than in terms of "waves", and this preference is mirrored in the theoretical approaches which work with "particle" creation and destruction operators. But in some instances it is exceedingly helpful to have an alternative framework, particularly when one wants to focus attention on some underlying wave-like phenomena. Duality transformations can sometimes serve this purpose, since they exchange the particle creation operators for a new set of "dual" operators which typically create "collective" excitations such as solitons (in 1d) or vortices (in 2d). Moreover, duality is playing an increasingly important role in describing novel electronic ground states which support excitations which carry fractional quantum numbers. The best studied situation is the one-dimensional interacting electron gas, a "quantum wire" [5], which exhibits a novel "Luttinger liquid" phase [6]. The "bosonization" reformulation of 1d interacting electrons, discussed in detail in Chapter 4, is in fact closely related to the 1d duality transformations introduced below.

This chapter provides a brief yet self-contained introduction to duality transformations, focussing on simple quantum Hamiltonians with global Ising or XY symmetry. In Section II the Hamiltonian for the quantum Ising model in transverse field is discussed, and is dualized in one dimension (1d) and two dimensions (2d) in subsections A and B, respectively. Section III considers a model of interacting bosons formulated in terms of quantum "rotors" - nicely exhibiting the phase-number uncertainty and being readily dualized in both 1d and 2d. A Hamiltonian version of the Z_2 gauge theory of electron fractionalization is discussed in Section IV. Finally, Section V is devoted to a brief discussion of some experimental systems of current interest for which the theoretical ideas introduced above might prove helpful.

2. Quantum Ising models

Consider the quantum Ising model in a transverse field with lattice Hamiltonian [7],

$$\mathcal{H}_I = -J \sum_{\langle ij \rangle} S_i^z S_j^z - K \sum_i S_i^x, \tag{13.1}$$

where S_i^x and S_i^z are Pauli matrices defined on the sites of a 1d lattice or a 2d square lattice and the sum in the first term is over near-neighbor sites. Here and throughout the rest of the chapter the 1d and 2d lattices are assumed to be infinite. The Pauli "spins" satisfy $S_i^x S_i^x = 1$ and $S_i^z S_i^z = 1$, commute on different sites and anticommute on the same site,

$$S_i^z S_i^x = -S_i^x S_i^z. \tag{13.2}$$

In the absence of the transverse field, $K = 0$, it is most convenient to work in a basis diagonal in $S^z = \pm 1$. The model then reduces to a *classical* Ising model with ferromagnetic exchange interaction J, and the ground state is the ferromagnetically ordered state with $S^z = 1$ (or $S^z = -1$) on every site. This ground state spontaneously breaks the global spin-flip (or Z_2) symmetry. A small transverse field will cause the spins to flip and the ground state will be more complicated, but provided $K \ll J$ one expects the ferromagnetic order to survive, $\langle 0|S_i^z|0\rangle \neq 0$. In the opposite extreme, $K \gg J$, all the spins will point in the x- direction, $S^x = 1$, which corresponds to a quantum paramagnetic ground state with zero magnetization, $\langle S_i^z \rangle = 0$. Based on this reasoning, one expects a quantum phase transition between the ferro- and para-magnetic ground states when J is of order K.

2.1 Quantum Ising duality in 1d

Further insight into the quantum Ising model follows upon performing a duality transformation [8]. Generally, a duality transformation is simply a change of variables wherein the original fields - in this case the Ising spins - are exchanged for a new set of "dual" fields. For the Ising model in 1d the Pauli spin operators, S^z and S^x are exchanged for a new set of (dual) Pauli matrices, σ^z and σ^x, defined on the *sites* of the dual lattice (which are the links of the original lattice),

$$S_i^z = \prod_{j \leq i} \sigma_j^x, \tag{13.3}$$

$$S_i^x = \sigma_i^z \sigma_{i+1}^z. \tag{13.4}$$

The product runs over a semi-infinite "string" of sites j on the dual lattice which satisfy $j \leq i$. One can readily check that provided the σ_i^μ fields obey the Pauli matrix algebra, then so do the spin fields S_i^μ. These expressions can be inverted,

$$\sigma_i^z = \prod_{j<i} S_j^x, \tag{13.5}$$

$$\sigma_i^x = S_i^z S_{i-1}^z, \tag{13.6}$$

taking essentially the same form.

The Hamiltonian in Eq. (13.1) when re-expressed in terms of the dual operators has precisely the same form as originally,

$$\mathcal{H}_I = -K \sum_i \sigma_i^z \sigma_{i+1}^z - J \sum_i \sigma_i^x, \tag{13.7}$$

except with an interchange of the coupling constants, $J \leftrightarrow K$. Then, $\langle \sigma^z \rangle$ serves as a *disorder* parameter, being non-zero in the paramagnetic ground state, and vanishing in the ferromagnetic state. The Hamiltonian is self-dual when $J = K$, and this point corresponds to the quantum phase transition separating the two phases.

The duality transformation is also useful in identifying the excitations above the ground state. Consider first the ferromagnetically ordered state with $J \gg K$ where $S^z = 1$. The lowest energy *classical* excitation (when $K = 0$) consists of a domain wall separating two domains with $S^z = \pm 1$. Notice that such an excitation can be created by acting with the operator σ_i^z on the classical ground state: $\sigma_i^z |0\rangle$. When K is small but non-zero this "domain-wall" excitation is no longer an eigenstate of the Hamiltonian, since acting with the first term in the dual Hamiltonian Eq. (13.7) can be seen to move the location of the domain-wall. All of these domain-wall excitations have the same energy when $K = 0$, but this low energy manifold of states is split by non-zero K. One can use standard degenerate perturbation theory to calculate the energy splitting of this degenerate manifold to leading (first) order in small K, and obtain the associated eigenstates. One thereby obtains a set of states in which the domain wall is propagating along, and behaves like a particle. Indeed, since a single domain wall is topologically protected, this "particle" will not decay. But *two* domain-wall "particles" can annihilate another and disappear altogether. For this reason one says that such domain walls carry an Ising or Z_2 "charge". The fact that domain walls are point-like objects in one spatial dimension and can propagate like particles is exploited in the bosonization approach to 1d interacting systems and underlies the physics of 1d particle "fractionalization".

The paramagnetic ground state when $J \ll K$ also supports particle-like excitations. These excitations correspond to domain walls in the ordered state of the dual Ising model, i.e. walls separating the two phases with $\langle \sigma_i^z \rangle \approx \pm 1$. The operator which creates this "particle" is simply the Ising spin itself, S_i^z, as is readily apparent from Eq. (13.3). By treating the original Ising spin Hamiltonian perturbatively to first order in small $J \ll K$, one can construct these gapped particle-like excitations in the paramagnetic phase of the Ising model, and obtain their dispersion relation. As one increases J towards K from below, the energy gap for creating these Ising-spin excitations vanishes, and in the ferromagnetic phase these particles "condense", exhibiting long-ranged order,

$$\langle S_i^z S_j^z \rangle \neq 0; \qquad |i - j| \to \infty. \tag{13.8}$$

2.2 Quantum Ising Duality in 2d

We next turn to the transverse field quantum Ising model in two spatial dimensions, which for simplicity we place on a 2d square lattice with nearest-neighbor exchange interaction J. As in 1d, this model is expected to have two quantum ground states as the couplings are varied, a ferromagnetic ground state when $J \gg K$, a paramagnetic state in the opposite limit $J \ll K$, and an intervening quantum phase transition when J is comparable to K. As we shall see, the duality transformation in 2d relates the quantum Ising model (with *global* Z_2 symmetry under $S^z \to -S^z$) to a dual *gauge* theory - specifically a gauge theory with a *local* Z_2 symmetry [8]. At the operator level, the duality transformation is implemented by re-expressing S^x and S^z directly in terms of the dual gauge fields, σ_{ij}^μ - a set of Pauli matrices defined on the *links* of the dual square lattice,

$$S_i^x = \prod_{pl(i)} \sigma_{jl}^z, \tag{13.9}$$

and

$$S_i^z = \prod_{jl=i}^{\infty} \sigma_{jl}^x. \tag{13.10}$$

Here, the first product is taken around an elementary four-sided plaquette on the dual square lattice (which encircles the site i of the original lattice). The second product involves an infinite string which connects sites of the original (direct) lattice, emanating from the site S_i^z and running to spatial infinity. For every bond of the dual lattice which is bisected by this string, a factor of σ_{ij}^x is present in the product. To assure that this definition is independent of the precise path taken by the string requires imposing the constraint that the product of σ_{ij}^x on all bonds connected to each site on the dual square lattice is set equal to unity,

$$\mathcal{G}_i = \prod_{j \in i} \sigma^x_{ij} = 1, \tag{13.11}$$

where j labels the nearest-neighbor sites of i. These local Z_2 gauge constraints must be imposed on the Hilbert space of the dual theory. (Note that because there are two bonds for every site of the 2d square lattice, the unconstrained dual Hilbert space is larger than the original Hilbert space, so it is reasonable that the dual Hilbert space be constrained.) In the resulting dual gauge theory, these constraints are analogous to Coulomb's law ($\nabla \cdot E = 0$) in conventional electromagnetism.

When re-expressed in terms of the dual fields, the Hamiltonian for the 2d quantum Ising model in a transverse field becomes,

$$\mathcal{H}_I = -K \sum_{pl} \prod_{pl} \sigma^z_{ij} - J \sum_{\langle ij \rangle} \sigma^x_{ij}. \tag{13.12}$$

In the first term products are taken around the elementary square plaquettes of the dual square lattice which surround the sites of the original lattice. These products measure "magnetic flux" in the dual gauge fields, that is plaquettes with $\prod_{pl} \sigma^z = -1$.

One can readily verify that the operators which implement a local gauge transformation, $\mathcal{G}_i$ in Eq. (13.11), commute with this dual Hamiltonian. Equivalently, since $\mathcal{G}_i \sigma^z_{ij} \mathcal{G}_i = -\sigma^z_{ij}$, the dual Hamiltonian is invariant under the general Z_2 gauge transformation,

$$\sigma^z_{ij} \to \epsilon_i \sigma^z_{ij} \epsilon_j, \tag{13.13}$$

with arbitrary $\epsilon_i = \pm 1$. We thus end up with a Z_2 gauge theory.

To gain some intuition for the behavior of this gauge theory we first focus on the limit $J \gg K$, where the global Ising model is ferromagnetically ordered. In this limit, the ground state of the gauge theory is simply $\sigma^x_{ij} = 1$ for all links ij. The low energy excitations about the ferromagnetically ordered state are droplets of $S^z = -1$ in the background of up spins ($S^z = 1$), and the "domain walls" are 1d closed paths (or "strings") which encircle the droplet (in contrast to the point-like domain walls for the 1d Ising model). To create such a droplet excitation from the ground state requires flipping all of the spins inside the droplet, that is,

$$|drop\rangle = \prod_{i \in drop} S^x_i |0\rangle, \tag{13.14}$$

where $|0\rangle$ denotes the ferromagnetically ordered ground state. This can be re-expressed in terms of the dual gauge fields as,

$$|drop\rangle = \prod_{\langle ij\rangle \in C} \sigma^z_{ij}, \tag{13.15}$$

where C denotes the closed path that encircles the droplet. The energy of this droplet excited state is roughly JL, where L is the linear dimension (circumference) of the droplet. This is called the "confining" phase of the gauge theory, since two "test Z_2-charges" placed on sites i and j of the dual lattice (with $\prod_{\ell \in i} \sigma^x_{i\ell} = \prod_{\ell \in j} \sigma^x_{j\ell} = -1$), will cost an energy linear in their separation - the two particles are "confined" together in much the same way that the quarks are confined inside the mesons and hadrons in the standard model of the strong interaction (QCD).

Consider next the paramagnetic phase of the Ising model with $K \gg J$. In this limit, the gauge theory ground state corresponds to a state with $\prod_{pl} \sigma^z_{ij} = 1$ for all plaquettes. Excited states correspond to making a single plaquette with $\prod_{pl} \sigma^z_{ij} = -1$, a plaquette with a penetrating Z_2 "magnetic flux". This point-like excitation is reminiscent of a vortex in a 2d superconductor, and has been christened a "vison" due to it's Ising-like character (see below). To study the dynamics of the vison, it is simplest to return to the original global Ising model, where the paramagnetic ground state corresponds to all sites having $S^x_i = 1$. As is clear from the definition in Eq. (13.10), the vison excitation can be created by acting on the ground state with S^z_i, where i is the site of the original lattice which is in the center of the corresponding dual plaquette. Thus, in terms of the original Ising spins, a vison simply consists of a site with $S^x = -1$.

When $J = 0$ there is a large manifold of degenerate single vison states (with energy $2K$), since the vison can occupy any site of the original lattice. This degeneracy will be split by a small non-zero J, and these single vison states will broaden into a dispersing band.

The paramagnetic phase of the global Ising model corresponds to the "deconfined" phase of the gauge theory. In this phase, "test Z_2-charges" introduced into the theory (with $\prod_{j \in i} \sigma^x_{ij} = -1$ at the sites i of the "test charges") cost a finite energy to create. In particular, the energy to separate two such particles does not grow linearly with separation, but saturates at some finite value even as the separation is taken to infinity. One of the key signatures of such a deconfined phase of the Z_2 gauge theory is the presence of the vison as a finite energy excitation. (In the context of high-temperature superconductivity, an experiment was recently proposed to detect whether or not the vison was present in the underdoped region of the phase diagram. Detection of the vison would establish the existence of electron fractionalization (or spin-charge separation). Upon increasing J and approaching the transition into the ferromagnetic phase of the Ising model the energy cost of the vison is reduced. In the ferromagnetically ordered phase the vison has condensed, with $\langle S^z \rangle \neq 0$, since S^z is the vison creation operator.

3. Quantum XY or Rotor models

We next turn attention to quantum Hamiltonians in 1d and 2d which have a conserved $U(1)$ symmetry [9, 10]. In particular, we focus on bosons hopping on a 1d or a 2d lattice with boson creation operators, $b_i^\dagger$ satisfying the usual Bose commutation relations, $[b_i, b_j^\dagger] = \delta_{ij}$. A simple Hamiltonian which conserves the total number of bosons is,

$$\mathcal{H}_{boson} = -t \sum_{\langle ij \rangle} b_i^\dagger b_j + h.c. + U \sum_i (b_i^\dagger b_i - \bar{n})^2. \tag{13.16}$$

The first term describes the hopping of bosons between nearest-neighbor sites, and the second term is an on-site repulsive interaction. Here, $\bar{n}$ plays the role of a chemical potential in setting the mean number of bosons, $\langle n_i \rangle$. This Hamiltonian is invariant under the global $U(1)$ symmetry: $b_i \to e^{i\Phi} b_i$, with a site-independent phase Φ. This global symmetry reflects the conservation of the total boson number.

Often it is convenient to consider a slight modification of this model, working with "rotor" variables rather than boson operators. In particular, we replace the boson creation operator by the exponential of a phase $\varphi_i \in [0, 2\pi]$: $b_i^\dagger \to e^{i\varphi_i}$, and the boson density by a number operator, n_i, which has integer eigenvalues, $b_i^\dagger b_i \to n_i$. The phase of the "rotor", φ_i, and the number operator are taken to satisfy the commutation relations,

$$[n_i, e^{i\varphi_j}] = \delta_{ij} e^{i\varphi_j}, \tag{13.17}$$

so that n_i can be thought of an "angular momentum" which is conjugate to the rotor phase. This commutation relation is directly analogous to,

$$[b_i^\dagger b_i, b_j^\dagger] = \delta_{ij} b_i^\dagger, \tag{13.18}$$

and indeed the operator $e^{i\varphi_i}$ increases the (boson) number n_i by one. In contrast to the operator $b_i^\dagger b_i$, which has non-negative eigenvalues, the eigenvalues of n_i span all the integers.

The "rotor" or XY Hamiltonian analogous to $\mathcal{H}_{boson}$ is,

$$\mathcal{H}_{XY} = -t \sum_{\langle ij \rangle} \cos(\varphi_i - \varphi_j) + U \sum_i (n_i - \bar{n})^2. \tag{13.19}$$

Notice that for large U the states with negative number $n_i < 0$ are up at high energy and can generally be neglected. Let us consider briefly the ground state phases of this quantum rotor or quantum XY model. When $U = 0$, the model reduces to a classical XY model, and the ground state is an ordered state with spatially constant rotor phases, $\varphi_i = \phi$ for all sites i. There is an associated non-vanishing order parameter, $\langle e^{i\varphi_i}\rangle \neq 0$. This ground state corresponds to the superfluid phase of the bosons, and exhibits off-diagonal long-ranged order,

$$G_{ij} = \langle e^{i\varphi_i} e^{-i\varphi_j}\rangle = |\langle e^{i\varphi_i}\rangle|^2 \neq 0 \quad |r_i - r_j| \to \infty. \tag{13.20}$$

For small but non-zero $U \ll t$ the ground state will be more complicated since the interaction term will induce some quantum fluctuations in the phases, but the off-diagonal long-range order and superfluidity should survive. (Actually, in 1d there will only be off-diagonal quasi-long-ranged order, and the correlator G_{ij} will vanish algebraically in the spatial separation.) The low-energy excitations above this ground state are the gapless Goldstone modes associated with the spontaneous breaking of the continuous $U(1)$ symmetry. (In 1d these should perhaps be called "quasi-Goldstone" modes, since the symmetry is not truly broken.) An effective Hamiltonian for these modes is obtained by expanding the cosine for small phase gradients,

$$\mathcal{H}_{Gold} = \frac{t}{2}\sum_{\langle ij\rangle}(\varphi_i - \varphi_j)^2 + U\sum_i (n_i - \bar{n})^2. \tag{13.21}$$

This Hamiltonian is quadratic in the conjugate variables (φ and n) and can be readily diagonalized to obtain the Goldstone modes. One can then evaluate the off-diagonal correlator G_{ij} in the ground state, and show that it decays algebraically in 1d but is infinitely long-ranged in 2d.

The behavior of the ground state in the opposite strong-interaction limit with $U \gg t$ depends sensitively on the average boson occupancy, $\langle n_i\rangle$ (which is only equal to $\bar{n}$ in the opposite $U \ll t$ limit). To understand this, consider the extreme limit with $t = 0$. For integer filling, such as $\langle n_i\rangle = 1$ say, the ground state will be unique with one boson on each site, and excited states with zero or two bosons on a site will cost a large energy of order U. This is a "Mott insulating" state with a large gap to charged excitations, and will be robust at integer filling, provided that $t \ll U$. Away from integer filling the ground states at $t = 0$ will be strongly degenerate, since the bosons can be arranged in many different ways on the lattice. In this case, non-zero hopping t will lift this degeneracy, leading to superfluidity. Henceforth, we focus primarily on the more interesting situation with integer boson filling.

3.1 Quantum XY Duality in 1d

Here we focus first on duality for the 1d rotor model. In close analogy with the Ising duality in 1d (Eq. (13.3) and 13.4), consider the change of variables,

$$e^{i\varphi_i} = \prod_{j\leq i} e^{iE_j}, \tag{13.22}$$

$$n_i = \theta_{i+1} - \theta_i, \tag{13.23}$$

where the dual "phase" field $E_i \in [0, 2\pi]$ and the integer-eigenvalue operator θ_i occupy the sites of the dual lattice. The dual operators are taken to satisfy

$$[e^{iE_i}, \theta_j] = \delta_{ij} e^{iE_i}, \tag{13.24}$$

which enables one to establish the desired commutator between n_i and $e^{i\varphi_i}$. In terms of these new fields the rotor Hamiltonian becomes,

$$\mathcal{H}_{XY} = -t\sum_i \cos(E_i) + U\sum_i (\theta_{i+1} - \theta_i - \bar{n})^2. \tag{13.25}$$

While formally exact, this dual Hamiltonian is often rather difficult to work with due to the integer constraints on the field θ_i. For this reason, it is both convenient and illuminating to modify the model by "softening" this integer constraint, allowing θ_i to take all real values and then adding a "potential term" acting on θ which favors integer values: $V(\theta) = -t_v \cos(2\pi\theta)$. Once this dual "angular momentum" θ_i is no longer quantized, it is legitimate to extend the "phase" field E_i to all real values, and to expand the cosine potential. In this way we arrive at an approximation to the dual Hamiltonian of the rotor model which should describe the same physics,

$$\tilde{\mathcal{H}}_{XY} = \sum_i \{\frac{t}{2}E_i^2 + U(\theta_{i+1} - \theta_i - 2\pi\bar{n})^2 - t_v\cos(2\pi\theta_i)\}, \tag{13.26}$$

where θ_i and E_i are now generalized coordinates and momenta which satisfy the canonical commutation relations,

$$[\theta_i, E_j] = i\delta_{ij}. \tag{13.27}$$

For integer boson occupancy, $\bar{n}$ can be eliminated from the theory by shifting the fields $\theta_j \to \theta_j + j(2\pi\bar{n})$, and $\tilde{\mathcal{H}}_{XY}$ reduces to a lattice sine-Gordon Hamiltonian. The associated Euclidian Lagrangian follows from,

$$\tilde{S}_{XY} = i\int d\tau \sum_i E_i\partial_\tau\theta_i + \int d\tau \tilde{\mathcal{H}}_{XY}, \tag{13.28}$$

and after integrating over the conjugate momenta becomes,

$$\tilde{S}_{XY} = \int d\tau \sum_i \{\frac{1}{2t}(\partial_\tau \theta_i)^2 + U(\theta_{i+1} - \theta_i - 2\pi\bar{n})^2 \quad (13.29)$$

$$- \quad t_v \cos(2\pi\theta_i)\}. \quad (13.30)$$

When $U \ll t$ the field θ is very soft and strongly fluctuating, and the cosine term becomes ineffective - this is the superfluid phase. After discarding the cosine term the Euclidian action (or Hamiltonian) is quadratic, and can be diagonalized to obtain the gapless "quasi-Goldstone" modes of the superfluid phase.

In the Mott insulating phase with $U \gg t$ (for integer boson density $\bar{n}$), the θ fluctuations are very "stiff" and become "pinned" in the minima of the cosine potential. In this limit one expects that the modes will become gapped. This can be verified by expanding the cosine potential to quadratic order for small θ and diagonalizing the resulting Hamiltonian to show that the normal-mode dispersion is gapped.

In addition to the gapped sound waves, the sine-Gordon theory will support "soliton"-like excitations separating regions in which the θ field is trapped in neighboring minima of the cosine potential. These correspond to single-boson excitations above the Mott ground state.

3.2 Quantum XY Duality in 2d

Finally, we consider dualizing the 2d quantum rotor model [10]. As for the Ising duality in 2 + 1d, the duality transformation for the 2d rotor model with global $U(1)$ (or XY) symmetry will take one to a gauge theory - but now a gauge theory with a (local) $U(1)$ gauge symmetry. Specifically, we re-express φ_i and n_i in terms of gauge fields defined on the links of the dual square lattice,

$$n_i = \Delta_x a_i^y - \Delta_y a_i^x \equiv \vec{\Delta} \times \vec{a}, \quad (13.31)$$

$$e^{i\varphi_i} = \overrightarrow{\prod_{j,\alpha=i}^{\infty}} e^{iE_j^\alpha}, \quad (13.32)$$

where a_i^α and E_i^α with $\alpha = x, y$ are vector fields defined on the links of the dual square lattice (a_i^x lives on the link running from site i to the site $i + \hat{x}$, and similarly for a_i^y). As above, E_i^α is a "phase" field defined on the interval $[0, 2\pi]$ and the operators a_i^α have integer eigenvalues. Here Δ_α denotes a discrete difference, $\Delta_x f_i = f_{i+\hat{x}} - f_i$. As for the Ising duality in 2+1d, the product above is along an infinite string - the string links sites of the original lattice starting at site i and running to spatial infinity, and for every link of the dual lattice bisected by the string a factor of $e^{iE_i^\alpha}$ is present in the product. The dual "vector potential" and "electric fields" are canonically conjugate variables, as in ordinary quantum electromagnetism,

$$[a_i^\alpha, e^{iE_j^\beta}] = \delta_{ij}\delta_{\alpha\beta}e^{iE_i^\alpha}. \quad (13.33)$$

To assure path-independence we must impose a constraint on the dual Hilbert space,

$$\mathcal{G}(\Lambda) = \prod_i e^{i\Lambda_i \vec{\Delta}\cdot\vec{E}_i} = 1, \quad (13.34)$$

for arbitrary integers Λ_i. Equivalently, the divergence of the "electric field" $\vec{\Delta}\cdot\vec{E}_i$ must equal $2\pi N_i$ for some integer N_i at each site of the dual lattice.

These integer "charges" actually correspond to vortices - point-like singularities around which the phase field φ_i winds by $2\pi N$. To see this, note that we can relate spatial gradients in the phase φ to the electric field

$$e^{i\Delta_\alpha \varphi_i} = e^{i\epsilon_{\alpha\beta}E_i^\beta}. \quad (13.35)$$

This is the discrete lattice version of $\nabla\varphi = \hat{z}\times\vec{E}$, and implies that $\nabla\times\nabla\varphi = \nabla\cdot\vec{E}$. As for the Ising case, the gauge constraints are generators of the local gauge transformations,

$$\mathcal{G}^\dagger a_i^\alpha \mathcal{G} = a_i^\alpha + \Delta_\alpha\Lambda_i. \quad (13.36)$$

In terms of the dual variables the 2d quantum XY model takes the form,

$$\mathcal{H}_{XY} = -t\sum_{i\alpha}\cos(E_i^\alpha) + U\sum_i(\vec{\Delta}\times\vec{a}_i - \bar{n})^2. \quad (13.37)$$

As in $1+1$d we now soften up the integer constraint on a_i^α, defining E_i^α in the range $[-\infty,\infty]$. Upon expanding the cosine term one obtains,

$$\tilde{\mathcal{H}}_{XY} = \sum_i\{\frac{t}{2}\vec{E}_i^2 + U(\vec{\Delta}x\vec{a}_i - \bar{n})^2\} \quad (13.38)$$

$$- \quad t_v\sum_{i\alpha}\cos(\Delta_\alpha\theta_i - 2\pi a_i^\alpha), \quad (13.39)$$

where we have explicitly displayed the longitudinal part of the gauge field, $2\pi\vec{a}_i^\ell \equiv \vec{\Delta}\theta_i$. After softening this constraint the appropriate local gauge symmetry becomes,

$$\tilde{\mathcal{G}}(\Lambda) = \prod_i e^{i\Lambda_i(\vec{\Delta}\cdot\vec{E}_i - 2\pi N_i)} = 1, \quad (13.40)$$

where N_i is a vortex number *operator* with integer eigenvalues which satisfies,

$$[N_i, e^{i\theta_j}] = \delta_{ij} e^{i\theta_i}, \tag{13.41}$$

so that

$$\tilde{\mathcal{G}}^\dagger a_i^\alpha \tilde{\mathcal{G}} = a_i^\alpha + \Delta_\alpha \Lambda_i \tag{13.42}$$

$$\tilde{\mathcal{G}}^\dagger \theta_i \tilde{\mathcal{G}} = \theta_i + 2\pi\Lambda_i. \tag{13.43}$$

We can now interpret the physics of the final dual Hamiltonian, $\tilde{\mathcal{H}}_{XY}$. The field, $e^{i\theta_i}$, is a vortex creation operator since it's action raises the vortex number, N_i, by one. The last term in the dual Hamiltonian thus describes the vortex kinetic energy, and the vortices are seen to be minimally coupled to the dual "electromagnetic" field. Thus the dual field mediates a logarithmic interaction between vortices. The dual $U(1)$ gauge symmetry can be interpreted as the conservation of vorticity.

When the vortices are absent from the ground state, with $t_v = 0$, the remaining terms in the Hamiltonian are quadratic and can be diagonalized to obtain the Goldstone modes of the 2d superfluid phase. In terms of the dual "electromagnetic field", this is nothing other than the massless "photon". Since the original boson density is equal to the curl of the dual "vector potential", this Goldstone mode is a longitudinal density (or sound) wave.

To describe the Mott insulating state we have to allow the proliferation of vortices and anti-vortices. Since the vortices are bosons, when present at zero temperature they will condense so that the ground state can be considered as a condensation of vortices, $\langle e^{i\theta} \rangle \neq 0$. Since the vortices are minimally coupled to the dual "electromagnetic" field, their condensation will lead to an expulsion of this dual "flux". For integer boson densities (i.e. integer $\bar{n}$) this phase will be the dual analog of the Meissner state in a superconductor. The gapless Goldstone mode of the superfluid (the dual "photon") will become gapped. To see this explicitly, one can expand the cosine to quadratic order in the dual "vector potential", and after choosing a convenient gauge ($\vec{\Delta} \cdot \vec{a_i} = 0$) diagonalize the resulting quadratic Hamiltonian.

After condensing the vortex, an externally applied dual "magnetic field" will be quantized into dual "flux quanta", analogous to the Abrikosov vortices in type-II superconductors. However, in this dual representation, a single quantized flux actually correspond to a boson excitation in the Mott insulating state. The dual "Abrikosov flux lattice" would then be a crystal of bosons.

This illustrates an appealing feature of dualizing to a vortex description when considering systems of 2d bosons: vortex-condensation gives one an order parameter for insulating (non-superfluid) phases of the bosons. It is very interesting to consider the possibility of *pairing* vortices, and condensing the pair, leaving single vortices uncondensed. As recently argued, this procedure leads to an exotic insulating state of bosons which supports fractionalized excitations - a gapped "half-boson" excitation (the dual quantized flux in the vortex pair condensate) and a gapped vison excitation (essentially an unpaired vortex).

4. Chargons, spinons and the Z_2 gauge theory of 2d electron fractionalization

The quantum Ising and XY Hamiltonians studied in Sections 2 and 3 are the simplest examples of quantum Hamiltonians which can be fruitfully analyzed by "duality" - re-expressing them in terms of a new set of "dual" operators. But many important models relevant to the quantum behavior of solids involve the fermionic electron creation and destruction operators, rather than the commuting bosonic operators entering in the quantum Ising and XY models. The classic example is the Hubbard model, which describes electrons hopping on the sites of a lattice interacting via a short-ranged (on-site) screened Coulomb repulsion. For the 1d Hubbard model and other 1d interacting electron models, a reformulation in terms of new operators - the so-called "bosonization" [6] technique - is possible and well understood. But "dualizing" models of 2d and 3d interacting electrons appears to be much more challenging. Nevertheless, some progress has been made in 2d, usually involving a "spin-charge" decomposition of the electron creation operator into a product of an operator which creates the spin of the electron - a "spinon" - and another which creates the charge of the electron - a "holon" or "chargon". These reformulations invariably involve a gauge field, which strongly couples together the spinons and chargons, and effectively "glues" them back together [11]. But in some situations the effects of the "gauge glue" can be weak, and exotic quantum ground states emerge within which the spinons and chargons can propagate as "deconfined" particle excitations. In effect, the electron is splintered into two fragments. A theory of such 2d "electron fractionalization" has recently been developed which involves a Z_2 gauge field [11]. The fractionalized state corresponds to the deconfined phase of the Z_2 gauge theory, and therefore supports a vison excitation precisely as discussed in Section IIB.

Here, a simple Hamiltonian version of the Z_2 gauge theory of 2d electrons is briefly presented. In the usual formulation, the $s = 1/2$ spinons carry the Fermi statistics of the electron, and the chargons are bosonic.

The full gauge theory Hamiltonian is [11],

$$\begin{aligned} H &= H_c + H_\sigma + H_s, && (13.44)\\ H_c &= -t\sum_{\langle ij\rangle}\sigma^z_{ij}\left(b^\dagger_i b_j + h.c\right) + U\sum_i \left(b^\dagger_i b_i - 1\right)^2, && (13.45)\\ H_\sigma &= -K\sum_{pl}\prod_{pl}\sigma^z_{ij} - J\sum_{\langle ij\rangle}\sigma^x_{ij}, && (13.46)\\ H_s &= -\sum_{\langle ij\rangle}\sigma^z_{ij}\left[t_s\left(f^\dagger_{i\alpha}f_{j\alpha} + h.c\right)\right.\\ &+ \left.\Delta_{ij}\left(f_{i\uparrow}f_{j\downarrow} - f_{i\downarrow}f_{j\uparrow} + h.c.\right)\right]. && (13.47) \end{aligned}$$

Here $b^\dagger_i$ creates a chargon at site i while $f^\dagger_{i\alpha}$ creates a spinon with spin $\alpha =\uparrow,\downarrow$ at site i. The operator $b^\dagger_i b_i$ measures the number of bosonic chargons at site i. For simplicity, we have specialized to half-filling, i.e. to an average of one boson per site. The constant Δ_{ij} contains the information about the pairing symmetry of the spinons. The $\sigma^z_{ij}, \sigma^x_{ij}$ are Pauli spin matrices which are defined on the links of the lattice, and H_σ is in fact identical to the Z_2 gauge theory Hamiltonian discussed in Section IIB.

The full Hamiltonian is invariant under the Z_2 gauge transformation $b_i \to -b_i, f_{i\alpha} \to -f_{i\alpha}$ at any site i of the lattice accompanied by $\sigma^z_{ij} \to -\sigma^z_{ij}$ on all the links connected to that site. This Hamiltonian must be supplemented with the constraint equation

$$\mathcal{G}_i = \prod_{j\in i}\sigma^x_{ij}e^{i\pi\left(f^\dagger_{i\alpha}f_{i\alpha}+b^\dagger_i b_i\right)} = 1. \qquad (13.48)$$

Here the product over σ^x_{ij} is over all links that emanate from site i. The operator $\mathcal{G}_i$, which commutes with the full Hamiltonian, is the generator of the local Z_2 gauge symmetry. Thus the constraint $\mathcal{G}_i = 1$ simply expresses the condition that the physical states in the Hilbert space are those that are gauge invariant.

When $J \gg K$ the gauge theory is deep within it's confining phase, and the chargon and spinon are confined back together to form the electron, with destruction operator $c_{i\alpha} = b_i f_{i\alpha}$. On the other hand, the fractionalized insulating phase is described as the deconfined phase of this gauge theory. This is obtained when $K \gg J, U \gg t$. A conventional superconducting state follows when the chargons condense, which occurs when $t \gg U$, or alternatively by doping away from half-filling. Note that the "pairing" symmetry of the superconductor is determined by Δ_{ij}.

5. Physics and Duality

This section provides a brief discussion of several strongly correlated electronic materials which exhibit unusual and in some cases poorly understood behavior, and considers how the theoretical ideas introduced above might provide a framework for gaining further insight into their properties.

5.1 One-dimensional systems

A number of complex molecular crystals exhibit highly anisotropic electrical properties. For crystals comprised of long (often organic), chain-like molecules, the conductivity along the chains can be many orders of magnitude larger than the transverse conductivity. In such cases, progress can be made by focusing on the properties of a single conducting chain. Modern lithographic techniques honed in the semiconductor industry provide another means to access one dimensional conductors, by controlling gates which further restrict the motion of electrons confined at the interface between two semiconductor materials [5]. However, carbon nanotubes - tube-shaped single molecules of carbon a nanometer in diameter and many microns long - provide the cleanest and most accessible example of a one-dimensional conductor [12].

It turns out that the strong effects of the interactions between the electrons moving up and down such nanotubes leads to exotic new behavior which is qualitatively different from the behavior of electrons in an ordinary conductor such as a copper wire [6, 13]. In particular, an electron added to a nanotube, for example by tunnelling from a metallic electrode, effectively splinters into fragments as it is propagates along the tube [14]. More precisely, the added spin and charge of the electron propagate in several "packets", one carrying the spin only and the others some fraction of the electronic charge. These exotic new "particles" are correctly considered as "solitons" in the background 1d fluid of electrons. They are quite similar to the "solitons" discussed in Section IIA in the context of the 1d quantum Ising model, which were domain walls between ferromagnetic domains which propagate as 1d "particles". The fractionally charged carriers in the nanotubes are even more closely related to the solitons mentioned in the context of the 1d quantum XY duality in Section IIIA (the solitons connecting different minima of the sine-Gordon cosine potential).

The central theoretical approach used to describe the physics of 1d interacting electron systems such as those occurring in carbon nanotubes is known as "bosonization" [15, 6]. In the bosonization approach the electron creation operator is exchanged for *two* bosonic fields, often denoted θ and φ. These two fields are essentially the same as the two fields employed in the discussion of 1d XY duality in Section IIIA, and provide two complementary (dual) descriptions of the same physics.

5.2 Two-dimensional systems

Quasi-two-dimensional layered materials occur both naturally (as with mica or graphite) and can also be grown, either out of the melt or layer-by-layer (in the case of semiconductors) using molecular beam epitaxy. Layered materials which exhibit strongly correlated electronic behavior typically have partially filled conduction band states, which can either lead to conduction or, when the band is very narrow, to self-localization. In this latter case, the residual electron spin degrees of freedom comprise a very interesting and challenging many-body system [16, 7]. The canonical examples are provided by the transition-metal oxides, where the 3d or 4d electrons form the localized interacting spin moments. The dynamics of such two-dimensional quantum spin systems can often be captured by (deceptively) simple lattice spin-Hamiltonians [17, 16]. For spin one-half moments, the spin Hamiltonians are in fact quite similar to the Hamiltonians in Eq. (13.1) and (13.16), the main difference being that the physical spin-systems have approximate spin-rotational symmetry rather than the extreme Ising-like "easy-axis" or XY-like "easy-plane" models considered here. Nevertheless, considerable insight can often be gained by appropriately dualizing the spin Hamiltonians [18]. Of interest are the myriad of possible quantum ground states that such many-body systems can possess, ranging from states with spontaneously broken spin-rotational symmetry (i.e. magnetic order) or broken translational symmetry ("spin-Peierls" order) [17] to exotic ground states with hidden "topological order" and fractionalized excitations [19].

Electrically conducting 2d layered materials offer an even more challenging arena of complicated many-body behavior. The high temperature cuprate superconductors [20, 1] offer the classic example. After more than 15 years of intensive effort (and at least tens of thousands of experimental publications), the underlying physics of these fascinating materials remains poorly understood and shrouded with theoretical controversy. The 2d electron system formed near the surface of an oxidized and gated silicon crystal (metal-oxide-semiconductor field-effect transistors or MOSFETS for short) provides another example of a well characterized material which exhibits strange behavior - an apparent 2d "metal-insulator" transition - which continues to defy theoretical consensus [21]. While the *fermionic* character of the conducting electrons is surely central to gaining an understanding of these materials, the 2d duality transformations discussed in this paper (which involve bosonic fields, commuting on different sites) might nevertheless be rather useful. One concrete approach was mentioned in Section IV, where a theory of 2d interacting electrons was reformulated in terms of spin-charge separated variables and a Z_2 gauge field - the same gauge theory shown to be dual to the 2d quantum Ising model in Section IIB. The 2d quantum XY duality transformation of Section IIIB has also been employed to access a new approach to 2d strongly correlated electrons [22, 23]. In this work, the vortices which appear in the dualized model of Section IIIB, are identified with the familiar vortices of a 2d superconductor. Very recent work [24] has exploited such a dual representation to obtain the first example of a genuine 2d "non-Fermi liquid phase" - a quantum ground state of 2d interacting electrons with no broken symmetries which has gapless charge and spin excitations but is *not* connected adiabatically to the free Fermi gas - in contrast to the familiar Fermi liquid phase. This novel quantum phase can apparently be accessed only by looking through a pair of "dual glasses". Determining whether such exotic states actually underlie the mysterious behavior of the cuprates or other 2d strongly correlated materials remains as one of the central challenges in contemporary theoretical physics.

Over the past 20 years my knowledge and appreciation of the wave-particle dualism of quantum mechanics in general and duality transformations of field theories in particular have been greatly aided by intensive and beneficial interactions and collaborations with (among others), Leon Balents, Daniel Fisher, Steve Girvin, Geoff Grinstein, Charlie Kane, Dung-Hai Lee, Chetan Nayak, T. Senthil and A. Peter Young - and I am deeply grateful and indebted to them all. This work has been generously supported by the National Science Foundation under grants DMR-0210790 and PHY-9907947.

References

[1] See, for example, J.C. Campuzano, "Angle resolved photemission in the high temperature superconductors", Chapter 3.

[2] *The Theory of Quantum Liquids*, by D. Pines and P. Nozieres, (Benjamin, New York, 1966).

[3] *Fermi-Liquid Theory: Concepts and Applications*, by G. Baym and C. Pethick (Wiley and Sons, New York, 1991).

[4] See for example, *Statistical Mechanics* by R.P. Feynman (Benjamin, Reading, 1972).

[5] See the article by A. Yacoby, "Transport in Quantum Wires", in Chapter 10.

[6] K. Schönhammer, "Luttinger Liquids: the basic concepts", Chapter 4.

[7] See *Quantum Phase Transitions*, by S. Sachdev, (Cambridge University Press, 1999), for a nice introduction to the transverse field quantum Ising model, and other more complicated quantum spin Hamiltonians.

[8] For an early review paper that discusses Ising lattice duality in both 1d and 2d see, J.B. Kogut, Rev. Mod. Phys. **51**, 659 (1979) and references therein.

[9] For an early paper on duality transformations in the *classical* 2d XY model see, J.V. Jose, L.P. Kadanoff, S. Kirkpatrick and D.R. Nelson, Phys. Rev. B**16**, 1217 (1978), and references therein. Duality for the *classical* 3d XY model is discussed in C. Dasgupta and B.I. Halperin, Phys. Rev. Lett. **47**, 1556 (1981).

[10] Duality for the 2d relativistic quantum XY model can be found in M. Peskin, Ann. Phys. **113**, 122 (1978) and P.O. Thomas and M. Stone, Nucl. Phys. B**144**, 513 (1978). For a discussion of duality for non-relativistic bosons see, M.P.A. Fisher and D.H. Lee, Phys. Rev. B**39**, 2756 (1989) and X.G. Wen and A. Zee, Int. J. Mod. Phys. B **4**, 437 (1990).

[11] See T. Senthil and M.P.A. Fisher, Phys. Rev. B **62**, 7850 (2000); T. Senthil and M.P.A. Fisher, Phys. Rev. B**63**, 134521 (2001) and references therein.

[12] For a review of carbon nanotube physics see Cees Dekker in "Physics Today" **52**, nr.5, 22-28 (May 1999).

[13] M. Bockrath, *et al.*, Nature **397**, 598 (1999); Z. Yao, H. Postma, L. Balents, C. Dekker, Nature **402**, 273 (1999) and H. Postma, M. de Jonge, Z. Yao, C. Dekker, Phys. Rev. B **62**, 10653 (2000).

[14] C.L. Kane, L. Balents, M.P.A. Fisher, Phys. Rev. Lett. **79**, 5086 (1997) and R. Egger, A. Gogolin, Phys. Rev. Lett. **79**, 5082 (1997).

[15] For various approaches to bosonization see, for example, V. Emery, in *Highly conducting one-dimensional solids*, edited by J. Devreese, R. Evrard, and V. Van Doren (Plenum Press, New York, 1979), p.247; A.W.W. Ludwig, Int. J. Mod. Phys. **B8**, 347 (1994); R. Shankar, Acta Phys. Polonica B **26**, 1835 (1995); M.P.A. Fisher and L.I. Glazman, Mes. Elec. Transp., ed. by L.L. Sohn, L.P. Kouwenhoven, and G. Schön, NATO Series E, Vol. 345, 331 (Kluwer Academic Publishing, Dordrecht, 1997).

[16] See, for example, C. Broholm and G. Aeppli, "Dynamic correlations in quantum magnets", Chapter 2.

[17] A. Auerbach, *Interacting Electrons and Quantum Magnetism,* (Springer-Verlag, New York, 1994).

[18] See for example, L. Balents, M.P.A. Fisher and S.M. Girvin, Phys. Rev. B**65**, 224412 (2002).

[19] For a discussion of topological order in the context of the fractional quantum Hall effect, see X.G. Wen and Q.Niu, Phys. Rev. **B41**, 9377 (1990).

[20] For a recent survey of various competing theories of the cuprate materials see "Physics in Canada", a special issue on High Temperature Superconductivity, edited by C. Kallin and J. Berlinsky, **56**, 242 (2000).

[21] For a review on the 2d metal-insulator transition in silicon MOSFETS see, E. Abrahams, S. V. Kravchenko and M. P. Sarachik, Rev. Mod. Phys. **73**, 251-266 (2001).

[22] L. Balents, M.P.A. Fisher and C. Nayak, Int. J. Mod. Phys. B**12**, 1033 (1998) and Phys. Rev. B**60**, 1654 (1999).

[23] See M.P.A. Fisher in, "Topological Aspects of Low Dimensional Field Theories", in Les Houches Lecures Session LXIX, edited by A. Comtet, T. Jolicoeur, S. Ouvry and F. David, (Springer, 1999).

[24] L. Balents and M.P.A. Fisher, unpublished (2002).

SUBJECT INDEX

MATERIALS INDEX

MIX
Papier aus verantwortungsvollen Quellen
Paper from responsible sources
FSC® C105338

If you have any concerns about our products,
you can contact us on
ProductSafety@springernature.com

In case Publisher is established outside the EU,
the EU authorized representative is:
Springer Nature Customer Service Center GmbH
Europaplatz 3, 69115 Heidelberg, Germany

Printed by Libri Plureos GmbH
in Hamburg, Germany